The Urban Book Series

The Urban Book Series is a resource for urban studies and geography research worldwide. It provides a unique and innovative resource for the latest developments in the field, nurturing a comprehensive and encompassing publication venue for urban studies, urban geography, planning and regional development.

The series publishes peer-reviewed volumes related to urbanization, sustainability, urban environments, sustainable urbanism, governance, globalization, urban and sustainable development, spatial and area studies, urban management, transport systems, urban infrastructure, urban dynamics, green cities and urban landscapes. It also invites research which documents urbanization processes and urban dynamics on a national, regional and local level, welcoming case studies, as well as comparative and applied research.

The series will appeal to urbanists, geographers, planners, engineers, architects, policy makers, and to all of those interested in a wide-ranging overview of contemporary urban studies and innovations in the field. It accepts monographs, edited volumes and textbooks.

Indexed by Scopus.

P. K. Joshi • Susanta Mahato • Anton Shkaruba
Editors

Geospatial Science for Urban Ecosystems

Insights from India and Beyond

Springer

Editors
P. K. Joshi
School of Environmental Sciences
Jawaharlal Nehru University
New Delhi, Delhi, India

Susanta Mahato
Department of General and Applied Geography, School of Applied Sciences
Dr. Harisingh Gour Vishwavidyalaya
Sagar, Madhya Pradesh, India

Anton Shkaruba
Environmental Protection and Landscape Management
Estonian University of Life Sciences
Tartu, Estonia

ISSN 2365-757X ISSN 2365-7588 (electronic)
The Urban Book Series
ISBN 978-3-032-07154-5 ISBN 978-3-032-07155-2 (eBook)
https://doi.org/10.1007/978-3-032-07155-2

This Springer imprint is published by the registered company Springer Nature Switzerland AG
The registered company address is: Gewerbestrasse 11, 6330 Cham, Switzerland

Contents

Case Studies

About the Editors

P. K. Joshi is a Professor School of Environmental Sciences (SES) and Former Chairperson Special Centre for Disaster Research (SCDR), Jawaharlal Nehru University (JNU), New Delhi. His research is focused on environmental studies, with specialization in landscape characterization and climate change analysis. He has immensely contributed in understanding geosphere and biosphere interactions. His present research work contributes to assessing impact and vulnerability of climate change, and disaster risk reduction. Now, he has spent more than two decades in actively contributing to a wide range of training, research and academic programmes in the field of geospatial analysis. He possesses a scholastic record coupled with strong leadership skills and networking capabilities. He has made significant contribution as a 'Scientific Administrator', and in promoting 'Scientific Research' *inter alia* through education, research publications and outreach activities.

Susanta Mahato is an Assistant Professor in the Department of General and Applied Geography at Dr. Harisingh Gour Vishwavidyalaya (A Central University), India. Dr. Mahato is a Dr. D.S. Kothari Postdoctoral Fellow (India), a Fellow of the Helmholtz Information & Data Science Academy (Germany), and an International Excellence Fellow at Karlsruhe Institute of Technology (Germany). His expertise spans land-atmosphere interactions, remote sensing and urban climate dynamics. His work integrates advanced remote sensing techniques with a deep understanding of urban and environmental systems, offering valuable insights into climate adaptation and policy development. He has contributed in understanding the complex interactions between extreme heat and air pollution. Beyond research, he is deeply committed to education and capacity building, developing innovative approaches to enhance interdisciplinary collaboration.

Anton Shkaruba is a Senior Researcher at Chair of Landscape Management and Environmental Protection at Estonian University of Life Sciences (Tartu, Estonia). His research areas are land use, forestry and biodiversity management and governance, as well as integration of urban ecosystem services and disservices to planning and management practices. This was acquired through the involvement in a number of EU Horizon Europe, Horizon 2020, FP5-7, Asia-Pacific Network and Swedish Institute projects. He was lead author of two IPBES reports, and collaborated with UNDP and European Environment Agency as consultant. Alongside research and teaching, he is also actively involved in a broad range of capacity building activities supporting research and higher education development in Eastern Europe and Asia.

Fundamental Concepts

Geospatial Science for Urban Ecosystems: Insights from India and Beyond

P. K. Joshi, Susanta Mahato, and Anton Shkaruba

Abstract This chapter delineates the content and significance of the book, Geospatial Science for Urban Ecosystems: Insights from India and Beyond, which meticulously examines the evolution and pivotal role of urban ecosystem research. It underscores the symbiotic relationship between nature and human-created environments within contemporary Indian cities, employing geospatial science as a crucial tool. The book accentuates the indispensable role of geospatial science in orchestrating a complex interplay of social, economic, and environmental factors that mold urban environments in India and globally. Illustrative applications of geospatial tools are showcased through a diverse array of case studies, encompassing various facets such as transportation planning and climate change adaptation, with a specific focus on the Indian context. At its essence, this chapter endeavors to furnish a comprehensive comprehension of the dynamic essence of urban environments, adept at navigating demographic shifts, technological advancements, and environmental challenges. It also presents an outline of the book's chapters. By accentuating urban ecosystems as the linchpin of sustainable urban development, the chapter endeavors to instill a deeper appreciation for this interdisciplinary realm of study. Ultimately, it beckons readers to embark on a captivating journey through the amalgamation of urban ecosystems and geospatial science, wherein innovative paradigms shape the trajectory of cities, both in India and beyond.

P. K. Joshi (✉)
School of Environmental Science, Jawaharlal Nehru University, New Delhi, India
e-mail: pkjoshi@mail.jnu.ac.in

S. Mahato
Department of General and Applied Geography, School of Applied Sciences, Dr. Harisingh Gour Vishwavidyalaya, Sagar, Madhya Pradesh, India
e-mail: ap_susantam@dhsgsu.edu.in

A. Shkaruba
Estonian University of Life Sciences, Tartu, Estonia
e-mail: anton@mespom.eu

P. K. Joshi et al. (eds.), *Geospatial Science for Urban Ecosystems*, The Urban Book Series, https://doi.org/10.1007/978-3-032-07155-2_1

Keywords Geospatial science · Urbanization · Biodiversity patterns · Climate change adaptation · Spatial analysis · Sustainable Development Goals · Environmental sustainability

1 Introduction

Geospatial science emerges as an invaluable tool for understanding and analyzing the components and interactions within urban ecosystems. The utilization of remote sensing, Geographic information system (GIS), and spatial analysis allows visualization, mapping, and quantification of urban landscapes. Geospatial technologies provide crucial data on land cover, biodiversity patterns, habitat fragmentation, and ecosystem services (ESs), facilitating informed decision-making in urban planning and management (Batty, 2013). Moreover, the conceptual framework of urban ecosystems underscores the dynamic nature of cities, constantly evolving under the influence of diverse factors. Urbanization, population growth, land use changes, and socioeconomic activities are among the major drivers shaping urban ecosystems. Alberti (2008) emphasizes the impacts of land use changes associated with urbanization on ecological patterns and processes, leading to habitat fragmentation and loss of biodiversity. Environmental stressors, such as pollution and climate change, further exacerbate these alterations. The Intergovernmental Panel on Climate Change (IPCC, 2021) reports rising temperatures and changing precipitation patterns, impacting urban areas and necessitating adaptive strategies for urban ecosystems. Comprehending the dynamics of urban ecosystems also involves recognizing ecological processes specific to urban environments. McDonnell and Pickett (1990) elucidate the modified nutrient cycles, altered hydrological patterns, and energy flow dynamics characteristic of urban landscapes. Urban areas exhibit unique ecological processes, influenced by human–nature interactions and resulting in novel ecosystems and hybrid ecologies. These processes shape the resilience and adaptive capacity of urban ecosystems, allowing organisms and ecological functions to respond to changing conditions (Pickett, 2010; Elmqvist et al., 2003).

The progression of yearly publications in the field of Urban Ecosystem and Geospatial Science from 2005 to 2024 paints a compelling picture of the field's sustained growth and increasing relevance (Fig. 1). The total number of publications, reaching 1135 by 2023, demonstrates a clear upward trend, indicating a burgeoning interest and scholarly activity over the years. The early years, marked by modest numbers in the 40 s, gave way to a more pronounced growth phase, notably from 2016 onward. Thereafter, the annual publications nearly doubled from 62 in 2016 to a peak of 110 in 2023, reflecting an intensified engagement with geospatial research within the urban ecosystem context. Several factors contribute to this observed growth. First and foremost is the rapid evolution of geospatial technologies, providing researchers with advanced tools and methodologies for analyzing urban landscapes. The intersection of geospatial science with urban ecosystem studies has likely prompted interdisciplinary collaborations, bringing together expertise

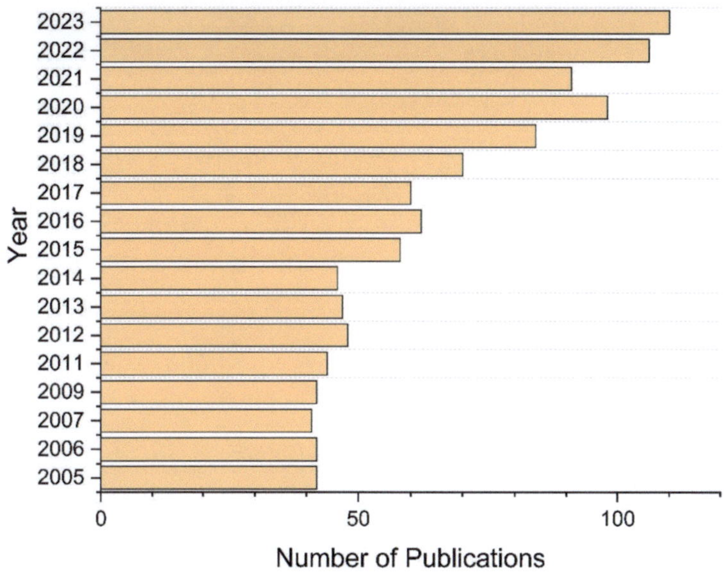

Fig. 1 Annual publication frequency in the field of Urban Ecosystem and Geospatial Science

from geography, ecology, urban planning, and related fields. Moreover, the increasing awareness of the critical role geospatial data plays in addressing urban challenges, such as sustainable development, environmental management, and infrastructure planning, has likely fueled the surge in research activity. The peak in 2022 with 106 publications suggests a particularly fruitful period, possibly influenced by external factors such as funding opportunities, policy emphasis on urban sustainability, or technological breakthroughs. The decline in 2023 might be a natural fluctuation or indicative of a shift in research priorities. The trajectory of geospatial research in the urban ecosystem demonstrates not only quantitative growth but also a maturation of the field. This reflects a vibrant and expanding landscape, with researchers actively contributing insights that are increasingly crucial for navigating the complexities of urban environments.

The co-occurrence analysis of keywords in the field of Urban Ecosystem and Geospatial Science over the past decade provides valuable insights into the evolving themes and priorities within this research domain (Fig. 2). In 2015, the focus was on fundamental aspects, such as management, patterns, conservation, urbanization, growth, ecosystem services, GIS, and spatial analysis, reflecting a broad interest in understanding the dynamics of urban ecosystems. Subsequent years saw an expansion of research themes, with keywords like climate change, remote sensing, biodiversity, impervious surfaces, land use change, and volunteered geographic information gaining prominence in 2016 and 2017. The years 2018 and 2019 introduced keywords related to sustainability, impacts, green infrastructure, air pollution, and green spaces, indicating an increased emphasis on the environmental and social aspects of urban ecosystems. The inclusion of terms like "climate change,"

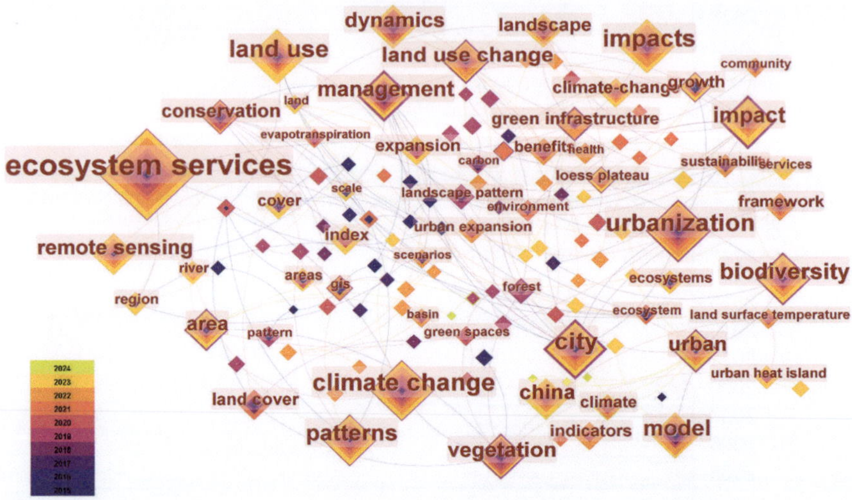

Fig. 2 Keyword co-occurrence network related to Urban Ecosystem and Geospatial Science in the past decade

"urban expansion," and "land surface temperature" in 2020 reflects a global perspective, possibly influenced by the increasing recognition of urbanization challenges worldwide. The subsequent years, from 2021 to 2024, witnessed a diverse range of keywords, including outdoor recreation, health, resilience, ecosystem services, and sustainable development goals, suggesting a broadening of the research focus to encompass interdisciplinary and holistic approaches to urban ecosystem and geospatial science. The keywords in 2024, such as "carbon emissions," "economic urbanization," "carbon neutrality," and "land urbanization," underscore a growing interest in the environmental consequences of urbanization, aligning with the global discourse on climate change and sustainable development. The mention of specific pollutants like "$PM_{2.5}$," "ammonia emissions," and "nitrate deposition" highlights a finer granularity in understanding the environmental impacts of urban areas. This co-occurrence analysis showcases the dynamic evolution of research themes in Urban Ecosystem and Geospatial Science over the past decade, reflecting an increasing awareness of the complex interactions between urbanization, environmental sustainability, and human well-being. The inclusion of diverse keywords suggests a multidisciplinary and integrated approach to addressing the challenges posed by urban ecosystems, positioning this field at the forefront of contemporary environmental research.

2 Understanding Urban Ecosystems

2.1 Conceptual Framework of Urban Ecosystems

Urban ecosystems represent complex, adaptive systems shaped by human activities, ecological processes, and socioeconomic dynamics. The conceptual framework of urban ecosystems involves a synthesis of ecological principles with the realities of urbanization. Studies (e.g., McPhearson et al., 2016; Pickett, 2010) have shown that cities exhibit unique ecological characteristics, where biotic and abiotic components interact in intricate ways. This framework acknowledges the multifaceted nature of urban spaces, considering factors like land use, socio-demographics, and environmental changes. The expansion of cities globally has led to alterations in ecological patterns and functions. For instance, the research by Seto et al. (2012) highlights that urbanization has profound impacts on biodiversity and ecosystem services. Urban ecosystems experience modifications in species composition, habitat fragmentation, and altered biogeochemical cycles due to human-induced changes.

2.2 Components and Interactions within Urban Ecosystems

Studies (e.g., Grimm et al., 2008; McKinney, 2006) indicate the diverse array of biotic elements in urban areas, ranging from native and non-native species to microorganisms. Urbanization influences species distribution, with urban-adapted species often dominating. The interactions between these species—competition, predation, and mutualism—shape biodiversity patterns within cities. Urban ecosystems comprise abiotic elements, such as air, water, soil, and built infrastructure. The interaction between these components and living organisms influences ecological processes. For instance, research (e.g., Pataki et al., 2011) demonstrates the urban heat island (UHI) effect, where built structures and altered landscapes lead to elevated temperatures, affecting microclimates and ecological dynamics. Understanding the interactions within urban ecosystems requires a holistic approach integrating ecological science, geospatial analysis, and socioeconomic factors. Geospatial technologies play a pivotal role in mapping, modeling, and analyzing these interactions by providing spatially explicit data on land cover, biodiversity, and ecosystem services.

3 Dynamics of Urban Ecosystems

3.1 Ecological Processes in Urban Environments

Urban landscapes harbor a spectrum of ecological processes unique to their settings, characterized by altered nutrient cycles, modified hydrological patterns, and distinctive energy flows. The research by McDonnell and Pickett (1990) underscores the significance of understanding these modified processes along urban–rural gradients. Urban areas exhibit distinct biogeochemical cycles influenced by human activities, such as increased nutrient runoff from impervious surfaces and altered nutrient inputs from atmospheric sources (Pataki et al., 2011). These modifications significantly affect the functioning of urban ecosystems, impacting ecosystem services and biodiversity. Furthermore, the human–nature interactions in urban settings result in novel ecosystems and hybrid ecologies. Urbanization creates opportunities for the emergence of adaptive capacities and self-organizing systems within these novel ecosystems. Elmqvist et al. (2003) discuss the importance of response diversity and ecosystem change, showcasing how urban ecosystems display resilience through adaptive processes.

3.2 Anthropocene and Urban Ecosystems

The Anthropocene epoch signifies the period in which human activities have significantly impacted Earth's ecosystems and geology. Urban ecosystems, being the focal points of human habitation, exhibit substantial changes reflective of the Anthropocene era. This shift is characterized by complex dynamics, reshaping ecological processes, biodiversity, and sustainability within cities. Urbanization has surged, with over half of the global population residing in urban areas, projected to reach 68% by 2050 (United Nations, 2018). This rapid urban expansion has altered landscapes, impacting ecosystems by fragmenting habitats, reducing biodiversity, and influencing resource consumption patterns (Grimm et al., 2008). The Anthropocene has ushered in various environmental challenges within urban ecosystems. For instance, air pollution levels in cities have surged due to increased industrialization and transportation. According to the World Health Organization (WHO), around 91% of the world's population lives in places where air quality exceeds WHO guideline limits, primarily in urban areas (WHO, 2018). Furthermore, urbanization has led to significant changes in hydrological cycles, affecting water availability and quality. Impermeable surfaces, such as concrete and asphalt, in cities disrupt natural water infiltration, exacerbating issues like urban flooding and water pollution (Pickett, 2010). However, amid these challenges, cities have also become hubs for innovation and sustainability initiatives. Urban planners and policymakers are implementing green infrastructure strategies, such as rooftop gardens, urban parks, and green spaces, to mitigate the adverse effects of urbanization (Ahern, 2013).

In the past two decades, India has experienced rapid urbanization (Onda et al., 2019; Shekhar, 2021). Cities are afflicted by unplanned urban sprawl, which has resulted in a scarcity of green and blue spaces. Many researchers have examined urban dynamics in cities, such as Delhi (Sharma et al., 2020; Mahato et al., 2020), Mumbai (Moghadam & Helbich, 2013; Yu et al., 2021), Kolkata (Sahana et al., 2018; Mandal et al., 2019), Bengaluru (Verma et al., 2017; Chandrashekar et al., 2021), Chennai (Aithal & Ramachandra, 2016; Devendran & Lakshmanan, 2019), Ahmedabad (Munshi et al., 2014; Yatoo et al., 2019), and Hyderabad (Ul Din & Mak, 2021). The urban area of Delhi (1977–2014) grew by 26.79% inside the National Capital Territory of Delhi and 8.19% in the outlying areas (Dutta et al., 2019). From 1973 to 2018, Mumbai's population increased by 3.91 times (Yu et al., 2021). Pal et al. (2022) found that Greater Mumbai has lost 2.5% of green space and 1.4% of waterbodies between 1999 and 2019. The spatial extension of urban growth and pattern has been studied using many methodologies, and future expansion is being studied using a variety of simulation models across time.

3.3 *Factors Shaping Urban Ecosystems*

Urban ecosystems are influenced by a multitude of factors that collectively shape their structure and function. Understanding these influences is pivotal in comprehending the dynamics of urban environments. Urbanization stands as a significant driver, altering landscapes and ecosystems. Studies (Alberti, 2008; Grimm et al., 2008) highlight that land use changes associated with urbanization modify ecological patterns, leading to habitat fragmentation and loss of biodiversity. The exponential growth of cities globally, coupled with land use transformations, poses challenges to the conservation of natural habitats and ecological connectivity. Moreover, environmental stressors, such as pollution, habitat destruction, and climate change, pose substantial pressures on urban ecosystems. These stressors exacerbate the alterations brought by urbanization, affecting biodiversity, ecosystem services, and overall ecological health. Understanding the complexities of factors shaping urban ecosystems is fundamental for devising strategies aimed at mitigating negative impacts, fostering sustainable urban development, and preserving ecological integrity.

3.4 *Resilience and Adaptation in Urban Ecosystems*

Urban ecosystems demonstrate remarkable resilience and adaptation capacities in response to environmental changes and disturbances. Pickett (2010) emphasized the adaptive nature of urban ecosystems, where organisms, communities, and ecological processes dynamically adjust to new conditions. Strategies such as green infrastructure, urban rewilding, and adaptive management play crucial roles in enhancing resilience in urban ecosystems. These strategies focus on enhancing ecosystem functions, restoring degraded habitats, and promoting biodiversity in urban landscapes. Understanding the dynamics of resilience and adaptation is essential for developing effective conservation and management approaches. The integration of geospatial science, ecological principles, and urban planning can aid in identifying and implementing strategies that enhance the adaptive capacity of urban ecosystems, promoting their sustainability amid ongoing urbanization and environmental changes. The dynamics of urban ecosystems are influenced by various interconnected factors, including urbanization, environmental stressors, and altered ecological processes. Understanding these dynamics is critical for devising strategies that support the resilience and adaptation of urban ecosystems, ensuring their sustainability and ecological integrity in the face of ongoing urban development and environmental challenges.

Urban ecological processes are influenced by human activities (Anthropocene) such as pollution, land use changes, and urbanization. These alterations impact species interactions, nutrient cycling, and energy flow within urban environments (Fig. 3). Conversely, changes driven by the Anthropocene reshape ecological

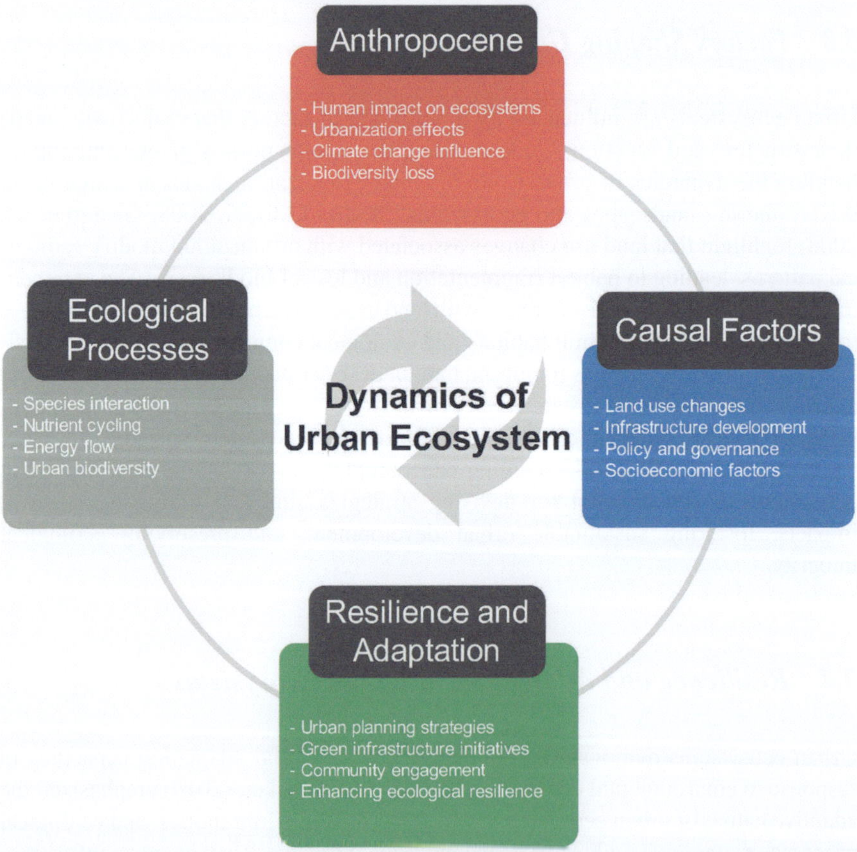

Fig. 3 This figure illustrates the interlinked relationships between the different aspects of urban ecosystem dynamics

processes. Urbanization, for instance, alters the landscape and species compositions, influencing how these ecological processes function. Human activities in the Anthropocene significantly shape the factors that define urban ecosystems. Urbanization, policy decisions, and socioeconomic factors are driven by human actions and consequently influence urban ecosystems. Simultaneously, the factors shaping urban ecosystems can influence human behaviors and activities. For instance, urban planning policies aimed at sustainability can directly impact the human-driven aspects of the Anthropocene. The factors shaping urban ecosystems, such as land use changes, infrastructure development, and governance policies, lay the groundwork for strategies aiming at resilience and adaptation. These strategies often directly address challenges posed by those shaping factors. Conversely, strategies for resilience and adaptation can impact and modify the factors shaping urban ecosystems. Initiatives like green infrastructure development or community engagement can alter policies and socioeconomic aspects, consequently influencing urban dynamics. Understanding these bidirectional linkages highlights the complexity of urban ecosystems, emphasizing the need for integrated approaches. Actions in one aspect can lead to cascading effects across the urban ecosystem, underscoring the importance of considering reciprocal relationships in managing and sustaining the urban environment.

4 Role of Geospatial Science in Urban Studies

Urban ecosystem research in India predominantly relies on remote sensing and geographic information system (GIS) tools. Moderate-resolution satellite data is commonly preferred for analyzing urban and peri-urban changes. However, the bulk of studies in India are focused on urban dynamics and characterization, exhibiting a bias toward larger cities. Additionally, the emphasis tends to be on analyzing changes within single cities rather than comparative studies across multiple urban centers. Urban growth modeling has the potential to shape future urban development policies and foster sustainable cities. However, its exploration within the Indian context remains relatively limited and is still in its early stages. Research on the thermal environment often centers on understanding the nonlinear spatial relationships between various factors, with less attention given to their interactions. In terms of green spaces, there is a notable absence of consideration for landscape connectivity and multifunctionality. While much research addresses the availability of green spaces, the aspect of accessibility remains poorly understood.

4.1 Overview of Geospatial Technology

Geospatial technology encompasses a suite of tools and methodologies for collecting, analyzing, and visualizing spatial data. These tools include geographic information system (GIS), remote sensing, Global Navigation Satellite System (GNSS), and spatial analysis software. The application of geospatial technology enables the acquisition of precise and detailed information about urban landscapes. Remote sensing, for instance, employs satellite imagery or aerial photography to capture detailed spatial data, facilitating the monitoring of land cover changes, urban growth, and environmental conditions with high spatial resolution (Jensen, 2009). Furthermore, GIS allows for the integration and analysis of various spatial datasets, enabling urban planners and researchers to explore relationships among different layers of information. This technology plays a fundamental role in understanding the complex interconnections within urban ecosystems by overlaying socioeconomic, environmental, and infrastructural data, providing a comprehensive understanding of urban landscapes (Longley et al., 2015).

4.2 Remote Sensing Techniques and Urban Mapping

Satellite-based remote sensing techniques offer invaluable capabilities for urban mapping, facilitating the detailed analysis and interpretation of urban environments. These techniques rely on sensors aboard satellites to capture data in different wavelengths of the electromagnetic spectrum. High-resolution satellite imagery, utilizing multispectral and hyperspectral sensors, provides detailed spatial information essential for urban mapping (Hebel et al., 2013). Multispectral imaging collects data in multiple spectral bands, allowing discrimination between different land cover types, such as vegetation, impervious surfaces, water bodies, and bare soil. This capability is fundamental in urban mapping, aiding in the delineation of land use patterns and the identification of urban features. Hyperspectral imaging, with a more extensive range of spectral bands, offers even finer discrimination, allowing for the detailed analysis of surface materials and vegetation health in urban areas (Lillesand et al., 2015). Light Detection and Ranging (LiDAR) technology, commonly mounted on satellites or aircraft, provides highly accurate three-dimensional (3D) data about urban terrain and structures. It enables the creation of precise digital elevation models (DEMs) and assists in urban modeling, infrastructure planning, and flood risk assessment by identifying topographic features and variations in the urban landscape (Hebel et al., 2013). Synthetic Aperture Radar (SAR), another remote sensing technology, operates in different frequency bands and is capable of penetrating clouds and darkness, providing consistent data, irrespective of weather conditions. SAR aids in monitoring changes in urban areas, such as ground subsidence, surface deformation, and infrastructure stability assessment, crucial for urban planners and disaster management (Lillesand et al., 2015).

4.3 Satellite Data for Monitoring Urban Changes

Satellite imagery is instrumental in monitoring urban changes, offering the ability to observe, analyze, and interpret alterations in urban landscapes over time. Time series analysis of satellite data enables the detection and assessment of various urban changes, including land use transformations, infrastructure development, and urban expansion (Seto et al., 2012).

By comparing satellite images captured at different times, it becomes possible to track the progression of urban growth and assess the impacts of human activities on the landscape. Studies leveraging satellite data, like those conducted by Kundu et al. (2024), showcase the capacity to monitor urban expansion, analyze sprawl patterns, and evaluate the consequences of urbanization on ecosystems and natural landscapes. Utilizing advanced image-processing techniques and machine learning algorithms enhances the analysis of satellite data, allowing for automated detection and quantification of changes in urban environments. Integration with GIS enables a spatially explicit understanding of urban dynamics, aiding policymakers and urban planners in making informed decisions for sustainable urban development (Anees et al., 2023). The temporal and spatial consistency of satellite data facilitates comprehensive analysis, providing a robust foundation for understanding urban dynamics and supporting evidence-based decision-making in urban planning, infrastructure development, and environmental management.

4.4 Importance of Geospatial Analysis in Urban Planning

Geospatial analysis is instrumental in urban planning, as it provides valuable insights into spatial patterns, trends, and relationships within cities. Urban planners leverage geospatial analysis to assess land use changes, identify suitable locations for infrastructure development, and optimize resource allocation. For instance, studies by Clarke (1997) and de Rathi et al. (2023) illustrate how GIS-based spatial analysis aids in identifying areas prone to flooding or other natural hazards, assisting in the formulation of effective risk mitigation strategies in urban planning. Moreover, geospatial analysis supports the evaluation of environmental impacts, facilitating sustainable urban development. It aids in assessing ecological footprints, analyzing transportation networks, and optimizing urban design to promote efficient resource utilization and minimize environmental degradation (Mitchel 2005).

Several case studies highlight the multifaceted applications of geospatial science in urban contexts. For instance, the City of Singapore has extensively utilized geospatial technology in its urban planning initiatives. By integrating GIS and remote sensing, Singapore effectively manages its limited land resources, monitors urban growth, and plans for infrastructure development while preserving green spaces and biodiversity (Tan et al., 2023). Similarly, in London, geospatial analysis has been integral in transport planning. Transport for London (TfL) utilizes GIS to manage the city's transportation networks, optimize public transport routes, and improve commuter experiences. Real-time spatial data analysis aids in predicting traffic congestion, optimizing traffic flow, and enhancing the efficiency of public transportation (Miyoshi & Torrell, 2019). Another notable case is the use of geospatial technology in slum mapping and urban poverty alleviation in cities like Nairobi, Kenya. NGOs and government agencies employ GIS and remote sensing to map informal settlements, assess living conditions, and plan for infrastructure improvements to enhance the quality of life for residents (Arif et al., 2022).

4.5 Unmanned Aerial Vehicles (UAVs) in Urban Analysis

UAVs equipped with remote sensing payloads have transformed the landscape of urban analysis, providing high-resolution imagery and data collection capabilities. These aerial platforms offer a cost-effective and flexible solution for acquiring detailed spatial information in urban areas (Colomina & Molina, 2014). UAVs capture high-resolution imagery with precision, enabling the creation of detailed orthophotos and three-dimensional models of urban environments. This data aids in land use classification, urban planning, environmental monitoring, and disaster management. The versatility of UAVs allows for the rapid and frequent acquisition of data, facilitating temporal analysis and monitoring of dynamic urban landscapes (Si Salah et al., 2019). Multispectral and thermal sensors mounted on UAVs capture various spectral bands, facilitating the assessment of vegetation health, surface temperatures, and environmental quality. Urban greenery, impervious surfaces, and infrastructure can be analyzed at high spatial resolutions, aiding in assessing urban heat islands, monitoring green spaces, and identifying pollution hotspots (Li et al., 2020).

UAVs play a vital role in urban infrastructure monitoring and maintenance. Equipped with LiDAR sensors or high-resolution cameras, UAVs capture detailed 3D models and high-definition imagery of urban infrastructure, such as buildings, roads, bridges, and utilities (Kim et al., 2019). These aerial platforms assist in asset inspection, allowing for the assessment of infrastructure conditions without disrupting traffic or causing significant disturbances. Inspection and monitoring of structural elements, identification of defects or damages, and assessment of maintenance needs are streamlined through UAV-based surveys (Kim et al., 2019). Moreover, UAVs contribute to disaster risk management by providing rapid post-disaster assessments. In the aftermath of natural disasters or emergencies, UAVs can quickly survey affected areas, assess damages, and support rescue and recovery efforts by providing detailed imagery and information to aid decision-makers (Mdaopa et al., 2023). The integration of UAV technology with advanced image-processing algorithms and artificial intelligence (AI) further enhances the capabilities of these platforms. Automated feature extraction, object recognition, and change detection algorithms leverage UAV-collected data to analyze urban environments more efficiently and accurately (Si Salah et al. 2019). UAVs have become indispensable tools in geospatial science for urban analysis. Their capabilities in remote sensing, data collection, and infrastructure monitoring offer a unique vantage point for understanding urban dynamics and supporting various applications in urban planning, environmental management, and disaster response.

4.6 Artificial Intelligence (AI) in Urban Studies

AI models have revolutionized urban planning by simulating complex urban scenarios and aiding in decision-making processes. Machine learning algorithms and AI-driven models enable planners to analyze and forecast urban development scenarios, simulate potential impacts, and optimize resource allocation (Batty, 2013). These models utilize historical and real-time urban data, incorporating socioeconomic, environmental, and infrastructural variables. For instance, UrbanSim, an AI-based urban simulation model, integrates demographic, economic, and land use data to forecast urban growth patterns, evaluate transportation infrastructure, and assess the potential impacts of policy interventions (Waddell, 2002). Similarly, land-use prediction models use machine learning techniques to forecast land use changes based on historical spatial data, aiding in long-term urban planning strategies (Saboori et al., 2022). AI models support planners in understanding and visualizing the consequences of different development scenarios, facilitating informed decision-making for sustainable urban growth. These models assist in optimizing land use plans, infrastructure development, and resource management strategies, contributing to more efficient and resilient urban landscapes.

Predictive analysis powered by AI plays a pivotal role in understanding and forecasting urban dynamics. AI algorithms process vast amounts of data, including historical trends, demographic changes, mobility patterns, and environmental factors, to predict future urban developments (Geertman et al., 2019). For instance, AI-based predictive models analyze transportation patterns, forecast traffic congestion, and optimize urban mobility. These models use real-time data from various sources, such as GPS devices, traffic cameras, and sensors, to predict traffic flow, identify congestion-prone areas, and propose efficient transportation strategies (Geertman et al., 2019). Moreover, AI-driven predictive analytics aids in urban risk assessment and resilience planning. These models assess vulnerabilities, such as those related to climate change, natural disasters, or infrastructure failures, predicting potential risks and enabling proactive measures for urban resilience (Sheladiya & Patel, 2023). By leveraging AI for predictive analysis in urban dynamics, planners and policymakers gain insights into future urban challenges and opportunities, enabling proactive planning and decision-making to address complex urban issues.

5 Challenges and Opportunities

5.1 *Sustainable Urban Development*

Geospatial science offers a paradigm for sustainable urban development by providing data-driven insights and facilitating evidence-based decision-making. Leveraging this science effectively can unlock numerous opportunities for creating resilient and livable urban environments. Geospatial technology enables smart urban planning by integrating diverse datasets to create comprehensive urban models. GIS-based planning assists in optimizing land use, transportation networks, and infrastructure development for sustainable growth (Li et al., 2016). Geospatial tools aid in monitoring environmental changes and managing ecological footprints. Satellite imagery and remote sensing facilitate real-time monitoring of environmental quality, deforestation, and air and water pollution, enabling timely interventions to safeguard ecosystems (Yan et al., 2020). Geospatial analysis enhances mobility solutions by optimizing transportation networks, predicting traffic flow, and facilitating efficient public transportation systems. Real-time data analysis helps in addressing congestion, improving transit services, and reducing carbon emissions (Sonaviya & Tandel, 2019). Geospatial technology plays a pivotal role in disaster risk reduction. UAVs, satellite imagery, and GIS aid in pre- and post-disaster assessments, facilitating quick response, evacuation planning, and rehabilitation efforts in the aftermath of disasters (Zhu et al., 2022). Geospatial tools promote community engagement in urban development. Citizen science initiatives, coupled with geospatial technology, empower communities to participate in data collection and decision-making processes, thereby fostering a sense of ownership in urban management (Kobayashi & Kinoshita, 2023).

5.2 *Addressing Challenges in Urban Ecosystem Management*

Urban ecosystems face multifaceted challenges stemming from rapid urbanization, environmental degradation, resource constraints, and socioeconomic disparities. Managing these challenges necessitates a holistic approach, integrating geospatial science, policy interventions, and community engagement. One challenge lies in preserving urban green spaces (UGSs) amid rapid urban expansion. Geospatial analysis using satellite imagery and GIS helps identify and prioritize green spaces for preservation, ensuring ecological balance and recreational opportunities for residents (Escobar & Britz, 2021). Urban heat islands exacerbate temperature differences between urban and rural areas due to human activities and infrastructure. Geospatial technology aids in identifying heat-prone zones, enabling urban planners to implement strategies like green roofs, reflective surfaces, and strategic vegetation to mitigate the heat island effects (Geng et al., 2021). Increasing urbanization intensifies the risk of flooding. Geospatial analysis assists in flood risk mapping, aiding in the identification of vulnerable areas and formulating resilient infrastructure strategies like green infrastructure and stormwater management systems (Zhou, 2020). Addressing socioeconomic disparities within cities poses a significant challenge. Geospatial data enables the identification of marginalized communities and supports targeted interventions for equitable resource allocation, improved access to amenities, and enhanced social services (Graham et al., 2021). While urban ecosystems face significant challenges, geospatial science stands as a powerful ally in addressing these challenges and unlocking opportunities for sustainable urban development. Integrating geospatial technology with effective policies and community engagement holds the key to building resilient, inclusive, and environmentally sustainable cities.

6 Scope and Structure of the Book

6.1 Overview of Chapters and Themes Covered

The book is conceptualized in two broad sections: Part I covering the fundamental concepts and Part II covering the case studies. The first part covers the broader theme of urban ecosystems. For example, Chap. 2 helps in understanding that urban form and dynamics are essential for promoting urban sustainability, yet the impact of various urban configurations on city growth and suburbanization remains poorly understood. This introduces the Urban Form Index (UFI), which evaluates urban land density, urban forms, and their expansion effects and applies it to 151 metropolises across Africa, China, and India (ACI). African and Indian cities predominantly exhibit compact or highly compact forms, while Chinese cities tend toward sprawled layouts. Examination of the UFI data spanning from 1985 to 2015 identifies 16 dynamic urban forms. African and Indian cities tend to maintain or transition toward compact forms, whereas Chinese cities shift from compact to sprawled forms. The rate of urban expansion, distribution of land density, spatial coverage, and built-up density are all influenced by these urban forms, with variations across regions. Recognizing the diverse patterns of urban expansion in the Global South challenges universal planning theories and underscores the necessity for context-specific urban planning to facilitate sustainable urban development in the future.

Chapter 3 reflects on the multitude of environmental challenges being experienced by the global population, predominantly residing in urban areas. While issues like smog, waste accumulation, and water contamination have long plagued cities worldwide, new challenges stemming from climate change, pandemics, and water scarcity are emerging. Ensuring access to safe water, effective disaster management, and robust health infrastructure are among the pressing concerns. Rapid and unplanned urbanization exacerbates these crises, necessitating global cooperation to tackle air, water, and noise pollution, as well as hazardous waste. The health impacts, both physical and psychological, demand immediate attention. Monitoring, evaluation, and public awareness are crucial initial steps, alongside adequate housing and strict energy and land use policies to prevent slum formation and maintain sanitation. Urban resource consumption also poses long-term threats to ecosystems beyond city limits. Effective mitigation requires comprehensive stakeholder involvement and understanding of all facets. This chapter delves into these urban environmental challenges in detail, utilizing case studies to illustrate the complexities involved.

Chapter 4 illustrates the dependence of the urban environment on biodiversity, encompassing diversity within species, between species, and across ecosystems. Biodiversity is prevalent in urban settings, with green spaces interspersed among cities and densely populated areas. Humans rely entirely on biodiversity, as it plays a critical role in ecosystem functioning. Disruptions to these processes directly affect ecosystem services. Of particular concern is land use fragmentation, which exacts a high cost in terms of biodiversity loss. This chapter highlights the essential relationship between urban environments and biodiversity, emphasizing the urgent need for conservation efforts to mitigate the detrimental effects of human activities on natural ecosystems.

Chapter 5 discusses how urban watersheds are encountering hydrological disruptions, including reduced infiltration rates, water volume loss, changes in water chemistry, destabilization of natural processes, human-made interferences, and escalating pollution levels. Historically, watersheds were managed systematically for sustainable development, but recent flood events worldwide indicate shortcomings in the scientific approach to watershed management. Urbanized watersheds globally suffer from decreased streamflow and degraded stream quality, posing significant threats to in-stream and near-stream habitats and water supply. It emphasizes that watersheds are crucial for livelihoods, especially in developing nations, yet unchecked agriculture and urbanization pose unsustainable demands on land and water resources, jeopardizing human life and civilization.

Chapter 6 on the urban heat island (UHI) effect widely acknowledges the phenomenon wherein urban areas exhibit higher temperatures compared to surrounding nonurban regions, primarily due to altered landscape morphology and materials. Rapid global urbanization has amplified concerns regarding UHI‘s adverse effects on human health, energy consumption, urban ecosystems, precipitation patterns, air quality, and climate change. This demonstrates that developing countries face exacerbated challenges due to rapid urban population growth without commensurate infrastructure development. Understanding land surface temperature (LST) changes driven by land use land cover (LULC) transformations is vital for managing urban microclimates. Implementing solutions like cool roofs, pavements, and increasing vegetation cover can mitigate UHIs and their impacts, with nature-based solutions (NbS) gaining prominence. Spatially and temporally mapping UHI intensity aids in identifying hotspots and high-intensity periods, informing urban planning and policymaking to alleviate UHIs' detrimental effects.

Chapter 7 draws the importance of urban natural ecosystems that possess valuable climate adaptation capabilities essential for building socio-ecological resilience and promoting the well-being of urban dwellers. Nature-based solutions

(NbS) have demonstrated effectiveness in bolstering urban resilience by providing sociocultural security, economic benefits, and essential ecosystem services. This chapter offers a comprehensive examination of the links between human health, urban green spaces, ecosystem services, and socio-ecological dynamics, emphasizing the relevance of urban–rural gradients. Key scientific concepts such as environmental neuroscience, biophilia, and environmental affordances are explored to integrate NbS options into urban planning, particularly in rapidly urbanizing regions of the Global South. Drawing on evidence and experiences from various countries, the chapter underscores the imperative of scaling up NbS to address urban stress, enhance psychological well-being, and mitigate diverse health risks, fostering the development of resilient cities in a warming and rapidly urbanizing world.

Chapter 8 reveals that urban ecosystems, which accommodate the world's highest population density, are increasingly susceptible to climate-related hazards like floods and heat waves. Vulnerability within these systems is influenced not solely by frequent hazards but also by evolving socioeconomic, political, and governance dynamics. Given the concentration of assets and the population at risk, addressing climate change vulnerability in urban ecosystems is imperative. This chapter offers a comprehensive exploration of climate-related hazards and vulnerability in urban ecosystems. It begins by elucidating fundamental theories and frameworks of vulnerability, followed by documenting vulnerability to climatic and social hazards to underscore critical factors. Finally, the chapter outlines proposed risk reduction strategies involving nature-based solutions and transitions toward urban sustainability to foster resilient urban ecosystems.

Chapter 9 is on urban sprawl and expansion that pose significant challenges to assessing a city's sustainable development. To address this complexity, advanced analysis and monitoring of urban growth processes, management, and planning practices are advocated. Modern urban geography employs remote sensing data and AI to track urban expansion and spatial patterns. Growth maps and categorized urban structures derived from remote sensing data aid planners and policymakers in understanding their cities' trajectories and underlying systems. Various analytical approaches and predictive models, utilizing remote sensing and GIS techniques, facilitate mapping, monitoring, analysis, and forecasting of urban growth. Efforts are increasingly focused on addressing environmental challenges, urban planning complexities, and long-term urban growth forecasting. Despite advancements in technology, there remains a gap between research-focused results in urban remote sensing and theoretical understanding in urban development studies. This chapter tries to bridge that gap, providing policymakers, local governments, and planners with a deeper understanding of urbanization through theoretical and research-based insights.

The second part of the book covers geospatial science for urban studies, and it takes the help of case studies to illustrate the concepts covered in the previous section. Chapter 10 proposes practical solutions and highlights the potential for creating a "waste-to-wealth" value chain to address the issue sustainably. It discusses that urban growth, driven by economic ambitions, sometimes overlooks sustainability practices, particularly in managing nondegradable solid waste generated in highly populated cities. Although Municipal Waste Management Rules have mandated proper waste segregation and disposal, effective implementation has been lacking, especially regarding plastic waste, which poses a significant challenge. This chapter focuses on the journey of plastic waste in densely populated Indian cities, from generation to disposal, and discusses its implications for the urban environment and ecology. It proposes practical solutions and highlights the potential for creating a "waste-to-wealth" value chain to address the issue sustainably.

Chapter 11 examines short-term fluctuations in the shallow water table in Cochin, a highly urbanized city in Kerala, India. Data from 97 well locations over sixteen months were analyzed to create monthly water table fluctuation and seasonal recharge maps. Recharge depths correlated with precipitation, with significant recharge occurring during the southwest monsoon. Monthly groundwater discharge flux rates indicated a predominant east-to-west discharge pattern, with higher seepage rates in the midland. Spatial distribution of total dissolved solids (TDS), electrical conductivity (EC), and pH concentrations suggested flush-out patterns influenced by rainfall recharge, with better water quality in the midland. The aquifer system in the midland was found to be more active due to laterite formation and steeper hydraulic gradients. Enriched groundwater from the midland drained rapidly after the monsoon, leading to water scarcity in the non-monsoon period. The study underscores the importance of implementing tailored aquifer management strategies to ensure sustainable groundwater resources based on local aquifer dynamics.

Chapter 12 aims to assess the effects of LULC changes on the hydrological regime of the Teesta River basin using the Soil and Water Assessment Tool (SWAT) model. Calibration and validation of the model were conducted using observed discharge data on a monthly time scale, showing good performance metrics. The northern part of the basin, characterized by glaciers, forests, and scrublands in the Himalayan ranges, contributes significantly to water yield. However, changes in LULC over the past decade have resulted in spatial and temporal shifts in water balance components at the subbasin level. Increased surface runoff from glacier melting and forest conversion to agriculture in the upper catchment may elevate the risk of flooding and landslides in lower-lying regions of the basin.

Chapter 13 provides a thorough analysis of the spatiotemporal dynamics of urban green spaces (UGSs) in Delhi, India, spanning from 1998 to 2018. Using Landsat satellite data, GIS tools, and census data, the study investigates how UGS changes impact urban residents' well-being. The analysis of LULC indicates a steady increase in human-dominated land use, particularly dense and sparse built-up areas, leading to a decline in essential green spaces like forests and water bodies. Morphological Spatial Pattern Analysis (MSPA) reveals the fragmentation and complexity of UGSs, especially along the Aravalli hills. Mapping the distribution of UGSs highlights decreasing trends over time, with unequal availability across different regions. The study introduces UGS per capita distribution and Greenness Index (GI) to assess the impact on residents, revealing concerning declines in UGS per capita, particularly in certain districts. The chapter concludes with a discussion contextualizing the findings within the broader urban development context, acknowledging study limitations and emphasizing the importance of continuous monitoring and nuanced exploration of socioeconomic factors. The insights gleaned offer valuable guidance for urban planners, policymakers, and residents in crafting sustainable strategies to preserve and enhance UGSs, promoting a healthier and more ecologically balanced urban environment.

Chapter 14 examines seasonal changes in green spaces and water bodies and their influence on LST from 1992 to 2023 using Google Earth Engine (GEE). Results show a significant increase in green spaces across all seasons but a decrease in water bodies, except during the post-monsoon season. The LST of green spaces, water bodies, and the Delhi region as a whole has steadily risen over the past three decades, indicating a substantial contribution of anthropogenic activities to the UHI effect, overshadowing the cooling impacts of green spaces and water bodies. Landscape metrics for urban green spaces, such as mean patch size (MPS), percent cover (PC), and largest patch index (LPI), are highest during the post-monsoon season and lowest in the pre-monsoon season. These findings are pertinent to smart cities, suggesting a need for specifying green space and water body areas per population. The study highlights the diminishing efficacy of green spaces and water bodies due to anthropogenic and climatic influences, underscoring the necessity for more extensive green space and water body coverage to mitigate future challenges.

Chapter 15 suggests a cost-effective method utilizing remote sensing data to analyze green space distribution, particularly beneficial for dynamically evolving urban areas in developed countries. Despite limitations such as the lack of high-resolution satellite images and diverse social data, the study highlights the importance of continuous improvement in methodology. The urban area surrounding the Siliguri Municipal region exhibits a complex blend of land use and socioeconomic characteristics, undergoing significant transformations from 1990 to 2020. These changes have had profound impacts on both the environment and socioeconomic conditions in Siliguri. Efforts to increase green spaces should focus not only on quantity but also on the layout and design to mitigate UHI effects effectively. However, the analysis reveals a significant decline in UGSs in the Siliguri municipal area after 2005 due to rapid urbanization, indicating a pressing need for effective planning and management policies to address this encroachment.

Chapter 16 investigates alterations in the LULC impact on ecosystem services (ESs) at a microlevel in selected subdistricts of the Central National Capital Region (CNCR) from 1991 to 2021. Multi-temporal LULC maps were then used to quantify ecosystem services for different land cover classes using a global coefficient. The evaluation of LULC changes' impact on ecosystem service valuation (ESV) employed elasticity approaches. The ESV in CNCR increased by 11.85% (from 1991 to 2021), with notable increases in Gurgaon, Faridabad, and Ghaziabad subdistricts. Croplands exhibited the greatest decrease in ESV in Gurgaon, Gautama Buddha Nagar, and Najafgarh. Changes in ecosystem services in urban areas affect air quality, the urban heat island effect, water infiltration, and recreational opportunities. These findings offer valuable insights for planners and scientists to develop sustainable management plans for local ecosystems, considering the effects of LULC changes on ecosystem services.

Chapter 17 provides a comprehensive evaluation of Urban Ecological Security (UES), analyzing the impact of urbanization pressure and ecological sensitivity within the context of the urban ecology and development nexus. Assessing UES is integral to sustainable urban development. The primary objective is to assess the ecological security status of rapidly growing metropolitan cities using advanced models, contributing to methodological advancements in UES. The findings offer scientifically robust insights to optimize land use planning strategies, potentially enhancing ecological security in urban areas. Moreover, the study emphasizes the importance of considering ecological security as a fundamental aspect of urban development, aligning with broader sustainability and environmental preservation goals.

Chapter 18 delves into the issues and challenges of land acquisition processes for urban housing projects in a rapidly expanding city in North India, focusing on the Aligarh City Master Plan. The study examines the realities of land acquisition policies, which often mirror the Land Acquisition Act of 1894. Utilizing a multi-stakeholder analysis approach, the study investigates issues such as lengthy acquisition processes, delayed land transfers, ownership disputes, landowners' unwillingness, and competition from the real estate market. The findings highlight a significant gap between housing demand and government supply mechanisms, with only 5216 housing plots developed over four decades, despite numerous planned projects. The study raises questions about the effectiveness and suitability of current land acquisition policies in addressing the housing needs of growing urban populations.

Chapter 19 employs the Random Forest (RF) approach to identify unrecognized slums and predicts slum expansion using the Cellular Automata-Markov Chain (CA-MC) technique in Hyderabad. Training samples collected from Google Earth are used to delineate slum boundaries, validated against original boundaries. The study presents a detailed methodology for informal slum delineation in cities. The chapter advocates that slums pose urban threats due to overcrowding, inadequate housing, and services, often neglected due to inadequate urban laws and management policies. The absence of sufficient slum maps hampers assessment and management efforts.

Chapter 20 examines the spatiotemporal patterns of LST to analyze the extent and characteristics of UHI in Herat, a major city in Afghanistan. Seasonal LST and UHI patterns are correlated with LULC changes over a decade (winter and summer; 2008–2009 and 2018–2019). Findings reveal rapid urban expansion during the study period, with barren land and vegetation being converted to built-up areas, expanding Herat's built-up area by 61.7%. The UHI effect is most pronounced in the summer, with the highest intensity observed in the city center due to dense urbanization and high anthropogenic activities. This study aids city planners in formulating sustainable land management policies.

Chapter 21 explores the dynamics of LST and its influencing factors in Kolkata Urban Agglomeration. The chapter examines different development zones to assess whether selected determinants exert varying effects on LST. The findings indicate that impervious surfaces like roads and buildings notably influence LST, with areas featuring higher impervious surface density exhibiting elevated LST. Air pollutants such as SO_2, nitrogen dioxide (NO_2), and carbon monoxide (CO) are positively correlated with nighttime LST. Decreased sky view leads to heightened heat retention within buildings and increased UHI effect. These results underscore the necessity for thoughtful urbanization and economic development planning to mitigate the UHI effect and alleviate LST.

Chapter 22 revisits the scenarios during the COVID-19 lockdown, when cities showed unexpected local climatic variations. Motivated by this observation, the chapter demonstrates temperature variations in cities during the pandemic years (2019–2021). It investigates factors such as pollutant concentration that contributed to these temperature changes over four years (2019–2022). Analysis revealed that, except for New Delhi and Kolkata, all cities experienced lower temperatures than usual during the summer months of the pandemic, despite the presence of urban heat effects. In New Delhi, aerosol index showed a positive correlation (0.29) with land surface temperature, while in Bangalore, carbon monoxide (CO) and nitrogen dioxide (NO_2) exhibited the highest correlations (0.57 and 0.69, respectively) with land surface temperature.

7 Roadmap for Understanding Urban Ecosystems

Urban ecosystems are complex and dynamic systems comprising a network of interactions between humans, flora, fauna, and the environment. Understanding these ecosystems is crucial for sustainable urban development and the well-being of urban residents. One of the foremost issues is urbanization and ecosystem dynamics, wherein urbanization alters natural landscapes, leading to changes in biodiversity, habitat loss, and ecosystem services. Understanding the dynamics of urban ecosystems involves studying the patterns of land use, vegetation cover, and wildlife distribution. Urban ecosystems provide essential services such as clean air, water regulation, climate regulation, and recreational opportunities. Evaluating these services helps quantify the benefits that ecosystems offer to urban residents. Often, urban areas are also habitats for diverse wildlife species, leading to interactions with humans. One of the fundamental reasons is that urbanization leads to habitat fragmentation and loss, affecting wildlife populations and ecosystem connectivity. Addressing these challenges requires urban planning strategies that prioritize green spaces, wildlife corridors, and ecological connectivity. Understanding these interactions is vital for managing conflicts, conserving biodiversity, and promoting coexistence between humans and wildlife. Urban areas are also known to face pollution from sources such as vehicle emissions, industrial activities, and waste disposal. Mitigating pollution and environmental degradation is essential for safeguarding urban ecosystems and public health. Access to green spaces and ecosystem services is often unequal across urban communities, leading to environmental injustice. Promoting social equity and inclusivity in urban planning ensures that all residents benefit from healthy and resilient urban ecosystems.

Geospatial tools encompassing remote sensing technologies, GPS, and information communication technology, including detailed field data, facilitate the mapping and monitoring of urban ecosystems. These tools provide valuable data on land cover changes, vegetation health, and urban heat islands. Such studies, supported by systematic field surveys and monitoring programs, allow researchers to assess biodiversity, habitat quality, and species distributions within urban areas. These surveys provide insights into the ecological health and functioning of urban ecosystems. However, integrating social and ecological perspectives helps understand the interactions between human activities and ecosystem dynamics. Socio-ecological research methods involve stakeholder engagement, participatory mapping, and community-based monitoring.

Understanding urban ecosystems is a multidisciplinary endeavor that requires collaboration between scientists, policymakers, urban planners, and local communities. By adopting holistic approaches and innovative methodologies, we can develop effective strategies for managing urban ecosystems sustainably (Table 1). This roadmap for understanding urban ecosystems lays the foundation for creating resilient, biodiverse, and livable cities in the face of rapid urbanization and environmental change.

Table 1 Description of different steps for the effective strategies for managing urban ecosystems sustainably

Major stakes	Description
Contextual analysis	Begin by providing a brief overview of the urbanization trends and environmental challenges faced by cities worldwide. Highlight the importance of understanding urban ecosystems in the context of sustainable development, human Well-being, and biodiversity conservation.
Conceptual framework	Define key concepts such as urban ecosystems, ecosystem services, and human–wildlife interactions. Discuss the interconnectedness of social, ecological, and economic factors shaping urban ecosystems.
Methodological approaches	Outline different research methodologies and tools used to study urban ecosystems, including remote sensing, GIS, ecological surveys, and socio-ecological approaches. Provide examples of how these methodologies can be applied to assess land cover changes, biodiversity, ecosystem services, and human impacts on urban ecosystems.
Key components of urban ecosystems	Identify and discuss the main components of urban ecosystems, such as green spaces, water bodies, built infrastructure, and wildlife habitats. Explore the functions and services provided by each component and their contributions to urban resilience and sustainability.
Drivers of change	Examine the drivers of change affecting urban ecosystems, including urbanization, land use change, pollution, climate change, and socioeconomic factors. Discuss the interactions and feedback loops between these drivers and their impacts on ecosystem structure and function.

Table 1 (continued)

Major stakes	Description
Challenges and opportunities	Identify the key challenges facing urban ecosystems, such as habitat fragmentation, pollution, inequitable access to green spaces, and loss of biodiversity. Discuss opportunities for addressing these challenges through sustainable urban planning, green infrastructure development, and community engagement.
Case studies and best practices	Present case studies highlighting successful examples of urban ecosystem management and conservation initiatives from different cities around the world. Analyze the factors contributing to their success, and identify lessons learned that can be applied in other urban contexts.
Policy implications and recommendations	Provide policy recommendations for policymakers, urban planners, and stakeholders to promote the conservation and sustainable management of urban ecosystems. Emphasize the importance of integrating ecological considerations into urban planning processes and fostering multi-stakeholder collaboration.
Future directions	Discuss emerging research areas and innovative approaches for advancing our understanding of urban ecosystems, such as ecosystem-based adaptation, NbS, and green infrastructure planning. Highlight the need for interdisciplinary research and continued monitoring of urban ecosystems to inform evidence-based decision-making and promote urban sustainability.
Conclusion	Summarize the key insights and findings from the framework. Reiterate the importance of understanding and conserving urban ecosystems for promoting sustainable and resilient cities in the face of global environmental challenges.

8 Conclusion

Geospatial Science for Urban Ecosystems: Insights from India and Beyond intricately traces the evolution of urban ecosystem research, spotlighting the harmonious relationship between nature and human-built environments within contemporary Indian cities and beyond. Geospatial science, encompassing cutting-edge technologies such as GIS, remote sensing, satellite imagery, and GPS, emerges as the cornerstone in understanding, managing, and thriving within urban landscapes in the Indian context. Tailored for students, academics, and urban planners alike, this book underscores geospatial science‘s pivotal role in urban studies, heralding an era of data-driven decision-making in Indian urban planning, environmental preservation, and public policy. Embarking on a comprehensive exploration of urban ecosystems, the book delves into the intricate interplay of social, economic, and environmental dynamics that shape urban spaces in India and globally. It meticulously examines how geospatial science facilitates the monitoring of Indian urban landscapes, the efficient allocation of resources, and the mitigation of city-specific challenges. Through a myriad of global case studies, it vividly illustrates the practical applications of geospatial tools, spanning from transportation planning to climate change adaptation, with a focus on the Indian context. At its core, the book strives to impart a holistic comprehension of the fluid nature of urban environments in India, adept at adapting to demographic shifts, technological innovations, and environmental exigencies. By underscoring urban ecosystems as the linchpin of sustainable urban growth in India and beyond, it seeks to foster a deeper appreciation for this interdisciplinary field of inquiry within the Indian context. Assembled through collaborative efforts from diverse contributors, the book amalgamates a spectrum of perspectives, theories, and methodologies to stimulate critical analysis and innovative problem-solving in the Indian urban context. It serves as both a scholarly reference, a practical handbook for Indian urban professionals, and a wellspring of inspiration for the burgeoning cohort of Indian urban scholars. Recognizing the myriad challenges confronting Indian cities, the book accentuates the potential for resilience, sustainability, and livability through the application of geospatial tools and knowledge in the Indian context. It extends an invitation to Indian readers to embark on an exhilarating expedition through the nexus of Indian urban ecosystems and geospatial science, where groundbreaking ideas shape the Indian urban landscape of tomorrow, one stride at a time.

References

Ahern, J. (2013). Urban landscape sustainability and resilience: The promise and challenges of integrating ecology with urban planning and design. *Landscape Ecology, 28*, 1203–1212.

Aithal, B. H., & Ramachandra, T. V. (2016). Visualization of urban growth pattern in Chennai using geoinformatics and spatial metrics. *Journal of the Indian Society of Remote Sensing, 44*(4), 617–633.

Alberti, M. (2008). *Advances in urban ecology: Integrating humans and ecological processes in urban ecosystems*. Springer.

Anees, M.M., Mann, D. and Mahato, S. (2023). Urban ecosystems research in India: advances and opportunities. *Current Landscape Ecology Reports, 8*(1), pp.34–48.

Arif, M. M., Ahsan, M., Devisch, O., & Schoonjans, Y. (2022). Integrated approach to explore multidimensional urban morphology of informal settlements: The case studies of Lahore, Pakistan. *Sustainability, 14*(13), 7788.

Batty, M. (2013). *The new science of cities*. MIT press.

Chandrashekar, M. C., Gupta, N., & Aithal, B. H. (2021). Analyzing and predicting urban expansion and its effects on surface temperature for two Indian megacities: Bengaluru and Chennai. In *Geospatial technology and smart cities* (pp. 93–116). Springer.

Clarke, K. C. (1997). *Getting started with geographic information systems*. Prentice Hall.

Colomina, I., & Molina, P. (2014). Unmanned aerial systems for photogrammetry and remote sensing: A review. *ISPRS Journal of Photogrammetry and Remote Sensing, 92*, 79–97.

Devendran, A. A., & Lakshmanan, G. (2019). Analysis and prediction of urban growth using neural-network-coupled agent-based cellular automata model for Chennai Metropolitan Area, Tamil Nadu, India. *Journal of the Indian Society of Remote Sensing, 47*(9), 1515–1526.

Dutta, D., Rahman, A., Paul, S. K., & Kundu, A. (2019). Changing pattern of urban landscape and its effect on land surface temperature in and around Delhi. *Environmental Monitoring and Assessment, 191*(9), 1–15.

Elmqvist, T., Folke, C., Nyström, M., Peterson, G., Bengtsson, J., Walker, B., & Norberg, J. (2003). Response diversity, ecosystem change, and resilience. *Frontiers in Ecology and the Environment, 1*(9), 488–494.

Escobar, N., & Britz, W. (2021). Metrics on the sustainability of region-specific bioplastics production, considering global land use change effects. *Resources, Conservation and Recycling, 167*, 105345.

Geertman, S., Allan, A., Zhan, Q., & Pettit, C. (2019). Computational urban planning and management for smart cities: An introduction. *Computational Urban Planning and Management for Smart Cities, 16*, 1–14.

Geng, S., Yang, L., Sun, Z., Wang, Z., Qian, J., Jiang, C., & Wen, M. (2021). Spatiotemporal patterns and driving forces of remotely sensed urban agglomeration heat islands in South China. *Science of the Total Environment, 800*, 149499.

Graham, E. B., Averill, C., Bond-Lamberty, B., Knelman, J. E., Krause, S., Peralta, A. L., Shade, A., Smith, A. P., Cheng, S. J., Fanin, N., & Freund, C. (2021). Toward a generalizable framework of disturbance ecology through crowdsourced science. *Frontiers in Ecology and Evolution, 9*, 588940.

Grimm, N. B., Faeth, S. H., Golubiewski, N. E., Redman, C. L., Wu, J., Bai, X., & Briggs, J. M. (2008). Global change and the ecology of cities. *Science, 319*(5864), 756–760.

Hebel, M., Arens, M., & Stilla, U. (2013). Change detection in urban areas by object-based analysis and on-the-fly comparison of multi-view ALS data. *ISPRS Journal of Photogrammetry and Remote Sensing, 86*, 52–64.

IPCC. (2021). *Climate change 2021: The physical science basis. Contribution of Working Group I to the Sixth Assessment Report of the Intergovernmental Panel on Climate Change.*

Jensen, J. R. (2009). *Remote sensing of the environment: An earth resource perspective 2/e.* Pearson Education India.

Kim, S., Park, S., Han, J., Son, S., Lee, S., Han, K., Kim, J., & Kim, J. (2019). Feasibility of UAV photogrammetry for coastal monitoring: A case study in Imlang Beach, South Korea. *Journal of Coastal Research, 90*(SI), 386–392.

Kobayashi, Y., & Kinoshita, T. (2023). Abandoned farmland detection using single-year satellite images in Japan. *Journal of Applied Remote Sensing, 17*(1), 014517–014517.

Kundu, S., Rana, N.K. and Mahato, S. (2024). Unravelling blue landscape fragmentation effects on ecosystem services in urban agglomerations. *Sustainable Cities and Society, 102*, p.105192.

Li, S., Dragicevic, S., Castro, F. A., Sester, M., Winter, S., Coltekin, A., Pettit, C., Jiang, B., Haworth, J., Stein, A., & Cheng, T. (2016). Geospatial big data handling theory and methods: A review and research challenges. *ISPRS Journal of Photogrammetry and Remote Sensing, 115*, 119–133.

Li, X., Levin, N., Xie, J., & Li, D. (2020). Monitoring hourly night-time light by an unmanned aerial vehicle and its implications to satellite remote sensing. *Remote Sensing of Environment, 247*, 111942.

Lillesand, T., Kiefer, R. W., & Chipman, J. (2015). *Remote sensing and image interpretation*. Wiley.

Longley, P. A., Goodchild, M. F., Maguire, D. J., & Rhind, D. W. (2015). *Geographic information science and systems*. Wiley.

Mahato, S., Pal, S. and Ghosh, K.G. (2020). Effect of lockdown amid COVID-19 pandemic on air quality of the megacity Delhi, India. *Science of the total environment, 730*, p.139086.

Mandal, J., Ghosh, N., & Mukhopadhyay, A. (2019). Urban growth dynamics and changing land-use land-cover of megacity Kolkata and its environs. *Journal of the Indian Society of Remote Sensing, 47*(10), 1707–1725.

McDonnell, M. J., & Pickett, S. T. (1990). Ecosystem structure and function along urban-rural gradients: An unexploited opportunity for ecology. *Ecology, 71*(4), 1232–1237.

McKinney, M. L. (2006). Urbanization as a major cause of biotic homogenization. *Biological Conservation, 127*(3), 247–260.

McPhearson, T., Pickett, S. T., Grimm, N. B., Niemelä, J., Alberti, M., Elmqvist, T., Weber, C., Haase, D., Breuste, J., & Qureshi, S. (2016). Advancing urban ecology toward a science of cities. *Bioscience, 66*(3), 198–212.

Mdaopa, B. E., Gumbo, T., Moyo, T., & Musonda, I. (2023, September). Functional applications of unmanned aerial vehicle Technology in Urban Planning Practice: A PRISMA systematic review. In *LET IT GROW, LET US PLAN, LET IT GROW. Nature-based solutions for sustainable resilient smart green and blue cities. Proceedings of REAL CORP 2023, 28th international conference on Urban Development, Regional Planning and Information Society* (pp. 305–314). CORP–Competence Center of Urban and Regional Planning.

Mitchel, A. (2005). The ESRI guide to GIS analysis, volume 2: Spartial measurements and statistics. *ESRI Guide to GIS Analysis, 2.*

Miyoshi, C., & Torrell, P. P. (2019). Flying economy–The economic impact of new propulsion technology: A cost-benefit analysis of the geared turbo fan engine on the London Heathrow and Frankfurt Route. In *Airline economics in Europe* (pp. 201–228). Emerald Publishing Limited.

Moghadam, H. S., & Helbich, M. (2013). Spatiotemporal urbanization processes in the megacity of Mumbai, India: A Markov chains-cellular automata urban growth model. *Applied Geography, 40*, 140–149.

Munshi, T., Zuidgeest, M., Brussel, M., & van Maarseveen, M. (2014). Logistic regression and cellular automata-based modelling of retail, commercial and residential development in the city of Ahmedabad, India. *Cities, 39*, 68–86.

Onda, K., Sinha, P., Gaughan, A. E., Stevens, F. R., & Kaza, N. (2019). Missing millions: Undercounting urbanization in India. *Population and Environment, 41*(2), 126–150.

Pal, S., Mahato, S., Mandal, I., Das, P. and Joshi, P.K. (2022). Evaluating aerosols concentration and air quality of Indian urban agglomerations over nationwide and regional lockdown. *Atmospheric Pollution Research, 13*(10), p.101567.

Pataki, D. E., Carreiro, M. M., Cherrier, J., Grulke, N. E., Jennings, V., Pincetl, S., Pouyat, R. V., Whitlow, T. H., & Zipperer, W. C. (2011). Coupling biogeochemical cycles in urban environments: Ecosystem services, green solutions, and misconceptions. *Frontiers in Ecology and the Environment, 9*(1), 27–36.

Pickett, S. T. A. (2010). Urban ecological systems: Linking terrestrial ecological, physical, and socioeconomic components of metropolitan areas. *Journal of Environmental Management, 92*, 331–362.

Rathi, N., Mahato, S., Pandey, D.N. and Joshi, P.K. (2023). Identifying the Flood Hazard Zones in Urban Area Using Flood Hazard Index (FHI)—A Case of Capital City of India. In Advancements in Urban Environmental Studies: Application of Geospatial Technology and Artificial Intelligence in Urban Studies (pp. 341–357). Cham: Springer International Publishing.

Saboori, M., Homayouni, S., Shah-Hosseini, R., & Zhang, Y. (2022). Optimum feature and classifier selection for accurate urban land use/cover mapping from very high resolution satellite imagery. *Remote Sensing, 14*(9), 2097.

Sahana, M., Hong, H., & Sajjad, H. (2018). Analyzing urban spatial patterns and trend of urban growth using urban sprawl matrix: A study on Kolkata urban agglomeration, India. *Science of the Total Environment, 628*, 1557–1566.

Seto, K. C., Reenberg, A., Boone, C. G., Fragkias, M., Haase, D., Langanke, T., Marcotullio, P., Munroe, D. K., Olah, B., & Simon, D. (2012). Urban land teleconnections and sustainability. *Proceedings of the National Academy of Sciences, 109*(20), 7687–7692.

Sharma, S., Nahid, S., Sharma, M., Sannigrahi, S., Anees, M. M., Sharma, R., Shekhar, R., Basu, A. S., Pilla, F., Basu, B., & Joshi, P. K. (2020). A long-term and comprehensive assessment of urbanization-induced impacts on ecosystem services in the capital city of India. *City and Environment Interactions, 7*, 100047.

Shekhar, S. (2021). Urbanization in India. In *Slum development in India* (pp. 1–20). Springer.

Sheladiya, K. P., & Patel, C. R. (2023). Leveraging Urban Growth Models (UGM) for sustainable urban planning and climate resilient cities: A bibliometric analysis. *Journal of Scientometric Research, 12*(3), 585–595.

Si Salah, H., Ait-Aoudia, S., Rezgui, A., & Goldin, S. E. (2019). Change detection in urban areas from remote sensing data: A multidimensional classification scheme. *International Journal of Remote Sensing, 40*(17), 6635–6679.

Sonaviya, D. R., & Tandel, B. N. (2019). A review on GIS based approach for road traffic noise mapping. *Indian Journal of Science and Technology, 12*(14), 1–6.

Tan, S. Y., Taeihagh, A., & Pande, D. (2023). Data sharing in disruptive technologies: Lessons from adoption of autonomous systems in Singapore. *Policy Design and Practice, 6*(1), 57–78.

Ul Din, S., & Mak, H. W. L. (2021). Retrieval of Land-Use/Land Cover Change (LUCC) maps and urban expansion dynamics of Hyderabad, Pakistan via Landsat datasets and support vector machine framework. *Remote Sensing, 13*(16), 3337.

United Nations. (2018). 68% of the world population projected to live in urban areas by 2050, says UN. https://www.un.org/development/desa/en/news/population/2018-revision-of-world-urbanization-prospects.html.

Verma, S., Chatterjee, A., & Mandal, N. R. (2017). Analysing urban sprawl and shifting of urban growth centre of Bengaluru city, India using Shannon's entropy method. *Journal of Settlements and Spatial Planning, 8*(2), 89–98.

Waddell, P. (2002). UrbanSim: Modeling urban development for land use, transportation, and environmental planning. *Journal of the American Planning Association, 68*(3), 297–314.

World Health Organization. (2018). 9 out of 10 people worldwide breathe polluted air, but more countries are taking action. https://www.who.int/news/item/02-05-2018-9-out-of-10-people-worldwide-breathe-polluted-air-but-more-countries-are-taking-action.

Yan, Y., Feng, C. C., Huang, W., Fan, H., Wang, Y. C., & Zipf, A. (2020). Volunteered geographic information research in the first decade: A narrative review of selected journal articles in GIScience. *International Journal of Geographical Information Science, 34*(9), 1765–1791.

Yatoo, S. A., Sahu, P., & Kalubarme, M. H. (2019). *Sustainable approaches for urban sprawl management of Ahmedabad city* (p. 130). Advance and Innovative Research.

Yu, S., Zhang, Z., Liu, F., Wang, X., & Hu, S. (2021). Urban expansion in the megacity since 1970s: A case study in Mumbai. *Geocarto International, 36*(6), 603–621.

Zhou, G. (2020). *Urban high-resolution remote sensing: Algorithms and modeling*. CRC Press.

Zhu, W., Cao, Z., Luo, P., Tang, Z., Zhang, Y., Hu, M., & He, B. (2022). Urban flood-related remote sensing: Research trends, gaps and opportunities. *Remote Sensing, 14*(21), 5505.

Patterns of Urban Growth: Megacities in the Developing World

Saurav Chakraborty

Abstract Studying urban forms and dynamics is crucial for achieving urban sustainability. However, the influence of different urban forms on city expansion and suburbanization remains largely unknown. This study introduces an Urban Form Index (UFI) based on urban land density and applies it to 151 metropolises in Africa, China, and India (ACI) to assess the urban forms and their effects on expansion. The UFI categorizes cities into four types: compact, highly compact, sprawled, and highly sprawled. African and Indian metropolises are mainly compact/highly compact, while China leans toward sprawled forms. Analyzing the UFI values from 1985 to 2015 reveals 16 dynamic urban forms. African and Indian cities maintain or transition toward compact forms, while Chinese cities shift from compact to sprawled forms. Urban expansion rate, land density distribution, spatial extent, and built-up density are influenced by these forms, with regional variations. Recognizing diverse urban expansion trends in the Global South questions universal planning theories and emphasizes the need for context-specific urban planning to ensure sustainable urbanization in the future.

Keywords Metropolises · Urban form · Global South · Suburbanization · Urban sustainability · Urban land density

1 Introduction

As of 2018, 43% of the worldwide urban population resided in 548 metropolises, each with a population exceeding one million people (United Nations, 2018). These metropolises, particularly in the "Global South," play a crucial role in national and global economies, driving socioeconomic development (Dadashpoor & Ahani,

S. Chakraborty (✉)
Department Geography, Faculty of Natural and Mathematical Sciences, Presidency University, Kolkata, India
e-mail: saurav.geog@presiuniv.ac.in

P. K. Joshi et al. (eds.), *Geospatial Science for Urban Ecosystems*, The Urban Book Series, https://doi.org/10.1007/978-3-032-07155-2_2

2021). Around 815 million people resided in 254 metropolises within Africa, China, and India (ACI) (United Nations, 2018). ACI metropolises are rapidly expanding due to population growth (Adhikari & de Beurs, 2017; Cobbinah et al., 2015), which is expected to continue until 2050 (United Nations, 2018). This expansion poses irreversible environmental changes, impacting land use, landscape structure, biodiversity, and ecosystem services (van Vliet, 2019; Huilei et al., 2017; Chakraborty, et al., 2021a; He et al., 2014; Das et al., 2020, 2021). Understanding evolving urban forms and expansion in ACI metropolises is crucial to mitigate these impacts and inform urban planning (Dadashpoor et al., 2019; Fei & Zhao, 2019; Xu et al., 2019c), fostering sustainable urban futures (Nagendra et al., 2018; Cobbinah et al., 2015; Tong et al., 2019; Zhang & Su, 2016).

Numerous studies have concentrated on examining the spatial expansion and growth patterns of cities (Bagheri & Tousi, 2018; Chakraborty, et al., 2021a; Wu et al., 2021; Sahana et al., 2018), with the aim of characterizing their urban macro-patterns. These analyses often delve into the concept of urban forms and their dynamics (Xu, et al., 2019a, b). Urban form describes the physical characteristics of different parts of a city, i.e., the city core, its periphery, and their relationship (Anderson et al., 1996; Živković, 2020). Usually, a landscape metrics-based approach (i.e., built-up patch shape, size, aggregation, and fragmentation) is used to identify the different typologies of urban forms (Chakraborty et al., 2024; Lemoine-Rodríguez et al., 2020; Huang et al., 2007; Xia et al., 2022). Nevertheless, the urban land density attribute, specifically, is acknowledged as a significant factor influencing city forms (Angel et al., 2010, 2020; Jiao, 2015; Schneider & Woodcock, 2008). Both the macroscale (city-wide) and microscale (neighborhood) built-up density attributes are employed to assess and analyze urban forms, along with their temporal changes (e.g., Chakraborty, et al. 2021a and Shukla et al. 2021).

The decline in urban land density from the city center to its outskirts is characterized by the inverse sigmoid curve function introduced by Jiao (2015). This curve has been utilized in recent studies (e.g., Sumari et al. (2020) and Wu et al. (2021)) to comprehend city forms and expansion, informing decision-making processes. For instance, Xu, et al. (2019a) used it to analyze 25 African cities, finding that larger cities are more compact. Wu et al. (2021) did a similar analysis for Latin American cities. Urban form arises from urban core and suburban growth rates (Jiao, 2015). Dong et al. (2019) identified four form typologies using centrality and compactness attributes derived from the curve fitting statistic. However, considering the "degree of urban form" (Jabareen, 2006; Schneider & Woodcock, 2008) is vital. Tsai (2005) and others examined this degree in terms of population, employment distribution, and spatial extents, affecting land use, living conditions, and ecology. This study emphasizes comparing the urban core and suburban areas' sizes, affecting the inverse sigmoid curve slope (Jiao, 2015). While urban growth inequality is documented, quantitatively analyzing unequal territorial extents remains unexplored. We compute the ratio between urban core and suburban areas to gauge cities' "degree of urban form."

Urban forms can be broadly categorized into compact or sprawled. Compact forms have denser and extensive urban cores, while sprawled forms have larger suburban areas (Dong et al., 2019). This distinction impacts urban expansion traits like hinterland land density and expansion rates (Xu, et al., 2019b). For instance, Xu, et al. (2019a) noted growing hinterland density over time, even in more compact cities, with spatial extents varying across similar forms. Schneider and Woodcock (2008) found increasing outskirt density, mainly in cities with smaller cores (sprawled forms). Xu et al. (2019b) linked expansion traits to the city size, based on population. Despite urban forms' impact on expansion, few studies comprehensively compare form attributes—curve shape, expansion rate, spatial extent, and hinterland density (Jiao, 2015; Schneider & Woodcock, 2008). Urban forms change as cities grow, creating temporal dynamics. Identifying dynamic forms and their traits enhances the understanding of expansion patterns. Regional variations of dynamic forms and associated suburban traits remain underexplored, adding importance to such research.

Comparisons of urban forms and patterns of spatial expansion are particularly useful in discerning global diversities in urbanization (Acuto et al., 2018; Angel et al., 2020). Rapid city growth is one of the most striking aspects of contemporary urbanization in the Global South (Angel et al., 2011; Güneralp et al., 2020). However, before this growth can be denoted as urban sprawl (frequently referred to as low-density development—e.g., Ewing (1997)), a quantitative assessment of these cities' urban forms and their characterization is required. Studies by Huang et al. (2007) and Jiao et al. (2021) have reported that cities in the Global South are more compact compared to those in the Global North. On the contrary, empirical studies have compared urban expansion patterns between the Global North and South, surmising that in the Global South, cities have developed a sprawled form and have also witnessed rapid urban expansion (e.g., He et al. (2019), Kuang et al. (2014), Lemoine-Rodríguez et al. (2020), and Schneider and Woodcock (2008)). We believe that the reasons behind these paradoxical claims need careful reassessment.

Urban forms vary across fast-growing Global South cities, leading to distinct expansion patterns (Wu et al., 2021; Xu, et al., 2019a). To grasp the nuances, an in-depth cross-country/regional analysis is essential (Güneralp et al., 2018; Nagendra et al., 2018). Although comparisons between cities in the Global North and Global South - particularly those involving Chinese cities (Dong et al. (2019) - are relatively common, intra-Global South comparisons remain underexplored (United Nations, 2014). Analyzing dynamic urban forms in ACI metropolises addresses whether generalized assumptions about urban forms in fast-growing Global South regions hold true, given the unique nature of these cities. Examining intra-regional urban form differences helps understand contemporary Global South urbanization (Randolph & Storper, 2020), refining urban planning theories for this region.

This study thus has two interrelated objectives. The first is to determine the degree of urban form using an *Index of Urban Form* (IUF), based on the area present in a city under its two broad spatial territories, i.e., the urban core and the suburban area, as demarcated using the urban land density function (Jiao, 2015), and discern their changes over time to identify the inherent dynamic urban forms. The second is to compare the occurring macro-urban expansion characteristics (the rate of urban expansion, land density distribution and its alterations, changing spatial extents, and the presence of low-density built-up areas beyond the urban fringe) across different dynamic urban forms. For this, we have examined a total of 151 metropolises situated across the ACI in this study.

2 Materials and Methods

2.1 Selection of Metropolises

The United Nations' World's Cities Data Booklet 2018 offers details about metropolitan cities globally (United Nations, 2018). According to this source, there were 548 metropolises worldwide in 2018, with a population of one million people or more. The provided information encompasses the name, country or area, population, and statistical concept (the spatial extent of the city) for each of these metropolises.

Out of the 548 metropolises worldwide, 254 are situated in ACI (Africa: 63, China: 130, and India: 61). However, due to insufficient built-up area data in 1985, this analysis has focused on only 151 cities (Africa: 46, China: 89, and India: 16) (Fig. 1; see Sect. 2.4 for the reason). In China and Africa, most of these cities are located along the coastal areas, while they are mostly situated inland (along principal river courses) in India (Fig. 1).

2.2 Built-Up Layers

The freely available Global Annual Urban Dynamics (GAUD) (Liu et al., 2020) dataset was used to elicit and frame the raster layers of the built-up extents for all the studied cities. This dataset encompasses global urban expansion from 1985 to 2015, determined through the analysis of Landsat images (at a spatial resolution of 30 m) on the Google Earth Engine platform. The creation of the GAUD dataset involved a two-step process. Initially, maps/layers representing global human settlement, global urban land, global artificial area, and global urban footprint (all of similar resolution) were integrated to outline the boundaries of urban lands between 1985 and 2015. During this process, particular attention was given to pixels that transitioned from nonurban in 1985 to urban in 2015. Additionally, the Normalized Urban Areas Composite Index was utilized to identify and extract built-up lands from Landsat images, forming an annual time series from 1985 to 2015. The accuracy of the final urban extent raster dataset was notably high compared to other available global built-up layers (Liu et al., 2020).

In this investigation, the GAUD dataset underwent reclassification for three distinct study years: 1985, 2000, and 2015. Initially, we utilized the built-up cells from

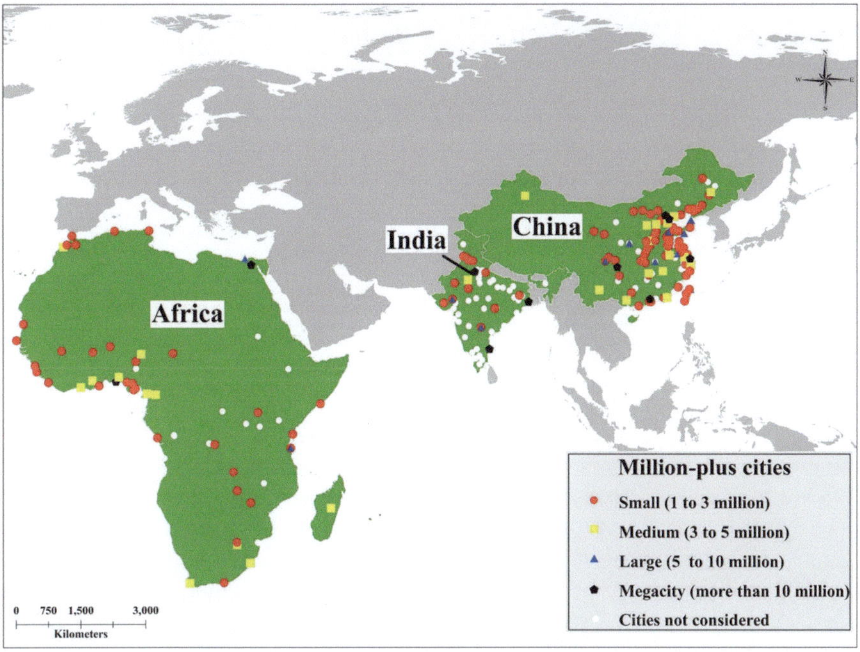

Fig. 1 Distribution of metropolis cities in Africa, China, and India by city sizeNote: For regional divisions of Africa, China, and India, see Supplementary Fig. S1. A total of 212 metropolises are present in ACI. Of these, 151 metropolises were studied to identify their respective dynamic urban forms due to the absence of the urban core territory in the other metropolises in 1985. For visualization purposes only, we have divided these 151 metropolises into four distinct classes according to their 2018 population

1985 as our foundational dataset. Subsequently, the annual built-up cells spanning from 1985 to 2000 were amalgamated into a single layer, constituting the built-up layer for the year 2000. Finally, we reclassified the entire 30-year dataset into a single class to prepare the built-up extents for 2015 (Supplementary Fig. S2).

2.3 *Urban Land Density and Its Function*

Urban land density was computed by creating a set of concentric ring buffers, each 1-km wide, originating from the designated city centers (Jiao, 2015; Schneider & Woodcock, 2008). The city centers for each metropolis were determined through the Google Earth platform and OpenStreetMap (e.g., Bosch et al. (2020)). To validate these locations, we cross-referenced them with the city locations available in the Atlas of Urban Expansion database (http://www.atlasofurbanexpansion.org).

Urban built-up density is characterized as the proportion of built-up land within each of the concentric rings (see Fig. 2). To ensure accuracy, protected lands, such

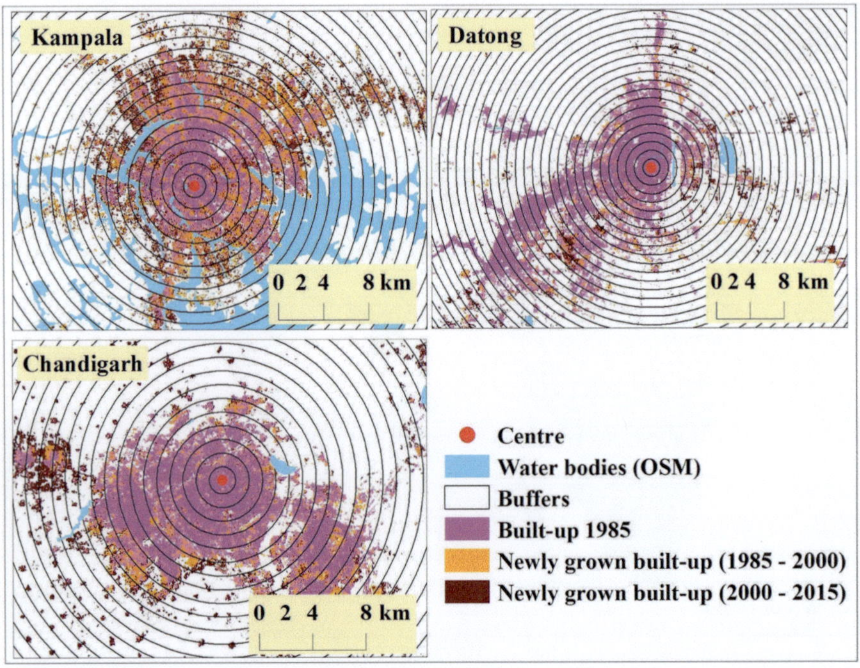

Fig. 2 Distribution of built-up land within concentric rings drawn around the city center (for monocentric cities). Note: For polycentric cities, see supplementary Figs. S3 and S4

as water bodies, must be excluded from the overlain ring area before determining the non-built-up area within each concentric ring, as outlined by Jiao (2015). Therefore, the urban land density for each ring is calculated using the following formula:

$$\text{Urban land density} = \frac{T_B}{T - T_w} \tag{1}$$

where T_B represents the total built-up land under each ring, T represents the total area under each ring, and T_W is the total area under water bodies that are present in each ring.

We used the available OpenStreetMap water layer to calculate the area under water bodies and excluded the same from each ring (see Fig. 2).

Urban land density tends to be higher in the city center, gradually decreasing toward the periphery of the urban core. Following this trend, it experiences a rapid decline within the inner and suburban areas, before gradually decreasing in the peripheral regions. Building upon this concept, Jiao (2015) introduced an "inverse sigmoid curve function" designed to capture the distance-decay nature of built-up land density. The function is expressed as follows:

$$f(r) = \frac{1-c}{1+e^{\alpha((2r/D)-1)}} + c \quad (2)$$

In the equation, f represents urban land density as a function of the distance from the city center (*r*), where e is Euler's number, and α, c, and D are parameters reflecting the key characteristics of the distribution of urban land density. The α parameter gauges the slope between the urban core and suburban areas, indicating the degree of compactness in the city—higher values signify a compact urban form, while lower values suggest a more sprawled nature. Parameter c represents built-up density in the hinterland, and parameter D denotes the estimated boundary distinguishing the city's urban fringe from its hinterland (refer to Jiao (2015) for details). In the case of polycentric cities, buffers were generated around multiple city centers, and curve fittings were conducted based on calculated land density within these buffers (see Jiao (2015) and Supplementary Table S1 for information on the spatial structure—monocentric or polycentric—of all the studied metropolises in ACI).

2.4 *Index of Urban Form*

An important aspect of the proposed inverse sigmoid function by Jiao (2015) is the calculation of the radius under the broad spatial territories, i.e., urban core (built-up density >=75%), inner-urban (built-up density >=50%), and suburban area (built-up density >=25%). These computations can only be done after fitting the parameters e, α, c, and D using the following equation (e.g., Dong et al. (2019) and Jiao (2015)):

$$r = \frac{D}{2}\left(\frac{1}{\alpha}\ln\frac{1-x}{x-c} + 1\right) \quad (3)$$

where *r* represents the radius of a given land density of *x*, i.e., urban core (>= 75%) or suburban (25–50%), and α, *c*, and *D* are parameters for the fitted function.

Utilizing the provided equation, Dong et al. (2019) introduced a two-dimensional urban form, considering centrality and compactness. This classification is based on the ratio of the area under the urban core to the combined area under the inner and suburban regions. A city can be said to be central when its urban core occupies more than 1/3 of its area and can be denoted as compact when its suburban zone occupies less than 1/3 of the total city area. Furthering this idea, here we propose an *Index of Urban Form* (IUF), which is simply a ratio between the proportion of area present under a city's urban core and that within its suburbs. This is written as follows:

$$\text{IUF} = \frac{\text{Proportion of area occupied by urban core}}{\text{Proportion of area occupied by suburban area}} \quad (4)$$

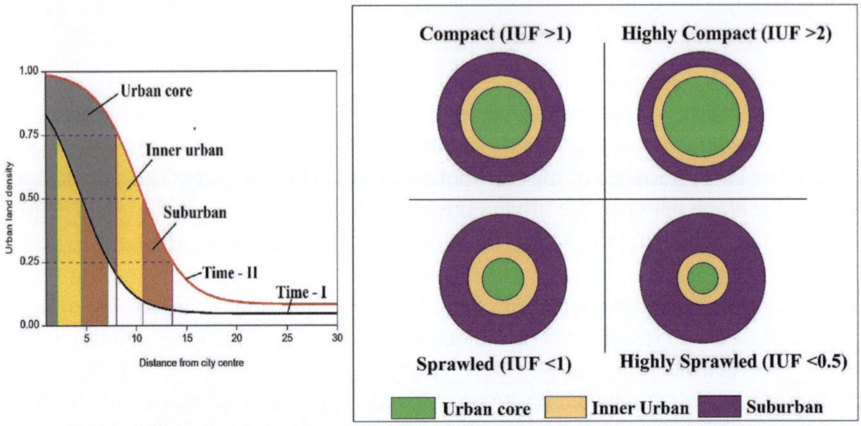

Fig. 3 Graphical depiction of four different urban form types. (Based on Dong et al., 2019)

To perform the aforementioned computation, we initially determined the respective areal proportions for the urban core, inner-urban, and suburban zones. Subsequently, we utilized the obtained figures for the urban core and suburban zone to compute the Index of Urban Form (IUF). When the ratio is greater than 1, the city has a compact form, while a value of IUF <1 represents a sprawled form. An IUF value of 1 signifies a balanced urban form, where the proportion of area under the urban core equals that of the suburban zone. Additionally, we introduce specific typologies within these broad forms: highly compact, characterized by an IUF greater than 2, indicating that the urban core area is more than twice that of the suburban zone, and highly sprawled, with an IUF less than 0.5, indicating that the suburban area is more than twice that of the urban core. Such parsing of the IUF score elicits four distinct types of urban forms, which can be particularly useful to quantitatively state the degree of compaction or sprawling for a city (see Fig. 3).

In 1985, only 151 metropolises in ACI had an urban land density of 75% or above. So, the area under the respective urban cores and suburban zones was computed only for these cities. The changing areas under their urban cores and suburban zones from 1985 to 2015 helped identify 16 different types of dynamic urban forms.

3 Results

3.1 Goodness of Fit Statistics of the Fitted Function

We fitted the nonlinear least squares urban land density function using the Origin 2018 (v 9.5) software on the prepared dataset for the 212 metropolises studied across ACI. Overall, the fitting achieved a high level of accuracy (median values of R^2 were higher than 0.95). However, it varied among countries/continents and over the years as well. Supplementary Table S2 lists the goodness of fit statistics and estimated parameters for each city.

3.2 Temporal Changes in Urban Forms for the Studied Cities

By calculating the ratio of the area proportion under the urban cores to that in suburban zones, we derived the Index of Urban Form (IUF) to categorize metropolises across ACI into four distinct types of urban forms (Table 1). In 1985, a total of 100 cities across ACI had either compact or highly compact urban forms, and 60 of them were located in China, followed by Africa (31) and India (09). Out of the 51 cities that had either sprawled or highly sprawled forms in 1985, 29 were located in China. Over time, the number of compact or highly compact cities in China declined, and cities with sprawled urban forms increased (see Table 1). Contrarily, the number of highly compact cities increased both in Africa and India. In particular, a rise in the number of highly compact cities in Africa was quite noticeable from 1985 to 2015. Moreover, the number of cities with sprawled forms declined in both Africa and India over this time period, and particularly in India, this decline was quite notable.

3.3 Dynamic Urban Forms and Their Spatial Distribution

To better understand these changing dynamics in urban forms, we compared the calculated IUF values of 1985 for each of the four urban forms with their 2015 IUF values and identified 16 different dynamic urban forms for these cities (Fig. 4). These are *remained sprawled, sprawled to compact, sprawled to highly compact, sprawled to highly sprawled, remained highly sprawled, highly sprawled to sprawled, highly sprawled to highly compact, highly sprawled to compact, remained compact, compact to sprawled, compact to highly compact, compact to highly sprawled, remained highly compact, highly compact to compact, highly compact to sprawled,* and *highly compact to highly sprawled*. We found that only 40% of the total 151 studied metropolises had maintained the same urban form from 1985 to 2015, and it was in the highly compact cities where such inertia in the urban form was most notable—e.g., 60% of the highly compact cities of 1985 remained highly compact in 2015. Moreover, both sprawled and highly sprawled cities in 1985 had either become more compact or their degree of sprawling had reduced by 2015. For example, nearly 62% of cities that were sprawled in 1985 had become compact or highly compact by 2015, whereas 50% of the highly sprawled cities continued to maintain this form even in 2015. On the contrary, cities with a compact form in

Table 1 Number of cities by different urban forms in 1985 and 2015

	1985			2015		
Urban forms	Africa	China	India	Africa	China	India
Compact	15	40	4	11	35	6
Highly compact	16	20	5	25	15	8
Sprawled	9	21	5	7	29	2
Highly sprawled	6	8	2	3	10	0

1985 continued to maintain a compact form in 2015 as well, but a growing suburban area around them was also apparent, which had resulted in the development of sprawled forms by 2015. In all, 30% of the studied cities exhibited such a transition from a previous compact form to a sprawled scenario (Fig. 4).

Figure 5 shows the spatial distribution of 151 metropolises having 16 dynamic urban forms across ACI. In Africa, although a number of cities with sprawled forms remained the same during the study period, their transformation from sprawled cities to compact entities was also seen (Fig. 5a). This was particularly noted for cities in central and southern Africa. Similarly, most of the sprawled cities in India had transformed into either compact or highly compact cities during 1985–2015, and these are geographically concentrated in the central, eastern, southern, and western parts of the country (see Supplementary Fig. S5). Most of the cities in China, on the other hand, had maintained their sprawled forms across the study years, and these cities are distributed all over the country. Those that converted and became compact and highly compact during 1985–2015 are specifically seen in the western and central parts of China.

In Africa, most of the compact cities in 1985 were concentrated in the western portion, and these remained compact or became highly compact. This was also true for metropolises located in northern and southern Africa (Fig. 5b). Indian metropolises with compact forms in 1985 also maintained their compact nature over the examined time period, while those that became highly compact were mostly located

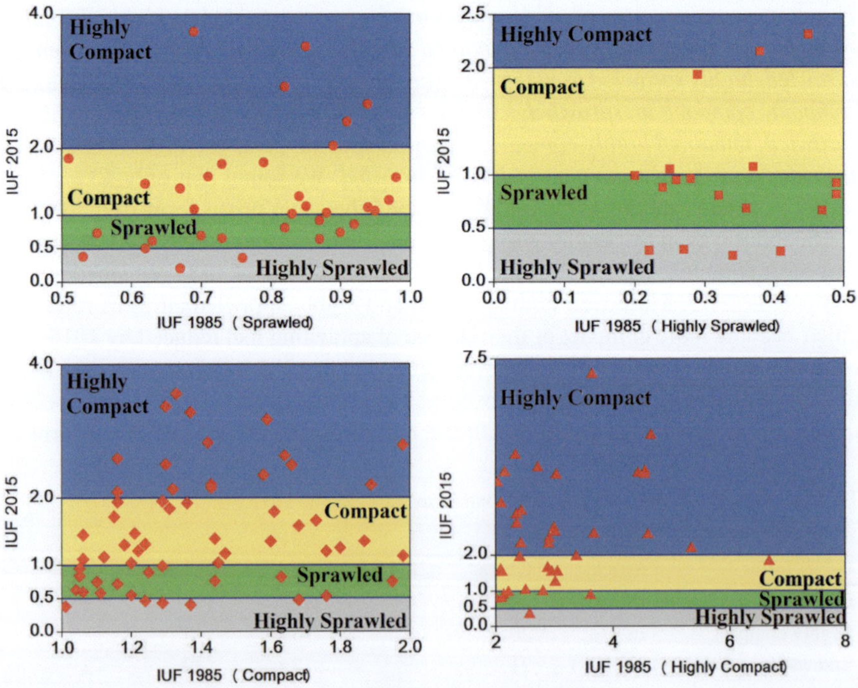

Fig. 4 IUF-based correlation for examining urban form dynamicity

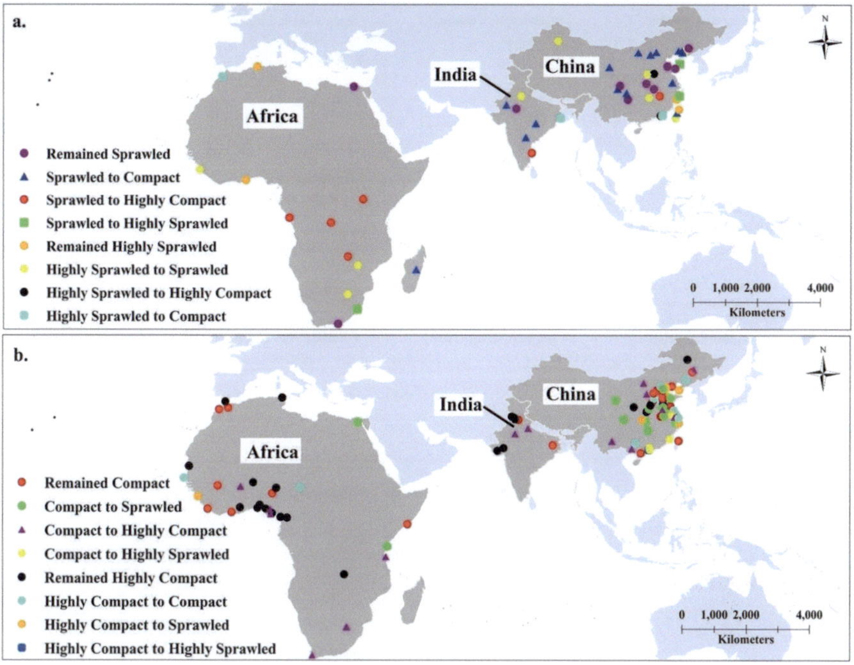

Fig. 5 Spatial distribution of metropolises by their dynamic urban forms across ACI. (**a**) Metropolises with sprawled and highly sprawled urban forms in 1985; (**b**) metropolises with compact and highly compact urban forms in 1985Note: Tables S3 and S4 for detailed IUF values and dynamic urban forms for 153 metropolises across ACI. Also see Supplementary Fig. S5 for the spatial distribution of these metropolises across different regional divisions in ACI

in northern and western India. Contrary to Africa and India, compact Chinese cities in 1985 had transformed into sprawled entities by 2015. Such cities are mostly situated in eastern and central China. Chinese cities that were able to maintain their compact forms, on the other hand, are mostly concentrated in the central region. It is important here to mention that the two extreme types of urban form transformations, i.e., from highly sprawled to highly compact and vice versa, were only noticed for Chinese metropolises and that only one city (Jiangmen) had transformed from being highly compact in 1985 to being highly sprawled by 2015 (see Fig. 5a, b).

3.4 Urban Expansion Characteristics Across Different Dynamic Urban Forms

3.4.1 Rate of the Urban Expansion and Land Density Distribution

Varied urban expansion patterns are mirrored in dynamic urban forms. To comprehend the distinct expansion patterns within the identified 16 dynamic urban forms, we generated generalized urban land density curves for each type (see Fig. 6).

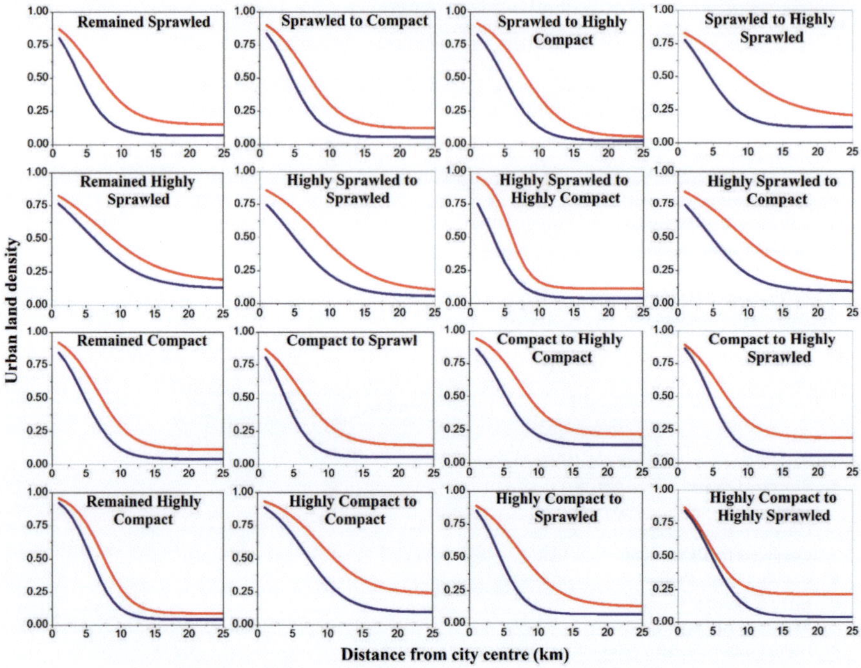

Fig. 6 Generalized land density distribution across 16 dynamic urban forms (fitted curves)

Initially, we categorized cities undergoing similar transformation paths and subsequently calculated the average land density within their concentric rings. Cities transitioning from initial sprawled or highly sprawled forms to compact or highly compact forms exhibited an inverse S-shaped curve in their land density distribution in 2015. In these cities, a phenomenal development of the urban core territory (land density ≥ 75%) was noticed during 1985–2015, wherein the core area (on average) increased from 3.1 km^2 in 1985 to 5.7 km^2 in 2015, with an annual average growth rate of 2.8%. This is quite the opposite of that which occurred in those cities that had remained sprawled or had become sprawled over time, wherein the built-up density increased more rapidly in their suburban areas compared to that within their urban cores. For example, in such cities, their suburban area (on average) doubled over time (from 3.2 km^2 in 1985 to 6.4 km^2 in 2015) compared to the only 1.1-km^2 (on average) increase in their urban core area (from 2.1 km^2 in 1985 to 3.2 km^2 in 2015). Besides this, a linear distance decay of urban land density distribution was also noted for cities within this group; i.e., land densities for sprawled cities declined sharply from their respective city center in a linear fashion and then became stable.

The rate of urban expansion, measured in terms of the interval between the two curves (1985 and 2015), also varies among these different dynamic urban forms (Fig. 6). Cities that either remained or had become sprawled and highly sprawled recorded a much rapid built-up growth in the concentric rings around their city center. In particular, the curve interval was exceptionally high in their suburban areas.

This also reflects an unbalanced growth trajectory between the urban core and suburban zones. Conversely, urban expansion was not as rapid for cities with initial compact/highly compact forms (the paucity of space in the inner city and high land value/rents being the probable causes behind this), so its curve intervals were smaller (Fig. 6). As a result, the urban core territory got consolidated further within the urban landscape, and presently, in these cities, the urban core extends up to a wider distance away from the city center as well.

Supplementary Figs. S6, S7, S8 and S9 show the land density distribution separately for metropolises in Africa, China, and India, based on the different dynamic urban forms. Generally, in China, metropolises that have remained sprawled or have become sprawled over time have grown more rapidly compared to African and Indian metropolises. As an illustration, the suburban areas of Chinese metropolises witnessed an average annual growth rate of 3.1%, in contrast to the 0.9% observed in African and Indian metropolises (see Supplementary Table S5). Besides this, a two-stage linear decay was noted for these cities in China. Contrarily, in African and Indian sprawling cities, this was more of a linear decay trend. Compact cities in China also exhibited faster growth compared to those in Africa and India. As a result, the city core had extended over a much wider area for compact metropolises in China (see Supplementary Table S5) and contributed toward the formation of the "inverted S-shaped" land density curves. The average built-up density in the core areas was also high for these cities (see Figs. S8 and S9). Contrary to this, the land density distribution of compact cities in both Africa and India approximated an "inverted S shape."

3.4.2 Suburbanization Characteristics Across Dynamic Urban Forms

Figure 7 shows the land density in the hinterlands (parameter c of the fitted curve) and the growing urban fringe boundary (parameter D of the fitted curve) of metropolises across different dynamic urban forms. It was no surprise that cities with existing sprawled forms or those that had become sprawled over time had the highest built-up land density in 1985 and recorded maximum growth over the examined time period as well. For example, cities that had transformed from being compact to highly sprawled had witnessed the highest growth in their hinterland built-up density, wherein the mean land density increased from 0.05% to 0.195% during 1985–2015 (annual average growth rate of 9.7%). On the contrary, cities that remained compact or had become compact over time had lower land density in their hinterlands (Fig. 7a) and recorded lower growth compared to sprawled cities. For example, cities that were compact in 1985 and had become highly compact in 2015 had the lowest land density (average) in the hinterland during the entire study period (0.012 in 1985 and 0.068 in 2015).

Urban extents were found to be the highest for cities that remained sprawled or had become highly sprawled and had also witnessed rapid growth in their urban extents over time. For such cities, the average annual growth rate was 2%. On the contrary, the spatial extents of cities that had either become compact or highly

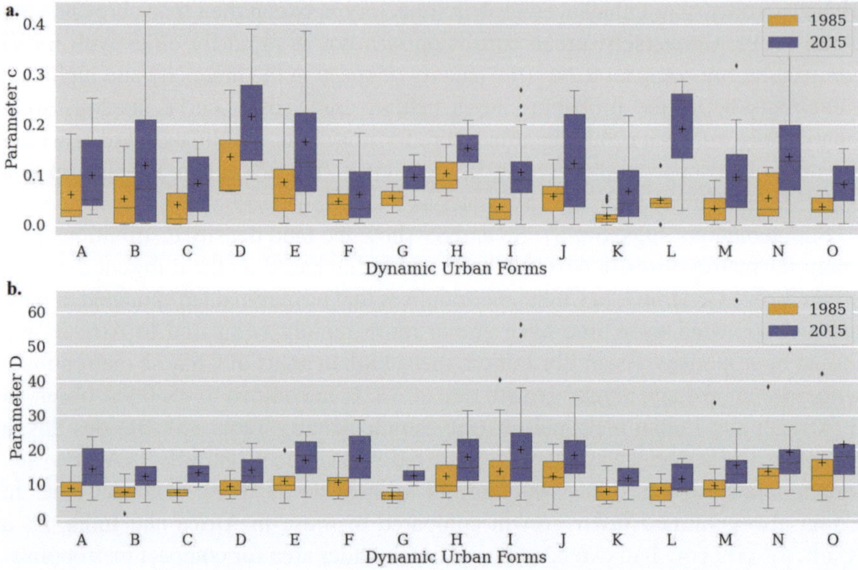

Fig. 7 Box plots for parameters C and D by different dynamic urban formsNote: *A* remained sprawled, *B* sprawled to compact, *C* sprawled to highly compact, *D* sprawled to highly sprawled, *E* remained highly sprawled, *F* highly sprawled to sprawled, *G* highly sprawled to highly compact, *H* highly sprawled to compact, *I* remained compact, *J* compact to sprawled, *K* compact to highly compact, *L* compact to highly sprawled, *M* remained highly compact, *N* highly compact to compact, and *O* highly compact to sprawled. Since there was only one metropolis under the highly compact to highly sprawled category, it was not included in Fig. 6

compact had remained small and grew relatively slowly, while their average annual growth rate was about 1.3%. However, cities that transformed from being highly sprawled to compact or had remained compact over the study period displayed different trends. In 1985, we found the highest spatial extents for these two groups of cities (Fig. 7b). Besides, their average annual growth rates were also the highest compared to all other forms, being 3.3% and 2.2%, respectively.

4 Discussion

Amid rapid urbanization in the Global South, sustainability challenges loom large, especially in major metropolises (Seto et al., 2011; Güneralp et al., 2020; van Vliet, 2019). Quantitative methods aid in understanding urban compaction and sprawl degrees, influencing metropolis development (Tsai, 2005). Urban form analysis is the key for informed urban planning to foster sustainability (Chakraborty et al., 2022; Jiao et al., 2021; Mahtta et al., 2019; Xu, et al., 2019b). Leveraging a freely accessible annual built-up layer (Liu et al., 2020), we introduce the "Index of Urban Form" (IUF) based on urban land density distribution (Jiao, 2015). This assesses urban forms and dynamics across 151 metropolises in Africa, China, and India (ACI), comparing their expansion and suburbanization characteristics (1985–2015) to illuminate ongoing Global South urbanization (Nagendra et al., 2018). By evaluating core-to-suburb ratios, we identified four urban forms for 1985 and 2015. Results diverge from prior studies (Dong et al., 2019; Gao et al., 2016; Xu, et al., 2019a), suggesting quantitative assessment's importance. China's trend contrasts with Africa's and India's compact or highly compact growth.

Comparing IUF values in 1985 and 2015 revealed 16 dynamic urban forms in the studied metropolises. While Dong et al. (2019) noted inertia in Chinese urban forms and Lemoine-Rodríguez et al. (2020) found increased complexity in African and Indian cities, our findings differ in two ways. First, most ACI cities experienced form changes during 1985–2015. Second, Chinese cities showed more dynamicity, with eastern China's compact cities turning sprawling due to investments and land leasing (Bian et al., 2021; Deng et al., 2020; Zhang, 2000; Yue et al., 2013). On the flip side, metropolises in Africa and India displayed a greater degree of form inertia, sustaining highly compact structures. Despite the prevalence of informal urbanization in the immediate peripheral zones of large metropolises in Africa and India, the surrounding high-density built-up areas experienced rapid growth, leading to the development of even more compact cities in these regions (e.g., Acheampong et al. (2017), Chakraborty, et al. (2021a), Jain (2018), Kantakumar et al. (2016), and Toure et al. (2020)).

The generalized land density curves for different dynamic urban forms revealed further insights about urban expansion characteristics as well. We found that sprawled cities had expanded more rapidly compared to cities with compact forms. However, such rapid expansion is very much concentrated in their suburban areas, and the development of their urban core was not very prominent. This ultimately contributed to a more linear urban land density decay curve for these cities. The growing low-density built-up zones in their hinterlands also indicated landscape fragmentation beyond the urban fringe of these sprawling cities (Asabere et al., 2020; Li et al., 2017). In contrast, compact cities exhibited relatively moderate rates of urban expansion, tending toward "inverse S-shaped" land density decay curves. Consequently, this phenomenon contributed to a diminished spatial growth of the urban fringe area and resulted in low-density built-up areas in the hinterlands.

Dynamic urban forms exert distinct control over expansion characteristics in ACI. In compact Chinese metropolises, wider urban cores and higher core density indicate simultaneous urban densification and outward core expansion. Leaving aside socio-political and financial aspects, compact cities impact suburban zones through hinterland density and urban fringe boundary. However, compact Chinese metropolises show stronger suburbanization compared to Africa and India (Jiao, 2015). Our approach highlights suburbanization‘s connection to urban forms and changes, emphasizing its context-specific complexity in rapidly urbanizing Global South (Dadashpoor & Ahani, 2021; Nagendra et al., 2018; Jiao et al., 2021).

The implications of our findings extend to urban and regional sustainability in ACI. Urban form plays a crucial role in influencing land use efficiency and sustainability, with sprawled forms notably decreasing per capita land consumption, as demonstrated in previous studies (Xu et al., 2020; Dong et al., 2019). While not directly linked to land consumption, our results reveal higher prevalence of sprawled forms in China compared to India and Africa (Li & Li, 2019; Deng et al., 2020). Many Chinese metropolises that were compact in 1985 have evolved into sprawled forms, especially in the east. Uncontrolled sprawl exacerbates urbanization's negative effects, including increased land consumption (Sumari et al., 2020; Dong et al., 2019, 2021). In China, the densification and suburbanization of compact cities require careful management by city authorities. Conversely, compact metropolises in Africa and India have a relatively low built-up density in their urban cores. Given the susceptibility of compact forms to become sprawled (Dong et al., 2019; Xu, et al., 2019a), enhancing urban core density is essential. However, this must be balanced with considerations for the urban microclimate and provision of green spaces (Chakraborty, et al., 2021b).

Our findings challenge the applicability of Global North–originated planning theories to contemporary Global South urbanization (Randolph & Storper, 2020), which differs significantly (Chakraborty et al., 2022). Existing cross-country/regional studies contrast urban expansion between the Global North and South (He et al., 2019; Kuang et al., 2014), rarely examining the intra–Global South variation. Recognizing diverse trends within the Global South questions universal planning theories (Watson, 2009, 2016). Moreover, our results caution against generalizing the Global South urban expansion ideas without proper context. Compact urban forms' prevalence differs from assumptions (Huang et al., 2007; Jiao et al., 2021). Compact forms and related expansion are more common in African and Indian metropolises, whereas they are outliers in Chinese ones. Chinese metropolises display distinct dynamics affecting expansion and suburban traits, differing from their African and Indian counterparts. Researchers must rethink the Global South urban planning theories to accommodate cross-country/regional expansion dynamics.

5 Conclusion

In this study, urban land density was employed, and the Index of Urban Form (IUF) was introduced to examine the forms and dynamics of urban areas in 151 metropolises spanning Africa, China, and India (ACI) from 1985 to 2015. Four urban form types—compact, highly compact, sprawled, and highly sprawled—were identified, revealing 16 dynamic urban forms. African and Indian metropolises trend toward compactness and slower growth, while many Chinese metropolises shift from compact to sprawled forms with faster expansion. Various expansion and suburban traits vary significantly among these forms. This study challenges misconceptions about prevalent compact urban forms in the Global South. Spatial differences exist in how forms impact expansion and suburbanization, deviating from previous assumptions. Recognizing diverse trends prompts rethinking universal planning theories, calling for context-specific urban planning theories that consider expansion patterns, forms, and dynamics.

While this study provides comprehensive insights into urbanization in the Global South cities, some limitations exist. We primarily used the GAUD dataset and hardly compared its accuracy against other datasets. Our focus was on cities with over a million people, neglecting different city classes with differing expansion dynamics. Broadening the analysis to the entire dataset and comparing expansion patterns between large and smaller cities could offer more insights. Investigating how urban form dynamics influence land consumption is also relevant. Furthermore, an examination of various urban expansion strategies, such as infilling, edge expansion, and outlying expansion, along with an understanding of the sociopolitical factors shaping diverse patterns, can contribute to the development of sustainable urban policy.

Supplementary Materials

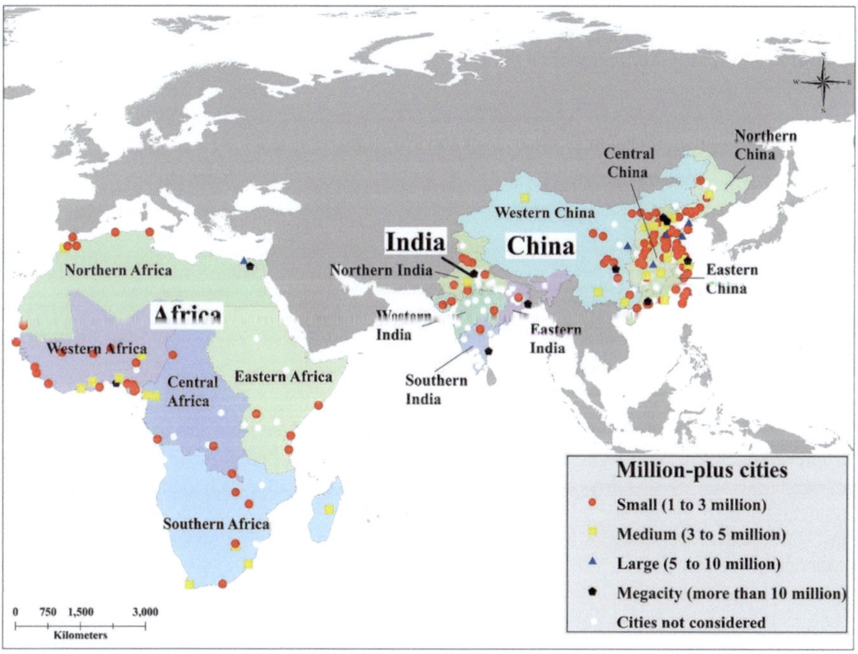

Supplementary Fig. S1 Distribution of metropolis cities in Africa, China, and India by city size (Regional divisions of Africa, China, and India also included)

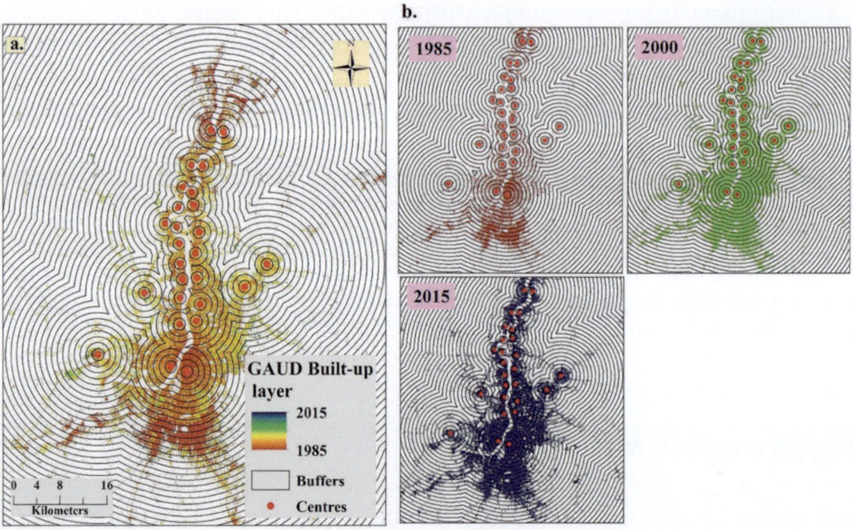

Supplementary Fig. S2 Preparation of built-up raster for different study years; (**a**) GAUD built-up layer (1985–2015); (**b**) reclassified built-up raster for 1985, 2000, and 2015

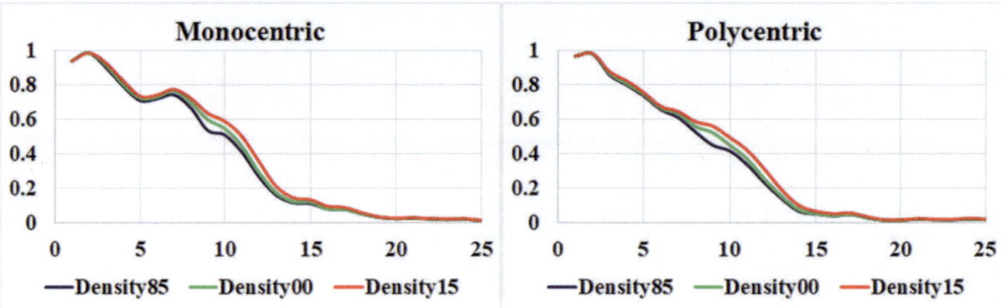

Supplementary Fig. S3 A graphical depiction of the selection of polycentric spatial structure of metropolisesNote: We first drew buffers around a single city center for each metropolis and prepare urban land density curves accordingly. We then checked for peaks in those curves (Supplementary Fig. S1); if peaks were present, we adopted a polycentric city structure; otherwise, we stuck with a monocentric city structure

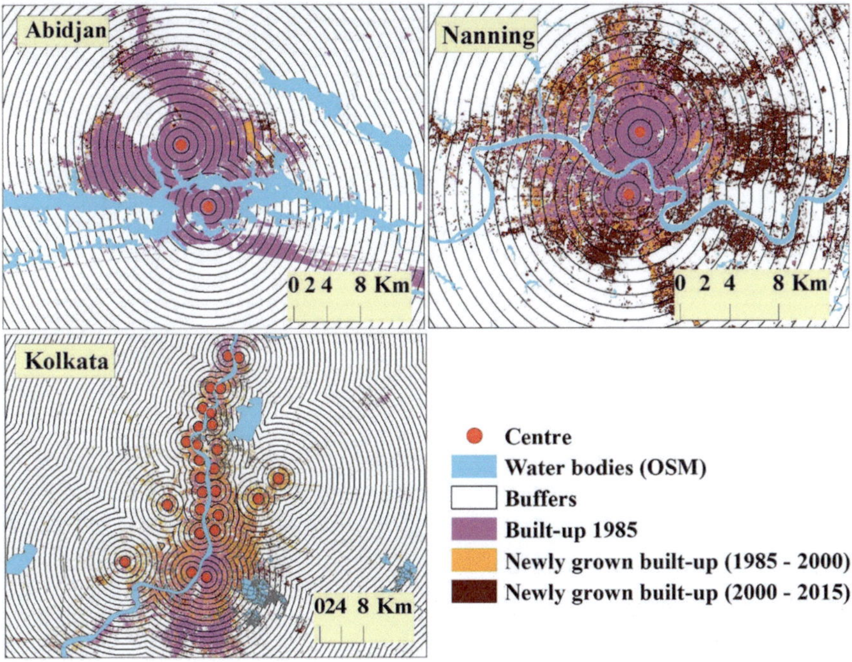

Supplementary Fig. S4 Distribution of built-up land within concentric rings drawn around the city center (for polycentric cities)

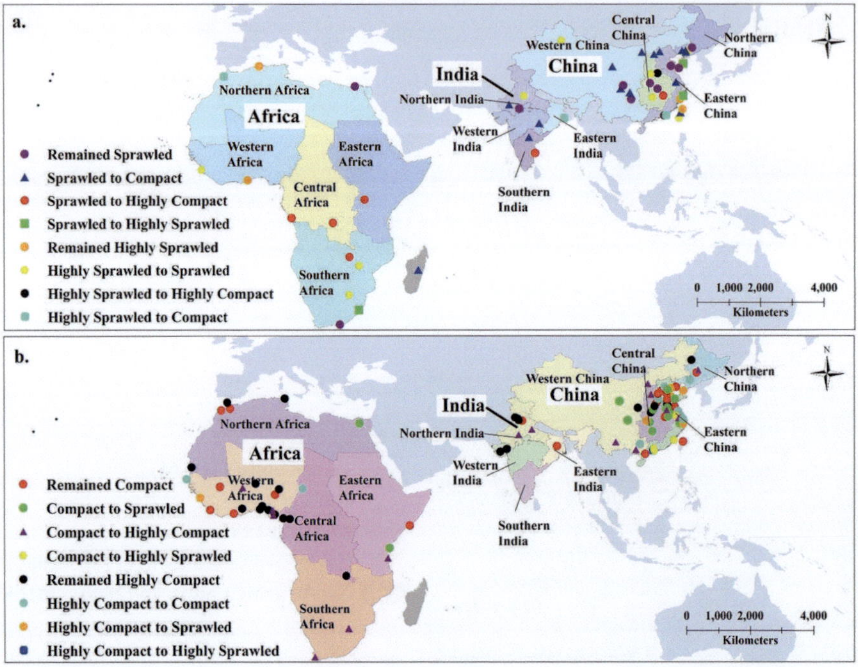

Supplementary Fig. S5 Spatial distribution of metropolises by their dynamic urban forms across ACI and their regional divisions. (**a**) Metropolises with sprawled and highly sprawled urban forms in 1985; (**b**) metropolises with compact and highly compact urban forms in 1985

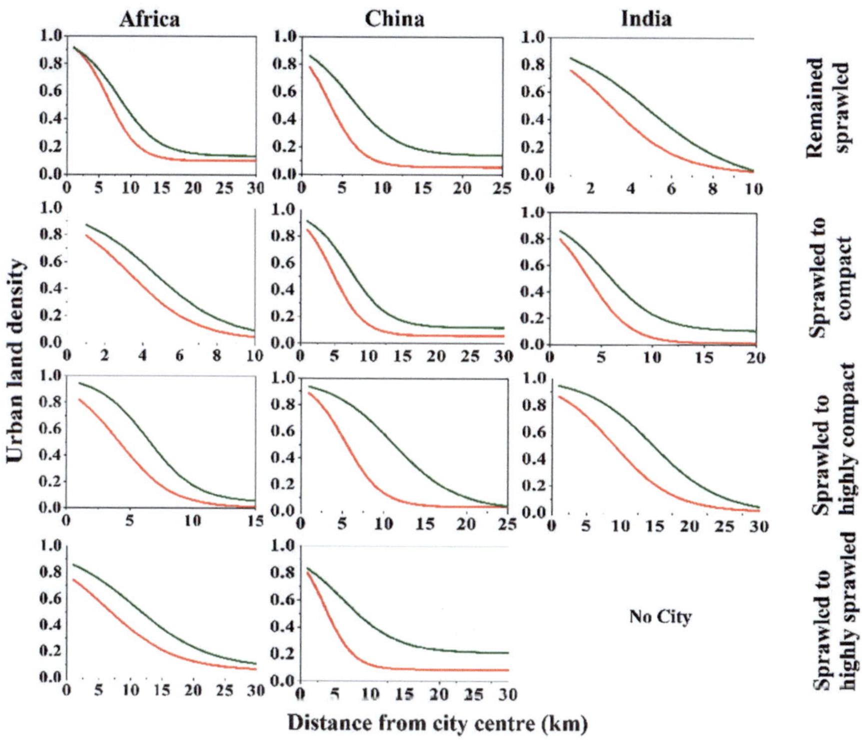

Supplementary Fig. S6 Generalized land density distribution across dynamic urban forms (A–D) by country/continent

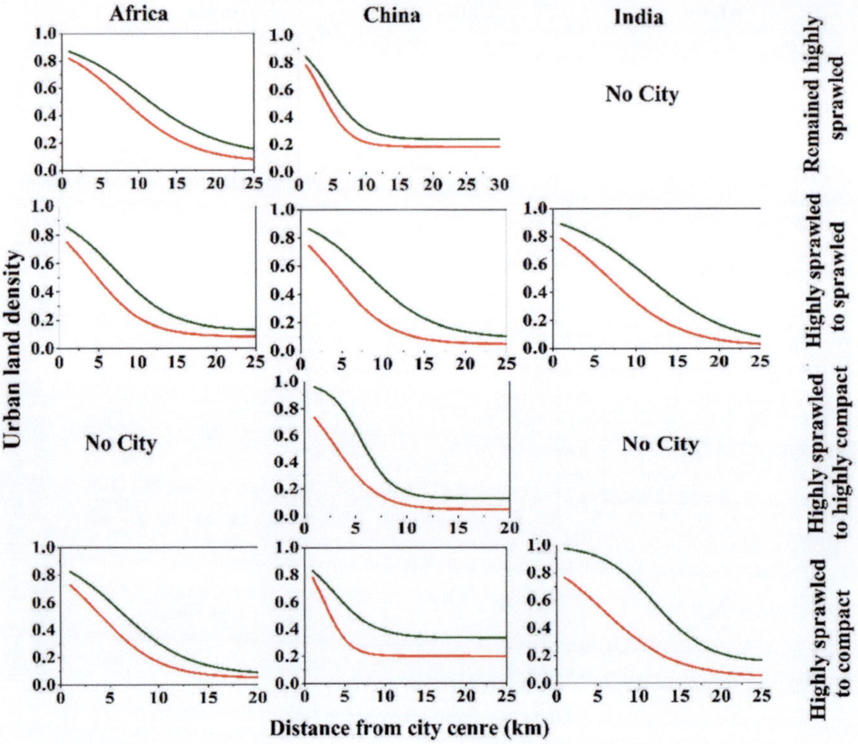

Supplementary Fig. S7 Generalized land density distribution across dynamic urban forms (E–H) by country/continent

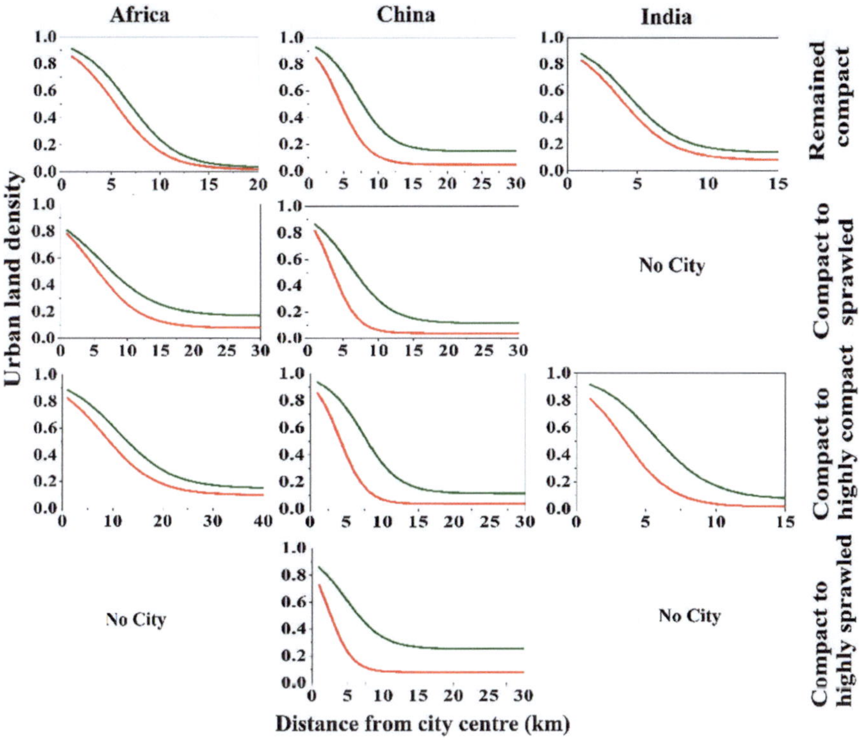

Supplementary Fig. S8 Generalized land density distribution across dynamic urban forms (I–L) by country/continent

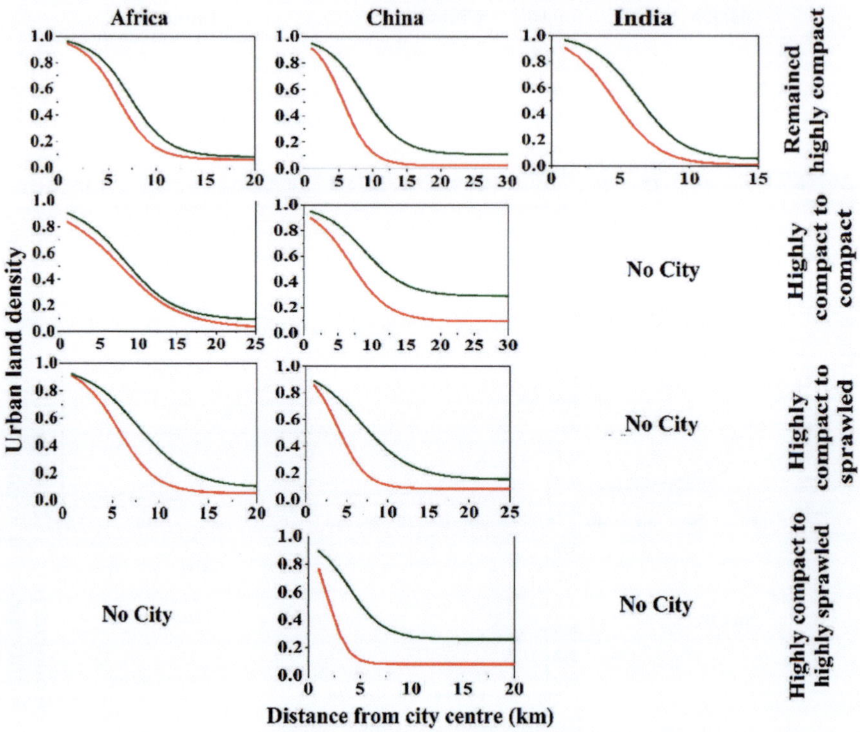

Supplementary Fig. S9 Generalized land density distribution across dynamic urban forms (M–P) by country/continent

Supplementary Table S1 Nature of spatial structure of metropolis cities in Africa, China, and India (ACI)

City	Country/continent	Nature of spatial structure
Aba	Africa	Monocentric
Abidjan	Africa	Polycentric
Abuja	Africa	Polycentric
Accra	Africa	Polycentric
Addis Ababa	Africa	Monocentric
Al-Iskandariyah (Alexandria)	Africa	Monocentric
Al-Qahirah (Cairo)	Africa	Polycentric
Antananarivo	Africa	Monocentric
Bamako	Africa	Polycentric
Benin City	Africa	Monocentric
Brazzaville	Africa	Polycentric
Cape Town	Africa	Monocentric
Conakry	Africa	Polycentric

Supplementary Table S1 (continued)

City	Country/continent	Nature of spatial structure
Dakar	Africa	Monocentric
Dar El Beida (Casablanca)	Africa	Monocentric
Douala	Africa	Monocentric
Durban (eThekwini)	Africa	Polycentric
Ekurhuleni	Africa	Monocentric
El Djazaïr (Algiers)	Africa	Monocentric
Fès	Africa	Monocentric
Freetown	Africa	Monocentric
Harare	Africa	Monocentric
Ibadan	Africa	Monocentric
Johannesburg	Africa	Monocentric
Kaduna	Africa	Monocentric
Kampala	Africa	Monocentric
Kano	Africa	Monocentric
Kigali	Africa	Polycentric
Kinshasa	Africa	Polycentric
Kisangani	Africa	Monocentric
Kumasi	Africa	Monocentric
Lagos	Africa	Polycentric
Lomé	Africa	Monocentric
Lubumbashi	Africa	Monocentric
Lusaka	Africa	Monocentric
Maputo	Africa	Monocentric
Mbuji-Mayi	Africa	Monocentric
Mombasa	Africa	Polycentric
Monrovia	Africa	Polycentric
Muqdisho (Mogadishu)	Africa	Monocentric
N'Djaména	Africa	Monocentric
Nairobi	Africa	Monocentric
Niamey	Africa	Monocentric
Nouakchott	Africa	Monocentric
Onitsha	Africa	Monocentric
Ouagadougou	Africa	Monocentric
Pointe-Noire	Africa	Monocentric
Port Elizabeth (Nelson Mandela Bay)	Africa	Monocentric
Port Harcourt	Africa	Monocentric
Pretoria	Africa	Polycentric
Rabat	Africa	Monocentric
Tanger	Africa	Monocentric
Tunis	Africa	Monocentric
Uyo	Africa	Monocentric

(continued)

Supplementary Table S1 (continued)

City	Country/continent	Nature of spatial structure
Yaoundé	Africa	Monocentric
Anshan	China	Monocentric
Anyang	China	Monocentric
Baoding	China	Monocentric
Baotou	China	Polycentric
Beijing	China	Monocentric
Binzhou	China	Monocentric
Cangzhou	China	Monocentric
Chángchun	China	Monocentric
Chángsha	China	Polycentric
Chaozhou	China	Monocentric
China, Hong Kong SAR	China	Polycentric
Chongqing	China	Polycentric
Cixi	China	Polycentric
Dàlián	China	Polycentric
Daqing	China	Polycentric
Datong	China	Monocentric
Dongguan	China	Polycentric
Dongying	China	Polycentric
Fóshan	China	Monocentric
Fushun, Liaoning	China	Polycentric
Fuzhou	China	Polycentric
Ganzhou	China	Polycentric
Gaoxiong, Taiwan	China	Monocentric
Guangzhou, Guangdong	China	Polycentric
Guìyáng	China	Monocentric
Haikou	China	Monocentric
Hándan	China	Monocentric
Hangzhou	China	Polycentric
Harbin	China	Polycentric
Hefei	China	Polycentric
Hengyang	China	Monocentric
Hohhot	China	Monocentric
Huai'an	China	Polycentric
Huaibei	China	Polycentric
Huainan	China	Polycentric
Huizhou	China	Polycentric
Jiangmen	China	Polycentric
Jiaxing	China	Monocentric
Jieyang	China	Polycentric
Jilin	China	Polycentric
Jinan	China	Monocentric

Supplementary Table S1 (continued)

City	Country/continent	Nature of spatial structure
Jingzhou, Hubei	China	Monocentric
Jinhua	China	Monocentric
Jining, Shandong	China	Monocentric
Jinzhou	China	Polycentric
Kaifeng	China	Monocentric
Kunming	China	Monocentric
Lanzhou	China	Monocentric
Lianyungang	China	Monocentric
Linyi, Shandong	China	Polycentric
Lu′an	China	Monocentric
Liuyang	China	Monocentric
Liuzhou	China	Monocentric
Luoyang	China	Polycentric
Maoming	China	Polycentric
Mianyang, Sichuan	China	Monocentric
Nanchang	China	Monocentric
Nanchong	China	Polycentric
Nanning	China	Polycentric
Nantong	China	Monocentric
Nanyang, Henan	China	Monocentric
Ningbo	China	Polycentric
Panjin	China	Monocentric
Pingdingshan, Henan	China	Polycentric
Puning	China	Monocentric
Putian	China	Monocentric
Qingdao	China	Polycentric
Qinhuangdao	China	Monocentric
Qiqihar	China	Monocentric
Quanzhou	China	Polycentric
Rizhao	China	Polycentric
Ruian, Wenzhou, Zhejiang, China	China	Polycentric
Shanghai	China	Monocentric
Shantou	China	Polycentric
Shaoxing	China	Monocentric
Shenyang	China	Monocentric
Shenzhen	China	Polycentric
Shijiazhuang	China	Monocentric
Shiyan	China	Monocentric
Suzhou, Anhui	China	Monocentric
Suzhou	China	Monocentric
Taian, Shandong	China	Monocentric

(continued)

Supplementary Table S1 (continued)

City	Country/continent	Nature of spatial structure
Taibei	China	Polycentric
Taiyuan	China	Monocentric
Taizhong	China	Monocentric
Taizhou, Jiangsu	China	Polycentric
Taizhou, Zhejiang	China	Monocentric
Tangshan	China	Monocentric
Taoyuan	China	Monocentric
Tengzhou	China	Monocentric
Tianjin	China	Polycentric
Ürümqi	China	Monocentric
Weifang	China	Polycentric
Weihai	China	Polycentric
Wenzhou	China	Monocentric
Wuhan	China	Polycentric
Wuhu, Anhui	China	Monocentric
Wuxi	China	Polycentric
Xi'an	China	Monocentric
Xiamen	China	Monocentric
Xiangtan, Hunan	China	Polycentric
Xiangyang	China	Monocentric
Xining	China	Monocentric
Xinxiang	China	Monocentric
Xuzhou	China	Monocentric
Yancheng, Jiangsu	China	Monocentric
Yangzhou	China	Monocentric
Yantai	China	Monocentric
Yichang	China	Monocentric
Yingkou	China	Monocentric
Yiwu	China	Monocentric
Zaozhuang	China	Monocentric
Zhangjiakou	China	Monocentric
Zhanjiang	China	Monocentric
Zhengzhou	China	Polycentric
Zhenjiang, Jiangsu	China	Polycentric
Zhongshan	China	Polycentric
Zhuzhou	China	Polycentric
Zibo	China	Monocentric
Zunyi	China	Monocentric
Allahabad	India	Monocentric
Amritsar	India	Monocentric
Asansol	India	Monocentric
Bareilly	India	Monocentric

Supplementary Table S1 (continued)

City	Country/continent	Nature of spatial structure
Chandigarh	India	Monocentric
Chennai (Madras)	India	Polycentric
Delhi, Connaught Place	India	Polycentric
Dhanbad	India	Monocentric
Durg-Bhilainagar	India	Polycentric
Guwahati (Gauhati)	India	Polycentric
Indore	India	Monocentric
Jaipur	India	Monocentric
Jalandhar	India	Monocentric
Jamshedpur	India	Monocentric
Jodhpur	India	Monocentric
Kanpur	India	Monocentric
Kolkata	India	Polycentric
Kota	India	Monocentric
Ludhiana	India	Monocentric
Meerut	India	Monocentric
Moradabad	India	Monocentric
Mumbai	India	Polycentric
Nagpur	India	Monocentric
Patna	India	Monocentric
Pune	India	Polycentric
Rajkot	India	Monocentric
Srinagar	India	Monocentric
Varanasi	India	Monocentric
Vijayawada	India	Monocentric
Visakhapatnam	India	Polycentric

Supplementary Table S2 Statistics for the R^2 value of the fitted function

Country/continent	Year	Median	Q1	Q3
Overall	1985	0.986	0.986	0.983
	2000	0.970	0.975	0.972
	2015	0.992	0.993	0.992
Africa	1985	0.984	0.984	0.985
	2000	0.968	0.970	0.970
	2015	0.992	0.992	0.993
India	1985	0.985	0.987	0.985
	2000	0.960	0.973	0.976
	2015	0.988	0.992	0.992
China	1985	0.987	0.987	0.982
	2000	0.974	0.977	0.972
	2015	0.993	0.993	0.991

Supplementary Table S3 Dynamic urban forms for metropolises with sprawled and highly sprawled urban forms in 1985

City	Country/ continent	IUF 1985	Urban form 1985	IUF 2015	Urban form 2015
Chéngdu	China	0.94	Sprawled	1.10	Compact
Hyderabad	India	0.85	Sprawled	1.12	Compact
Gaoxiong, Taiwan	China	0.32	Highly sprawled	0.80	Sprawled
Taizhong	China	0.51	Sprawled	1.85	Compact
Taoyuan	China	0.34	Highly sprawled	0.24	Highly sprawled
Anshan	China	0.49	Highly sprawled	0.92	Sprawled
Baotou	China	0.67	Sprawled	1.39	Compact
Datong	China	0.98	Sprawled	1.55	Compact
Fushun, Liaoning	China	0.63	Sprawled	0.61	Sprawled
Jinzhou	China	0.79	Sprawled	1.78	Compact
Kaifeng	China	0.38	Highly sprawled	2.15	Highly compact
Luoyang	China	0.26	Highly sprawled	0.95	Sprawled
Mianyang, Sichuan	China	0.62	Sprawled	0.50	Sprawled
Panjin	China	0.84	Sprawled	1.27	Compact
Puning	China	0.45	Highly sprawled	2.31	Highly compact
Ruian	China	0.27	Highly sprawled	0.30	Highly sprawled
Taizhou, Zhejiang	China	0.67	Sprawled	0.20	Highly sprawled
Xiangyang	China	0.87	Sprawled	0.91	Sprawled
Xining	China	0.95	Sprawled	1.05	Compact
Yangzhou	China	0.88	Sprawled	1.02	Compact
Yantai	China	0.76	Sprawled	0.35	Highly sprawled
Yingkou	China	0.73	Sprawled	1.76	Compact
Zhangjiakou	China	0.71	Sprawled	1.57	Compact
Zunyi	China	0.73	Sprawled	0.65	Sprawled
Durg-Bhilainagar	India	0.83	Sprawled	1.01	Compact
Chángsha	China	0.24	Highly sprawled	0.88	Sprawled
Nanchang	China	0.89	Sprawled	2.03	Highly compact
Qingdao	China	0.55	Sprawled	0.73	Sprawled
Shantou	China	0.25	Highly sprawled	1.05	Compact
Wuhan	China	0.90	Sprawled	0.73	Sprawled

Supplementary Table S3 (continued)

City	Country/ continent	IUF 1985	Urban form 1985	IUF 2015	Urban form 2015
Ürümqi	China	0.36	Highly sprawled	0.68	Sprawled
Zibo	China	0.82	Sprawled	0.80	Sprawled
Delhi, Connaught Place	India	0.49	Highly sprawled	0.81	Sprawled
Chennai (Madras)	India	0.91	Sprawled	2.39	Highly compact
Chongqing	China	0.62	Sprawled	1.46	Compact
Kolkata	India	0.29	Highly sprawled	1.93	Compact
Antananarivo	Africa	0.69	Sprawled	1.08	Compact
Kampala	Africa	0.82	Sprawled	2.91	Highly compact
Lusaka	Africa	0.69	Sprawled	3.74	Highly compact
Harare	Africa	0.28	Highly sprawled	0.96	Sprawled
Pointe-Noire	Africa	0.85	Sprawled	3.51	Highly compact
Mbuji-Mayi	Africa	0.94	Sprawled	2.65	Highly compact
El Djazaïr (Algiers)	Africa	0.22	Highly sprawled	0.29	Highly sprawled
Al-Iskandariyah (Alexandria)	Africa	0.70	Sprawled	0.69	Sprawled
Rabat	Africa	0.37	Highly sprawled	1.07	Compact
Durban (eThekwini)	Africa	0.53	Sprawled	0.38	Highly sprawled
Port Elizabeth (Nelson Mandela Bay)	Africa	0.92	Sprawled	0.85	Sprawled
Pretoria	Africa	0.20	Highly sprawled	0.99	Sprawled
Accra	Africa	0.41	Highly sprawled	0.28	Highly sprawled
Freetown	Africa	0.47	Highly sprawled	0.66	Sprawled
Jodhpur	India	0.97	Sprawled	1.21	Compact
Kota	India	0.87	Sprawled	0.63	Sprawled

Supplementary Table S4 Dynamic urban forms for metropolises with compact and highly compact urban forms in 1985

City	Country/ continent	IUF 1985	Urban form 1985	IUF 2015	Urban form 2015
Dar es Salaam	Africa	1.43	Compact	2.15	Highly compact
Ahmedabad	India	2.35	Highly compact	3.13	Highly compact
Taibei	China	1.73	Compact	1.66	Compact
Anyang	China	1.45	Compact	1.03	Compact
Baoding	China	1.44	Compact	1.39	Compact
Binzhou	China	2.80	Highly compact	0.97	Sprawled
Cixi	China	1.68	Compact	0.47	Highly sprawled
Hohhot	China	1.66	Compact	2.49	Highly compact
Huai'an	China	1.05	Compact	0.95	Sprawled
Huaibei	China	1.25	Compact	0.89	Sprawled
Jiangmen	China	2.57	Highly compact	0.33	Highly sprawled
Jiaxing	China	1.60	Compact	1.35	Compact
Jilin	China	1.15	Compact	1.72	Compact
Jingzhou, Hubei	China	1.06	Compact	0.60	Sprawled
Jining, Shandong	China	3.01	Highly compact	1.24	Compact
Lianyungang	China	1.98	Compact	1.13	Compact
Lu′an	China	1.22	Compact	1.21	Compact
Liuzhou	China	2.40	Highly compact	1.93	Compact
Maoming	China	1.18	Compact	1.30	Compact
Nanchong	China	1.20	Compact	0.55	Sprawled
Nantong	China	3.37	Highly compact	1.95	Compact
Nanyang, Henan	China	2.99	Highly compact	2.75	Highly compact
Pingdingshan, Henan	China	1.63	Compact	0.82	Sprawled
Putian	China	1.01	Compact	0.38	Highly sprawled
Qinhuangdao	China	1.76	Compact	1.20	Compact
Qiqihar	China	2.42	Highly compact	2.32	Highly compact
Shaoxing	China	1.76	Compact	0.53	Sprawled
Taian, Shandong	China	1.47	Compact	1.17	Compact
Taizhou, Jiangsu	China	1.20	Compact	1.03	Compact
Tengzhou	China	2.33	Highly compact	4.80	Highly compact

Supplementary Table S4 (continued)

City	Country/ continent	IUF 1985	Urban form 1985	IUF 2015	Urban form 2015
Xiangtan, Hunan	China	1.05	Compact	0.84	Sprawled
Xinxiang	China	1.68	Compact	1.58	Compact
Xuzhou	China	1.44	Compact	0.76	Sprawled
Yancheng, Jiangsu	China	2.50	Highly compact	1.01	Compact
Yichang	China	2.21	Highly compact	0.95	Sprawled
Zhenjiang, Jiangsu	China	1.29	Compact	0.43	Highly sprawled
Bareilly	India	1.30	Compact	2.50	Highly compact
Jalandhar	India	2.15	Highly compact	4.31	Highly compact
Cangzhou	China	1.61	Compact	1.79	Compact
Chángchun	China	1.59	Compact	3.17	Highly compact
Dàlián	China	2.13	Highly compact	0.88	Sprawled
Hándan	China	2.95	Highly compact	1.52	Compact
Hefei	China	1.10	Compact	0.75	Sprawled
Jinan	China	1.87	Compact	1.35	Compact
Kunming	China	1.58	Compact	2.34	Highly compact
Lanzhou	China	1.29	Compact	0.98	Sprawled
Nanning	China	1.32	Compact	2.13	Highly compact
Ningbo	China	2.01	Highly compact	0.76	Sprawled
Shenyang	China	3.05	Highly compact	1.52	Compact
Shijiazhuang	China	1.80	Compact	1.25	Compact
Taiyuan	China	1.98	Compact	2.79	Highly compact
Tangshan	China	2.10	Highly compact	0.78	Sprawled
Weifang	China	1.16	Compact	0.72	Sprawled
Wuxi	China	1.16	Compact	2.59	Highly compact
Xi'an	China	2.35	Highly compact	2.87	Highly compact
Zhengzhou	China	2.90	Highly compact	2.31	Highly compact
Zhongshan	China	1.24	Compact	0.46	Highly sprawled

(continued)

Supplementary Table S4 (continued)

City	Country/ continent	IUF 1985	Urban form 1985	IUF 2015	Urban form 2015
Shanghai	China	2.10	Highly compact	1.52	Compact
Beijing	China	2.08	Highly compact	1.58	Compact
Guangzhou, Guangdong	China	1.37	Compact	0.40	Highly sprawled
Lagos	Africa	5.34	Highly compact	2.17	Highly compact
Tianjin	China	1.95	Compact	0.75	Sprawled
Mombasa	Africa	1.11	Compact	0.58	Sprawled
Muqdisho (Mogadishu)	Africa	1.21	Compact	1.47	Compact
Douala	Africa	3.03	Highly compact	4.24	Highly compact
Yaoundé	Africa	13.69	Highly compact	4.09	Highly compact
N'Djaména	Africa	2.88	Highly compact	1.63	Compact
Lubumbashi	Africa	4.64	Highly compact	5.35	Highly compact
Al-Qahirah (Cairo)	Africa	1.04	Compact	0.63	Sprawled
Dar El Beida (Casablanca)	Africa	1.16	Compact	1.94	Compact
Fès	Africa	1.06	Compact	1.45	Compact
Tanger	Africa	2.43	Highly compact	3.23	Highly compact
Tunis	Africa	3.67	Highly compact	2.59	Highly compact
Cape Town	Africa	1.37	Compact	3.28	Highly compact
Johannesburg	Africa	1.16	Compact	2.09	Highly compact
Ekurhuleni	Africa	1.89	Compact	2.19	Highly compact
Ouagadougou	Africa	1.42	Compact	2.83	Highly compact
Abidjan	Africa	1.36	Compact	1.92	Compact
Kumasi	Africa	2.09	Highly compact	3.44	Highly compact
Conakry	Africa	3.61	Highly compact	0.86	Sprawled
Monrovia	Africa	1.24	Compact	1.31	Compact
Bamako	Africa	1.29	Compact	1.95	Compact
Nouakchott	Africa	2.91	Highly compact	2.46	Highly compact
Niamey	Africa	3.64	Highly compact	7.06	Highly compact

Supplementary Table S4 (continued)

City	Country/ continent	IUF 1985	Urban form 1985	IUF 2015	Urban form 2015
Aba	Africa	4.54	Highly compact	4.34	Highly compact
Benin City	Africa	4.42	Highly compact	4.28	Highly compact
Ibadan	Africa	4.59	Highly compact	2.56	Highly compact
Kaduna	Africa	1.12	Compact	1.12	Compact
Kano	Africa	2.71	Highly compact	4.44	Highly compact
Onitsha	Africa	1.64	Compact	2.63	Highly compact
Port Harcourt	Africa	1.30	Compact	3.37	Highly compact
Dakar	Africa	6.69	Highly compact	1.80	Compact
Amritsar	India	3.01	Highly compact	2.64	Highly compact
Chandigarh	India	1.06	Compact	1.09	Compact
Dhanbad	India	1.31	Compact	1.84	Compact
Jaipur	India	1.43	Compact	2.20	Highly compact
Ludhiana	India	2.03	Highly compact	4.02	Highly compact
Rajkot	India	4.53	Highly compact	4.25	Highly compact
Suzhou, Anhui	China	1.33	Compact	3.56	Highly compact

Supplementary Table S5 Average radii of two broad territories and their growth rates for cities across ACI by urban forms

Country/continent	Urban forms	Urban core		Suburban		Annual growth rate (%)	
		1985	2015	1985	2015	Urban core	Suburban
Africa	Overall	4.1	5.5	3.0	3.8	1.1	0.9
	Compact cities	4.5	6.1	2.4	2.7	1.2	0.4
	Sprawling cities	2.7	4.3	4.8	7.1	2.0	1.6
China	Overall	2.9	5.4	2.5	4.8	2.9	3.1
	Compact cities	3.5	6.8	2.6	4.1	3.1	1.9
	Sprawling cities	2.2	4.3	2.4	7.0	3.2	6.4
India	Overall	2.5	4.2	2.3	2.9	2.3	0.9
	Compact cities	2.6	4.4	2.3	2.7	2.3	0.6
	Sprawling cities	1.9	3.4	2.6	4.5	2.6	2.4

Note: Clubbed different dynamic urban forms related to compact or becoming compact into compact cities and sprawling and becoming sprawling into sprawling cities

References

Acheampong, R. A., Agyemang, F. S. K., & Abdul-Fatawu, M. (2017). Quantifying the spatio-temporal patterns of settlement growth in a metropolitan region of Ghana. *GeoJournal, 82*, 823–840.

Acuto, M., Parnell, S., & Seto, K. C. (2018). Building a global urban science. *Nature Sustainability, 1*, 2–4.

Adhikari, P., & de Beurs, K. M. (2017). Growth in urban extent and allometric analysis of West African cities. *Journal of Land Use Science, 12*, 105–124.

Anderson, W. P., Kanaroglou, P. S., & Miller, E. J. (1996). Urban form, energy and the environment: A review of issues, evidence and policy. *Urban Studies, 33*, 7–35.

Angel, S., Parent, J., & Civco, D. L. (2010). Ten compactness properties of circles: Measuring shape in geography. *Canadian Geographic, 54*, 441–461.

Angel, S., Parent, J., Civco, D.L. and Blei, A.M. (2011). *Making room for a planet of cities*. https://center4affordablehousing.org/wp-content/uploads/2018/12/Angel-Making-Room-for-a-Planet-of-Cities-2011.pdf. Last accessed 30 Aug 2021.

Angel, S., Arango Franco, S., Liu, Y., & Blei, A. M. (2020). The shape compactness of urban footprints. *Progress in Planning, 139*, 0–1.

Asabere, S. B., Acheampong, R. A., Ashiagbor, G., Beckers, S. C., Keck, M., Erasmi, S., Schanze, J., & Sauer, D. (2020). Urbanization, land use transformation and spatio-environmental impacts: Analyses of trends and implications in major metropolitan regions of Ghana. *Land Use Policy, 96*, 104707.

Bagheri, B., & Tousi, S. N. (2018). An explanation of urban sprawl phenomenon in Shiraz Metropolitan Area (SMA). *Cities, 73*, 71–90.

Bian, H., Gao, J., Wu, J., Sun, X., & Du, Y. (2021). Hierarchical analysis of landscape urbanization and its impacts on regional sustainability: A case study of the Yangtze River Economic Belt of China. *Journal of Cleaner Production, 279*, 123267.

Bosch, M., Jaligot, R., & Chenal, J. (2020). Spatiotemporal patterns of urbanization in three Swiss urban agglomerations: insights from landscape metrics, growth modes and fractal analysis. *Landscape Ecology, 35*, 879–891.

Chakraborty, S., Maity, I., Patel, P. P., Dadashpoor, H., Pramanik, S., Follmann, A., Novotný, J., & Roy, U. (2021a). Spatio-temporal patterns of urbanization in the Kolkata Urban Agglomeration: A dynamic spatial territory-based approach. *Sustainable Cities and Society, 67*, 102715.

Chakraborty, S., Pramanik, S., Follmann, A., Giri, B., Mondal, B., Patel, P. P., Maity, I., Das, J., Punia, M., & Sahana, M. (2021b). Dominant urban form and its relation to nighttime land surface temperature in the rapidly urbanizing National Capital Region of India. *Urban Climate, 40*, 101002.

Chakraborty, S., Maity, I., Dadashpoor, H., Novotný, J., & Banerji, S. (2022). Building in or out? Examining urban expansion patterns and land use efficiency across the global sample of 466 cities with million+ inhabitants. *Habitat International, 120*, 102503.

Chakraborty, S., Novotný, J., Maity, I., Lemoine-Rodríguez, R. and Follmann, A. (2024). Same planet but different worlds! Diverging convergence pattern of urban form typologies across 413 cities with million+ inhabitants and their sustainability tradeoffs. *Habitat International, 145*, p.103024.

Cobbinah, P. B., Erdiaw-Kwasie, M. O. & Amoateng, P. (2015). Rethinking sustainable development within the framework of poverty and urbanisation in developing countries. *Environmental Development, 13*, pp.18–32.

Das, M., Das, A., & Mandal, A. (2020). Research note: Ecosystem Health (EH) assessment of a rapidly urbanizing metropolitan city region of eastern India—A study on Kolkata Metropolitan Area. *Landscape and Urban Planning, 204*, 103938.

Das, S., Shit, P. K., & Patel, P. P. (2021). Ecosystem services value assessment and forecasting using integrated machine learning and CA-Markov model: an empirical investigation of an Asian megacity. *Geocarto International*. https://doi.org/10.1080/10106049.2021.2002424

Dadashpoor, H., Azizi, P., & Moghadasi, M. (2019). Land use change, urbanization, and change in landscape pattern in a metropolitan area. *Science of the Total Environment, 655*, 707–719.

Dadashpoor, H., & Ahani, S. (2021). Explaining objective forces, driving forces, and causal mechanisms affecting the formation and expansion of the peri-urban areas: A critical realism approach. *Land Use Policy, 102*, 105232.

Deng, Y., Qi, W., Fu, B., & Wang, K. (2020). Geographical transformations of urban sprawl: Exploring the spatial heterogeneity across cities in China 1992–2015. *Cities, 105*, 102415.

Deng, H., Zhang, K., Wang, F., & Dang, A. (2021). Compact or disperse? Evolution patterns and coupling of urban land expansion and population distribution evolution of major cities in China, 1998–2018. *Habitat International, 108*, 102324.

Dong, T., Jiao, L., Xu, G., Yang, L., & Liu, J. (2019). Towards sustainability? Analyzing changing urban form patterns in the United States, Europe, and China. *Science of the Total Environment, 671*, 632–643.

Ewing, R. (1997). Is Los Angeles-style sprawl desirable? *Journal of the American Planning Association, 63*(1), 107–126.

Fei, W., & Zhao, S. (2019). Urban land expansion in China's six megacities from 1978 to 2015. *Science of the Total Environment, 664*, 60–71.

Gao, B., Huang, Q., He, C., Sun, Z., & Zhang, D. (2016). How does sprawl differ across cities in China? A multi-scale investigation using nighttime light and census data. *Landscape and Urban Planning, 148*, 89–98.

Güneralp, B., Lwasa, S., Masundire, H., Parnell, S., & Seto, K. C. (2018). Urbanization in Africa: Challenges and opportunities for conservation. *Environmental Research Letters, 13*.

Güneralp, B., Reba, M., Hales, B. U., Wentz, E. A., & Seto, K. C. (2020). Trends in urban land expansion, density, and land transitions from 1970 to 2010: a global synthesis. *Environmental Research Letters, 15*(4), 044015.

He, C., Liu, Z., Tian, J., & Ma, Q. (2014). Urban expansion dynamics and natural habitat loss in China: A multiscale landscape perspective. *Global Change Biology, 20*, 2886–2902.

He, Q., Zeng, C., Xie, P., Tan, S., & Wu, J. (2019). Comparison of urban growth patterns and changes between three urban agglomerations in China and three metropolises in the USA from 1995 to 2015. *Sustainable Cities and Society, 50*, 101649.

Huang, J., Lu, X. X., & Sellers, J. M. (2007). A global comparative analysis of urban form: Applying spatial metrics and remote sensing. *Landscape and Urban Planning, 82*(4), 184–197.

Huilei, L., Jian, P., Yanxu, L., & Yi'na, H. (2017). Urbanization impact on landscape patterns in Beijing City, China: A spatial heterogeneity perspective. *Ecological Indicators, 82*, 50–60.

Jabareen, Y. R. (2006). Sustainable urban forms: Their typologies, models, and concepts. *Journal of Planning Education and Research, 26*(1), 38–52.

Jain, M. (2018). Contemporary urbanization as unregulated growth in India: The story of census towns. *Cities, 73*, 117–127.

Jiao, L. (2015). Urban land density function: A new method to characterize urban expansion. *Landscape and Urban Planning, 139*, 26–39.

Jiao, L., Dong, T., Xu, G., Zhou, Z., Liu, J., & Liu, Y. (2021). Geographic micro-process model: Understanding global urban expansion from a process-oriented view. *Computers, Environment and Urban Systems, 87*, 101603.

Kantakumar, L. N., Kumar, S., & Schneider, K. (2016). Spatiotemporal urban expansion in Pune metropolis, India using remote sensing. *Habitat International, 51*, 11–22.

Kuang, W., Chi, W., Lu, D., & Dou, Y. (2014). A comparative analysis of megacity expansions in China and the U.S.: Patterns, rates and driving forces. *Landscape and Urban Planning, 132*, 121–135.

Lemoine-Rodríguez, R., Inostroza, L., & Zepp, H. (2020). The global homogenization of urban form. An assessment of 194 cities across time. *Landscape and Urban Planning, 204*, 103949.

Li, G., & Li, F. (2019). Urban sprawl in China: Differences and socioeconomic drivers. *Science of the Total Environment, 673*, 367–377.

Li, H., Peng, J., Yanxu, L., & Yi'na, H. (2017). Urbanization impact on landscape patterns in Beijing City, China: A spatial heterogeneity perspective. *Ecological Indicators, 82*, 50–60.

Liu, X., Huang, Y., Xu, X., Li, X., Li, X., Ciais, P., Lin, P., Gong, K., Ziegler, A. D., Chen, A., Gong, P., Chen, J., Hu, G., Chen, Y., Wang, S., Wu, Q., Huang, K., Estes, L., & Zeng, Z. (2020). High-spatiotemporal-resolution mapping of global urban change from 1985 to 2015. *Nature Sustainability, 3*, 564–570.

Mahtta, R., Mahendra, A., & Seto, K. C. (2019). Building up or spreading out? typologies of urban growth across 478 cities of 1 million+. *Environmental Research Letters, 14*.

Nagendra, H., Bai, X., Brondizio, E. S., & Lwasa, S. (2018). The urban south and the predicament of global sustainability. *Nature Sustainability, 1*, 341–349.

Randolph, G. F., & Storper, M. (2020). Is urbanisation in the Global South fundamentally different? Comparative global urban analysis for the 21st century. *Urban Studies*, 00420980211067926.

Sahana, M., Hong, H., & Sajjad, H. (2018). Analyzing urban spatial patterns and trend of urban growth using urban sprawl matrix: A study on Kolkata urban agglomeration, India. *Science of the Total Environment, 628–629*, 1557–1566.

Schneider, A., & Woodcock, C. E. (2008). Compact, dispersed, fragmented, extensive? A comparison of urban growth in twenty-five global cities using remotely sensed data, pattern metrics and census information. *Urban Studies, 45*, 659–692.

Seto, K. C., Fragkias, M., Güneralp, B., & Reilly, M. K. (2011). A meta-analysis of global urban land expansion. *PLoS One, 6*(8), e23777.

Shukla, A., Jain, K., Ramsankaran, R. A. A. J., & Rajasekaran, E. (2021). Understanding the macro-micro dynamics of urban densification: A case study of different sized Indian cities. *Land Use Policy, 107*, 105469.

Sumari, N. S., Cobbinah, P. B., Ujoh, F., & Xu, G. (2020). On the absurdity of rapid urbanization: Spatio-temporal analysis of land-use changes in Morogoro, Tanzania. *Cities, 107*, 102876.

Tong, L., Hu, S., & Frazier, A. E. (2019). Hierarchically measuring urban expansion in fast urbanizing regions using multi-dimensional metrics: A case of Wuhan metropolis, China. *Habitat International, 94*, 102070.

Toure, S. I., Stow, D. A., Clarke, K., & Weeks, J. (2020). Patterns of land cover and land use change within the two major metropolitan areas of Ghana. *Geocarto International, 35*, 209–223.

Tsai, Y. H. (2005). Quantifying urban form: compactness versus' sprawl'. *Urban Studies, 42*(1), 141–161.

United Nations. (2014). *World urbanization prospects*. United Nations.

United Nations. (2018). The world's cities in 2018. Data Booklet (ST/ESA/SER. A/392). https://www.un.org/en/development/desa/population/publications/pdf/urbanization/the_worlds_cities_in_2018_databooklet.pdf.

van Vliet, J. (2019). Direct and indirect loss of natural area from urban expansion. *Nature Sustainability*.

Watson, V. (2009). Seeing from the South: Refocusing urban planning on the globe's central urban issues. *Urban Studies, 46*(11), 2259–2275.

Watson, V. (2016). Shifting approaches to planning theory: Global North and South. *Urban Planning, 1*(4), 32–41.

Wu, S., Sumari, N. S., Dong, T., Xu, G., & Liu, Y. (2021). Characterizing urban expansion combining concentric-ring and grid-based analysis for Latin American cities. *Land, 10*(5), 444.

Xia, C., Zhang, A., & Yeh, A. G. (2022). The varying relationships between multidimensional urban form and urban vitality in Chinese megacities: Insights from a comparative analysis. *Annals of the American Association of Geographers, 112*(1), 141–166.

Xu, G., Jiao, L., Liu, J., Shi, Z., Zeng, C., & Liu, Y. (2019a). Understanding urban expansion combining macro patterns and micro dynamics in three Southeast Asian megacities. *Science of the Total Environment, 660*, 375–383.

Xu, G., Dong, T., Cobbinah, P. B., Jiao, L., Sumari, N. S., Chai, B., & Liu, Y. (2019b). Urban expansion and form changes across African cities with a global outlook: Spatiotemporal analysis of urban land densities. *Journal of Cleaner Production, 224*, 802–810.

Xu, G., Jiao, L., Liu, J., Shi, Z., Zeng, C. & Liu, Y. (2019c). Understanding urban expansion combining macro patterns and micro dynamics in three Southeast Asian megacities. *Science of the Total Environment, 660*, pp.375–383.

Xu, G., Zhou, Z., Jiao, L., & Zhao, R. (2020). Compact urban form and expansion pattern slow down the decline in urban densities: A global perspective. *Land Use Policy, 94*, 104563.

Yue, W., Liu, Y., & Fan, P. (2013). Measuring urban sprawl and its drivers in large Chinese cities: The case of Hangzhou. *Land Use Policy, 31*, 358–370.

Zhang, T. (2000). Land market forces and government's role in sprawl. The case of China. *Cities, 17*, 123–135.

Zhang, Q., & Su, S. (2016). Determinants of urban expansion and their relative importance: A comparative analysis of 30 major metropolitans in China. *Habitat International, 58*, 89–107.

Živković, J. (2020). Urban form and function. *Climate Action*, 862–871.

Urban Environmental Challenges

Saloni Sharma and Amit Kumar Mishra

Abstract The whole world is facing various environmental challenges owing to its large human population. As most of the world's population now lives in urban areas, the pressure on the urban environment has significantly increased. The urban environmental crisis is not new to the world; several cities worldwide have experienced dense smog, solid waste accumulation, water contamination, and poor air quality, both outdoors and indoors. But now, more severe, new challenges are arising due to climate change, the pandemic rise, and the water availability crisis. Ensuring safe potable water, good disaster management, and well-developed health infrastructure are some of the new urban environment challenges. Areas where rapid and unplanned urbanization is taking place are the worst hit by these environmental crises. The magnitude of these crises has significantly increased, necessitating global involvement in formulating effective strategies to address urban pollution in the form of poor air and water quality, noise pollution, and hazardous waste. Physical and psychological health impacts of these crises are the greatest and need immediate response from the stakeholders. Proper monitoring, evaluation, and planning that involve awareness and education among the common people are some of the beginning steps. Availability of enough housing and stringent policies for energy and land use are required to avoid slum formation and maintain sanitation. The enormous amount of resource consumption in urban areas has far-reaching and long-term effects on the natural ecosystem, not only on urban land but outside of it as well. These threatened and deteriorated natural ecosystems are also part of urban environmental challenges. Effective participation from stakeholders for the mitigation of these urban environmental challenges requires knowledge of all aspects. In this chapter, we will be highlighting these urban environmental challenges in detail using some case studies.

Keywords Urban environmental crisis · Climate change · Urbanization · Sustainable development · Pollution mitigation

S. Sharma · A. K. Mishra (✉)
School of Environmental Sciences, Jawaharlal Nehru University, New Delhi, India
e-mail: amitmishra@mail.jnu.ac.in

P. K. Joshi et al. (eds.), *Geospatial Science for Urban Ecosystems*, The Urban Book Series, https://doi.org/10.1007/978-3-032-07155-2_3

1 Introduction

Urban areas are characterized by high population density. Continuous inhabitation of land by a large number of people implies an enormous amount of resource utilization that generates large quantities of waste and other by-products. The urban environment includes the built environment with a large human population, along with the natural ecosystem in which it is located. Urban environment is exposed to a lot of stress and burden from the hazardous chemicals and gases generated from various anthropogenic activities. Vehicular and industrial exhausts, releasing gaseous and particulate pollutants into the air, along with municipal and industrial discharges into water bodies, including groundwater, and solid waste generation from households, industries, and hospitals, are significant sources that degrade the urban environment. The very basic structure of urban settlement implies intense anthropogenic activities; therefore, the various challenges faced by urban environments all around the world are inevitable. Only with proper planning and management, we can mitigate those to maintain a quality and sustainable human life. Unplanned urban settlements with poor management have always been showing the world the extremities of urban environment challenges; e.g. London smog (1952), Ontario Minamata disease (1970), and electronic waste in Guiyu (1990) are some of the deadliest urban environmental crises. The incident of Guiyu, China, where informal e-waste recycling has led to elevated blood lead levels in children, illustrates the severe health impacts of urban pollution (Huo et al., 2007).

Currently, more than 50% of the global population resides in urban centres, which is projected to rise to approximately 68% by the year 2050 as per UN Department of Economic and Social Affairs (Ritchie & Roser, 2018). With such a significant population at stake, comprehensive knowledge about the urban environment challenges and their mitigation have become a matter of concern for the governance body, urban planners, policymakers, and scientists all around the world. Specifically, in the developing countries, where urbanization is taking place more quickly, a properly planned approach is of utmost importance.

Unplanned and rapid urbanization in developing regions leads to an increase in environmental and health hazards. Such regions will soon become overcrowded and create more challenges to the urban environment in the absence of proper housing/residence, with the formation of slums and a lack of clean water supply, drainage, roads, toilets, and medical facilities. The quality of the urban environment has a substantial influence on public health, sanitation, access to clean water, and the availability of fresh air. A lack of sound policies for transport and urban land use and their implementation are the most important reasons for an unsustainable urban environment. Urban environmental challenges are interlinked with energy, infrastructure, and land use and land change patterns. We can ensure a sustainable future in cities only with proper recognition, sensitization, and participation of all stakeholders of the urban ecosystem. In view of the importance of the urban ecosystem and the associated spurring of environmental issues, this chapter is going to highlight these urban environmental challenges (in terms of pollution) in detail using some case studies as examples.

2 Air Pollution

The phenomenon of adding physical, chemical, and biological agents, which are known as air pollutants, to air, which makes it harmful for human health, is known as air pollution. Air pollution is now a global threat, and its impacts are more evident in urban regions. Smog, produced as a result of the chemical reaction between primary air pollutants, is causing serious health problems in indoor as well as outdoor environments (Gilbert, 2013). A recent World Health Organization (WHO) report showed that now the entire world's population is now exposed to air quality levels that do not meet WHO guidelines, with developing parts of the world facing the most critical level of exposures (WHO, 2022a, b). In urban areas, both indoor and outdoor air pollution together cause increased morbidity in the form of various heart diseases and pulmonary diseases and have become the cause of millions of untimely deaths. Urban areas contribute to the majority of air pollutant emissions due to a comparatively higher population density than rural and suburban areas. Such a high level of pollution is causing an adverse effect on the population dwelling in these densely populated areas (~50% of the world's population).

According to a study by the WHO (2016), toxic air is regularly inhaled by 93% of children globally. The study found that 1.8 billion children globally breathe polluted air, which presents a significant risk to their health and development. In 2016, air pollution-associated acute lower respiratory illnesses were responsible for the deaths of roughly 600,000 children. The study further uncovered that over 40% of the world's population, encompassing one billion children under the age of 15, is significantly exposed to household air pollution. This exposure arises from the use of polluting fuels and appliances for cooking. Women in economically disadvantaged nations frequently resort to coal and biomass for cooking and heating, posing a threat to the health of their families (Aroud et al., 2020).

Various WHO reports highlight the close association between air pollution and climate change, particularly with regard to global warming, and urban areas are major contributors to this phenomenon. UN-Habitat reports that urban agglomerations, occupying less than 2% of the Earth's surface, consume over two-thirds of the world's energy and are responsible for more than sixty per cent of greenhouse gas emissions (Ruiz, 2021). The concentrated population density in urban areas relying on fossil fuels renders them particularly vulnerable to the impacts of climate change. This vulnerability is compounded by the limited availability of green spaces (United Nations Human Settlements Programme (UN-Habitat), 2010). The intergovernmental panel on climate change (IPCC) report underscores the need for substantial and rapid changes in energy usage, land use, infrastructure (including transportation and buildings), and industrial systems to restrict global warming to 1.5 °C. Urban environments are a major source of pollution, which is closely associated with climate change. In general, combustion sources contribute most significantly to air pollution (Liang & Gong, 2020) and also lead to elevated CO_2 emissions and serve as the primary driver of global warming (Waisman et al., 2019). Additionally, the specifics of air pollution, encompassing sources, types of pollutants, and their health impacts can be explored independently for outdoor and indoor environments, as outlined below.

2.1 Outdoor Air Pollution

In urban areas, industrial emissions and vehicle exhaust are the primary contributors to outdoor or ambient air pollution. Outdoor air pollution encompasses a wide range of contaminants. Among the most prevalent are ground-level ozone (O_3); carbon monoxide (CO), produced by the incomplete combustion of fossil fuels; NO_2, emitted from vehicle emissions and industrial activities; SO_2, generated from burning fossil fuels containing sulfur, particularly coal; particulate matter (PM), a mixture of tiny solid particles and liquid droplets that can include dust, dirt, soot, and smoke; and lead (Pb), a toxic metal released from industrial processes and previously from leaded gasoline (WHO, 2021).

Further, the outdoor air pollutants are categorized into primary and secondary pollutants based on their origin. Primary pollutants are directly emitted from sources into the atmosphere, while secondary pollutants are formed by chemical reactions in the atmosphere. Examples of primary air pollutants include sulfur oxides (SO_X), nitrogen oxides (NO_X), CO, and volatile organic compounds (VOCs), which are emitted through activities such as vehicle combustion, industrial processes, and natural events like volcanic eruptions (EPA, 2021). On the other hand, secondary pollutants are created when primary pollutants react chemically in the atmosphere. Examples of secondary pollutants include ozone, which forms when NO_X and VOCs react in the presence of sunlight, and photochemical smog, a complex mixture resulting from the interaction of various air pollutants under sunlight (WHO, 2024).

These outdoor air pollutants pose significant health risks and environmental challenges. Ground-level ozone can cause respiratory problems and exacerbate conditions like asthma (EPA, 2019). Carbon monoxide affects the oxygen-carrying capacity of blood, leading to cardiovascular issues (CDC, 2020). NO_2 and SO_2 can affect the respiratory system and also contribute to the formation of acid rain, which harms ecosystems (EPA, 2021). Particulate matter, depending on its size, can penetrate deep into the lungs and even enter the bloodstream, causing cardiovascular and respiratory diseases (WHO, 2021). Lead exposure can result in neurological damage, particularly in children (CDC, 2018). The sources and impacts of these pollutants are interconnected. For instance, emissions from vehicles and industrial activities contribute significantly to the levels of NO_2, SO_2, and particulate matter in the air. These primary pollutants, in turn, play crucial roles in the formation of secondary pollutants like ozone and photochemical smog (WHO, 2024). The widespread presence of these criteria air pollutants in the air necessitates comprehensive monitoring and regulatory measures to protect public health and the environment.

Efforts to reduce emissions, improve air quality standards, and develop cleaner technologies are essential in addressing the challenges posed by outdoor air pollution (EPA, 2021).

Environmental health is significantly affected by outdoor air pollution, impacting individuals across low-, middle-, and high-income countries alike. In India, around 39 cities rank among the top 50 most polluted places in the world, according to a report by IQAir (2022). The cities that fare the worst in terms of air pollution are Delhi, Bhiwandi, Gurugram, Ghaziabad, Kanpur, Patna, and Chapra (Guttikunda et al., 2014; IQAir, 2022). As per Saini's and Sharma's (2020) report, air pollution, including in both urban and rural areas, led to approximately 4.2 million premature deaths globally in 2016. The primary culprit behind these fatalities was the exposure to fine particulate matter (PM2.5.), a pollutant associated with cancer, cardiovascular disease, and respiratory issues. Notably, populations in low- and middle-income countries bear a disproportionate burden, with outdoor air pollution being responsible for 91% of premature deaths worldwide. Southeast Asia and Western Pacific areas bear the majority of this burden. Recent burden figures highlight the severe impact of air pollution on heart disease and mortality. Increasing amounts of data, including studies from highly polluted areas, demonstrate the links between ambient air pollution and the likelihood of getting heart disease (WHO, 2009). A significant portion of outdoor air pollution is contributed by traffic or vehicular emissions, which is discussed in the following section.

2.1.1 Traffic-Induced Air Pollution

Combustion from fossil fuels (petrol, diesel, and CNG) is the main cause of traffic-related pollution as it increases amounts of NO_x, VOCs, and carbonaceous particulate matter (PM). Occasionally, gases (e.g. SO_2) and particulate matter (like fly ash, which includes metals like selenium, vanadium, mercury, and nickel) from fuel combustion are also released (Aslam et al., 2006). Moreover, combustion can generate nitrogen oxides (NO_x), which, in conjunction with other airborne chemicals, can give rise to detrimental pollutants like ozone and particulate matter. These pollutants can have adverse health effects, particularly for those with respiratory problems. It is important to ensure proper ventilation and maintenance of combustion appliances to minimize the release of these pollutants into indoor and outdoor air. Some of these potentially harmful organic species, such as dioxins, quinones, and polycyclic aromatic hydrocarbons (PAHs), are known or believed to cause cancer. The sort of fuel and the circumstances of combustion affect the relative abundance of the constituents. Spark-ignition engines historically produced more CO and VOCs than diesel but usually less PM and NO_x. Complexity abounds in the carbonaceous PM produced by burning. Not all car pollutants come from the tailpipe; brake and tyre wear can result in copper and other metals (Mwangi et al., 2015).

2.2 Indoor Air Pollution

Indoor air quality (IAQ) is a significant health concern, with people spending over 90% of their time indoors, where pollutant levels can be twice as high as outdoors. IAQ issues arise from indoor pollution sources that release gases or particles into the air, posing greater health risks than outdoor pollution, especially for vulnerable groups like the young, elderly, and those with chronic illnesses (Cincinelli & Martellini, 2017; EPA, 2018). Indoor pollutants can cause immediate health issues, such as eye, nose, and throat irritation; headaches; dizziness; and fatigue. Prolonged exposure to these pollutants can lead to serious health conditions, including respiratory diseases, heart problems, and cancer. Approximately 6.1% of total global deaths, or 4.31 million annually, are attributed to indoor pollution, with the highest rates in low-income countries (Bennitt et al., 2021).

Poor ventilation is a major contributor to indoor pollution, allowing pollutants to accumulate (WHO, 2016). High humidity and temperature can significantly increase contaminant levels (Wolkoff, 2018). Common sources of indoor pollution include fuel-burning appliances, tobacco products, building materials, furnishings, and household maintenance items (Cincinelli & Martellini, 2017). Biological pollutants, such as animal dander, bacteria, moulds, and viruses, can cause significant health effects, including the spread of diseases like influenza in poorly ventilated environments (EPA, 2019).

Improving IAQ necessitates a comprehensive approach that involves identifying pollution sources, monitoring pollutant levels, and implementing effective control measures. Utilizing both mechanical and natural enhanced ventilation systems is crucial for maintaining IAQ, as these systems help dilute indoor pollutants and introduce fresh air. Additionally, innovative technologies like smart monitoring systems and advanced material-based sensors present promising solutions for better IAQ management (Tran et al., 2020). Traditional physical and chemical technologies for indoor air quality improvement, like filters and ozonizers, often fail to address the full range of indoor pollutants cost-effectively, necessitating complex and expensive sequential solutions (González-Martín et al., 2021). In contrast, biotechnologies, which use plants, bacteria, fungi, and micro-algae, offer a sustainable and efficient alternative. These biological systems not only improve air quality but also enhance energy efficiency and provide aesthetic and psychological benefits. For example, plants can absorb volatile organic compounds (VOCs) through their leaves and roots, while microorganisms can break down pollutants into less harmful substances (Yue et al., 2021). Additionally, emerging materials like activated carbon and metal-organic frameworks (MOFs) show promise in adsorbing VOCs and particulate matter (PM2.5). Photocatalytic oxidation, using materials such as titanium dioxide (TiO2), can effectively eliminate pollutants under light exposure.

In conclusion, efforts to mitigate indoor air pollution must prioritize innovative solutions that address the diverse sources and complex nature of indoor pollutants. Biotechnologies, with their ability to harness natural processes like plant and microbial action, offer a sustainable pathway towards improving indoor air quality. By integrating these biological systems into building designs and ventilation strategies, we can not only reduce pollutant levels but also enhance energy efficiency and promote healthier indoor environments. Continued research and development in biotechnologies, alongside advancements in materials science for adsorption and photocatalytic oxidation, are crucial to advancing effective strategies for indoor air purification. These efforts are essential to safeguarding public health, particularly in densely populated areas and low-income settings, where indoor pollution poses the greatest risks.

3 Water Pollution

Water pollution occurs when substances contaminate water sources, such as rivers, lakes, streams, oceans, and aquifers. These substances degrade the water quality, making it toxic and unsuitable for various human activities, including drinking, cooking, cleaning, swimming, and other purposes (Owa, 2014). Various substances such as chemicals, garbage, germs, and parasites can pollute water. All types of pollution ultimately lead to water contamination (EPA, 2021). Air pollution, for instance, can cause contamination in lakes and oceans through dry and wet deposition. Similarly, land contamination can pollute underground water resources, rivers, and eventually, the ocean. This means that waste dumped in an empty lot can potentially contaminate a water source (EPA, 2021). The impact of human activities on the surface water and groundwater quality is increasingly intricate due to diverse water resource characteristics. Challenges persist in forecasting and mitigating contaminants, particularly heavy metals. Pollutants from human activities (industrial, urban, and agricultural) and natural sources (climate change, natural disasters, and geological factors) contribute major inorganic substances like nitrogen, fluoride, and heavy metals (Akhtar et al., 2021). Nutrients, sediments, and toxic pollutants from non-point sources pose significant threats to water quality (Chaudhry & Malik, 2017). The existence of harmful substances in water systems can have negative repercussions on the local plant and animal life. These organisms may accumulate these chemicals in their bodies, and when humans consume them, it can result in mild or severe toxic symptoms. Additionally, plants and animals may suffer from reduced reproduction or even death as a result of water pollution (Anju et al., 2010). Water pollution in India has escalated to a critical level, as nearly all of the country's water systems are severely contaminated. The research conducted by the National Environmental Engineering Research Institute (NEERI) in Nagpur indicates that almost 70% of India's water is polluted (Dwivedi, 2017).

Water is highly susceptible to contamination due to its ability to dissolve more compounds than any other liquid on Earth, earning it the nickname "universal solvent" (Tripathi & Hussain, 2022). This property of water also explains why it is easily contaminated, as it readily dissolves and mixes with toxic substances from factories, towns, and fields, resulting in water pollution. Point-source pollutants enter water bodies through defined pathways, such as pipes or conduits, during dry weather, whereas non-point-source pollutants originate from diffuse pathways. Storm drainage, despite eventually reaching watercourses through pipes or channels, is categorized as non-point-source pollution. Other non-point sources include farm runoff, construction sites, and land disturbances. Point-source pollution primarily comes from industrial facilities and municipal wastewater treatment plants, with the types of pollutants discharged into water systems varying based on drainage contents (Loague & Corwin, 2006).

3.1 *Wastewater*

Wastewater denotes water that has been used for various purposes and is no longer suitable for use without undergoing treatment. It can originate from diverse sources, including industries, households, and agriculture. The release of untreated wastewater into the environment can result in polluting water bodies and create a substantial risk to human health and the environment (Edokpayi et al., 2017). Eliminating pollutants from water to make it acceptable for release or reuse is a crucial step in the treatment of wastewater. This process employs a variety of physical, chemical, and biological methods to eliminate pollutants such as pathogens, heavy metals, organic waste, and nutrients. In numerous economically poor neighbourhoods within cities and towns in developing nations, a significant amount of wastewater is occasionally inadequately treated and released straight into the nearest surface water drain or informal drainage channel. Both public health and the environment are seriously threatened by this practice. In urban areas, hospitals and various small- and large-scale industries often release hazardous chemicals and medical waste into the wastewater system, alongside household effluent and human waste (Apha et al., 2012).

Depending on the sort of company that produces it, the quality and quantity of industrial effluent might vary significantly (Amoatey & Bani, 2011). It may contain materials that are biodegradable or not, as well as those that are resistant to treatment, such as heavy metals or organic synthetic materials. Industrial wastewater in developing nations may differ drastically from industrial wastewater in industrialized nations. Industrial wastewater contains a rising amount of synthetic substances, which raises serious concerns because it can be challenging to detect and manage them. In order to solve this problem, it is advised that industries switch to safer production techniques by substituting dangerous recalcitrant substances with safer substitutes. Increasing public awareness can also aid in lowering the use of dangerous substances. A growing push for businesses to decrease their wastewater and treat it before discharge is the result of societal and environmental demands (NRC, 2012; Kiselev et al., 2019). The utilization of wastewater after proper treatment can bring about economic and financial benefits. Wastewater is increasingly recognized as a possible source; through the concept of "industrial symbiosis," wastewater can be utilized within the industry or by various firms. There is a compelling financial motivation to use wastewater locally and internally as a means to reduce costs.

Processed water, for instance, can be utilized for cooling or heating, while rainwater collected from roofs can be used for irrigation, flushing toilets, or washing cars.

The heightened reliance on pesticides and chemical fertilizers in agriculture to meet escalating demands has raised concerns about potential environmental pollution. In numerous developing countries, the practice of using untreated or insufficiently treated wastewater in irrigation has resulted in the contamination of groundwater. Improved management of wastewater is necessary to reduce the risk of pathogen exposure, particularly in the agricultural sector, where workers are at a high risk of infection. Farmers are exploring unconventional water sources, such as wastewater, recognizing it as a valuable reservoir of water and nutrients. When used safely, this approach has the potential to enhance water and food security and improve livelihoods (UN, 2017).

Overall, efficient wastewater management is crucial for shielding both the environment and public health. The efficiency of treatment can vary, even in cities where wastewater is collected and treated, depending on the specific treatment techniques applied. One of the most popular techniques for wastewater treatment uses the activated sludge process, which relies on microorganisms to break down organic matter (Singh et al., 2022). Additionally, man-made wetlands, oxidation ponds, and biological filters are also employed to treat wastewater (Amoatey & Bani, 2011). Effluent discharge standards have been set up by regulatory agencies and organizations all over the world to guarantee that the treated wastewater satisfies particular quality requirements before it is released into the environment. Depending on how the water is going to be used, different parameters may apply for these standards (Chen et al., 2012). Ensuring the proper management of wastewater is vital to safeguard human health and the environment from the adverse impacts of wastewater discharge. Treating wastewater to a quality level suitable for its intended purpose not only enhances the potential for cost recovery but also contributes to sustainability.

Investing in the right treatment technologies and implementing sustainable wastewater management practices can help developing nations. Additionally, effective wastewater treatment and reuse can promote economic growth, increase agricultural output, and lessen water scarcity. Leveraging sustainable technologies and innovative wastewater management systems can mitigate the environmental consequences of wastewater discharge. Through the adoption of effective wastewater management practices, we can promote the sustainable utilization of water resources and diminish the risk of waterborne diseases.

3.2 Thermal Water Pollution

Any divergence in the temperature of a habitat that can potentially harm the living organisms inhabiting it is defined as thermal pollution. The temperature can rise or fall due to diverse human activities, such as industrial cooling and the discharge of cold water into the stream (Dodds & Whites, 2010). As urbanization and industrialization continue to rise, thermal pollution in metropolitan areas is becoming a bigger concern (Chen, et al., 2021). Various sources of thermal pollution include nuclear, coal, and electric power plants; crude oil refineries; smelting industries; etc. (Speight, 2019). These industries release substantial amounts of heat into water bodies, impacting the physical, chemical, and biological characteristics of the receiving environments. Among these sources, nuclear power plants stand out as the largest contributors, requiring 30%–100% more cooling water than other power plants (Kirillin et al., 2013). While noticeable temperature increases have clear effects, even minor changes can have significant biological consequences. Additionally, climate change serves as a non-point source of thermal pollution, already influencing a wide range of freshwater habitats. The documented global temperature rise of 0.4–0.8 °C over the past 15 years is a reality. The specific effects of climate change on freshwater habitats remain uncertain, but experts anticipate significant alterations in the structure and function of streams, lakes, and wetlands, particularly in higher latitudes (Rose et al., 2023; Alahuhta, et al., 2019).

The introduction of heat into surface water can have harmful effects on the living organisms within them, thus making it a form of water pollution. This type of pollution can occur due to heated industrial waste products or human activities that alter stream bank vegetation, resulting in increased water temperatures from solar radiation. As a result, the warmer water temperature can decrease the solubility of oxygen in the water and speed up the decomposition of organic matter, leading to a reduction in dissolved oxygen. Even a small increase in water temperature can have negative impacts on aquatic life (Kumar et al., 2021). Increased water temperature can lead to a reduction in the oxygen levels within the water, adversely affecting the reproductive cycles, respiratory and digestive rates, and other physiological processes of aquatic life. This can pose challenges to their survival (Speight, 2019). Thermal pollution often occurs as a result of power plants and industrial manufacturers using water as a coolant. Thermal pollution can cause sudden and fatal changes in water temperature that can harm fish and other aquatic organisms that have adapted to specific temperature ranges. These sudden changes in temperature, known as thermal shock, can be lethal for these creatures. The use of geothermal fields often results in thermal pollution of both air and water, and the excess heat released as steam can affect cloud development and alter local weather patterns.

3.3 Water Quality Assessment

Evaluating water quality is a crucial step in ensuring the safety of water for various purposes, such as drinking, irrigation, and recreational activities (Gorde & Jadhav, 2013). This assessment encompasses the analysis of various physical, chemical, and biological parameters, including temperature, dissolved oxygen and solids, pH, turbidity, heavy metals, and pathogens. Any levels exceeding the set limits indicate inadequate water quality. If these parameters' levels exceed the recommended limits, it can adversely affect human health. As previously noted, population growth, urbanization, and modernization are factors contributing to the increasing challenge of sewage disposal and pollution of water bodies. Natural water sources are also subject to contamination from activities such as mining, soil leaching, and rock weathering. The accumulation of nutrients in lakes, attributed to various factors, has been recognized as a significant issue. This can result in eutrophication and a decline in biodiversity in rivers, ponds, and lakes (Gorde, & Jadhav, 2013). The application of fertilizers in response to land use changes and extended growing seasons can worsen this issue by amplifying the leaching of nutrients into water sources. Many organizations and agencies globally have set forth guidelines and standards to protect water quality. The World Health Organization (WHO) has developed guidelines for drinking water quality that define limits for various parameters to ensure human health protection. Likewise, the US Environmental Protection Agency (EPA) has implemented regulations to maintain water quality within the country. To assess water quality, several methods are utilized, including laboratory testing, remote sensing, and modelling (Ammenberg et al., 2002). Laboratory testing involves the collection of water samples from different sources, followed by analyses of various parameters. Remote sensing utilizes satellites and other sensors to remotely monitor water quality parameters. Modelling applies mathematical models to estimate water quality parameters based on various inputs. Different methods, including the water quality index and hazard quotient, have been employed to evaluate water quality. It is advisable to implement pollution prevention, water reuse, and nutrient recycling in controlled urban agriculture to address these concerns.

4 Industrial Pollution

Different environmental issues have arisen in developing nations like India as a result of the unprecedented development of industrialization and urbanization, which pollute both water and air of the urban environment. Some of the most polluting sectors are categorized as "Red Category" by the Central Pollution Control Board (CPCB), including sugar, pulp and paper, textiles, fertilizers and pesticides, pharmaceuticals, oil refineries, thermal power plants, petrochemicals, and metal smelters (Arif, et al., 2021). Through the release of liquid waste, hazardous solid waste, and exhaust, these sectors have a significant negative impact on the water, soil, noise, and air quality (Baranwal & Agrawal, 2022). Heavy metals from these industries are released into water, soil, and air, making them not just unfit for any use but also a threat to biodiversity and human health. A large amount of CO_2 is released by power plants that produce electricity from fossil fuels (coal, oil, and gas). Approximately 1 MWh of power results in the production of about 990 kg of CO_2 from coal, 839 kg from petroleum, and 540 kg from natural gas (Reed & Renner, 1995).

Chemical pollution is caused by the discharge of various chemicals into the environment. Chemicals such as ammonia, hydrogen sulfide, arsenic, cadmium, and mercury, with many other organic compounds, pollute the groundwater, soil, and rivers (Gupta & Roy, 2007), while chemicals in gaseous form (carbon dioxide, hydrogen sulfide, oxides of nitrogen, volatile organic compounds, and ozone) make the atmosphere poisonous. Certain chemicals like pesticides and plastics can remain in the environment for a significantly long time and adversely affect the ecosystems within it (Zhang & Wang, 2020). Along with the water scarcity condition, chemical pollution severely impacts the biological dynamics of an ecosystem (Sánchez et al., 2016). Prolonged discharge of chemicals in water bodies is the leading cause of eutrophication. The contamination of soil and water bodies with heavy metals leads to the accumulation of these elements in crops and vegetables at potentially hazardous concentrations, adversely affecting human health. Increase in diseases (such as cancer, organ damage, cardiovascular disease, and nervous system disorders) has been found to be linked with these contaminations (Kumar et al., 2021).

5 Noise Pollution

Noise is generally referred to as sound that is unwanted or disturbs individuals. Unwanted noise becomes a disturbance when it hinders routine activities like conversation or sleep, thereby detrimentally impacting the quality of life. Noise pollution, however, has not garnered as much attention as other forms of pollution, such as air or water pollution, possibly due to its lack of visibility, tangibility, or odour. Despite being constantly surrounded by sounds, most people would not describe their environment as noisy. Nevertheless, the continuous and escalating sources of sound can be regarded as a nuisance to some individuals and can have severe effects on one's overall health and well-being (Raja et al., 2022). Maintaining acceptable ambient noise levels in public spaces requires vigilant monitoring and regulation of noise-emitting sources. Elevated noise levels stemming from diverse sources like industrial and construction activities, generator sets, loudspeakers, public address systems, music systems, vehicular horns, and various mechanical devices can detrimentally impact both human mental and physical health. Indeed, millions of individuals experience the deleterious effects of noise pollution, leading to both short-term and long-term health issues (Baro, 2023). Although noise-induced hearing loss is a prevalent and extensively discussed health consequence of noise pollution, research indicates that exposure to high or continuous noise levels can lead to various adverse health effects beyond hearing impairment. Issues such as stress-related illnesses, elevated blood pressure, sleep disturbances, cardiovascular effects, decreased productivity, and hearing loss are just a few of the negative outcomes associated with noise exposure (Wokekoro, 2020). Noise can affect some populations more than others (Swift, 2010); for example, children are more exposed to nocturnal noise because they spend more time in bed than adults do. The elderly and individuals with long-term illnesses are more sensitive to disturbance. Shift employees are more vulnerable because their sleep patterns are stressed. Additionally, the less wealthy are likely to suffer disproportionately because they cannot afford to live in quiet neighbourhoods or have properly insulated houses. There are numerous actions people can take to safeguard themselves from the negative impacts of noise pollution. If individuals must be around deafening noises, hearing protection is an option (Bistrup et al., 2001).

There are many ways to deal with noise in your house, place of education, place of employment, and neighbourhood. As per the WHO guidelines for neighbourhood noise (Berglund et al., 2000), it is recommended to maintain noise levels below 30 A-weighted decibels (dB) in bedrooms at night to support quality sleep. Additionally, classrooms should ideally have noise levels below 35 dB to foster conducive teaching and learning environments. In order to avoid negative health effects from night noise, the WHO guidelines for night noise suggest an annual average of less than 40 dB outside of bedrooms. The WHO defines a quiet place as one where the sound pressure level is less than or equal to 55 dB (Hurtley, 2009). Numerous sources of noise pollution in urban areas are detailed in the following subsections.

5.1 *Transport Noise*

Noise pollution, particularly in urban centres with a high density of road and railway vehicles, is significantly influenced by transportation activities. While road transportation is often the main focus, noise is produced in all modes of transportation and can harm the health of living beings (Griefahn et al., 2000). The main contributors to transport noise encompass road traffic noise, emanating from engines and horns of cars, trucks, buses, and various vehicles, along with aircraft and railway noise. Road traffic noise, largely stemming from the prevalence of internal combustion engine vehicles, stands out as the predominant urban noise source, constituting 80% of all communal noise sources (Basner et al., 2014). However, the integration of electric and hybrid cars, known for producing lower noise levels, is aiding in mitigating noise pollution in urban settings (Piotrowska et al., 2020). Studies indicate a substantial rise in traffic noise in urban areas since the early twenty-first century, driven by the increased transportation of people and goods. Addressing this surge in traffic noise is crucial to alleviating its detrimental impact on human health and well-being. There are many ways to mitigate transportation noise, from actions taken on the vehicle itself to the installation of natural or artificial sound barriers. Future research should focus on developing more effective noise mitigation strategies and understanding the complex interactions between different sources of noise, particularly in electric vehicles.

5.2 *Neighbourhood Noise*

The growing concern over the impact of neighbourhood noise on the physical and emotional health of a significant population is on the rise. Long-term loud sound exposure can damage ears, interfere with sleep, and even cause cardiovascular disease (Kalawapudi, et al., 2020). In addition, it may cause annoyance, worry, and a decline in quality of life (Miedema & Oudshoorn, 2001). Research has confirmed that noise pollution from various sources, including traffic, significantly and adversely affects public health. Exposure to noise levels exceeding 55 dB has been shown to disrupt sleep and increase the risk of cardiovascular disease (Cohen et al., 1986). Another study showed that long-term exposure to noise pollution can increase stress hormone levels and have a detrimental effect on mental health (Stansfeld & Matheson, 2003). Children's growth and schooling can be significantly impacted by neighbourhood noise as well. Children who experienced elevated noise levels at home exhibited poorer cognitive performance and achieved lower reading scores (Klatte et al., 2013). Sound insulation, landscaping, and zoning laws are a few strategies for reducing neighbourhood noise. Landscaping residential areas with trees and other vegetation (urban green spaces) can greatly lower noise levels (Kruize, et al., 2019). Noise pollution can also be decreased by zoning laws that restrict noise levels and the use of loud devices during specific hours (Ordoñez et al., 2018). Hence, neighbourhood noise stands as a substantial issue that can negatively impact the quality of life, education, and both physical and mental health. To mitigate noise pollution and protect public health, it is essential for individuals, communities, and governments to undertake appropriate measures, particularly in developing countries like India.

5.3 Industrial Noise

Globally, industrial noise poses serious work and public health risks. Occupational noise exposure contributes to at least 16% of adult instances of hearing loss (Nelson et al., 2005). Numerous studies have looked into how industrial noise affects employees' health. Manufacturing workers who were exposed to loud noise had significantly higher blood pressure and heart rates (Pouryaghoub et al., 2013). A different study revealed that employees who endured noise levels exceeding 85 dB for more than eight hours every day faced an increased risk of developing hearing loss (Zare et al., 2013). Industrial noise pollution can harm wildlife, including bird populations and other vulnerable species too (Ljungberg et al., 2015). Different tactics can be employed to lessen the effects of industrial pollution. Effective engineering measures include sound-absorbing materials and pollution barriers. Earmuffs and earplugs are two examples of personal safety equipment that can be used to lessen noise exposure (Wang et al., 2014). Hearing loss can be avoided with regular hearing tests and instruction on the dangers of noise exposure. In addition to affecting the health of workers, industrial noise can have adverse effects on the environment and nearby communities.

6 Urban Waste

Cities worldwide grapple with a significant challenge presented by urban waste, resulting in adverse effects on both the environment and human health. Cities all over the world are struggling with an increasing amount of urban garbage as a result of rising populations and consumption trends. Inadequate waste management can lead to various harmful consequences for both the environment and human health. Solid waste handling poses a significant challenge in addressing urban waste issues. Waste is not frequently collected or disposed of in a way that is secure for the environment. As a consequence, there might be an accumulation of trash in public areas, creating unhealthy conditions. Ineffective solid waste management practices were linked to an increase in cases of dengue fever, malaria, and diarrhoea in India (Sarkar et al., 2008). Urban regions also produce substantial amounts of wastewater in addition to solid waste. Wastewater improperly disposed of has a significant negative effect on public health because it can contaminate groundwater and surface water. According to a study done in Accra, Ghana, drinking tainted water for domestic reasons increases your risk of getting diarrhoea (Kyei et al., 2015).

Several approaches have been proposed for a more sustainable management of urban solid waste. These consist of greater recycling and composting, as well as waste reduction through better product design and packaging. The creation of the necessary infrastructure, such as garbage collection and treatment facilities, is also necessary for effective waste management (Cucchiella et al., 2016). The creation of circular economies is one creative strategy for handling urban waste. This entails developing closed-loop systems that utilize and recycle waste materials rather than disposing of them. Circular economy practices can boost the economy and greatly lessen the environmental effects of waste management (Su et al., 2013). To lessen the negative effects of waste and progress towards a more sustainable future, effective waste management strategies are needed (Otoma et al., 2019). The main component of urban waste, i.e. municipal waste, and its management techniques are discussed in the following subsections.

6.1 Municipal Waste

The quantity of municipal waste generated in a country is influenced by various factors like economic growth, urbanization rate, cultural practices, and climatic conditions (Hoornweg & Bhada-Tata, 2012). Municipal waste encompasses both liquid and solid waste. Solid refuse includes waste from urban and suburban areas, which can be either homogeneous or heterogeneous (Getahun et al., 2012). Most solid waste items, such as paper, glass, and metal, are recyclable; however, they require sorting either before or after collection. Organic leftovers, after decomposition, can enhance soil nutrients. Although there will inevitably be a portion that is neither recyclable nor easily decomposable, it can be utilized for energy production, necessitating special handling during disposal or treatment. The majority of the municipal garbage produced by cities should be recycled, with the leftovers being either burned or dumped in landfills. The diverse composition of urban solid waste, which includes organic wastes, batteries, packaging materials, paper, metal containers, glass, plastics, clothing, electronics, furniture, and more, complicates the management of these materials. Hazardous substance-filled industrial and pharmaceutical refuse is frequently disposed of with regular trash.

Municipal waste used to frequently be dumped in landfills without sufficient planning. Leachate was created as rain seeped through the landfills during strong downpours. Leachate composition can be affected by internal processes such as organic matter decomposition, landfill age, waste makeup, and hydrogeological circumstances (Mukherjee, et al., 2015). Leachate often contains hazardous concentrations of ions, heavy metals, chemicals, poisonous substances, organic compounds, and pathogens. Surface water and groundwater have become contaminated and unusable in many nations as a result of wastewater infiltration in the vadose zone. Unlined landfills have been connected to contaminated surface and groundwater in places, including Sant'Agostino, Italy; Zhoukou, China; and Ranital, India (Brindha & Schneider, 2019). To mitigate the adverse environmental impacts of landfilling, liners and leachate collection systems are incorporated into modern landfill design. These systems prevent the infiltration of leachate into groundwater and avoid the associated contamination of aquifers, which can render the water unsuitable for use. This approach is now widely adopted and has replaced the older practice of haphazardly dumping municipal waste in landfills without proper planning or protective measures (Rapti-Caputo & Vaccaro, 2006; Han et al., 2014; Samadder et al., 2017; Sivakarun et al., 2020). However, many older landfills were not built with such precautions, leading to groundwater contamination in several developed and developing nations. Hence, adopting sustainable waste management practices is essential to decrease the volume of generated municipal waste and minimize its environmental impact. Initiatives such as waste reduction, segregation, recycling, and the implementation of new technologies to recover valuable resources from waste are crucial components of such efforts. By taking proactive measures, we can create a cleaner and healthier future for ourselves and future generations.

6.2 Waste Management

Waste is handled through diverse techniques, including recycling, composting, incineration, and landfill disposal. Municipal waste management typically encompasses several processes, such as collection, separation, reuse, storage, transportation, recycling, and end-of-life actions like landfill disposal, land application, and energy production utilizing the caloric content of the waste. Municipal garbage is produced and thrown away from daily living and includes both hazardous and non-hazardous elements. Considering the global increase in the volume of municipal solid waste due to factors like population growth, urbanization, economic expansion, and human activities, it presents a significant threat to both public health and the environment. There are numerous approaches to deal with the hazardous components of municipal waste management, including landfill disposal, anaerobic digestion, fuel recovery, composting, and incineration (UNEP, 2010).

In many parts of the world, managing municipal waste is becoming more difficult due to manual scavenging. Manual scavenging, according to Fidelis et al. (2020), entails the risky and solitary process of gathering and recycling municipal solid refuse. Economic, environmental, and societal issues are raised by this practice. Sadly, discarded garbage frequently contains harmful or possibly hazardous materials. Municipal rubbish burning and the decomposition of organic materials in open dumps and landfills both discharge harmful pollutants into the atmosphere.

Managing solid waste is a crucial challenge experienced by both developed and developing countries, encompassing rural and urban regions. Municipal solid waste (MSW) management, notably in urban areas, poses particular difficulties, especially in developing nations (Ferronato & Torretta, 2019). Effective MSW management strategies must align with environmental sustainability, cost-effectiveness, social and legal acceptability, and scientific feasibility. The management of solid waste remains a significant challenge for municipalities, regardless of their size. The environmental repercussions of MSW management include greenhouse gas emissions, which can pose health risks for humans and contribute to pollution in soil, air, and water. These issues have significant implications for public health and aesthetics. This pertains to the contamination of soil, air, and water, along with its implications on aesthetics, potentially causing health issues for humans. Household hazardous waste differs from industrial waste and is discarded alongside regular domestic waste in landfills. The quantity, nature, and disposal methods of household hazardous waste are not thoroughly comprehended, resulting in uncertainty regarding potential risks. Some believe that since there are not many household hazardous wastes, there is no significant danger associated with their disposal (LaGrega et al., 2010). On the other hand, when industrial, municipal solid, and other wastes are disposed of separately, the presence of toxic and hazardous elements in such wastes becomes more significant (LaGrega et al., 2010). The widespread use of certain chemicals in household products has raised serious concerns. The environmental consequences and impacts of dumping household hazardous waste are also worrisome. To minimize environmental risks, household hazardous waste should be disposed of in accordance with the prevailing regulations in landfills (Slack et al., 2004).

Approximately 71% of MSW is dumped in landfills on a global scale. Batteries, paints, mercury-containing garbage, medications, and auto repair equipment are just a few of the dangerous elements found in these dumps. On the contrary, over 53% of the waste disposed of in landfills consists of materials that can be decomposed by anaerobic bacteria, including hardboard, paper, yard waste, and food. Consequently, landfilling remains the prevalent method of solid waste disposal for various waste types in Europe and the United States. In light of this, it is essential to grasp the fundamentals of landfill design (Nanda Berruti, 2021). For instance, the design and operation of landfills in the United States are subject to regulations outlined in the New Source Performance Standards of the Clean Air Act, Subtitle D of the Resource Conservation and Recovery Act, and various relevant state laws. Consequently, landfills have transformed from basic open-air refuse disposal sites into intricately engineered structures and facilities. They are sequestered from the environment, capturing contaminated leachate (water that comes into contact with the waste) and controlling gas migration (Louis, 2004). To protect groundwater, a landfill site is typically excavated and lined with a system of layers that minimizes leachate migration to the ground layers and collects such leachate for treatment.

7 Aesthetic Pollution

Aesthetic pollution, also known as visual pollution or environmental blight, is a term employed to describe the adverse effects of certain human activities on the visual appeal of the environment. According to a United Nations study, aesthetic pollution constitutes a significant issue in cities, exacerbating overall urban climatic conditions (United Nations, 2014). Aesthetic pollution can harm both people and societies in a variety of ways. For instance, it can lead to sensory overload or visual fatigue, which makes it challenging for people to enjoy the built or natural environment around them. Additionally, it can lower a community's quality of life by creating a feeling of discontent or alienation from its surroundings (Gobster et al., 2007, Fernandez-Montoya et al., 2014). Aesthetic pollution can take many forms, including billboards, graffiti, litter, industrial infrastructure, and other types of visual clutter. Billboard advertising is one of the most pervasive types of visible pollution. Billboards could be considered a form of visible clutter that detracts from the built or natural environment. Researchers from the University of California found a correlation between less favourable views of the beauty of nature and billboards in urban areas (Gobster et al., 2007).

Another type of aesthetic pollution that can harm the beauty of the ecosystem is litter. Litter was connected to poorer levels of perceived safety, cleanliness, and natural beauty in urban areas (Gobster et al., 2007). Cities and communities can use a variety of tactics to lessen the harmful impacts of aesthetic pollution. To prevent the development of visibly intrusive buildings or structures, they might, for instance, enact stringent zoning laws and design standards. Additionally, they can spend money on projects that will improve the aesthetic appeal of public areas, such as landscaping and public art. Aesthetic pollution is a serious issue that can harm both people and societies in a variety of ways. However, we can contribute to the development of more vibrant, liveable, and sustainable communities for both present and future generations by taking measures to decrease aesthetic pollution and improve the visual quality of our surroundings.

8 Radioactive Pollution

An important environmental issue that poses a risk to both the ecosystem and human health is radioactive pollution. Radon gas is one example of a natural source of radioactive pollutants. Human activities such as the operation of nuclear power plants, nuclear weapon testing, and the disposal of radioactive waste are other examples. The primary danger posed by radioactive pollutants is that they can ionize atoms and molecules, which can harm genetic material and living cells. High radioactive pollutant exposure can cause radiation sickness and even mortality. Extended exposure to radioactive pollutants can elevate the risk of developing cancer and other fatal diseases (WHO, 2012).

One of the primary contributors to radioactive pollution is nuclear power plants. Nuclear reactions are used in nuclear power facilities to heat water, which creates steam to drive turbines. The procedure also generates radioactive waste, which must be handled carefully in order to avoid environmental contamination. Accidents at nuclear power plants, like the Fukushima tragedy in 2011 and the Chernobyl disaster in 1986, highlighted the hazards of nuclear energy and underscored the importance of appropriately managing radioactive waste (International Atomic Energy Agency, 2014). Nuclear weapon testing is another source of radioactive pollution, with atmospheric testing during the Cold War significantly contaminating the ecosystem with radioactive material. While nuclear weapon testing has mostly been prohibited, its legacy persists.

Strong regulations must be in place to guarantee the safe disposal of radioactive waste and avoid accidents in order to reduce the risks associated with radioactive pollutants. To stop leaks and other accidents, nuclear power facilities and other sources of radioactive pollution need to be better monitored (United Nations Scientific Committee on the Effects of Atomic Radiation, 2016). To sum up, radioactive pollution is an important environmental concern that needs to be managed and monitored carefully in order to safeguard both human health and the ecosystem.

9 Conclusion

Increase in the frequency of urban migration, along with population growth in the current warming world, needs special attention in terms of environmental health of urban regions. Moreover, environmental indicators that are used to monitor the health of the system have a direct association with human health. The detrimental effects of environmental pollution on human health pose a significant concern for urban settlements. To explore the numerous environmental issues linked to urban areas, this chapter focuses on various concerns within urban ecosystems, including air pollution, water pollution, noise pollution, waste pollution, and their implications for human health. This chapter also provides a brief discussion on the possible preventive and management techniques for various urban environmental issues.

One of the important contributions of this chapter could be providing the answers to a few questions, such as the following: (1) How are the pollution of different spheres (atmosphere, geosphere, hydrosphere, and biosphere) interconnected? (2) What are the management practices and preventive measures to curb these environmental issues? We argue that the overall economic development, particularly in developing countries, leads to an escalation in the levels of environmental pollutants (such as in air, water, soil, and land) in urban cities if adequate preventive measures are not implemented. Waste management is a daunting task in urban areas, which are connected with almost all kinds of pollution forms in urban regions. Moreover, effective wastewater management, employing a blend of traditional and advanced techniques, is a critical process focused on eliminating pollutants from water to safeguard public health and the environment. Embracing diverse treatment methods and adhering to effluent discharge standards is pivotal for the secure disposal and reuse of wastewater. The implementation of sustainable wastewater management practices plays a crucial role in mitigating the impact of wastewater discharge and safeguarding the health and well-being of communities and the environment.

In summary, we provide a systematic introduction to various urban environmental issues using scientific literature as a dataset. The utilization of the above information by national and international agencies to shape environmental policies has become increasingly prevalent. Measuring the impact of environmental pollution on public health has become a pivotal element in policy discussions, as governments assess their options for pollution management.

Acknowledgements The authors would like to acknowledge the School of Environmental Sciences, Jawaharlal Nehru University, New Delhi, for providing the required logistics to complete this article.

References

Akhtar, N., Syakir Ishak, M. I., Bhawani, S. A., & Umar, K. (2021). Various natural and anthropogenic factors responsible for water quality degradation: A review. *Water, 13*(19), 2660.

Alahuhta, J., Erős, T., Kärnä, O. M., Soininen, J., Wang, J., & Heino, J. (2019). Understanding environmental change through the lens of trait-based, functional, and phylogenetic biodiversity in freshwater ecosystems. *Environmental Reviews, 27*(2), 263–273.

Ammenberg, P., Flink, P., Lindell, T., Pierson, D., & Strombeck, N. (2002). Bio-optical modelling combined with remote sensing to assess water quality. *International Journal of Remote Sensing, 23*(8), 1621–1638.

Amoatey, P., & Bani, R. (2011). *Wastewater management*. INTECH Open Access Publisher.

Anju, A., Ravi, S. P., & Bechan, S. (2010). Water pollution with special reference to pesticide contamination in India. *Journal of Water Resource and Protection, 2010.*

Apha, A., Wef, A. W., & Wpcf, A. (2012). *Standard methods for the examination of water and wastewater* (22nd ed.). American Public Health Association.

Arif, Z., Sethy, N. K., Swati, Mishra, P. K., & Verma, B. (2021). Grossly polluting industries and their effect on water resources in India. *Pollutants and Water Management: Resources, Strategies and Scarcity*, 47–65.

Aroud, R. A., Blasi, A. H., & Alsuwaiket, M. A. (2020). Intelligent risk alarm for asthma patients using artificial neural networks. *arXiv preprint arXiv:2009.12175.*

Aslam, M. U., Masjuki, H. H., Kalam, M. A., Abdesselam, H., Mahlia, T. M. I., & Amalina, M. A. (2006). An experimental investigation of CNG as an alternative fuel for a retrofitted gasoline vehicle. *Fuel, 85*(5–6), 717–724.

Baranwal, P., & Agrawal, S. (2022). A pollution assessment of grossly polluting industries in India. *International Journal of Environment and Waste Management, 30*(1), 30–54.

Baro, D. K. (2023). Pollution and it's effects on health. Societal Challenges and Sustainability: Socio-economic and Environmental Perspectives, 63.

Basner, M., Babisch, W., Davis, A., Brink, M., Clark, C., Janssen, S., & Stansfeld, S. (2014). Auditory and non-auditory effects of noise on health. *The Lancet, 383*(9925), 1325–1332.

Bennitt, F. B., Wozniak, S. S., Causey, K., Burkart, K., & Brauer, M. (2021). Estimating disease burden attributable to household air pollution: New methods within the Global Burden of Disease Study. *The Lancet Global Health, 9*, S18.

Berglund, B., Lindvall, T., & Schwela, D. H. (2000). New WHO guidelines for community noise. *Noise & Vibration Worldwide, 31*(4), 24–29.

Bistrup, M. L., Hygge, S., Keiding, L., & Passchier-Vermeer, W. (2001). *Health effects of noise on children and perception of the risk of noise*. National Institute of Public Health.

Brindha, K., & Schneider, M. (2019). Impact of urbanization on groundwater quality. In *GIS and geostatistical techniques for groundwater science* (pp. 179–196). Elsevier.

Centers for Disease Control and Prevention. (2018, updated 2024). Childhood Lead Poisoning Prevention: Guidelines and recommendations. U.S. Department of Health & Human Services. Retrieved from https://www.cdc.gov/lead-prevention/php/guidelines/index.html

Centers for Disease Control and Prevention, National Center for Environmental Health. (2020). Clinical guidance for carbon monoxide (CO) poisoning. U.S. Department of Health & Human Services. Retrieved from https://stacks.cdc.gov/view/cdc/122250

Chaudhry, F. N., & Malik, M. F. (2017). Factors affecting water pollution: A review. *Journal of Ecosystem and Ecography, 7*(1), 225–231.

Chen, X., Chen, G., & Chen, Y. (2012). Review of biological methods for wastewater treatment. *Renewable and Sustainable Energy Reviews, 16*(5), 3263–3277.

Chen, H. Y., Hodges, C. C., & Dymond, R. L. (2021). Modeling watershed-wide bioretention stormwater retrofits to achieve thermal pollution mitigation goals. *JAWRA Journal of the American Water Resources Association, 57*(1), 109–133.

Cincinelli, A., & Martellini, T. (2017). Indoor air quality and health. *International Journal of Environmental Research and Public Health, 14*(11), 1286.

Cohen, S., Evans, G. W., Stokols, D., & Krantz, D. S. (1986). *Behavior, health, and environmental stress*. Plenum Press.

Cucchiella, F., D'Adamo, I., & Lenny Koh, S. C. (2016). Circular economy: What does it mean for industrial ecology? *Resources, Conservation and Recycling, 114*, 32–39.

Dodds, W., & Whites, M. (2010). *Freshwater ecology: Concepts and environmental applications of limnology* (p. 829). Academic Press.

Dwivedi, A. K. (2017). Researches in water pollution: A review. *International Research Journal of Natural and Applied Sciences, 4*(1), 118–142.

Edokpayi, J. N., Odiyo, J. O., & Durowoju, O. S. (2017). Impact of wastewater on surface water quality in developing countries: A case study of South Africa. *Water Quality, 10*(66561), 10 5772.

Environmental Protection Agency. (2019). Indoor air quality (IAQ). Retrieved from https://www.epa.gov/indoor-air-quality-iaq/introduction-indoor-air-quality.

EPA. (2018). *An introduction to Indoor Air Quality (IAQ)*. United States Environmental Protection Agency. Retrieved from https://www.epa.gov/indoor-air-quality-iaq/introduction-indoor-air-quality.

Fernandez-Montoya, A., Fernandez-Morales, A., & Martinez-Ruiz, M. P. (2014). The impact of graffiti on residential property values. *Journal of Real Estate Research, 36*(3), 283–308.

Ferronato, N., & Torretta, V. (2019). Waste mismanagement in developing countries: A review of global issues. *International Journal of Environmental Research and Public Health, 16*(6), 1060.

Fidelis, R., Marco-Ferreira, A., Antunes, L. C., & Komatsu, A. K. (2020). Socio-productive inclusion of scavengers in municipal solid waste management in Brazil: Practices, paradigms and future prospects. *Resources, Conservation and Recycling, 154*, 104594.

Getahun, T., Mengistie, E., Haddis, A., Wasie, F., Alemayehu, E., Dadi, D., et al. (2012). Municipal solid waste generation in growing urban areas in Africa: Current practices and relation to socioeconomic factors in Jimma, Ethiopia. *Environmental Monitoring and Assessment, 184*, 6337–6345.

Gilbert, S. (2013). A small dose of air pollution or an introduction to the health effects of air pollution. In *A small dose of toxicology–The health effects of common chemicals*.

Gobster, P. H., Nassauer, J. I., Daniel, T. C., & Fry, G. (2007). The shared landscape: What does aesthetics have to do with ecology? *Landscape Ecology, 22*(7), 959–972.

González-Martín, J., Kraakman, N. J. R., Pérez, C., Lebrero, R., & Muñoz, R. (2021). A state–of–the-art review on indoor air pollution and strategies for indoor air pollution control. *Chemosphere, 262*, 128376.
Gorde, S. P., & Jadhav, M. V. (2013). Assessment of water quality parameters: A review. *International Journal of Engineering Research and Applications, 3*(6), 2029–2035.
Griefahn, B., Schuemer-Kohrs, A., Schuemer, R., Moehler, U., & Mehnert, P. (2000). Physiological, subjective, and behavioural responses during sleep to noise from rail and road traffic. *Noise and Health, 3*(9), 59.
Gupta, P. K., & Roy, S. (2007). *Environmental pollution and health*. APH Publishing.
Guttikunda, S. K., Goel, R., & Pant, P. (2014). Nature of air pollution, emission sources, and management in the Indian cities. *Atmospheric Environment, 95*, 501–510.
Han, D., Tong, X., Currell, M. J., Cao, G., Jin, M., & Tong, C. (2014). Evaluation of the impact of an uncontrolled landfill on surrounding groundwater quality, Zhoukou, China. *Journal of Geochemical Exploration, 136*, 24–39.
Hoornweg, D., & Bhada-Tata, P. (2012). *What a waste: A global review of solid waste management*. The World Bank.
Huo, X., Peng, L., Xu, X., Zheng, L., Qiu, B., Qi, Z., & Piao, Z. (2007). Elevated blood lead levels of children in Guiyu, an electronic waste recycling town in China. *Environmental Health Perspectives, 115*(7), 1113–1117. https://doi.org/10.1289/ehp.9697
Hurtley, C. (Ed.). (2009). *Night noise guidelines for Europe*. WHO Regional Office Europe.
International Atomic Energy Agency. (2014). *Management of radioactive waste*. Retrieved from https://www.iaea.org/topics/radioactive-waste-management.
IQAir. (2022). *World Air Quality Report*. https://www.iqair.com/in-en/world-most-polluted-cities.
Kalawapudi, K., Singh, T., Dey, J., Vijay, R., & Kumar, R. (2020). Noise pollution in Mumbai Metropolitan Region (MMR): An emerging environmental threat. *Environmental Monitoring and Assessment, 192*, 1–20.
Kirillin, G., Shatwell, T., & Kasprzak, P. (2013). Consequences of thermal pollution from a nuclear plant on lake temperature and mixing regime. *Journal of Hydrology, 496*, 47–56.
Kiselev, A. V., Magaril, E. R., & Rada, E. C. (2019). Energy and sustainability assessment of municipal wastewater treatment under circular economy paradigm. *WIT Transactions on Ecology and the Environment, 237*, 109–120.
Klatte, M., Bergstrom, K., & Lachmann, T. (2013). Does noise affect learning? A short review on noise effects on cognitive performance in children. *Frontiers in Psychology, 4*, 578.
Kruize, H., van der Vliet, N., Staatsen, B., Bell, R., Chiabai, A., Muiños, G., et al. (2019). Urban green space: Creating a triple win for environmental sustainability, health, and health equity through behavior change. *International Journal of Environmental Research and Public Health, 16*(22), 4403.
Kumar, A., Taxak, A. K., Mishra, S., & Pandey, R. (2021). Long term trend analysis and suitability of water quality of River Ganga at Himalayan hills of Uttarakhand, India. *Environmental Technology & Innovation, 22*, 101405.
Kyei, K. A., Danquah, L. O., & Mensah, C. M. (2015). Risk of diarrheal diseases associated with wastewater and fecal sludge reuse in urban farming in Accra, Ghana. *Environmental Science and Pollution Research, 22*(19), 15222–15231.
LaGrega, M. D., Buckingham, P. L., & Evans, J. C. (2010). *Hazardous waste management*. Waveland Press.
Liang, L., & Gong, P. (2020). Urban and air pollution: A multi-city study of long-term effects of urban landscape patterns on air quality trends. *Scientific Reports, 10*(1), 1–13.
Ljungberg, J. K., Neely, G., Sánchez-Donoso, R., & Soulsby, C. (2015). Industrial noise pollution and its impacts on birds: A case study from an urban riverine Island. *Journal of Environmental Management, 161*, 26–32.
Loague, K., & Corwin, D. L. (2006). Point and nonpoint source pollution. In *Encyclopedia of hydrological sciences*.

Louis, G. E. (2004). A historical context of municipal solid waste management in the United States. *Waste Management & Research, 22*(4), 306–322.

Miedema, H. M. E., & Oudshoorn, C. G. M. (2001). Annoyance from transportation noise: Relationships with exposure metrics DNL and DENL and their confidence intervals. *Environmental Health Perspectives, 109*(4), 409–416.

Mukherjee, S., Mukhopadhyay, S., Hashim, M. A., & Sen Gupta, B. (2015). Contemporary environmental issues of landfill leachate: Assessment and remedies. *Critical Reviews in Environmental Science and Technology, 45*(5), 472–590.

Mwangi, J. K., Lee, W. J., Chang, Y. C., Chen, C. Y., & Wang, L. C. (2015). An overview: Energy saving and pollution reduction by using green fuel blends in diesel engines. *Applied Energy, 159*, 214–236.

Nanda, S., & Berruti, F. (2021). Municipal solid waste management and landfilling technologies: A review. *Environmental Chemistry Letters, 19*, 1433–1456.

Nelson, D. I., Nelson, R. Y., Concha-Barrientos, M., & Fingerhut, M. (2005). The global burden of occupational noise-induced hearing loss. *American Journal of Industrial Medicine, 48*(6), 446–458.

NRC, National Research Council. (2012). *Water reuse: Potential for expanding the nation's water supply through reuse of municipal wastewater*. National Academies Press.

Ordoñez, C., Aspuru, I., & Begueria, S. (2018). Vegetation for noise abatement in urban areas: A review of the effectiveness. *Journal of Environmental Psychology, 60*, 145–154.

Otoma, S., Kubota, J., Matsubae, K., & Kameya, T. (2019). Circular economy initiatives and their environmental benefits in waste management. *Journal of Material Cycles and Waste Management, 21*(2), 246–259.

Owa, F. W. (2014). Water pollution: Sources, effects, control and management. *International Letters of Natural Sciences, 3*, 1.

Piotrowska, A., Kujawa, A., & Śliwiński, P. (2020). An overview of the current state of electric vehicle noise research. *Applied Acoustics, 163*, 107243. https://doi.org/10.1016/j.apacoust.2019.107243

Pouryaghoub, G., Mehrdad, R., Karami, K., & Soleimani, E. (2013). Occupational noise exposure and its effects on blood pressure and heart rate of workers in an automotive parts manufacturing factory. *International Journal of Environmental Research and Public Health, 10*(9), 3509–3519.

Raja, R. V., Sriraman, G., & Kameswaran, M. (2022). Noise pollution and associated hearing loss in a Metropolitan City—A preliminary report. *Indian Journal of Otolaryngology and Head & Neck Surgery*, 1–7.

Rapti-Caputo, D., & Vaccaro, C. (2006). Geochemical evidences of landfill leachate in groundwater. *Engineering Geology, 85*(1–2), 111–121.

Reed, M. J., & Renner, J. L. (1995). In F. S. Sterret (Ed.), *Alternative fuels and the environment*.

Ritchie, H., and Roser, M. (2018). *Urbanization*. Published online at OurWorldInData.org. Retrieved from: https://ourworldindata.org/urbanization [Online resource].

Rose, K. C., Bierwagen, B., Bridgham, S. D., Carlisle, D. M., Hawkins, C. P., Poff, N. L., Read, J. S., Rohr, J. R., Saros, J. E., & Williamson, C. E. (2023). Indicators of the effects of climate change on freshwater ecosystems. *Climatic Change, 176*(23). https://doi.org/10.1007/s10584-022-03457-1

Ruiz, F. J. D. (2021). Smart cities, big data, artificial intelligence and respect for the European Union Data Protection rules. *European Journal of Formal Sciences and Engineering, 4*(1), 90–108.

Saini, P., & Sharma, M. (2020). Cause and age-specific premature mortality attributable to PM2.5 exposure: An analysis for million-plus Indian cities. *Science of the Total Environment, 710*, 135230.

Samadder, S. R., Prabhakar, R., Khan, D., Kishan, D., & Chauhan, M. S. (2017). Analysis of the contaminants released from municipal solid waste landfill site: A case study. *Science of the Total Environment, 580*, 593–601.

Sánchez, A., Rico, A., & Vighi, M. (2016). Effects of water scarcity and chemical pollution in aquatic ecosystems: State of the art. *Science of the Total Environment, 572*, 390–403. https://doi.org/10.1016/j.scitotenv.2016.07.211

Sarkar, A., Mandal, R. N., & Chakrabarti, S. (2008). Impact of solid waste on health: A quantitative analysis in Kolkata, India. *Waste Management, 28*(2), 248–259.

Singh, R. K., Tirth, V., & Singh, M. (2022). Wastewater treatment processes with special reference to activated sludge process in Indian conditions for water use sustainability. In *Wastewater assessment, treatment, reuse and development in India* (pp. 227–238). Springer International Publishing.

Sivakarun, N., Udayaganesan, P., Chidambaram, S., Venkatramanan, S., Prasanna, M. V., Pradeep, K., & Panda, B. (2020). Factors determining the hydrogeochemical processes occurring in shallow groundwater of coastal alluvial aquifer, India. *Geochemistry, 80*(4), 125623.

Slack, R., Gronow, J., & Voulvoulis, N. (2004). Hazardous components of household waste. *Critical Reviews in Environmental Science and Technology, 34*(5), 419–445.

Speight, J. G. (2019). *Natural water remediation: Chemistry and technology*. Butterworth-Heinemann.

Stansfeld, S. A., & Matheson, M. P. (2003). Noise pollution: Non-auditory effects on health. *British Medical Bulletin, 68*(1), 243–257.

Su, B., Heshmati, A., & Geng, Y. (2013). A review of the circular economy in China: Moving from rhetoric to implementation. *Journal of Cleaner Production, 42*, 215–227.

Swift, H. (2010). A review of the literature related to potential health effects of aircraft noise. TG, G. K. *Harmonious cities: Issues and suggestions*.

Tran, V. V., Park, D., & Lee, Y. C. (2020). Indoor air pollution, related human diseases, and recent trends in the control and improvement of indoor air quality. *International Journal of Environmental Research and Public Health, 17*(8), 2927.

Tripathi, S., & Hussain, T. (2022). Water and wastewater treatment through ozone-based technologies. In *Development in wastewater treatment research and processes* (pp. 139–172). Elsevier.

United Nations Environment Programme (UNEP). (2010). *Waste and climate change: Global trends and strategy framework. UNEP International Environmental Technology Centre*. Retrieved from https://www.unep.org/resources/report/waste-and-climate-change-global-trends-and-strategy-framework

United Nations Human Settlements Programme (UN-Habitat). (2010). *State of the World's Cities 2010/2011: Bridging the Urban Divide.* Earthscan/UN-Habitat. Retrieved from https://unhabitat.org/state-of-theworlds-cities-20102011-cities-for-all-bridging-the-urban-divide

UNEP, United Nations Scientific Committee on the Effects of Atomic Radiation. (2016). *Sources and effects of ionizing radiation: UNSCEAR 2016 report.* Retrieved from https://www.unscear.org/unscear/en/publications/2016_1.html.

United Nations. (2014). *Global Environment Outlook 6: Regional assessments*. Retrieved from https://www.unep.org/resources/report/global-environment-outlook-6-regional-assessments

United Nations. (2017). *Sustainable development goal 6 synthesis report 2018 on water and sanitation*. United Nations.

United States Environmental Protection Agency. (2021). *National Ambient Air Quality Standards (NAAQS) Table*. Retrieved from https://www.epa.gov/criteria-air-pollutants/naaqs-table

Waisman, H., De Coninck, H., & Rogelj, J. (2019). Key technological enablers for ambitious climate goals: Insights from the IPCC special report on global warming of 1.5 C. *Environmental Research Letters, 14*(11), 111001.

Wang, S., Ji, Y., Li, R., Li, L., & Li, Y. (2014). Noise-induced hearing loss in China: A potentially costly public health issue. *Noise and Health, 16*(72), 223–229.

WHO, World Health Organization. (2009). *Global health risks: Mortality and burden of disease attributable to selected major risks*. World Health Organization.

WHO, World Health Organization. (2012). *Ionizing radiation, health effects and protective measures*. Retrieved from https://www.who.int/ionizing_radiation/chernobyl/WHO_Chernobyl_report_2012.pdf

WHO, World Health Organization. (2016). *Indoor air pollution: The world's largest environmental health risk*. WHO. https://www.who.int/news-room/fact-sheets/detail/indoor-air-pollution

WHO, World Health Organization. (2022a). *Billions of people still breathe unhealthy air: New WHO data*. WHO. Retrieved March 22, 2023, from https://www.who.int/news/item/04-04-2022-billions-of-people-still-breathe-unhealthy-air-new-who-data.

WHO, World Health Organization. (2022b). Billions of people still breathe unhealthy air: New WHO data.

World Health Organization. (2021). *WHO global air quality guidelines: Particulate matter (PM2.5 and PM10), ozone, nitrogen dioxide, sulfur dioxide and carbon monoxide*. WHO. Retrieved from https://www.who.int/publications/i/item/9789240034228

World Health Organization. (2024). *Ambient (outdoor) air quality and health*. WHO. Retrieved from https://www.who.int/news-room/fact-sheets/detail/ambient-(outdoor)-air-quality-and-health

Wokekoro, E. (2020). Public awareness of the impacts of noise pollution on human health. *World Journal of Research and Review, 10*(6), 27–32.

Wolkoff, P. (2018). Indoor air humidity, air quality, and health–An overview. *International Journal of Hygiene and Environmental Health, 221*(3), 376–390.

Yue, X., Ma, N. L., Sonne, C., Guan, R., Lam, S. S., Van Le, Q., et al. (2021). Mitigation of indoor air pollution: A review of recent advances in adsorption materials and catalytic oxidation. *Journal of Hazardous Materials, 405*, 124138.

Zare, M., Mehrdad, R., Pouryaghoub, G., & Akbari, H. (2013). The effects of exposure to occupational noise on blood pressure and heart rate of workers in a textile factory in Shiraz, Iran. *Journal of Environmental and Public Health, 2013*, 1–6.

Zhang, Q., & Wang, C. (2020). Natural and human factors affect the distribution of soil heavy metal pollution: A review. *Water, Air, & Soil Pollution, 231*(350). https://doi.org/10.1007/s11270-020-04728-2

Urban Biodiversity: Threats and Conservation Strategies

Neha Jaiswal, Ankita Roy Chowdhury, and S. Jayakumar

Abstract Urban biodiversity encompasses the variety of species, ecosystems, and habitats within urban and peri-urban areas, playing a crucial role in sustaining ecological balance. However, rapid urbanization, fuelled by economic growth and population expansion, poses significant challenges to biodiversity. The transformation of natural landscapes, habitat fragmentation, and rising greenhouse gas (GHG) emissions contribute to environmental degradation and climate change, threatening both urban ecosystems and human well-being. Despite these threats, urban biodiversity provides essential ecosystem services that benefit both nature and society. Cities can also serve as refuges for species at risk in their natural habitats, sometimes even supporting greater species density than surrounding rural areas. By integrating biodiversity into urban planning and policy frameworks, cities can create resilient, liveable, and sustainable environments that benefit both present and future generations.

Keywords Ecosystem services · Green infrastructure · Climate resilience · Habitat fragmentation · Sustainable urban planning

1 Introduction

The urban environment relies significantly on biodiversity, which encompasses variations within species, between species, and across ecosystems. We can find biodiversity almost anywhere; for example, green spaces can be found even within cities or densely populated areas (Andersson et al., 2014; Nilon, 2011). As we all know, humans are entirely reliant on biodiversity, which plays a vital role in the ecosystem's functioning. Therefore, any interference with these processes will have

N. Jaiswal · A. R. Chowdhury · S. Jayakumar (✉)
Department of Ecology & Environmental Sciences, School of Life Sciences, Pondicherry University, Puducherry, India
e-mail: s.jkumar1@pondiuni.ac.in

P. K. Joshi et al. (eds.), *Geospatial Science for Urban Ecosystems*, The Urban Book Series, https://doi.org/10.1007/978-3-032-07155-2_4

a direct impact on the ecosystem's services (Zari, 2018). One of the most concerning effects is land-use fragmentation, which comes at a higher cost to biodiversity loss (Jaiswal & Jayakumar, 2021), followed by degradation.

1.1 Urban Biodiversity

The intricate fusion of diverse urban species, ecosystems, and habitats constitutes urban biodiversity. This ecological concept is frequently interchanged with the terms "urban green space"and "urban green infrastructure," as noted by Müller et al. (2013) and Barhadiya and Ghosh (2022). There are multiple perspectives on defining urban biodiversity, as expressed in various statements. For instance, Muller et al. (2010) defines it as "the range of biodiversity encompassing genetic variations and variety of habitat present within and surrounding human communities." According to this definition, rock faces, agricultural landscapes, urban industrial landscapes, and remains of pristine ecosystems are all included in the category of urban biodiversity, covering centres, including city centres, residential zones, and industrial parks, suggesting that it extends from the rural to the urban core (Snep & Ottburg, 2008).

In other words, as the name implies, urban biodiversity refers to the diversity of animals and plants found in urban areas, or it can be the diversity of cities and towns that hold a large amount of biodiversity and protect it from extinction. Approximately half of the global population currently resides in urban environments, and this percentage is projected to potentially rise to 66% by the year 2050, as forecasted by the United Nations in 2018 (United Nations, 2019).

On the one hand, where urban biodiversity contains numerous biodiverse regions, on the other hand, it is facing various threats. The most detrimental peril to urban biodiversity stems from urbanization or the expansion of urban land, a phenomenon increasingly observed in recent decades (He & Zhou, 2014). This expansion is primarily attributed to population growth, economic development, and rural-to-urban migration, as highlighted by D'amour et al. (2017) and Chen et al. (2020). The impact of urbanization on the diversity and abundance of plant and animal species is evident, manifesting both positive and negative consequences. Interestingly, 70% of the world's anthropogenic greenhouse gas emissions originate from metropolitan regions, directly influencing climate change. As we can see, this is both subtle and significant, and it impacts urban life (Semerdzhieva & Borisova, 2021). Moreover, more than 80% of local natural habitats have been lost due to urbanization, as indicated by Faeth et al. (2011) and Hopkins et al. (2016). Illustratively, between 2009 and 2050, the growth of urban areas is projected to accommodate an additional 1.86 billion people, expanding at a rate twice as fast as the overall population, as noted by Ke et al. (2018).

The opportunity for individuals to engage with plants and animals for recreational enjoyment is a crucial dimension of the coexistence of diverse organisms in urban settings. This exemplifies ecosystem services within cities, encompassing the diverse benefits that humans derive from ecosystems, as emphasized by Clark et al. (2014) and Gunnarsson et al. (2017). However, as previously stated, urbanization poses a severe threat to biodiversity, resulting in species extinction and biotic homogenization. The process of urbanization not only disrupts the habitats of numerous endemic species but also creates favourable environments for the adaptation of a limited number of species to urban conditions (McKinney & Lockwood, 1999; Buczkowski & Richmond, 2012).

As per the Millennium Ecosystem Assessment, a minimum of 60% of ecosystem services have undergone degradation or unsustainable use, posing a threat to human communities (Millennium Ecosystem Assessment, 2005). Safeguarding and preserving urban ecosystems, including vital natural elements such as green and blue infrastructure, are crucial for the health and well-being of urban residents, as highlighted by Semerdzhieva and Borisova (2021). Contrarily, Yam et al. (2016) argue that urbanization should not always be perceived as a threat to local urban biodiversity because cities are distinct ecosystems capable of providing niches and habitats for various species of urban flora and fauna.

1.2 Importance of Urban Biodiversity

Although urban landscapes occupy a limited portion of the surface of Earth, they have a crucial impact on sustaining urban biodiversity, given that they accommodate over half of the global population (Liu & Opdam, 2014; Duke & Soulsbury, 2021). Additionally, green spaces within cities, such as parks, play a role in alleviating the impacts of water stress, heat, carbon emissions, and pollution cycles, as noted by Ossola and Livesley (2016). On the other hand, the socio-cultural context has uniquely shaped urban biodiversity, which is one of the main benefits of urban biodiversity. Citizens can, for example, connect with nature in urban green spaces, observe ecological processes in action, and potentially become scientifically minded citizens capable of making informed decisions about conservation programmes and policies (Lepczyk et al., 2017). Besides, urban biodiversity provides many benefits to humans and the ecosystem (Fig. 1).

1.2.1 Human Well-being

Urban landscapes include green and blue spaces, which link the environment and the human population. Understanding the connectivity of green spaces is particularly crucial as it plays a pivotal role in facilitating the movement of organisms and processes, contributing significantly to the conservation of biodiversity. Blue spaces, on the other hand, provide ecosystem services and serve a functional purpose. Various activities in urban areas gravitate towards blue spaces, making them significant focal points (Grellier et al., 2017). Several studies suggest that these spaces contribute to improved health and well-being. Individuals living in urban areas,

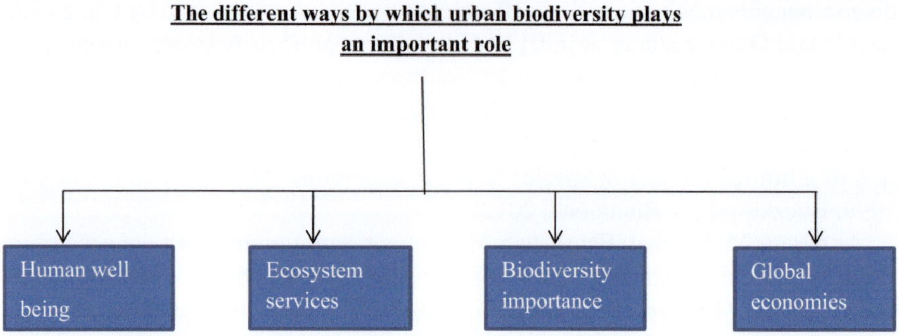

Fig. 1 The global importance of urban biodiversity

Fig. 2 Advantages of urban spaces: green and blue

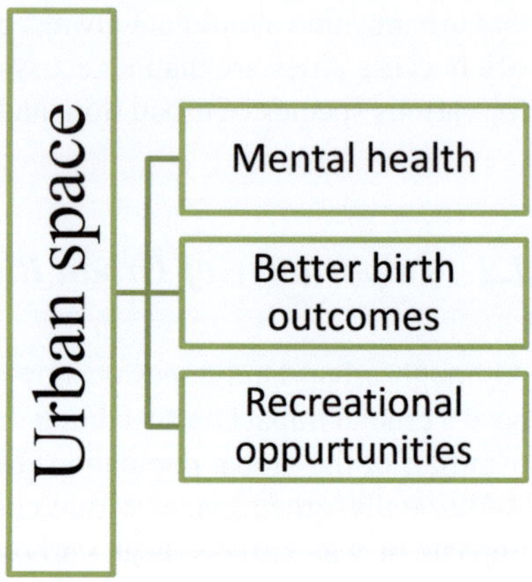

particularly those near green spaces, exhibit better physical fitness and mental health, as depicted in Fig. 2 (De Vries et al., 2013; Gascon et al., 2015). Similarly, blue spaces also demonstrate significance in terms of human well-being, particularly mental health (De Bell et al., 2017).

Although a minimal amount of research has been done on the effects of larger bodies of water on wildlife and humans living near them (Slaughter et al., 2010), recently, there has been a growing emphasis on recognizing the significance of blue spaces in urban areas (Oertli & Parris, 2019), as well as understanding the connectivity between green and blue spaces and their impact on biodiversity (Hyseni et al., 2021). For example, a study conducted by Garrett et al. (2019) suggests that the blue spaces in Hong Kong could serve as a valuable resource for the health of the elderly population.

The conservation of urban biodiversity is still a contentious issue due to the lack of a comprehensive plan and definitions on a large scale. Consequently, these linkages provide a chance to leverage human–wildlife interactions and blue spaces for engaging local residents, simultaneously addressing social and environmental concerns (Gearey et al., 2019). However, according to some studies, access to these green spaces can increase awareness of the need to protect and use them sustainably (Sarkar et al., 2014; Hejkal et al., 2017; Tian et al., 2017; Dong et al., 2020). Green spaces that are well-connected within the city not only embody greater ecological and sustainable qualities, enhancing residents' perceptions of well-being but also typically enhance accessibility to the overall green space system (Park, 2017).

Increasing green space connectivity, in other words, can improve microclimate, living environment, urban landscape quality, and so on, potentially increasing green space use. It also helps in the well-being of communities by providing various employment opportunities, such as gardening and horticulture (Anderson & Minor, 2017). The way individuals come into contact with and experience urban nature not only varies based on its form, structure, and function but also influences ethnocultural identities (Egerer et al., 2019).

1.2.2 Urban Ecosystem Services (UES)

Urban ecosystem services (UES) play a crucial role in sustaining and enhancing human livelihood and overall quality of life. These services, as categorized by the Common International Classification of Ecosystem Services (CICES, 2022), encompass a diverse array of functions. Firstly, *provisioning services* address essential needs like food, water, medicinal plants, and various resources. Secondly, *regulating services* contribute to the smooth functioning of urban environments by managing air and soil quality, as well as controlling floods, storm water, and disease. Thirdly, *habitat or supporting services* form the foundation for nearly all other services, including the preservation of plant and animal diversity. Lastly, *cultural services* encompass non-material, socio-ecological benefits that individuals derive from their interactions with the environment, spanning psychological and cognitive aspects. These include recreational opportunities, aesthetic appreciation, spiritual well-being, psychological benefits, and tourism (Cowling et al., 2008; Haase et al. 2014a, b). Altogether, urban ecosystem services not only withstand but also improve human livelihood and overall quality of life.

Broadly, the operation of these ecosystems provides urban landscapes with the adaptability necessary to build resilience and address various challenges, particularly the heightened risks of heatwaves and flooding. Although analyses of urban socio-ecological systems have proven effective in enhancing our understanding of how ecosystems can mitigate the effects of climate change, there are notable knowledge gaps that require attention and focused efforts.

1.2.3 Biodiversity

Urban areas can positively influence regional biodiversity in various ways. One example is how urban habitats can sustain populations that face threats or extinction in their natural habitats (Ives et al., 2016). For example, according to Mason (2000), in the United Kingdom, the song thrush (*Turdus philomelos*), a species of national conservation concern experiencing decline, is discovered in urban habitats with densities over three times greater than those in the surrounding rural environment. Urban areas often exhibit richness in species and can function as a focal point for local and regional biodiversity. Sanderson et al. (2006) argue that the pattern and dynamics of habitat patches in the area, as well as the local and regional species pool, influence biodiversity in urban environments; however, there are more definitions that go beyond these fundamental ideas (Angold et al., 2006; Nilon, 2011).

The decline in biodiversity can impact numerous services derived from it, particularly affecting plant-based medicines, especially in developing nations. The World Health Organization (WHO) has approximated that as much as 80% of the global population depends on plants for their fundamental health needs. According to Mexia et al. (2018), tree density and species composition improve carbon sequestration by urban vegetation. On the contrary, biodiversity loss leads to crop-land degradation and, ultimately, affects food systems.

Abundant evidence supports the direct impact of biodiversity on health. The reduction in biodiversity can lead to the rise and dissemination of infectious diseases, as well as health issues arising from insufficient food and nutritional diversity. Conversely, the enhancement of biodiversity can have direct positive effects on health. For instance, augmenting plant abundance, such as through the implementation of green roofs, can aid in mitigating air pollution, thereby lowering the prevalence of respiratory and cardiovascular diseases (Johnson et al., 2013; Clark et al., 2014).

1.2.4 Global Economies

Urbanization stimulates economic development through the expansion of industries and various activities. The positive correlation between urbanization, development, and economic growth is acknowledged (Njoh, 2003). To a certain degree, it is reasonable to affirm that urbanization is indeed positively connected with development, particularly in terms of economic progress.

2 Biodiversity Loss Assessment

Climate change and biodiversity loss are two major concerns and challenges in today's world, which have resulted in a slew of problems for our planet. Current estimates indicate that species extinction rates are 100–1000 times higher than the natural rate, a level significantly surpassing background predictions (May, 2011). This decline in diversity, encompassing genetic diversity, poses a critical threat to global food security by compromising the resilience of numerous agricultural systems against threats like pests, pathogens, and climate change.

To conserve biodiversity, it is important to comprehend the factors that drive biodiversity loss. Different researchers use different methods to assess biodiversity and the causes of the loss.

At local, national, and global levels, alterations in land use play a substantial role as the primary driver of biodiversity loss attributed to urbanization. These changes primarily stem from factors such as population growth, economic development, and rural-to-urban migration (Zebisch et al., 2004). Moreover, according to the Convention on Biological Diversity (CBD), biodiversity loss poses major challenges to the global environment that needs to be discussed and protected subsequently. Hence, it is obvious that human impact and the reduction in species richness are directly correlated.

Wildlife trade is another major challenge for biodiversity conservation. The trade in wildlife is mainly occurring because of medicinal or cultural purposes. Despite the number of policies and regulations, this trade is still a threat, especially for urban areas (Challender et al., 2015; Nguyen & Jones, 2021). Globally, over 25% of terrestrial bird, mammal, amphibian, and squamate reptile species are subject to international trade, and an anticipated 4064 species are projected to be incorporated into the future wildlife trading system (Scheffers et al., 2019). A further study mentioned that one out of every five species is impacted by trade in some way.

Substantial evidence indicates that biodiversity loss profoundly influences cultural belief systems, resulting in notable adverse effects on people's well-being, including anxiety, frustration, and stress. Nevertheless, only a limited number of studies have commenced exploring the connections within these cultural pathways, given our limited understanding of the repercussions for human well-being and, consequently, health. However, in recent years, evidence suggests that urban environments are much safer for some species than rural habitats and wildlands (Shochat et al., 2010).

According to a UN report in 2020, it was disclosed that over a million species are at risk of extinction in the near future (United Nations, 2020). Also, some studies (Raven & Wagner, 2021; Wagner et al., 2021) reported that these ongoing threats resulted in a sharp decline of around 80% in many mammalian species over the past decade, especially insects that behave as a link of the food chain. Similarly, they also reported that several butterfly species were found to be declining in Europe, mainly because of agricultural expansion.

Urbanization is increasing the difficulty of resilience and resistance because it is occurring at a scale and rate that ecosystems and biodiversity cannot handle. Even in scenarios with lower urbanization rates, urban areas worldwide face serious challenges, such as biodiversity loss and climate change. These occurrences are linked to a loss of biological diversity (Smith et al., 2018). Seto et al. (2012) also reported that urban expansion would be continued, and by 2030, it would be around three times compared to 2000. Likewise, the World Cities Report released in 2020 forecasted a continued global urbanization trend over the next decade, projecting an increase from the current 56.2% to 60.4% by 2030 (Zhongming et al., 2020). That would directly affect biodiversity and habitat loss at a considerable rate. In a study by Sol et al. (2014) on birds, it was determined that species loss resulting from urbanization varied from 19.5% to 76.5%, with the most urbanized areas showing the most substantial decline. Consequently, it is a valid conclusion that urbanization presents a significant threat to both biodiversity and the environment.

Biodiversity matters as it provides a variety of services and benefits to both nature and humanity. For example, crop genetic diversity boosts crop yields; plant species diversity improves grassland fodder and, ultimately, aids in climate regulation and resilience (Roe, 2019). Nonetheless, a recent study indicates that around 50–63% of the newly expanded urban land will be situated on existing croplands. Consequently, global crop production is projected to decline by 1–4%, equivalent to the annual food requirements of 122–1389 million people for a specific crop (Chen et al., 2020). Furthermore, human activities directly impact urban biodiversity through habitat loss, habitat fragmentation, and the introduction of new species. Indirectly, these impacts extend to alterations in urban climate, soils, hydrology, and biogeochemical cycles, as highlighted by Kowarik (2011).

Given the global decline in biodiversity, primarily attributed to intensified land management practices and the conversion of natural areas for agriculture, production forestry, and settlements (IPBES, 2019), the conservation of urban biodiversity should be designed to align with broader global conservation initiatives (Knapp et al., 2021).

Additionally, anthropogenic noise sources, such as vehicles, planes, construction activities, tools, and human interactions, all contribute to environmental noise. This noise has evolutionary implications as it impacts biodiversity by influencing the behavioural traits of various taxa that rely on acoustic communication, affecting aspects like habitat choice and mating patterns (Lampe et al., 2014). Besides, Cardinale et al. (2012) discussed in their review how the loss of biodiversity impacted the functionality of the ecosystem and human beings.

On the one hand, unplanned and unmanaged urbanization can endanger global environments. In contrast, on the other hand, sustainable urbanization is a method of restoring urban biodiversity that can be accomplished through proper planning and management (Ishi, 2004). Enhancing local species richness and increasing the likelihood of preserving our portion of global biodiversity can be achieved by prioritizing the planting of indigenous species. On the other hand, green roofs are a new strategy for restoring urban biodiversity that can be beneficial and well-accepted and will also improve connectivity (Braaker et al., 2014).

Regrettably, despite compelling justifications for preserving biodiversity and the growing understanding of urban biodiversity, which is increasingly incorporated into policy and planning, habitat reduction and the loss of biodiversity persist (Given, 2000; Knapp et al., 2021). In the present scenario, climate emergency and biodiversity crisis are two of the most difficult challenges we face, both of which we must take seriously and begin to address.

3 Invasive Species Spread

Invasive species, also known as non-native alien species, are something that can harm nature and its surroundings. Species are typically introduced into a new habitat for three primary reasons: unintentional introductions, often involving pathogens and invertebrates; species initially imported for a specific purpose but subsequently escaping; and deliberate introductions, commonly observed in the case of plants and vertebrates. Many of the upcoming introductions are motivated by people's desire to protect species that benefit them.

Species used for agricultural, forestry, and ornamental purposes face heightened vulnerability, particularly considering their significant roles. Across most regions globally, introduced species play a pivotal role in meeting essential dietary needs, as exemplified by the presence of species like cattle, goats, and maize in Africa. The introduction of species becomes integral to human welfare in nearly every corner of the world (McNeely, 2001). Interestingly, invasive non-native species exhibit a higher prevalence in urban settings compared to rural areas (Kowarik, 2011), influenced by various factors. The socio-economic ramifications of biological invasions are nuanced and challenging to evaluate using conventional monetary methods and market-based models (Binimelis et al., 2008). Notably, the species richness of a particular area can be increased by introducing some non-native species, but overall global biodiversity will be decreased (Given, 2000).

In an extensive exploration of the global trade in plants and animals, McNeely (2001) delved into the history and ecology of introducing non-native species into new habitats. His findings indicated that the primary motivation behind introducing non-native species is often rooted in financial considerations. Economic interests have driven the introduction of numerous non-native species, with introduced fish offering excellent sport fishing opportunities; introduced plants serving purposes such as providing food, fodder, timber, and energy; and introduced insects playing a role in biological control. An illustrative example of the scale of such introductions is reflected in the period between 2000 and 2008, when over one billion live non-native animals were imported into the United States for wildlife trade (Smith et al., 2008). Historical instances further underscore this trend, as seen with the intentional introduction of brushtail possums from Australia to New Zealand between 1858 and 1900, aiming to establish a fur trade. Similarly, the water hyacinth (*Eichhornia crassipes*) from South America was introduced to China in the 1930s. Its widespread distribution from the 1950s to the 1970s was driven by its utilization as an ornamental plant, as a livestock food source, and for pollution control through its capacity for heavy metal absorption. These examples highlight the multifaceted and often complex motivations behind introducing non-native species.

Hence, we can say that invasive species are suitable in some aspects (Di Castri, 1989) but cost numerous elements of biodiversity and ecology (Borden & Flory, 2021). For example, the impacts of introduced species include diminished crop and livestock production, the decline of native biodiversity, heightened production costs, and various other associated expenses. Introduced invasive species have significantly impacted some protected areas set up to protect native species (Bratton, 1982) as they out-compete the native species and spread in the area.

The proliferation of non-native plant species has witnessed a significant surge, primarily attributed to the deliberate introduction of numerous species into managed green spaces. An illustrative example is evident in the Shanghai area, where approximately 300 alien plant species were introduced between 1980 and 2005, contributing to the substantial diversification of the local flora. This trend is mirrored in the composition of woody plant species within the city's green spaces, where a total of 206 such species have been documented. Notably, only 64 of these species (31.1%) are native to Shanghai and its adjacent regions, underscoring the considerable prevalence of non-native woody plants in the urban landscape (Zhao et al., 2006). This influx of non-native species poses challenges to the ecological balance, necessitating a comprehensive understanding of the dynamics between introduced and native vegetation for effective environmental management. Overall, there are two sides of invasive species spread, which are illustrated in the given figure (Fig. 3).

3.1 Positive Effects of Invasive Species

Nevertheless, the drawbacks associated with invasive species outweigh their benefits in many cases. While some invasive species were intentionally introduced to offer specific ecosystem services, such as trees to prevent erosion and produce timber, their subsequent spread and proliferation often result in a net detrimental impact (Vilà et al., 2010). Conversely, investigations conducted in urban forests in Florida, for instance, have revealed that invasive trees exhibited notable success in sequestering carbon dioxide (CO_2), suggesting potential ecological benefits despite the overall challenges posed by invasive species (Escobedo et al., 2010). This underscores the nuanced nature of the ecological consequences associated with invasive species, where both positive and negative aspects must be carefully considered in environmental management strategies.

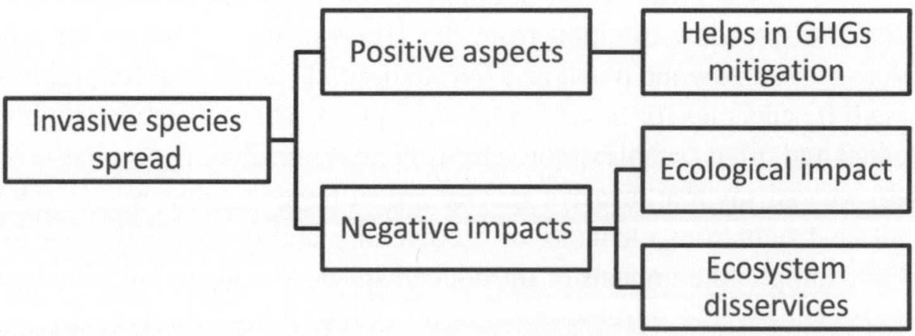

Fig. 3 The positive and negative impacts of invasive species

3.2 *Negative Impacts of Invasive Species Spread*

Invasive species are primarily harmful to the environment and surroundings in a number of ways. In urban environments, invasive species can disrupt crucial ecosystem services like water filtration, flood attenuation, and coastal protection (Pandey & Ghosh, 2023). The presence of invasive plants has the potential to obstruct streams and canals, leading to flooding, while also elevating the risk of increased fire severity and soil erosion (Van Wilgen & Scott, 2001). Beyond these ecological impacts, invasive species may exert adverse effects on urban biodiversity, directly impacting human well-being. This influence is particularly evident in changes to the aesthetics of the environment, as highlighted by recent studies (Britton, 2023; Hu et al., 2023; Kowarik, 2011). Consequently, the challenges posed by invasive species extend beyond ecological disruptions, encompassing alterations in the urban landscape, which can have tangible consequences for the quality of life and the overall experience of city dwellers.

In addition, many non-native plants are responsible for ecosystem disservices, which means that the ecosystem has negative impacts on humans (Potgieter et al., 2019), such as floods, allergenic substances, and aesthetic issues (Lyytimäki, 2014; Mclellan & Shackleton, 2019). Amid increasing apprehensions regarding invasive species, certain authors posit that, in specific cases, we might need to adapt to "cohabitating" with them in urban settings (Foster & Sandberg, 2004).

4 Impacts of Cities on Biodiversity

In its most comprehensive context, biodiversity is indispensable for the functioning and endurance of life on Earth, as well as for human existence and welfare. The urgent recognition of its swift decline necessitates immediate global attention (Given, 2000). As per the World Health Organization, around 80% of plants are employed in the production of medicines for human health, underscoring the significance of biodiversity. But people are migrating to cities for a variety of reasons, putting a significant strain on biodiversity. Besides, a city's ecological footprint contributes to the threat of biodiversity loss; for example, if one species is harmed, the entire ecosystem is harmed. Nevertheless, cities can exert both direct and indirect influences on biodiversity; for instance, food consumption indirectly affects biodiversity (McDonald et al., 2020).

As discussed earlier, urban growth is a significant concern for biodiversity conservation. As we all know, cities play an essential role in greenhouse gas emissions; approximately 70% of GHG emissions are anthropogenic (Hopkins et al., 2016), resulting in climate change and biodiversity threats. Cities play a pivotal role in the noteworthy escalation of global temperatures, primarily functioning as major contributors to the emission of greenhouse gases (Semerdzhieva & Borisova, 2021). The concentration of human activities in urban areas amplifies the release of pollutants, including carbon dioxide and other greenhouse gases, into the atmosphere, thereby intensifying the urban heat island effect and contributing to climate change. Simultaneously, the impact of human activities on global carbon dynamics extends beyond urban boundaries, encompassing widespread issues like deforestation and forest degradation. A substantial share of anthropogenic carbon dioxide emissions, estimated to range between 6% and 17%, is attributed to deforestation and forest degradation, with the tropics being the primary source of this ecological concern (Van der Werf et al., 2009). This dual challenge highlights the interconnectedness of urbanization and environmental issues, emphasizing the pressing need for comprehensive strategies to address climate change and preserve global biodiversity.

Biodiversity hotspots comprise a tiny portion of Earth, but cities within these hotspots or the junction of the ecosystems are increasing with years, in the form of urban expansion, which will ultimately affect the biodiversity. For example, Seto et al. (2012) projected growth in urban areas by 2030, and they found that urban expansion would increase to around 2% from 1% of all hotspots globally. Likewise, based on a projection of future urban land expansion, global crop production is anticipated to decrease by approximately 1–4%, equating to the annual food requirements for a specific crop serving 122–1389 million people (Chen et al., 2020).

Furthermore, global drivers of change, including land-use alterations, invasive species, and climate change, are modifying both biodiversity and ecosystem functions (Smith et al., 2018). The various major impacts of cities are illustrated in Fig. 4 and described in detail below.

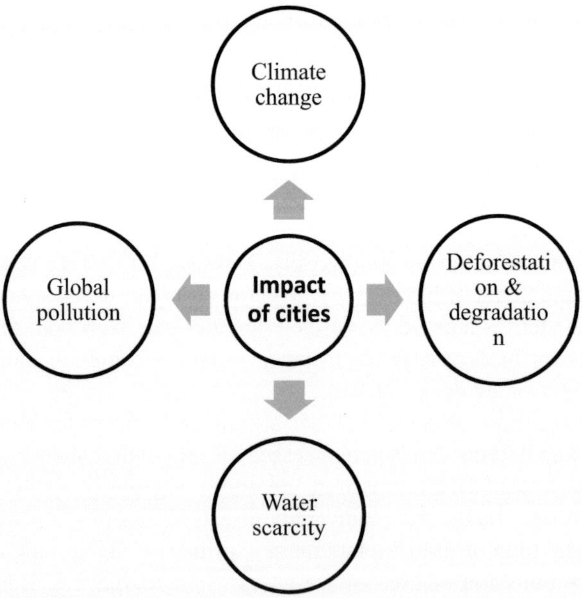

Fig. 4 Various impacts of cities on biodiversity

4.1 Land-Use Change and Degradation

Environmental conditions and human drivers are the primary drivers of land-use change, which multiple interacting factors can describe. Take, for example, the increase in agricultural land and the associated infrastructure development, leading to deforestation and placing substantial strain on biodiversity and ecosystem services. Moreover, deforestation contributes to the fragmentation of habitats, diminishing patch size and isolating suitable environments (Tantipisanuh & Gale, 2013; Trisurat et al., 2019). A case in point is a study that observed noteworthy declines in avian species abundances and overall biodiversity as green spaces became increasingly fragmented due to urbanization, as evidenced in Metro Manila and cities in Brazil (Nor et al., 2017). This underscores the multifaceted impact of human-induced changes on ecosystems and emphasizes the importance of sustainable practices in land use and development. This shows that the deterioration of landscape connectivity negatively impacts biodiversity conservation.

4.2 Climate Change at a Faster Rate

Climate change can be the most significant cause of biodiversity loss after habitat destruction, especially at the rate at which it is happening, and the future undoubtedly will be devastating (Jaiswal & Jayakumar, 2021). And it has already started affecting the different components of biodiversity, beginning from the species to the ecosystem level (Parmesan, 2006). The shifts in temperature causing global warming result in a serious problem, especially for biodiversity. Those shifts cause species to migrate and, ultimately, become extinct, and it can be even worse at a higher level than expected. Urban areas significantly contribute to global-scale climate change due to elevated levels of greenhouse gas emissions, influencing local-scale shifts in temperature and precipitation patterns (Grimm et al., 2008). Nevertheless, it is crucial to recognize the interconnectedness of the impact of climate change on biodiversity and the global economy (Fig. 5).

There are so many climate adaptation plans (Butt et al., 2018), which ultimately affect the global economies, as cities/megacities have to invest a lot to tackle climate change. For example, New York City alone spent £1.6 billion in 2014 and 2015 (Georgeson et al., 2016).

4.3 Global Pollution

It would be fair to say that cities can produce more pollution than other areas. Vehicular emissions, heat islands, air pollution, etc. are among the risk factors promoting biodiversity loss. Similarly, water pollution adds risks to the ecosystem and

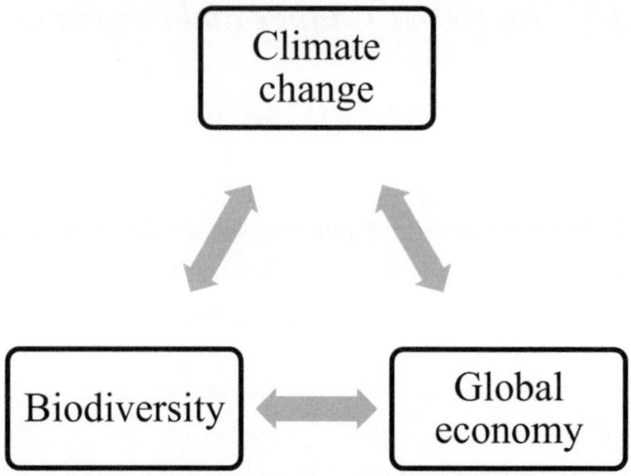

Fig. 5 Interrelations between climate change, biodiversity, and the economy

severely affects biodiversity, mainly in domestic or industrial discharge. In particular, the coastal cities have detrimental effects and damage the surrounding areas of biodiversity.

4.4 Natural Resources' Scarcity

The consequences of anthropogenic actions surpass every effort we could have made to save the planet and its natural resources. The cities have a high impact on natural resources, especially water (Smith et al., 2018), resulting in water scarcity in various places. Despite all of the mentioned drivers for diversity loss, urbanization, on the other hand, cannot always be viewed as a threat to local urban biodiversity, as cities are unique ecosystems in and of themselves, providing niches and habitats for a diverse range of urban flora and fauna (Yam et al., 2016). Yet, in recent years, there has been an emergence of innovative concepts for sustainable and adaptable urban governance, which integrate urban green infrastructure, ecosystem services, and nature-based solutions (Almenar et al., 2021; Vandermeulen et al., 2011). Therefore, if management and conservation are prioritized, the cities can help maintain biodiversity and the ecosystem.

5 Impact of Biodiversity on People

Degradation of natural environments, fragmentation, and habitat loss as a result of development have led to increased awareness (Irwin & Bockstael, 2007). Globally set performance targets have resulted in the need to implement sustainable development, with conservation and sustainable development of biodiversity as its necessary conditions. Biodiversity and its links to ecosystem services have different impacts on human lives.

5.1 Impact of Biodiversity on Ecosystem Services

Biodiversity plays an important role in different ecosystem services, such as the provisioning services, which include the supply of the products obtained from the natural environment that meet the basic necessities of mankind. These include food, water, and different natural resources. The resilience of agriculture plays an important role in the healthy functioning of ecosystems as it intensifies to meet growing food demands sustainably. Biodiversity in the agricultural production system makes essential contributions to meet the nutritional requirements of individuals. Loss of biodiversity from the agricultural system through the introduction of high-yielding varieties of genetically modified crops replacing the native breeds and the introduction of a mono-cropping system to get larger crop production is leading to a loss in genetic diversity and negative health consequences. Thus, increasing sustainable production will not only play an important role in providing better living but also help the ecosystem adopt the actions needed to cope with future changes and enhance the resilience of the ecosystem, mitigating its harmful consequences (Liu & Opdam, 2014). Regulating services: these services regulate and promote the proper functioning of ecosystems, such as the regulation of climate, water purification, disease control, and pollination. Supporting services: these include natural processes that help in the proper functioning of other ecosystem services, such as photosynthesis and nutrient cycling. Cultural services: these services promote healthy well-being and provide non-material benefits, such as recreation, tourism, spiritual values, and education. Developing a better understanding of the interdependent relations of ecological and social systems is important for the sustainable functioning of the ecosystem. Proper assessment of the impact of ecosystem services is important as it serves as a support tool for resource management.

5.2 Impact of Biodiversity on Mental, Physical, and Cultural Well-being

Biodiversity and human well-being are deeply interconnected and influence each other in many ways. Biodiversity provides us with two layers of protection: the microbiota, or the outer layer, which includes elements of the external environment an individual is a part of, and the inner layer, which comprises the gut microbiome, also known as the second genome. It colonizes our body from the outer layer. Interactions with the natural environment and green spaces have improved overall human health and well-being. Increased urbanization and other related factors are changing our environment and lifestyle (Romanelli et al., 2015). It reduces the adaptive immune-regulatory functions and dynamic homoeostasis, resulting in chronic inflammatory disorders (Haahtela, 2019). It has also resulted in an increased number of respiratory and cardiovascular diseases and cancer mortality. Studies suggest that short-term exposure to forests, urban parks, gardens, and other natural environments reduces stress and symptoms of depression, restores attention fatigue, increases positive emotions, and improves self-esteem, mood, and mental health. Long-term exposure to natural environments in residing areas with high greenness has been associated with reduced lifestyle diseases and particularly results in better respiratory and mental health. Exposure to green or natural environments is essential during prenatal development and early life, as it has a positive effect on the birth weight of their infants. Childhood exposure to natural environments reduces the risk of developing schizophrenia, obesity prevalence, and atopic sensitization in children and has a positive effect on blood pressure in adolescents. Better development of the immune system and the early exposure to nature amplify the potential beneficial effects of green spaces in later life, including the stress-reducing effects of therapeutic immersion in nature (Aerts et al., 2018; Bijnens et al., 2017; Lee et al., 2017). Access to natural environments also tends to enhance outdoor physical activity, improving physical health by reducing the prevalence of obesity and type 2 diabetes (Lachowycz & Jones, 2011). Medicinal and aromatic plants from the wild are used as traditional medicines; they are also used in various other industries. Sustainable use of such resources can provide benefits to biodiversity, livelihood, and human health. The introduction of biodiversity creates a balance in the ecosystem, and it has links to ecosystem services, which have different impacts on traditional, aesthetic, cultural, educational, and spiritual values that are important to society. Culture not only comprises the practical skills and languages of society but also the ideas and beliefs of the individuals of society. Culture evolves in and around the natural environment in which it grows. It includes sacred sites, burial grounds, native flora, and fauna, which are used for various ceremonial practices. Thus, spiritual and cultural beliefs of an individual are linked to nature. Changes in the natural environment and dislocation of individuals from their native environment cause distress and harm their well-being. Biodiversity has remarkable impacts on human lives. Its loss is a key environmental determinant of human health and well-being. For further understanding the effects of biodiversity on people, a better understanding of various complex linkages and interactions in biodiversity is required.

6 Incorporation of Biodiversity into Cities

With the increase in global biodiversity loss, humanity has shifted its habitat resulting in predominantly urban population, and today a significant percentage of the world's population resides in cities. Migration of many individuals to these areas, change in energy use, demand for food, overexploitation of natural resources, urbanization, and other factors have led to biotic exchange; change in habitat; increased competition and predation; nitrogen deposition in the aquatic systems, causing defoliation and adversely affecting the aquatic life; microclimatic change; loss of habitat; extinction; and land-use change due to fragmentation. All these factors collectively result in biodiversity loss. With industrialization, more migrants move to the cities, leading to more and more invasions at the city periphery. Such a lifestyle leads to automobile dependency and heavy use of fossil fuels. The early neighbourhoods are identified with substandard housing, a lack of open spaces, widespread illness due to overcrowding, and unhealthy living standards. Biodiversity loss in cities causes reduced resilience of the ecosystem, thereby resulting in reduced ecosystem services, reduction in the quality of the natural resources of the area, and changes in disease patterns among the residents, which collectively impact the physical, mental, societal, and economic well-being of people (Næss & Vogel, 2012). These adverse impacts on the health and well-being of people have led to the need to adopt the concept of sustainable living, which aims at preventing the exploitation of natural resources. Sustainable living can be achieved by (1) developing new technologies that promote the use of renewable energy and (2) adopting a lifestyle that conserves natural resources for the use of future generations. Most steps are taken towards using green technology or renewable resources for the production of energy, whereas implementing measures for sustainable living by adopting lifestyle changes for the conservation of the environment is far more challenging. The introduction of biodiversity into cities and its associations with urban life tend to suggest its role in (1) improved environmental quality and reduction in air, water, and soil pollution; (2) better social interactions; (3) promotion of physical activity; (4) restoration from lifestyle stress; and (5) promotion of spiritual and emotional well-being. A few aspects of urban design that incorporate biodiversity into cities and their impacts are as follows.

6.1 Walkable Space

Walkability is an important concept in sustainable design. Besides aiding in safe and easy walking, these pedestrian areas are beneficial to society in different ways. They promote physical activities, which help in reducing diseases that are a result of an urban lifestyle, and also manage problems of over-nutrition and obesity. Obesity is no longer referred to as a problem of the high-income economy; it is escalating and penetrating even the poorest of economies in the world. Comprehensive studies on walkability show that countries that report fewer cases of obesity and health problems related to over-nutrition rely more on walking and using public transport in their daily life. They have low motorization levels despite the current surge in owning personal vehicles in order to cope with fast-paced lives. Walkable communities are sustainable communities that are designed to retain a sense of history and place, not space. It connects people to nature; contributes to social, cultural, and environmental diversity; connects people to important destinations; and helps improve resource responsibility, safety, and social interactions. Increased walking helps develop a more attractive and functional community, thus improving individual and community health (Taylor & Hochuli, 2015). A few other impacts of walking include reducing automobile exhaust, which, in turn, reduces the emission of greenhouse gases; giving an opportunity for physical exercise; reducing crime by promoting social interactions among residents; and supporting the local economy by encouraging business in nearby areas. In addition to the physical and mental health benefits, providing walkable space can encourage walking instead of driving, and it serves as a cost-effective transport option that is environmentally friendly.

6.2 Community Space

These are open spaces that allow interaction of people with other residents, nature, and animals. Openness and accessibility are the two key elements of community spaces. Properly planned, open spaces in urban set-ups influence the quality of life in the city. Different kinds of interactions provide social support, which helps people cope with the potential stresses of an urban lifestyle and enhances their emotional well-being. The interactions between humans and the community spaces can be divided into the following. (1) Human–nature interactions, which can be further classified into three types: (a) natural, (b) aesthetic, and (c) recreational. Neighbourhood parks contribute mostly towards interactions with nature; such environments are designed to provide calmness and activities directed towards nature. According to the attention restoration and stress reduction theory, natural settings restore positive effects and reduce symptoms of fear, anger, and aggression (Ulrich et al., 1991). Aesthetic interactions are experienced in all types of open spaces, though it is dependent on scenic beauty, cleanliness, etc., factors that are influenced by the residents of the community. Recreational activities are provided by almost all open spaces in a community, such as local parks, playing fields, playgrounds, neighbourhood parks, and all other types of community spaces. (2) Human–human interactions: such interactions can be divided into (a) social interactions, (b) citizen participation, and (c) sense of community. A good design of open space should promote linking people belonging to all sections of society with the community. Playing fields and the participation of citizens in urban parks contribute the most towards social interactions. They help in strengthening relationships among citizens. Local parks contribute significantly to fostering a sense of social security and community, as they provide space for public gathering and outdoor activities that promote community identity and a sense of social security among their users.

6.3 Green Spaces

Green spaces include all natural spaces in urban areas, such as public or private gardens, trees along streets, playing fields, and urban agricultural fields. The use of green spaces mainly infers three types of interactions: (1) indirect, that is, visual green space engagement; (2) incidental, where physical interactions are a by-product of other activities, such as passing by the green spaces while going to work; and (3) intentional, where the aim is to experience nature through green spaces. The benefits of urban green spaces often overlap with walkable roads and community spaces. All these initiatives aim at reducing the ill effects of an urban lifestyle and promoting better health and well-being. Ecosystem resilience is increased by the presence of green spaces, thus helping it recover from changes caused by different environmental stresses due to urbanization. In addition to the above-mentioned aspects, biodiversity and ecosystem services, some other undefined ecosystem services, such as parks, gardens, and indoor plants, make significant contributions to human well-being. Though their contribution towards biodiversity and ecosystem services is very low, such small initiatives help build a healthy human and biodiversity relationship, where both are beneficial. While evidence points out the positive impacts of integrating nature in the cities, research tends to point out the assumptions that inevitably link the presence of green spaces with well-being. Such assumptions are considered limitations as they prevent researchers from finding out when and why people use green spaces and how personal factors influence our choices regarding the use of green spaces for well-being. Such assumptions prevent us from getting the optimum benefits from green spaces; ruling out such assumptions will help design community spaces where the green spaces are planned according to the needs of the residents of a community (Sarah L. Bell et al. (2014)). To understand such assumptions, opportunities should be examined, alongside aspects like understanding personal well-being priorities and how the presence of green space is perceived and experienced by individuals (Atkinson et al., 2020). Thus, ruling out assumptions helps understand community priorities and helps in incorporating nature in a better way in daily life.

7 Strategies for Improving Urban Biodiversity Through Naturalization Practices

7.1 Increasing Environmental Awareness Among Citizens

New environmental issues have developed with the increase of urbanization and people adapting to urban life. This has led to a rising demand for environmentally friendly business practices. Though there is a strong approach of different businesses adopting green technology, their level of implementation of environmentally friendly practices is considerably low. Environmental awareness is needed to understand the connection between the need for environmentally friendly measures and, in practice, the implementation of such measures. Environmental practices among people influence them to reduce waste and properly segregate waste before it is collected by municipalities, thus contributing towards solid waste management. It increases awareness among people about the ill effects of using chemical fertilizers and pesticides and promotes the use of organic products. It educates people about sustainable living and ways to achieve it. The knowledge of the effects of increasing urbanization should not only be limited to the experts, but public awareness should also be spread so as to implement the best planning strategies for ecological sustainability. Through environmental awareness, people become aware of the importance of the environment and its positive impacts on our lives; thus, they tend to protect their environment from exploitation, which in the long run helps in maintaining a balanced ecosystem.

7.2 Green Architecture Designs

In order to reduce urban metabolism and ecological footprints of cities, green architecture designs are being considered, where principles of ecological sustainability and environmental conservation are integrated with architecture. They attempt to use eco-friendly building materials and construction practices. They safeguard the natural resources, air, water, and soil from the harmful effects of changing the environment. Green designs mainly include construction practices such as (a) efficient ventilation system, (b) the use of energy-efficient household appliances and gadgets, (c) the development of water-saving systems, (d) an increase in the use of renewable energy sources, (e) the use of non-toxic materials, (f) the use of locally available materials, (g) the reuse of older buildings, and (h) the efficient use of available space. Green architecture is also known as sustainable development with eco-friendly architecture (USGBC, 2002). A green building design begins with a detailed understanding of the site, considering both its aesthetic value and the issues associated with the land. Its main aim is to integrate the design with the on-site ecological functions. Efforts are being made to develop designs that will mimic a particular function of the ecosystem, which has been compromised while building the urban settlement. These measures help in minimizing the harmful effects of urbanization. The major components of green buildings are the following: (1) management of the water system: the green buildings implement methods of rainwater harvesting, grey water systems, and living pools. Water-conserving systems, such as ultra-low flush toilets and low-flow showerheads and taps, can minimize wastewater generation. Initiatives are taken towards recycling water, thereby, using water in an efficient way. (2) Natural buildings and green building materials: natural buildings focus on the durability and use of renewable resources that are not highly processed, all those materials that promote a healthy living environment; these materials include clay blocks, earth bags, wood, straw, rice hulls, bamboo, etc. A wide variety of reused or recycled non-toxic materials are commonly used as building materials for natural buildings. (3) Living architecture encompasses features such as green roofs, vegetated walls, and other sustainable roof technologies including solar collectors and cool roofs. While green roofs integrate vegetation and technology on rooftops, green walls specifically introduce plants onto the building façade. Green walls can be of the hanging type, climbing type, or module type. The concept of green buildings is not only a simple trend but is also appropriately suited to meet the needs of an urban settlement. Its importance will continue to increase with time.

7.3 *Bioengineering Techniques and Urban Green Designs*

These practices mainly include conservation practices such as anti-erosion practices, which focus on the treatment of surface water and the creation of wetlands. They are design practices to attain a sustainable system that integrates human society with its natural environment, benefitting both (Dhital & Tang, 2014). Bioengineering works include (1) river bed modelling, the movement of sediments and their deposition, which has a profound significance to the rivers. Sediment deposition is a key factor that hinders river development and management. Increasing human activities have led to problems with sediment deposition in rivers. To solve this problem, bioengineering techniques are adopted, and models are developed to estimate the deposition thickness using hydrodynamics, sedimentology, and bed deformation processes. The results are useful not only in calculating the sediment thickness but also in calculating probable sediment deposition in the future. Such simulations help design the sediment transport process and act as a technical tool to restore the ecological balance of the region. (2) Flood control in different parts of the world uses soil bioengineering techniques to control soil erosion along riverbanks and slopes. In bioengineering practices, engineering practices with integrated ecological principles are used in flood control measures. Mostly, vegetation, along with other non-living materials, is used in these practices. These practices are cost-effective, simple, and flexible and require no elaborate civil engineering work. Planting vegetation along slopes and stream banks effectively controls soil erosion and stabilizes slopes. It increases water infiltration in soil and soil fixation along river banks, reduces near-bed shear stress, and controls erosion. Furthermore, large vegetation prevents the transfer of sediments through heavy rain run-offs. The selection of plants for such plantations is also important as the plants should have simple and fast propagation and should also be well adapted to the environment. In addition to vegetation, wire net check dams are also used to stabilize the banks of rivers and streams. (3) Anti–soil erosion: soil erosion increases sediments in river beds, increases risks of flood, and reduces soil fertility by removing the top layer of soil. Prolonged erosion reduces soil porosity, thus reducing water infiltration and increasing rainwater run-off. Some of the main bioengineering techniques used to reduce soil erosion are brush layering, live check dams, fascines, and vegetative stone pitching. These structures slow down run-offs, catch debris, and filter sediments. All these methods provide resistance to flow, thus strengthening the soil and stabilizing it.

References

Aerts, R., Honnay, O., & Van Nieuwenhuyse, A. (2018). Biodiversity and human health: Mechanisms and evidence of the positive health effects of diversity in nature and green spaces. *British Medical Bulletin, 127*(1), 5–22. https://doi.org/10.1093/bmb/ldy021

Almenar, J. B., Elliot, T., Rugani, B., Philippe, B., Gutierrez, T. N., Sonnemann, G., & Geneletti, D. (2021). Nexus between nature-based solutions, ecosystem services and urban challenges. *Land Use Policy, 100*, 104898.

Anderson, E. C., & Minor, E. S. (2017). Vacant lots: An underexplored resource for ecological and social benefits in cities. *Urban Forestry & Urban Greening, 21*, 146–152.

Andersson, E., Barthel, S., Borgström, S., Colding, J., Elmqvist, T., Folke, C., & Gren, Å. (2014). Reconnecting cities to the biosphere: Stewardship of green infrastructure and urban ecosystem services. *Ambio, 43*(4), 445–453.

Angold, P. G., Sadler, J. P., Hill, M. O., Pullin, A., Rushton, S., Austin, K., & Thompson, K. (2006). Biodiversity in urban habitat patches. *Science of the Total Environment, 360*(1–3), 196–204.

Atkinson, S., Bagnall, A. M., Corcoran, R., South, J., & Curtis, S. (2020). Being well together: Individual subjective and community Well-being. *Journal of Happiness Studies, 21*(5), 1903–1921.

Barhadiya, G., & Ghosh, C. (2022). Urban biodiversity conservation a basic concept of sustainable urbanism: A systematic review. *Sustainability, Agri, Food and Environmental Research, 11*(1).

Bell, S. L., Phoenix, C., Lovell, R., & Wheeler, B. W. (2014). Green space, health and well-being: Making space for individual agency. *Health & Place, 30*, 287–292.

Bijnens, E. M., Nawrot, T. S., Loos, R. J. F., Gielen, M., Vlietinck, R., Derom, C., & Zeegers, M. P. (2017). Blood pressure in young adulthood and residential greenness in the early-life environment of twins. *Environmental Health, 16*(1), 53. https://doi.org/10.1186/s12940-017-0266-9

Binimelis, R., Born, W., Monterroso, I., & Rodríguez-Labajos, B. (2008). Socio-economic impact and assessment of biological invasions. In *Biological invasions* (pp. 331–347). Springer.

Borden, J. B., & Flory, S. L. (2021). Urban evolution of invasive species. *Frontiers in Ecology and the Environment, 19*(3), 184–191.

Braaker, S., Ghazoul, J., Obrist, M. K., & Moretti, M. (2014). Habitat connectivity shapes urban arthropod communities: The key role of green roofs. *Ecology, 95*(4), 1010–1021.

Bratton, S. P. (1982). The effects of exotic plant and animal species on nature preserves. *Natural Areas Journal, 2*(3), 3–13.

Britton, J. R. (2023). Contemporary perspectives on the ecological impacts of invasive freshwater fishes. *Journal of Fish Biology, 103*(4), 752–764.

Buczkowski, G., & Richmond, D. S. (2012). The effect of urbanization on ant abundance and diversity: A temporal examination of factors affecting biodiversity.

Butt, N., Shanahan, D. F., Shumway, N., Bekessy, S. A., Fuller, R. A., Watson, J. E., & Hole, D. G. (2018). Opportunities for biodiversity conservation as cities adapt to climate change. *Geo: Geography and Environment, 5*(1), e00052.

Cardinale, B. J., Duffy, J. E., Gonzalez, A., Hooper, D. U., Perrings, C., Venail, P., & Naeem, S. (2012). Biodiversity loss and its impact on humanity. *Nature, 486*(7401), 59–67.

Challender, D. W., Harrop, S. R., & MacMillan, D. C. (2015). Understanding markets to conserve trade-threatened species in CITES. *Biological Conservation, 187*, 249–259.

Chen, G., Li, X., Liu, X., Chen, Y., Liang, X., Leng, J., & Huang, K. (2020). Global projections of future urban land expansion under shared socio-economic pathways. *Nature Communications, 11*(1), 1–12.

CICES. https://cices.eu/revision-highlights/. Accessed 30 Jan 2022.

Clark, N. E., Lovell, R., Wheeler, B. W., Higgins, S. L., Depledge, M. H., & Norris, K. (2014). Biodiversity, cultural pathways, and human health: A framework. *Trends in Ecology & Evolution, 29*(4), 198–204.

Cowling, R. M., Egoh, B., Knight, A. T., O'Farrell, P. J., Reyers, B., Rouget, M., & Wilhelm-Rechman, A. (2008). An operational model for mainstreaming ecosystem services for implementation. *Proceedings of the National Academy of Sciences, 105*(28), 9483–9488.

D'amour, C. B., Reitsma, F., Baiocchi, G., Barthel, S., Güneralp, B., Erb, K. H., & Seto, K. C. (2017). Future urban land expansion and implications for global croplands. *Proceedings of the National Academy of Sciences, 114*(34), 8939–8944.

De Bell, S., Graham, H., Jarvis, S., & White, P. (2017). The importance of nature in mediating social and psychological benefits associated with visits to freshwater blue space. *Landscape and Urban Planning, 167*, 118–127.

De vries, S., Van Dillen, S. M., Groenewegen, P. P., & Spreeuwenberg, P. (2013). Streetscape greenery and health: Stress, social cohesion and physical activity as mediators. *Social Science & Medicine, 94*, 26–33.

Dhital, Y. P., & Tang, Q. (2014). Soil bioengineering application for flood hazard minimization in the foothills of Siwaliks, Nepal. *Ecological Engineering, 74*, 468–472.

Di Castri, F. (1989). History of biological invasions with special emphasis on the Old World. In *Biological invasion: A global perspective* (pp. 1–30). Wiley.

Dong, Y., Liu, H., & Zheng, T. (2020). Does the connectivity of urban public green space promote its use? An empirical study of Wuhan. *International Journal of Environmental Research and Public Health, 17*(1), 297.

Duke, L., & Soulsbury, C. D. (2021). Human–wildlife interaction networks at urban blue spaces. *Area, 53*(1), 122–133.

Egerer, M., Ordóñez, C., Lin, B. B., & Kendal, D. (2019). Multicultural gardeners and park users benefit from and attach diverse values to urban nature spaces. *Urban Forestry & Urban Greening, 46*, 126445.

Escobedo, F., Varela, S., Zhao, M., Wagner, J. E., & Zipperer, W. (2010). Analyzing the efficacy of subtropical urban forests in offsetting carbon emissions from cities. *Environmental Science & Policy, 13*(5), 362–372.

Faeth, S. H., Bang, C., & Saari, S. (2011). Urban biodiversity: Patterns and mechanisms. *Annals of the New York Academy of Sciences, 1223*(1), 69–81.

Foster, J., & Sandberg, L. A. (2004). Friends or foe? Invasive species and public green space in Toronto. *Geographical Review, 94*(2), 178–198.

Garrett, J. K., White, M. P., Huang, J., Ng, S., Hui, Z., Leung, C., & Wong, M. C. (2019). Urban blue space and health and well-being in Hong Kong: Results from a survey of older adults. *Health & Place, 55*, 100–110.

Gascon, M., Triguero-Mas, M., Martínez, D., Dadvand, P., Forns, J., Plasència, A., & Nieuwenhuijsen, M. J. (2015). Mental health benefits of long-term exposure to residential green and blue spaces: A systematic review. *International Journal of Environmental Research and Public Health, 12*(4), 4354–4379.

Gearey, M., Church, A., & Ravenscroft, N. (2019). From the hydrosocial to the hydrocitizen; Water, place and subjectivity within emergent urban wetlands. *Environment and Planning E: Nature and Space, 2*(2), 409–428.

Georgeson, L., Maslin, M., Poessinouw, M., & Howard, S. (2016). Adaptation responses to climate change differ between global megacities. *Nature Climate Change, 6*(6), 584–588.

Given David, R. (2000). Biodiversity of the urban environment: The importance of indigenous species and the role urban environments can play in their preservation.

Grellier, J., White, M. P., Albin, M., Bell, S., Elliott, L. R., Gascón, M., & Fleming, L. E. (2017). BlueHealth: A study programme protocol for mapping and quantifying the potential benefits to public health and well-being from Europe's blue spaces. *BMJ Open, 7*(6), e016188.

Grimm, N. B., Faeth, S. H., Golubiewski, N. E., Redman, C. L., Wu, J., Bai, X., & Briggs, J. M. (2008). Global change and the ecology of cities. *Science, 319*(5864), 756–760.

Gunnarsson, B., Knez, I., Hedblom, M., & Sang, Å. O. (2017). Effects of biodiversity and environment-related attitude on perception of urban green space. *Urban Ecosystems, 20*(1), 37–49.

Haahtela, T. (2019). A biodiversity hypothesis. *Allergy, 2019*(74), 1445–1456.

Haase, D., Frantzeskaki, N., & Elmqvist, T. (2014a). Ecosystem services in urban landscapes: Practical applications and governance implications. *Ambio, 43*(407–412), 2.

Haase, D., Larondelle, N., Andersson, E., Artmann, M., Borgström, S., Breuste, J., & Elmqvist, T. (2014b). A quantitative review of urban ecosystem service assessments: Concepts, models, and implementation. *Ambio, 43*(4), 413–433.

He, C., & Zhou, Y. (2014). *How much of the world's land has been urbanized, really? A hierarchical framework for evading confusion (No. PNNL-SA-102088)*. Pacific Northwest National Lab.(PNNL).

Hejkal, J., Buttschardt, T. K., & Klaus, V. H. (2017). Connectivity of public urban grasslands: Implications for grassland conservation and restoration in cities. *Urban Ecosystems, 20*(2), 511–519.

Hopkins, F. M., Ehleringer, J. R., Bush, S. E., Duren, R. M., Miller, C. E., Lai, C. T., & Randerson, J. T. (2016). Mitigation of methane emissions in cities: How new measurements and partnerships can contribute to emissions reduction strategies. *Earth's Future, 4*(9), 408–425.

Hu, S., Jin, C., Liao, R., Huang, L., Zhou, L., Long, Y., & Yang, Y. (2023). Herbaceous ornamental plants with conspicuous aesthetic traits contribute to plant invasion risk in subtropical urban parks. *Journal of Environmental Management, 347*, 119059.

Hyseni, C., Heino, J., Bini, L. M., Bjelke, U., & Johansson, F. (2021). The importance of blue and green landscape connectivity for biodiversity in urban ponds. *Basic and Applied Ecology, 57*, 129–145.

IPBES. (2019). https://www.ipbes.net/global-assessment

Irwin, E. G., & Bockstael, N. E. (2007). The evolution of urban sprawl: Evidence of spatial heterogeneity and increasing land fragmentation. *Proceedings of the National Academy of Science of the USA, 104*, 20672–20677.

Ishii, S. (2004). *Urban air pollution and urban management: Applicability of ecosystem approach and the way forward*. UNU-IAS working paper no. 120

Ives, C. D., Lentini, P. E., Threlfall, C. G., Ikin, K., Shanahan, D. F., Garrard, G. E., & Kendal, D. (2016). Cities are hotspots for threatened species. *Global Ecology and Biogeography, 25*(1), 117–126.

Jaiswal, N., & Jayakumar, S. (2021). Covid-19 pandemic-changes in the context of global environment and lessons learned. In *Environmental resilience and transformation in times of COVID-19* (pp. 207–222). Elsevier.

Johnson, P. T., Preston, D. L., Hoverman, J. T., & Richgels, K. L. (2013). Biodiversity decreases disease through predictable changes in host community competence. *Nature, 494*(7436), 230–233.

Ke, X., Van Vliet, J., Zhou, T., Verburg, P. H., Zheng, W., & Liu, X. (2018). Direct and indirect loss of natural habitat due to built-up area expansion: A model-based analysis for the city of Wuhan, China. *Land Use Policy, 74*, 231–239.

Knapp, S., Aronson, M. F., Carpenter, E., Herrera-Montes, A., Jung, K., Kotze, D. J., & Hahs, A. K. (2021). A research agenda for urban biodiversity in the global extinction crisis. *Bioscience, 71*(3), 268–279.

Kowarik, I. (2011). Novel urban ecosystems, biodiversity, and conservation. *Environmental Pollution, 159*(8–9), 1974–1983.

Lachowycz, K., & Jones, A. P. (2011). Greenspace and obesity: A systematic review of the evidence. *Obesity Reviews, 12*(5), e183–e189. https://doi.org/10.1111/j.1467-789X.2010.00827.x

Lampe, U., Reinhold, K., & Schmoll, T. (2014). How grasshoppers respond to road noise: Developmental plasticity and population differentiation in acoustic signalling. *Functional Ecology, 28*(3), 660–668.

Lee, I., Choi, H., Bang, K.-S., Kim, S., Song, M. K., & Lee, B. (2017). Effects of forest therapy on depressive symptoms among adults: A systematic review. *International Journal of Environmental Research and Public Health, 14*(3), 321. https://doi.org/10.3390/ijerph14030321

Lepczyk, C. A., Aronson, M. F., Evans, K. L., Goddard, M. A., Lerman, S. B., & MacIvor, J. S. (2017). Biodiversity in the city: Fundamental questions for understanding the ecology of urban green spaces for biodiversity conservation. *Bioscience, 67*(9), 799–807.

Liu, J., & Opdam, P. (2014). Valuing ecosystem services in community-based landscape planning: Introducing a wellbeing-based approach. *Landscape Ecology, 2014*(29), 1347–1360.

Lyytimäki, J. (2014). Bad nature: Newspaper representations of ecosystem disservices. *Urban Forestry & Urban Greening, 13*(3), 418–424.

Mason, C. F. (2000). Thrushes now largely restricted to the built environment in eastern England. *Diversity and Distributions, 6*(4), 189–194.

May, R. M. (2011). Why should we be concerned about loss of biodiversity. *Comptes Rendus Biologies, 334*(5–6), 346–350.

McDonald, R. I., Mansur, A. V., Ascensão, F., Crossman, K., Elmqvist, T., Gonzalez, A., & Ziter, C. (2020). Research gaps in knowledge of the impact of urban growth on biodiversity. *Nature Sustainability, 3*(1), 16–24.

McKinney, M. L., & Lockwood, J. L. (1999). Biotic homogenization: A few winners replacing many losers in the next mass extinction. *Trends in Ecology & Evolution, 14*(11), 450–453.

Mclellan, V., & Shackleton, C. M. (2019). The relative representation of ecosystem services and disservices in South African newspaper media. *Ecosystems and People, 15*(1), 247–256.

McNeely, J. (2001). Invasive species: A costly catastrophe for native biodiversity. *Land Use and Water Resources Research, 1*(1732-2016-140260).

Mexia, T., Vieira, J., Príncipe, A., Anjos, A., Silva, P., Lopes, N., Freitas, C., Santos-Reis, M., Correia, O., Branquinho, C., & Pinho, P. (2018). Ecosystem services: Urban parks under a magnifying glass. *Environmental Research, 160*, 469–478.

Millennium Ecosystem Assessment. (2005). *Ecosystems and human well-being* (Vol. 5, p. 563). Island Press.

Muller, N., Werner, P., & Kelcey, J. G. (Eds.). (2010). *Urban biodiversity and design*. John Wiley & Sons.

Müller, N., Ignatieva, M., Nilon, C. H., Werner, P., & Zipperer, W. C. (2013). Patterns and trends in urban biodiversity and landscape design. In *Urbanization, biodiversity and ecosystem services: Challenges and opportunities* (pp. 123–174). Springer.

Næss, P., & Vogel, N. (2012). Sustainable urban development and the multi-level transition perspective. *Environmental Innovation and Societal Transitions, 4*(2012), 36–50.

Nguyen, M. H., & Jones, T. E. (2021). Exploring biodiversity mental constructs of urban residents in an emerging South-East Asian country.

Nilon, C. H. (2011). Urban biodiversity and the importance of management and conservation. *Landscape and Ecological Engineering, 7*(1), 45–52.

Njoh, A. J. (2003). Urbanization and development in sub-Saharan Africa. *Cities, 20*(3), 167–174.

Nor, A. N. M., Corstanje, R., Harris, J. A., & Brewer, T. (2017). Impact of rapid urban expansion on green space structure. *Ecological Indicators, 81*, 274–284.

Oertli, B., & Parris, K. M. (2019). Toward management of urban ponds for freshwater biodiversity. *Ecosphere, 10*(7), e02810.

Ossola, A., & Livesley, S. J. (2016). Drivers of soil heterogeneity in the urban landscape. *Urban Landscape Ecology*, 37–59.

Pandey, B., & Ghosh, A. (2023). Urban ecosystem services and climate change: A dynamic interplay. *Frontiers in Sustainable Cities, 5*, 1281430.

Park, S. (2017). A preliminary study on connectivity and perceived values of community green spaces. *Sustainability, 9*(5), 692.

Parmesan, C. (2006). Ecological and evolutionary responses to recent climate change. *Annual Review of Ecology, Evolution, and Systematics, 37*, 637–669.

Potgieter, L. J., Gaertner, M., O'Farrell, P. J., & Richardson, D. M. (2019). Perceptions of impact: Invasive alien plants in the urban environment. *Journal of Environmental Management, 229*, 76–87.

Raven, P. H., & Wagner, D. L. (2021). Agricultural intensification and climate change are rapidly decreasing insect biodiversity. *Proceedings of the National Academy of Sciences, 118*(2), e2002548117.

Roe, D. (2019). Biodiversity loss—More than an environmental emergency. *The Lancet Planetary Health, 3*(7), e287–e289.

Romanelli, C., Cooper, D. et al., (2015). Connecting global priorities: Biodiversity and human health. World Health Organization and Secretariat of the Convention on Biological Diversity.

Sanderson, M. G., Collins, W. J., Johnson, C. E., & Derwent, R. G. (2006). Present and future acid deposition to ecosystems: The effect of climate change. *Atmospheric Environment, 40*(7), 1275–1283.

Sarkar, C., Webster, C., & Gallacher, J. (2014). *Healthy cities: Public health through urban planning*. Edward Elgar Publishing.

Scheffers, B. R., Oliveira, B. F., Lamb, I., & Edwards, D. P. (2019). Global wildlife trade across the tree of life. *Science, 366*(6461), 71–76.

Semerdzhieva, L., & Borisova, B. (2021). Urban ecosystems assessment: An integrated approach to maintenance of habitats and their biodiversity. *Journal of the Bulgarian Geographical Society, 45*, 99.

Seto, K. C., Güneralp, B., & Hutyra, L. R. (2012). Global forecasts of urban expansion to 2030 and direct impacts on biodiversity and carbon pools. *Proceedings of the National Academy of Sciences, 109*(40), 16083–16088.

Shochat, E., Lerman, S. B., Anderies, J. M., Warren, P. S., Faeth, S. H., & Nilon, C. H. (2010). Invasion, competition, and biodiversity loss in urban ecosystems. *Bioscience, 60*(3), 199–208.

Slaughter, M. S., Pagani, F. D., Rogers, J. G., Miller, L. W., Sun, B., Russell, S. D., et al. (2010). Clinical management of continuous-flow left ventricular assist devices in advanced heart failure. *The Journal of Heart and Lung Transplantation, 29*(4), S1–S39.

Smith, K. F., Behrens, M. D., Max, L. M., & Daszak, P. (2008). U.S. drowning in unidentified fishes: Scope, implications, and regulation of live fish import. *Conservation Letters, 1*(2), 103–109.

Smith, W. S., Silva, F. L., Amorim, S. R., & Severino, M. S. (2018). Urban biodiversity: How the city can do its management? *Biodiversity International Journal, 2*(3), 272–277.

Snep, R., & Ottburg, F. G. (2008). The habitat backbone sites as a nature conservation strategy for industrial areas: Lesson from the natterjack toad (bufo calamita) in the Port of Antwerp (Belgium). *Landscape Ecology, 23*, 1277–1289.

Sol, D., González-Lagos, C., Moreira, D., Maspons, J., & Lapiedra, O. (2014). Urbanisation tolerance and the loss of avian diversity. *Ecology Letters, 17*(8), 942–950.

Tantipisanuh, N., & Gale, G. A. (2013). Representation of threatened vertebrates by a protected area system in Southeast Asia: The important of non-forest habitats. *The Raffles Bulletin of Zoology, 61*, 359–395.

Taylor, L., & Hochuli, D. F. (2015). Creating better cities: How biodiversity and ecosystem functioning enhance urban residents' well-being. *Urban Ecosystem, 18*, 747–762.

Tian, Y., Liu, Y., Jim, C. Y., & Song, H. (2017). Assessing structural connectivity of urban green spaces in metropolitan Hong Kong. *Sustainability, 9*(9), 1653.

Trisurat, Y., Shirakawa, H., & Johnston, J. M. (2019). Land-use/land-cover change from socio-economic drivers and their impact on biodiversity in Nan Province, Thailand. *Sustainability, 11*(3), 649.

Ulrich, R. S., Simons, R. F., Losito, B. D., Fiorito, E., Miles, M. A., & Zelson, M. (1991). Stress recovery during exposure to natural and urban environments. *Journal of Environmental Psychology, 11*(3), 201–230.

United Nations. (2019). Nature's dangerous decline 'unprecedented'; species extinction rates' accelerating'.

United Nations. (2020). https://unhabitat.org/sites/default/files/2020/10/wcr_2020_report.pdf

USGBC. (2002). *U.S. Green Building Council, building momentum: "National Trends and prospects for high-performance green buildings"*. Prepared for the U.S. Senate Subcommittee on Environmental and Public Works by the U.S. Green Building Council, November 2002.

Van der Werf, G. R., Morton, D. C., DeFries, R. S., Olivier, J. G., Kasibhatla, P. S., Jackson, R. B., & Randerson, J. T. (2009). CO2 emissions from forest loss. *Nature Geoscience, 2*(11), 737–738.

Van Wilgen, B. W., & Scott, D. F. (2001). Managing fires on the Cape Peninsula, South Africa: Dealing with the inevitable. *Journal of Mediterranean Ecology, 2*, 197–208.

Vandermeulen, V., Verspecht, A., Vermeire, B., Van Huylenbroeck, G., & Gellynck, X. (2011). The use of economic valuation to create public support for green infrastructure investments in urban areas. *Landscape and Urban Planning, 103*(2), 198–206.

Vilà, M., Basnou, C., Pyšek, P., Josefsson, M., Genovesi, P., Gollasch, S., et al. (2010). How well do we understand the impacts of alien species on ecosystem services? A pan-European, cross-taxa assessment. *Frontiers in Ecology and the Environment, 8*(3), 135–144.

Wagner, D. L., Grames, E. M., Forister, M. L., Berenbaum, M. R., & Stopak, D. (2021). Insect decline in the Anthropocene: Death by a thousand cuts. *Proceedings of the National Academy of Sciences, 118*(2).

Yam, K., Tan, T., Doyle, R., Clos, J., Gutiérrez, F., Chan, M., & Agar, B. (2016). Our Planet-October 2016: Urban solutions: Making cities strong, smart, sustainable.

Zari, M. P. (2018). The importance of urban biodiversity–an ecosystem services approach. *Biodiversity International Journal, 2*(4), 357–360.

Zebisch, M., Wechsung, F., & Kenneweg, H. (2004). Landscape response functions for biodiversity—Assessing the impact of land-use changes at the county level. *Landscape and Urban Planning, 67*(1–4), 157–172.

Zhao, S., Da, L., Tang, Z., Fang, H., Song, K., & Fang, J. (2006). Ecological consequences of rapid urban expansion: Shanghai, China. *Frontiers in Ecology and the Environment, 4*(7), 341–346.

Zhongming, Z., Linong, L., Wangqiang, Z., & Wei, L. (2020). World cities report 2020: The value of sustainable urbanization.

Urban Watershed Management

Abhishek Pandey Bharat, Ashwani Kumar Tiwari, Antonino Pisciotta, and Abhay Kumar Singh

Abstract Urban watersheds in India are currently facing significant hydrological disruptions due to a range of factors, including diminished infiltration capacities, reduced water volumes, altered water chemistry, instability in natural processes, anthropogenic encroachments, and escalating pollution levels. Historically, watersheds were conceptualised as integrated systems for the sustainable management of water resources within a region. However, recent flooding events worldwide underscore the complexities involved in contemporary watershed management practices. Globally, urbanized watersheds are increasingly experiencing stream-flow depletion and deteriorating water quality, which severely compromise the ecological integrity of both in-stream and riparian habitats, as well as the overall water supply. In developing nations, watersheds are essential for providing livelihoods through various products and services. Nonetheless, the unchecked expansion of agriculture and urbanization poses a grave threat to human life and civilization by creating unsustainable demands on land and water resources. By adopting a holistic approach to urban watershed management, cities can reduce the risks associated with urbanization and create more resilient and sustainable water systems. Furthermore, the implementation of nature-based solutions can provide numerous benefits, including biodiversity enhancement, improved water quality, and increased public health and well-being.

Keywords Sustainable water systems · Hydrological disruptions · Water quality degradation · Watershed management · Nature-based solutions

---

A. P. Bharat · A. K. Singh
CSIR-Central Institute of Mining and Fuel Research, Dhanbad, Jharkhand, India

Academy of Scientific and Innovative Research (AcSIR), Ghaziabad, Uttar Pradesh, India

A. K. Tiwari (✉)
School of Environmental Sciences, Jawaharlal Nehru University (JNU), New Delhi, India
e-mail: ashwaniktiwari@mail.jnu.ac.in

A. Pisciotta
Istituto Nazionale di Geofisica e Vulcanologia (INGV), Palermo, Italy

P. K. Joshi et al. (eds.), *Geospatial Science for Urban Ecosystems*, The Urban Book Series, https://doi.org/10.1007/978-3-032-07155-2_5

1 Introduction

A watershed, a primary and essential component of environmental planning, is a topographically delineated land area that channels rainfall and snow-melt water into common outlets such as creeks, streams, and rivers, which finally submerge into reservoirs, bays, or oceans (Wang et al., 2004; McGinnis, 1999). The watershed may be surrounded by a drainage divide that spans thousands of square miles, with surface runoff and groundwater beneath it (Forte & Whipple, 2018). Collaborative watershed management has become a new paradigm in environmental policy during the past few decades. The extraction of water resources in fulfilment of the demands generated due to the rapid growth of population has exacerbated the issues of water scarcity and pollution all around the world (Dangar et al., 2021). Planning for the watershed and its surrounding areas could address problems, including contaminated streams, contaminated agricultural land, and a lack of freshwater supplies, all of which have an influence on human populations. Reports say that over one-third of the global population is compelled to drink contaminated water, creating waterborne diseases in humans and resulting in premature deaths of more than nine million people worldwide (Mukherjee et al., 2019). The primary goal of watershed management is to benefit humanity in an integrated manner by conserving water resources of a catchment, plants, and soil, although fragile ecosystem management could lead to the degradation of environmental components (Kerr, 2002; Richter et al., 2003). Additionally, watershed management should be oriented towards long-term sustainable planning that could minimise the impacts on humans and natural resources, and avoid disaster-related aspects.

Urban watersheds, in the current scenario, have been experiencing degradation in both quality and quantity and are going to be invisible soon due to a lack of appropriate, sustainable, and scientific planning (Platt et al., 2008; Fletcher et al., 2013). A larger population density, increasing municipal wastes generated from the households, sewage generation, industrial effluent drainage, deficiency of treatment plants, high concretion, blockage of the soil surface due to unplanned infrastructures, and disturbances in the flow path of a river system in an urban watershed are the key factors responsible for remarkable changes in the hydrological behaviour of water regimes in a watershed (Sewnet & Abebe, 2018; Raphael et al., 2023; Bernhardt et al., 2008; Liao et al., 2020; Alam et al., 2006; Sánchez et al., 2007).

Approximately 55% of people reside in metropolitan regions globally, where there is interaction between natural and man-made elements. According to a report by the United Nations, around 68% of the global population will start to live in urban regions by 2050, and the impact of such a shift in population on the environment will also be intensified. The ongoing construction and reshaping of cities continuously modify urban watersheds. The neglect of appropriate scientific approaches in the management of watersheds will put great pressure not only on the water regimes but also on the other environmental components, such as soil, hydrology, human health, and ecology.

2 Evolution of Watershed Management

Initially, the idea of managing watersheds was centred on utilisation and conservation of water and soil entities, which was further integrated into social, economic, and environmental development (Katusiime & Schütt, 2020). Scientific watershed management, however, was cited in the Atharva Veda text from 800 BC, stating, "one should take proper measures to conserve the water resources associated with mountains, aquifers, rivers and rainwater for utilizing it in domestic, irrigation and industrial sectors" (Wang et al., 2016; Neary, 1998). The recent centuries have observed the first plan of watershed-based management in 1878 by John Wesley Powell, who proposed that political jurisdictions conform to watersheds in the American West (Worster, 2009; Powell, 1879). However, most recently, in the 1990s, the approach to watershed management in environmental sustainability gained popularity (Schueler, 1996). Further, the watersheds were categorised based on the shape, size, and land use pattern into five different types, such as mini-watershed (1<area<100 hect), micro-watershed (100<area<1000 hect), milli-watershed (1000<area<10,000 hect), sub-watershed (10,000<area<50,000 hect), and macro-watershed (area>50,000 hect) (Singh, 2017; Sen et al., 2007). The integrated management of watersheds evolved from understanding the complex relations among land use change, hydrological systems, human health, and ecological components, in association with scientific and political aspects, to sustain and enhance watershed functions. Through a consecutive change in watershed planning for many decades, a number of major goals were added, which include damage control and utilisation of runoff and, thereby, water and soil conservation to protect, conserve and rejuvenate the area of the watershed for sustainable development; identify the soil erosion mechanism for reducing the sediment yield; rehabilitate the deteriorating area; moderate the flood peaks; manage the increase in infiltration of rainwater for groundwater conservation; improve wildlife resources; and maintain the quality of water resources.

The scientific and technological facets of environmental management are increasingly intertwined with socio-economic elements, and when combined, they have the potential to address the complex and urgent issues being experienced by society (Cosgrove & Loucks, 2015; Lienert et al., 2006). Priority should be given to scientific planning that is cost-effective and enhanced for the current issues of society. A proper focus on size, shape, slope, drainage density, land use pattern, climate, geography, geology, and soil properties of the selected area should also be focused on the characterization of the watershed (Waikar & Nilawar, 2014). The watersheds can also be categorised on a hydrological basis into the following parts: urban, agricultural, forest, mountainous, desert, coastal, and wetland watersheds, which have their specific characteristics (Ouyang et al., 2018; Kaushal & Belt, 2012; Qi et al., 2009; Wolka et al., 2015). In this chapter, the urban watershed, its classes, environmental degradation, and remedial measures and management are discussed.

3 Urban Watershed

The diversity of the physical, political, ecological, and social spheres affects urban watersheds (Pickett et al., 2001; Larson & Lach, 2010). Different components of the urban watershed are shown in Fig. 1. The structure and sustainability of watershed systems are altered by the interaction of hydrology, geology, climate, soils, vegetation, and human activity (Sadeghi et al., 2019; Jacobson, 2011; Sobota et al., 2009). The availability of water is seen as an essential component of the natural world and human life and as a necessary precondition for social development (Li et al., 2023; Azhari et al., 2022). An urban watershed continuum paradigm recognises a continuum of artificial and natural hydrological flow routes that expand hydrologic networks in seldom-considered ways. It recognises that the kind of hydrologic connectivity influences downstream fluxes and transformations of nutrients, energy, pollutants, and carbon over four dimensions of space and time (Covino, 2017).

Urban infrastructure, such as drains, ditches, gutters, and pipelines, replaces first-order streams in watersheds, both longitudinally and laterally (Jarden et al., 2016; Potter et al., 2014; Napieralski & Welsh, 2016). In designed headwaters, organic carbon and nutrient retention have undergone substantial longitudinal and lateral change. The relationship between land use changes and hydrologic processes highlights the longitudinal downstream pulses observed in material and energy exports. Leaky pipes and groundwater interact vertically, affecting how stream solutes are transported (Hoard et al., 2020). The urban watershed continuum transforms and transports materials and energy based on hydrologic residence durations.

Fig. 1 Watershed and its interrelated aspects (after Veale & Cooke, 2018)

Biogeochemical cycles and ecosystem are evolving throughout time as land use and urban infrastructure change.

The report says that a significant section of the population will move to urban areas in the next 20 years; the twenty-first century may see problems with urban drinking water and sanitation (United Nations, 2010; Sincero & Sincero, 2002; Kareem et al., 2023). The primary source of toxins that harm hydrochemical properties and human health in watersheds is urban development. Urban development enhances hydrologic connectivity by constructing pipes, ditches, swales, and gutters (Haghighatafshar et al., 2018). The vastly larger flows and changed subsidies and fluxes of organic matter, organisms, energy, and nutrients set urban stream networks apart from those that drain rural environments. With the growing trend of cities becoming more sustainable and hygienic, developing models that integrate ecosystem function with infrastructure across space and time will be essential. A significant and sometimes overlooked biological role in stream networks is played by artificial headwaters, which are part of a continuum of natural and designed hydrologic flow routes that should be explicitly considered in studies of urban watersheds. Hydrological linkages and urban infrastructure affect downstream fluxes and transformations. Urban planning and the inevitable decay of infrastructure have an impact on the relationship between ecosystems and urban infrastructure.

4 Natural Vs. Urban Watersheds

Maintaining the quality of surface water and groundwater is addressed by natural resource management techniques that preserve water and soil together (Addisu et al., 2015; Zalidis et al., 2002). The management of natural resources has an effect on the quality of the soil and water; therefore, in order to apply corrective management strategies, these changes must be regularly monitored and assessed. In a rainfed agro-ecosystem, soil, water, and production systems are the most important natural resources and components of watersheds, and environmental harmony should be a prerequisite for sustainable industrial systems.

Urban watersheds, on the other hand, are essential for the management of water for a sizable population. Urban watersheds provide a novel and dynamic conceptual framework for analysing several research questions, including how storms and other disturbances affect patterns in ecological community succession and biogeochemical functions, how biodiversity and ecosystem functions may change along hydrologic flow paths and watershed environmental gradients (Samhouri et al., 2022; Yi et al., 2018), and how nutrients, ecological stoichiometry of carbon, metals, and hydrologic retention change along flow paths in urban watersheds across space and time (Shields et al., 2008).

5 Degradation of Urban Watershed

Urban watershed degradation has accelerated due to the synergy between building artificial structures, urban area expansion, and rapid population rise (Pauchard et al., 2006; Peng et al., 2011). Deforestation, wetland reclamation, river channel alteration, urban pollution, and impervious surfaces (such as roads, paving, and rooftops) have all contributed to the significant degradation of watersheds. This may limit the capacity of watersheds to deliver essential goods and services that support metropolitan areas and local lives, as well as their natural capacity to regulate and buffer against water-related risks, including floods, landslides, and drought (Zhang et al., 2022; Cui et al., 2009). The population of wealthy countries is more urbanised; for instance, the US population is about 80% urban. Human activities upstream and in peri-urban regions can worsen runoff by reducing ground infiltration, causing drainage systems to get clogged with silt and other wastes. Communities downstream are frequently compelled to make up for lost ecosystem benefits. Impervious surfaces critically impact how well urban watersheds operate (Zimmerman & Civco, 2020).

Ecological buildings have the potential to significantly alter the hydrology in metropolitan settings. For instance, less tree cover in metropolitan areas results in higher runoff rates and shorter times between the start of storms and the start of runoff (Carlyle-Moses et al., 2020; Yao et al., 2015; Hough, 1995). The increased runoff in metropolitan areas changed the shape of streams, drastically increasing the floodplains in the regions. Urbanization is becoming more often acknowledged as a disruptive force that alters but does not completely remove its physical and ecological environment. It is an important demographic trend as well as a major component of the global land transformation process. Urban ecology in the 1990s concentrated on the benefits of nature to human society, whether it be rural or urban. These benefits included soil formation, water purification, flood storage, aquifer replenishment, pollution reduction, and carbon sequestration. As a consequence, due to urbanisation, the less expensive natural services may have to be replaced by more expensive technological options, such as artificial fertilisers, air conditioning, sewage and water treatment, flood control, and air conditioning.

There are a number of man-made structures that impact the naturally available hydrological structures, such as local streams of the first order as well as the second order, with dense networks, runs, kills, brooks, and creeks, as per the commonly used Strahler categorisation system. Local watersheds generally feature a diverse array of hydrological and ecological characteristics, such as river channels, riverbanks, ponds, lakes, floodplains, aquatic as well as riparian habitats, different aquifer units, coastal areas with estuaries, wetlands, and many others. Natural land cover in urban-source watersheds contributes significantly to ecosystem that help keep the quality of the city's water sources stable. However, with the development of urban works, there is a possibility of increased pollutant load in suspended and dissolved forms from different domestic, sewage, or municipal activities. Urbanisation has a negative impact on watersheds because small tributary streams carry increased loads of silt and pollutants downstream to higher-order waterways. They also have lower base flows and greater peak flows. The loss of natural land cover could result in the deterioration of water quality, simultaneously with a reduction in the provision of ecosystem services. The major reasons behind these issues could be the results of a group of several activities, out of which converting the natural land cover into urban areas, agricultural activities, industrial activities, and disposing of sewage into natural rivers are important ones. The man-made impacts on urban watersheds can be categorised into the following kinds.

5.1 *Industrial Waste*

The degradation of urban watersheds by industrial waste is a critical environmental issue that significantly impacts water quality and ecosystem health. The majority of industrialised countries use water treatment plants that use physical, biological, and chemical processes to clean the wastewater (Ahmed et al., 2021; Sonune & Ghate, 2004); in contrast, these wastes—especially in developing nations—are disposed of straight into natural ecosystems without any prior treatment. After each phase, bioassays can be used to assess the treatment's effectiveness and compare the results to the input wastewater's toxicity. Industrial wastewaters typically require pretreatment because of their higher toxicity before being transported to a municipal wastewater treatment facility (Mantzavinos & Psillakis, 2004). High levels of biochemical oxygen demand (BOD), chemical oxygen demand (COD), total organic carbon, and variable pH are typical characteristics of industrial effluents. Examples of organic materials include proteins, lignin, lipids, soaps, artificial detergents, carbohydrates, and the by-products of their breakdown, in addition to many organic compounds from the process industries (Patwardhan, 2017). The three categories of strong, medium, and mild domestic wastewater are determined by the concentration of the constituent parts. For strong, medium, and mild wastewater, the chemical oxygen demand is 1000, 500, and 250 mg/L, respectively (Sincero & Sincero, 2002). In dry and semi-arid environments, domestic water use is often very low, leading to extremely strong sewage. Heavy metals like lead, cadmium, and mercury, often found in industrial effluents, can accumulate in the sediments and biota of water bodies, leading to toxic effects on aquatic organisms and posing health risks to humans through bioaccumulation and biomagnification (Dreshaj et al., 2022; Kareem et al., 2023).

Since the effluent carries large levels of iron (>20 times higher than ambient), which blankets the stream bed, it has an adverse effect on water quality by elevating in-stream temperatures to 13–18 °C during the colder months (Nedeau et al., 2003). The wastewater also change the substrate and flow conditions in the habitat by increasing the overall stream discharge by 50–150%. In order to gather benthic macroinvertebrates in depositional and erosional habitats and assess the relative significance of habitat quality and water quality changes, three different techniques are used. The main causes of pollution include waste discharges and industrial wastewater. Many urban areas lack adequate wastewater treatment facilities, allowing untreated or poorly treated industrial effluents to enter rivers and streams directly. The wastes come from homes, businesses, recreational venues, offices, and institutions, including hospitals and schools. In addition to paper, batteries, textiles, plastics, glass, excreta, aerosol cans, and many other materials, they also comprise food waste.

5.2 Unplanned Groundwater Extraction

Unplanned groundwater extraction in urban watersheds leads to significant degradation of water resources and associated ecosystems. As urban populations grow and industrial activities expand, the demand for groundwater increases, often outstripping the natural recharging rates of aquifers. This excessive abstraction can result in a variety of environmental issues. Lowering water tables can cause the intrusion of pollutants and saline water into freshwater aquifers, further degrading water quality and making it unsuitable for consumption and agriculture (Seyam et al., 2020; Odeh et al., 2019; Liu et al., 2003). Groundwater extraction from wells is a critical activity globally. Due to increased urbanisation and over-pumping of water from the wells, the groundwater salinity rises over time (Seyam et al., 2020; Odeh et al., 2019; Liu et al., 2003). Effective government action is required to mitigate these negative effects since rising salinity levels in water have a severe impact on the sustainability of the economy and society, particularly the agricultural sector. Hydrological problems, such as excessive groundwater extraction in certain areas, rising water tables in other areas, and generally declining groundwater quality, are limiting agricultural production all over the world. In different parts of the world, such as Peru, California, Iran, Chile, Bangladesh, India, China, and Southern Europe, groundwater withdrawals are rising, where a growing population and an extended drought are causing problems (Langridge & Ansell, 2018; Motagh et al., 2008; Hernandez-Mora et al., 2001; Schwarz & Mathijs, 2017). Alternative sources of freshwater and saltwater that can be used with desalination and/or treatment are deep groundwater aquifers. India is one of the water-stressed nations in the world, with 600 million people enduring high levels of water stress, according to a 2018 NITI Aayog report. The estimate predicts that by 2030, water demand may quadruple the present supply, leaving millions of people severely short of water and causing a 6% drop in the country's GDP.

5.3 Decreased Infiltration Rate

Decreased infiltration rate is a significant issue in urban watersheds, primarily caused by urbanisation and soil compaction. Urban soils often experience greater compaction due to construction activities, leading to reduced infiltration rates. Studies have shown that infiltration rates on compacted soils can be 70–99 per cent lower than those on non-compacted soils (Gregory et al., 2006). For example, a study in North Central Florida found that average infiltration rates on compacted soils ranged from 8.175 mm/hr. to 188 mm/hr., significantly lower than the design storm infiltration rate of 254 mm/hr. (Gregory et al., 2006). These changes include the large-scale creation of impervious areas (roads, concrete walkways, rooftops, and driveways), a decline in the natural depressions that store surface water, and a reduction in the soil's ability to absorb rainfall (Zhou et al., 2018; Barnes et al., 2001). Sealing is the process of covering soil with artificial materials that have a low porosity, like asphalt and concrete. Soil sealing has an impact on a number of processes that are hindered or slowed down by soil biodiversity, including habitat fragmentation, carbon sequestration, water infiltration, groundwater recharge, and loss of biodiversity (Geisen et al., 2019). These findings suggest that decreased infiltration rates can lead to increased storm water runoff, emphasising the need for effective storm water management strategies in urban watersheds. The measurement of the impact of impervious area is crucial since it increases runoff volume and water velocity, both of which tend to raise peak flows. Furthermore, research has also shown that afforestation efforts can lead to improved infiltration rates, highlighting the importance of urban green spaces in mitigating the effects of urbanisation (Tang et al., 2024). Overall, addressing decreased infiltration rates is crucial for managing storm water runoff and maintaining the health of urban watersheds.

5.4 Increased Flood Zones

Urban services and, in particular, the management of emergencies and disasters are under increasing pressure as a result of the unrelenting migration of people from rural to urban regions. The planning, support, and reconstruction of society after a natural or man-made disaster is known as emergency or disaster management. Rapid urbanisation can result in unwanted urban development due to the increased population density and prolonged land use. As a result, urban built-up regions are typically only found in geographically hazardous locations like riverbanks, floodways, and water basins that are susceptible to floods (Tabios III, 2023; Masoumi, 2022; Desalegn & Mulu, 2021). Climate change has also played a significant role in increasing flood zones, with rising global temperatures leading to more frequent and intense flooding events. The Intergovernmental Panel on Climate Change (IPCC) has found that climate change "has detectably influenced" several of the variables that contribute to floods, such as rainfall and snow-melt (Tabari, 2020). Furthermore, the IPCC projects that heavy precipitation events will increase by 50% to as much as three times the historical average in this century, leading to more frequent and severe flooding events (Tabari, 2020). The expansion of urban areas into flood zones, combined with the impacts of climate change, has resulted in a significant increase in flood-prone areas, highlighting the need for effective flood risk mitigation efforts and urban planning strategies that prioritise resilience and adaptation.

5.5 Water Quality Change

High concentrations of heavy metals are considered to be the main problem with urban runoff water quality, while other factors, including temperature, minerals, microorganisms, and pharmaceutical residues involving active compounds and metabolites, frequently play a part as well (Dodgen et al., 2017; Bekele et al., 2011). Water pollution in urban watersheds may occur due to the combined impacts of industrial processes, domestic sewage, septic tank wastewater, solid waste disposal and associated leachate effects, leakage in sewage systems, fertilisation impacts, etc. The most common causes of elevated heavy metal concentrations in soils are agrochemical impurities resulting from the use of fertilisers and pesticides, irrigation with contaminated water, surface runoff from adjacent industrial facilities, the extraction of mineral ore, and the ensuing disposal of waste, road dust, sewage sludge, waste disposal, livestock manures, and atmospheric deposition (Khan et al., 2022; Hou et al., 2017). The contaminated water circulates through different drinking water sources and has deteriorating impacts on human health.

5.6 Lack of Treatment Plants

Urbanisation alters the natural water cycle of rainfall by removing vegetation and replacing it with sealed surfaces, altering the aggregate arrangement, pore space, and hydraulic properties of the soil, and using artificial drainage systems to collect excess runoff from urban surfaces and transport it to stream channels. In India, around 62,000 million litres of sewage is generated per day in urban areas, but the treatment capacity of India is limited to 23,277 million litres per day (Chatterjee et al., 2016; Kaur et al., 2012). Sewage treatment plants (STPs) are expensive enterprises that can only be maintained when a thorough financial structure is available. To treat sewage, several components of an STP must operate properly. Typically, when untreated sewage is discharged into a water body, natural processes such as stream cleaning and self-purification can facilitate its cleansing. However, the accelerated discharge of sewage, fuelled by population growth and extensive urbanisation, has outpaced the natural purification rate. This imbalance poses significant environmental challenges. Similarly, India grapples with substantial environmental issues arising from extensive trash generation and shortcomings in waste collection, transportation, treatment, and disposal (Chariar, 2017). The existing waste management systems in India need to be equipped to handle the escalating volume of waste generated by the expanding urban population, impacting both ecology and public health.

5.7 Municipal Solid Waste

In urban regions, solid waste disposal is a serious and pervasive issue in many developed and developing nations. The amount of waste produced has increased as a result of high standards of living and an ever-growing population. Water with relatively low percentages of dissolved and suspended organic and inorganic particles makes up the majority of municipal wastewater. Municipal solid waste (MSW) is typically a mixture of domestic and commercial garbage produced by a living community (Bhat et al., 2018). The management of MSW and its effects on groundwater quality have recently risen to the top of the list of numerous environmental issues being faced by developing countries' urbanising cities. Leachate is produced when solid waste is placed in low-lying regions, where it may come into contact with runoff, rainwater, or groundwater (Bhalla et al., 2012). Such leachate comprises a variety of harmful and other detrimental substances that, if dispersed into groundwater, would affect the quality of the water or render it unusable for drinking and other household purposes (Abd El-Salam & Abu-Zuid, 2015; Fatta et al., 1999). Groundwater may become unsuitable for drinking and other uses once contamination enters the sub-surface ecosystem and is hidden for a long time. Municipal solid waste (MSW) landfills are recognised as major sources of groundwater contamination because of the leakage of leachate, a complex mixture of pollutants with a high chemical oxygen demand, a high ammonium nitrogen concentration, and long-lasting toxicological properties. In developed areas, waste should be disposed of following efficient solid waste management protocols.

5.8 Deforestation

Deforestation continued with an average of 14.6 million hectares worldwide from 1990 to 2000 (FAO, 2001). Deforestation is a common phenomenon and a major environmental problem in the period of rapid climate change. Deforestation in urban watersheds has profound and multifaceted impacts on the environment and water resources. When trees are removed, the natural water cycle is significantly disrupted. Forests play a crucial role in intercepting rainfall through their canopy, which slows down the fall of rainwater and allows for greater infiltration into the soil. This process is vital for maintaining groundwater recharge and reducing surface runoff. However, deforestation leads to a substantial decrease in interception and evapotranspiration, resulting in increased runoff and a higher likelihood of flooding (Hlásny et al., 2015).

The removal of forest cover also exacerbates soil erosion and sedimentation. Without the protective layer of vegetation, soil becomes more susceptible to erosion by rainwater, leading to increased sediment loads in nearby streams and rivers. Moreover, the loss of trees affects the overall stability of the watershed. Trees and their root systems help hold soil in place, preventing landslides and maintaining the structural integrity of the watershed. When deforestation occurs, the absence of these root systems makes the soil more prone to erosion and degradation, further destabilising the watershed and impacting its ability to support biodiversity and clean water. It is difficult to quantify the effects of deforestation on watershed dynamics and aquatic ecosystems downstream, such as reservoirs, and separate these from the direct effects of climate change, let alone estimate future changes to guide management (Bleich et al., 2014; Coe et al., 2009). Deforestation, on the other hand, results in the loss of important ecosystem processes and has long-term repercussions on the hydrosphere of catchments by altering runoff patterns or amplifying nutrient input from these catchments.

5.9 Soil Degradation and Its Effects on Water Entities

Urban areas face a heightened risk of environmental contamination due to the substantial release of pollutants from various sources, such as residential, commercial, transportation, and industrial activities, as well as improper waste disposal practices. Urban soils have been found to be associated with different pollutants, such as heavy metals, polychlorinated biphenyls (PCBs), polycyclic aromatic hydrocarbons (PAHs), and phthalate esters/phthalic acid esters (PAEs) (Ferreira et al., 2019; Wang et al., 2018). Activities like the application of road traffic, sewage sludge, and the use of wastewater for park irrigation can contribute to soil pollution caused by heavy metals. Sites in close proximity to traffic routes often exhibit heightened levels of heavy metals, attributed to factors like tyre wear particles and weathered street surfaces, with a direct correlation to the frequency of vehicular traffic (Cowan et al., 2021; Nawrot et al., 2020). Because they are persistent in the environment, heavy metals have come under more and more scrutiny. Therefore, they may have a negative impact on soil microbial activity, disturb biogeochemical cycles, and alter how the food web functions. Heavy metals can be harmful to people's health when they come into contact with them through ingestion or contact with suspended dust in urban parks, lawns, and urban woodland soils.

6 Watershed Management

The imminent surge in urban water demands and the imperative to fortify urban water infrastructure against climate change are poised to catalyse a revolution in urban water management. It is crucial to reconsider traditional paradigms, shifting away from managing discrete elements of the urban water cycle in isolation. Instead, an integrated approach supported by all stakeholders is necessary to address escalating competition, conflicts, shortages, wastage, and the degradation of water resources. Urban sustainability heavily depends on effective water management (Barroca et al., 2023). In most modern cities, water services are provided by underground pipe networks that link customers to wastewater treatment facilities and, eventually, to water sources (Babovic et al., 2002). To effectively manage a watershed and its natural resources, such as water and vegetation, a comprehensive grasp of the spatio-temporal hydrological characteristics of the watershed is imperative. The estimation of runoff resulting from rainfall holds particular significance in watershed management (Kumar et al., 2021). Accurate data on both the volume and distribution of runoff is indispensable for formulating efficient water management strategies. This includes decisions regarding the placement of artificial water abstraction points, the construction of water storage facilities, and the implementation of measures to prevent soil erosion. By adopting a holistic and inclusive approach, urban watershed management not only improves water quality and reduces flooding risks but also enhances the ecological integrity and liability of urban environments. The flow chart (Fig. 2) shows how different components of the watershed are interconnected to each other, how environmental impacts occur, and how a sustainable approach could help in mitigating these impacts.

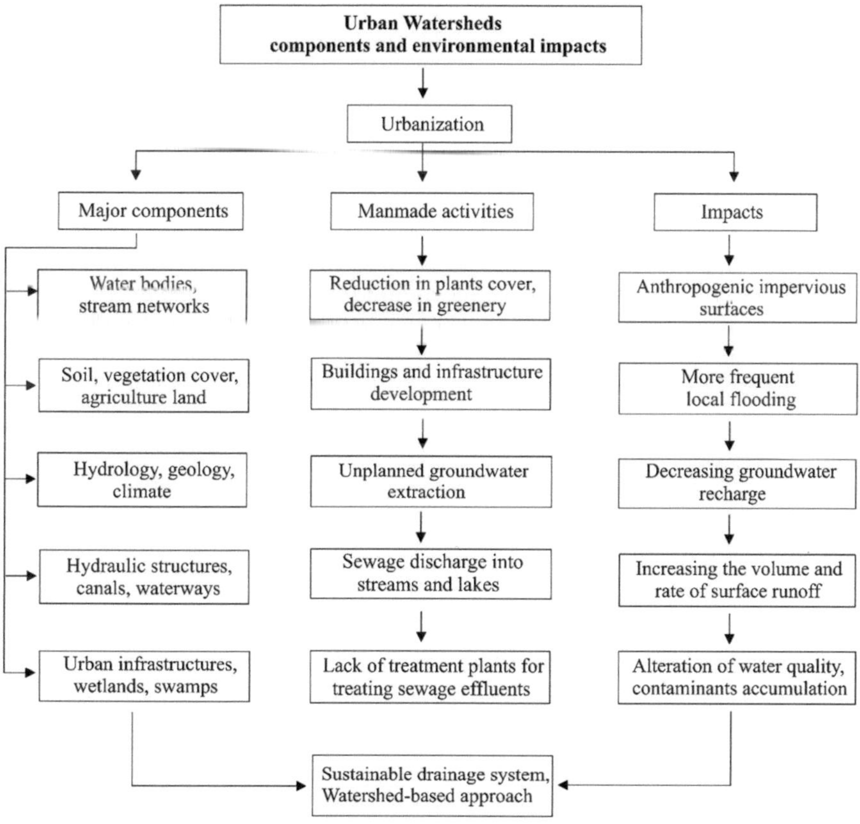

Fig. 2 Flow chart for reviewing watersheds, their components, environmental impacts, and sustainable approaches

6.1 Rainfall-Runoff Modelling

Precipitation runoff stands out as one of the most dynamic hydrologic variables in the hydrologic cycle (Sitterson et al., 2018). It serves as a crucial component for managing natural resources, specifically water from precipitation, and plays a pivotal role in mitigating the impact of natural disasters to achieve sustainable development. Surface runoff calculations play a major role in the planning of water conservation measures, the implementation of groundwater recharge structures, and the estimation of potential discharge for each watershed. Rainfall initiates intricate hydrological processes influenced by a multitude of factors, resulting in storm water generation. The transformation of rainfall into runoff involves two primary stages: firstly, hydrological losses occur through interception, depression storage, infiltration, and evapotranspiration; secondly, surface routing transforms the effective rainfall into overland flow, which is then collected by the drainage system.

The Precipitation-Runoff Modelling System (PRMS) is a physically based, distributed-parameter, deterministic modelling system designed to assess the impact of various climatic and agro-ecological configurations on stream-flow and watershed hydrology. The main objectives of PRMS are as follows: (i) make use of distributed meteorological data (precipitation, temperature, and solar radiation) and simulate hydrologic processes, such as evaporation, transpiration, runoff, infiltration, and interflow, based on the energy and water budgets of the soil zone, snowpack, and plant canopy; (ii) create models of hydrologic water budgets at the watershed scale over time spans ranging from days to centuries; (iii) establish linkages between natural resource management models and models from other scientific domains; and (iv) provide a modular architecture that enables the user to choose from a variety of hydrologic process algorithms in the PRMS module library by default. Rainfall-runoff models have developed during the past few decades, moving from more basic approaches with fewer parameters to more intricate and sophisticated ones. The process of choosing a suitable model involves first determining the issues and then addressing them using a sustainable strategy. Due to surplus nutrients and pesticides from agricultural lands being washed into streams by rain events, runoff is also a key factor in pollutant transfer (Tournebize et al., 2017). Knowing this information enables water resource managers to take runoff-related water contamination into account.

6.2 Sustainable Drainage System

Since the release of the Sustainable Drainage System (SuDS) manual in 2007, urban water management has undergone some changes. Instead of constructing traditional sewers, SuDS is now being used (Davis & Naumann, 2017). SuDS is considered to be among the best management practices in the USA and helps with water-sensitive urban design in Australia. In recent years, sustainable drainage systems have garnered growing public interest, resulting in positive impacts on water quality and quantity concerns. Additionally, these systems contribute to the incorporation of new recreational amenities into the urban landscape. Although precipitation is an unreliable source of water entering cities, it may also be quite beneficial. Surface water must flow through foul sewer systems in order to scour the pipes and keep them clear. In addition to watering gardens, rain also helps remove roadside dust.

SuDS provides a variety of advantages over traditional drainage methods (Green, 2019). It preserves and improves the biodiversity and water quality of urban streams. It defends people and property against flooding now and, possibly, in the future. It prevents nearby or far-off environmental pollution and safeguards urban watercourses against pollution brought on by unintentional spills. It allows long-term operations to be flexible enough to meet changing demands by minimising the use of natural resources. Improved drainage channels are created by smooth and straight channels which support to carry water more swiftly out of the metropolitan area. A decline in water quality, an increase in sediment yields and related pollutant fluxes, a decrease in hydrological amenity, and an increased risk of flooding are some of the cumulative repercussions of these changes. The SuDS approach places a strong emphasis on people and the receiving environment when making decisions about drainage design, construction, and maintenance. The idea can be broken down into three major groupings that seek to (i) minimise the quantity of runoff from the area by utilising source control measures and use source control strategies to lessen the amount of runoff from the location, (ii) decrease runoff velocity to allow settlements to infiltrate and filter water (permeable conveyance systems), and (iii) provide a treatment measure for the collected water prior to releasing water into land areas or into a water body. This may contribute to a region that comes within a watershed having a long-lasting and stable water capacity system.

6.3 *Watershed-Based Approach to Resource Management*

To achieve sustainable development through the planning and management of natural resources, it is essential to identify watershed boundaries within expansive drainage basins and subsequently prioritise these boundaries. The study of drainage basins is important in geomorphic studies because a watershed has a significant impact on how landforms develop (Rai et al., 2017). In the pursuit of sustainable development, a watershed serves as an ideal unit for the management of natural resources, including land and water, and for mitigating the impacts of natural disasters. A watershed-based approach to resource management in urban contexts is essential for addressing the multifaceted challenges of urbanisation, such as storm water management, water quality, and habitat protection. This approach involves comprehensive planning that integrates various sectors, including land use, housing, energy, and transportation, to avoid fragmented and inefficient resource management. Watershed management plans (WMPs) typically include flood assessment studies, water quality and storm water master plans, and the mapping of grey and green infrastructure. These plans help identify ecologically significant areas for protection and restoration, establish storm water management priorities, and set more stringent water quality design criteria for new developments.

Watershed management recognises the interconnections among land use, slope, geomorphology, lowland areas, uplands, and soil. Morphology, as defined by Agarwal (1998), involves the measurement and mathematical analysis of the Earth's surface configuration, as well as the shape, size, and dimensions of its landforms. Over the past few decades, geomorphology has witnessed significant attention towards the development of quantitative physiographic approaches to characterise the history and behaviour of surface drainage networks. The drainage system is a crucial element in watershed characterisation, and morphometric analysis provides a quantitative description of it.

6.4 Watershed Management Practices

Watersheds play a crucial role in the overall health of urban ecosystems, providing essential resources and supporting various life forms. A systematic approach is required to ensure sustainable management and conservation. The following are some important segments that could be followed as sequential steps in managing watershed characteristics sustainably.

Marking Out and Mapping Watershed Boundaries The first step in effective watershed management is to delineate its boundaries clearly. This involves identifying the geographical area that drains into a common water body, such as a river or a lake. Once boundaries are established, smaller drainage basins within the watershed must be identified. Modern mapping technologies, including geographic information system (GIS), play a vital role in accurately mapping and visualising these spatial features.

Creating Land Use and Land Cover Land use and land cover mapping provide insights into how human activities impact the watershed. This includes urban development, agriculture, and forestry. By studying changes in land use patterns over the years and decades, authorities can implement zoning regulations and land management practices that balance human needs with environmental conservation goals.

Classifying Soil Properties Soil composition plays a vital role in water retention, nutrient cycling, and overall ecosystem health. Mapping soil types and characteristics allows for informed decision-making in agriculture, construction, and conservation efforts. This knowledge helps prevent soil erosion and ensures sustainable land use practices.

Demarcating Natural and Man-Made Drainage Systems Understanding the drainage systems within a watershed is crucial for effective water management.

Mapping both natural and man-made drainage systems helps identify potential areas of flooding, erosion, and water quality concerns. This information is invaluable for designing infrastructure to mitigate these issues and promote sustainable development.

Identifying Watershed Resources A comprehensive inventory of the watershed's resources is essential for understanding its ecological dynamics. This includes vegetation, wildlife, wetlands, and other natural features. GIS technology enables the creation of detailed resource maps, aiding in the development of conservation strategies and sustainable land use practices.

Finding Erosion Areas Erosion poses a significant threat to watershed health. By identifying erosion-prone areas, such as stream banks and construction sites, preventive measures can be implemented. This may include vegetation restoration, erosion control structures, and sustainable construction practices to minimise negative impacts.

Evolution of Water Contaminants Impacting Water Quality Monitoring and assessing water quality are essential for establishing initial conditions for comparison over time. Regular sampling and analysis help identify trends, pollution sources, and potential threats to aquatic ecosystems. This baseline information serves as a foundation for developing water quality improvement strategies.

Source Apportionment of Pollutants Identifying both point and non-point sources of pollution is critical for effective pollution control. Mapping industrial discharge pipes, municipal storm water systems, failing septic systems, and illicit discharges provides a clear picture of potential pollution hotspots. This information guides regulatory measures and pollution prevention initiatives.

7 Conclusion

The significance of watershed management has gained increased attention in recent years due to the growing understanding that the long-term viability of water projects is a crucial condition for ensuring human welfare. The management of soil and water resources around the world is becoming more and more dependent on watershed initiatives. Research is required to make sure that new ventures learn from their forerunners' mistakes. Evaluation is challenging, however, due to the watershed initiatives' social and technical complexity. Communities may build on shared interests and overcome divisions when they design their management plans if they have a deeper knowledge of the similarities and variations in the watershed aims and concerns of stakeholders. Effective urban watershed management is crucial for mitigating the impacts of urbanisation on water resources and ensuring the sustainability of urban ecosystems. Building local partnerships and implementing best management practices such as non-point-source controls, stream restoration, and environmental education are essential for improving water quality and reducing pollution. Additionally, integrating nature-based solutions in urban water management can help maintain the natural hydrologic regime, mitigate flood and drought impacts, and support storm water and water supply management. Therefore, urban planners and policymakers need to prioritise urban watershed management and incorporate nature-based solutions into their strategies to ensure the long-term sustainability of urban water resources. By employing modern technologies and collaborative efforts, stakeholders can make informed decisions to ensure watersheds' long-term health and sustainability. Implementing these practices will contribute to the conservation of natural resources, the protection of water quality, and the overall well-being of ecosystems within the watershed.

Acknowledgements We thank the vice-chancellor of JNU, the director of CIMFR, and the director of INGV for providing the necessary support to carry out this work.

References

Abd El-Salam, M. M., & Abu-Zuid, G. I. (2015). Impact of landfill leachate on the groundwater quality: A case study in Egypt. *Journal of Advanced Research, 6*(4), 579–586.

Addisu, D. A., Husen, M. A., & Demeku, M. A. (2015). Determinants of adopting techniques of soil and water conservation in Goromti Watershed, Western Ethiopia. *Journal of Soil Science and Environmental Management, 6*(6), 168–177.

Agarwal, C. S. (1998). Study of drainage pattern through aerial data in Naugarh area of Varanasi district, UP. *Journal of the Indian Society of Remote Sensing, 26*, 169–175.

Ahmed, S. F., Mofijur, M., Nuzhat, S., Chowdhury, A. T., Rafa, N., Uddin, M. A., Inayat, A., Mahlia, T. M. I., Ong, H. C., Chia, W. Y., & Show, P. L. (2021). Recent developments in physical, biological, chemical, and hybrid treatment techniques for removing emerging contaminants from wastewater. *Journal of Hazardous Materials, 416*, 125912.

Alam, M. J., Alam, M. J. B., Rahman, M. H., Khan, S. K., & Munna, G. M. (2006). Unplanned urbanization: Assessment through calculation of environmental degradation index. *International Journal of Environmental Science & Technology, 3*, 119–130.

Azhari, H. E., Cherif, E. K., Sarti, O., Azzirgue, E. M., Dakak, H., Yachou, H., Esteves da Silva, J. C., & Salmoun, F. (2022). Assessment of surface water quality using the Water Quality Index (IWQ), Multivariate Statistical Analysis (MSA) and Geographic Information System (GIS) in Oued Laou Mediterranean Watershed, Morocco. *Water, 15*(1), 130.

Babovic, V., Drecourt, J. P., Keijzer, M., & Hansen, P. F. (2002). A data mining approach to modelling of water supply assets. *Urban water, 4*(4), 401–414.

Barnes, K. B., Morgan, J., & Roberge, M. (2001). *Impervious surfaces and the quality of natural and built environments*. Department of Geography and Environmental Planning, Towson University.

Barroca, B., Clemente, M. F., & Yang, Z. (2023). Application of "behind the barriers" model at neighbourhood scale to improve water management under multi-risks scenarios: A case study in Lyon, France. *International Journal of Environmental Research and Public Health, 20*(3), 2587.

Bekele, E., Toze, S., Patterson, B., & Higginson, S. (2011). Managed aquifer recharge of treated wastewater: Water quality changes resulting from infiltration through the vadose zone. *Water Research, 45*(17), 5764–5772.

Bernhardt, E. S., Band, L. E., Walsh, C. J., & Berke, P. E. (2008). Understanding, managing, and minimizing urban impacts on surface water nitrogen loading. *Annals of the New York Academy of Sciences, 1134*(1), 61–96.

Bhalla, B., Saini, M. S., & Jha, M. K. (2012). Characterization of leachate from municipal solid waste (MSW) landfilling sites of Ludhiana, India: A comparative study. *International Journal of Engineering Research and Applications, 2*(6), 732–745.

Bhat, R. A., Dar, S. A., Dar, D. A., & Dar, G. H. (2018). Municipal solid waste generation and current scenario of its management in India. *International journal of advance research in science and engineering, 7*(2), 419–431.

Bleich, M. E., Mortati, A. F., André, T., & Piedade, M. T. F. (2014). Riparian deforestation affects the structural dynamics of headwater streams in southern Brazilian Amazonia. *Tropical Conservation Science, 7*(4), 657–676.

Carlyle-Moses, D. E., Livesley, S., Baptista, M. D., Thom, J., & Szota, C. (2020). Urban trees as green infrastructure for stormwater mitigation and use. *Forest-Water Interactions*, 397–432.

Chariar, V., 2017. Waste management in India—Perspectives and challenges.

Chatterjee, P., Ghangrekar, M. M., & Rao, S. (2016). Low efficiency of sewage treatment plants due to unskilled operations in India. *Environmental Chemistry Letters, 14*, 407–416.

Coe, M. T., Costa, M. H., & Soares-Filho, B. S. (2009). The influence of historical and potential future deforestation on the stream flow of the Amazon River–Land surface processes and atmospheric feedbacks. *Journal of Hydrology, 369*(1–2), 165–174.

Cosgrove, W. J., & Loucks, D. P. (2015). Water management: Current and future challenges and research directions. *Water Resources Research, 51*(6), 4823–4839.

Covino, T. (2017). Hydrologic connectivity as a framework for understanding biogeochemical flux through watersheds and along fluvial networks. *Geomorphology, 277*, 133–144.

Cowan, N., Blair, D., Malcolm, H., & Graham, M. (2021). A survey of heavy metal contents of rural and urban roadside dusts: Comparisons at low, medium and high traffic sites in Central Scotland. *Environmental Science and Pollution Research, 28*, 7365–7378.

Cui, B., Wang, C., Tao, W., & You, Z. (2009). River channel network design for drought and flood control: A case study of Xiaoqinghe River basin, Jinan City, China. *Journal of environmental management, 90*(11), 3675–3686.

Dangar, S., Asoka, A., & Mishra, V. (2021). Causes and implications of groundwater depletion in India: A review. *Journal of Hydrology, 596*, 126103.

Davis, M., & Naumann, S. (2017). Making the case for sustainable urban drainage systems as a nature-based solution to urban flooding. In *Nature-based solutions to climate change adaptation in urban areas: Linkages between science, policy and practice* (pp. 123–137).

Desalegn, H., & Mulu, A. (2021). Mapping flood inundation areas using GIS and HEC-RAS model at Fetam River, Upper Abbay Basin, Ethiopia. *Scientific African, 12*, e00834.

Dreshaj, A., Shala, A., Hyseni, M., Millaku, B., & Gashi, A. (2022). Analysis of the impact of industrial waste on river water quality towards using the dynamics of land quality. *Journal of Ecological Engineering, 23*(4), 191–196.

Dodgen, L. K., Kelly, W. R., Panno, S. V., Taylor, S. J., Armstrong, D. L., Wiles, K. N., Zhang, Y., & Zheng, W. (2017). Characterizing pharmaceutical, personal care product, and hormone contamination in a karst aquifer of southwestern Illinois, USA, using water quality and stream flow parameters. *Science of the Total Environment, 578*, 281–289.

FAO (2001). https://www.fao.org/3/y4252e/y4252e08.htm. Accessed on 07 Aug 2023.

Fatta, D., Papadopoulos, A., & Loizidou, M. (1999). A study on the landfill leachate and its impact on the groundwater quality of the greater area. *Environmental Geochemistry and Health, 21*, 175–190.

Ferreira, C. S., Kalantari, Z., Salvati, L., Canfora, L., Zambon, I., & Walsh, R. P. (2019). Urban areas. In *Advances in chemical pollution, environmental management and protection* (Vol. 4, pp. 207–249). Elsevier.

Fletcher, T. D., Andrieu, H., & Hamel, P. (2013). Understanding, management and modelling of urban hydrology and its consequences for receiving waters: A state of the art. *Advances in Water Resources, 51*, 261–279.

Forte, A. M., & Whipple, K. X. (2018). Criteria and tools for determining drainage divide stability. *Earth and Planetary Science Letters, 493*, 102–117.

Geisen, S., Wall, D. H., & van der Putten, W. H. (2019). Challenges and opportunities for soil biodiversity in the anthropocene. *Current Biology, 29*(19), R1036–R1044.

Green, A. (2019). Sustainable drainage systems (SuDS) in the UK. In *Urban Stormwater and flood management: Enhancing the liveability of cities* (pp. 69–101).

Gregory, J. H., Dukes, M. D., Jones, P. H., & Miller, G. L. (2006). Effect of urban soil compaction on infiltration rate. *Journal of Soil and Water Conservation, 61*(3), 117–124.

Haghighatafshar, S., la Cour Jansen, J., Aspegren, H., & Jönsson, K. (2018). Conceptualization and schematization of mesoscale sustainable drainage systems: A full-scale study. *Water, 10*(8), 1041.

Hernandez-Mora, N., Llamas, R., & Cortina, L. M. (2001). Misconceptions in aquifer over-exploitation: Implications for water policy in southern Europe. In *Agricultural use of groundwater: Towards integration between agricultural policy and water resources management* (pp. 107–126). Springer.

Hlásny, T., Kočický, D., Maretta, M., Sitková, Z., Barka, I., Konôpka, M., & Hlavatá, H. (2015). Effect of deforestation on watershed water balance: Hydrological modelling-based approach. *Forestry Journal, 61*(2), 89–100. https://doi.org/10.1515/forj-2015-0017

Hoard, C. J., Haefner, R. J., Shuster, W. D., Pieschek, R. L., & Beeler, S. (2020). Full water-cycle monitoring in an urban catchment reveals unexpected water transfers (Detroit MI, USA). *JAWRA Journal of the American Water Resources Association, 56*(1), 82–99.

Hou, D., O'Connor, D., Nathanail, P., Tian, L., & Ma, Y. (2017). Integrated GIS and multivariate statistical analysis for regional scale assessment of heavy metal soil contamination: A critical review. *Environmental Pollution, 231*, 1188–1200.

Hough, M. (1995). *Cities and natural processes*. Routledge.

Jacobson, C. R. (2011). Identification and quantification of the hydrological impacts of imperviousness in urban catchments: A review. *Journal of Environmental Management, 92*(6), 1438–1448.

Jarden, K. M., Jefferson, A. J., & Grieser, J. M. (2016). Assessing the effects of catchment-scale urban green infrastructure retrofits on hydrograph characteristics. *Hydrological Processes, 30*(10), 1536–1550.

Katusiime, J., & Schütt, B. (2020). Linking land tenure and integrated watershed management—A review. *Sustainability, 12*(4), 1667.

Kaur, R., Wani, S. P., Singh, A. K., & Lal, K. (2012, May). Wastewater production, treatment and use in India. In *National report presented at the 2nd regional workshop on safe use of wastewater in agriculture* (pp. 1–13).

Kareem, H. A., Riaz, S., Sadia, H., & Mehmood, R. (2023). *Industrial-waste, types, sources, pollution potential, country wise comparisons. Waste problems and management in developing countries. Book.* https://doi.org/10.1201/9781003283621

Kaushal, S. S., & Belt, K. T. (2012). The urban watershed continuum: Evolving spatial and temporal dimensions. *Urban Ecosystems, 15*, 409–435.

Kerr, J. (2002). Watershed development, environmental services, and poverty alleviation in India. *World Development, 30*(8), 1387–1400.

Khan, M. Y. A., El Kashouty, M., Gusti, W., Kumar, A., Subyani, A. M., & Alshehri, A. (2022). Geo-temporal signatures of physicochemical and heavy metals pollution in groundwater of Khulais region—Makkah Province, Saudi Arabia. *Frontiers in Environmental Science, 9*, 800517.

Kumar, A., Kanga, S., Taloor, A. K., Singh, S. K., & Đurin, B. (2021). Surface runoff estimation of Sind River basin using integrated SCS-CN and GIS techniques. *HydroResearch, 4*, 61–74.

Langridge, R., & Ansell, C. (2018). Comparative analysis of institutions to govern the groundwater commons in California. *Water Alternatives, 11*(3).

Larson, K. L., & Lach, D. (2010). Equity in urban water governance through participatory, place-based approaches. *Natural Resources Journal*, 407–430.

Li, J., Zuo, Q., Yu, L., & Ma, J. (2023). A multi-dimensional relationship assessment framework for water resources, social economy and eco-environment: A case study of China's largest arid zone. *Environmental Impact Assessment Review, 102*, 107221.
Liao, H., Yen, J. Y., Guan, Y., Ke, D., & Liu, C. (2020). Differential responses of stream water and bed sediment microbial communities to watershed degradation. *Environment International, 134*, 105198.
Lienert, J., Monstadt, J., & Truffer, B. (2006). *Future scenarios for a sustainable water sector: A case study from Switzerland* (Vol. 40).
Liu, C. W., Lin, K. H., & Kuo, Y. M. (2003). Application of factor analysis in the assessment of groundwater quality in a Blackfoot disease area in Taiwan. *Science of the Total Environment, 313*(1–3), 77–89.
Mantzavinos, D., & Psillakis, E. (2004). Enhancement of biodegradability of industrial wastewaters by chemical oxidation pre-treatment. *Journal of Chemical Technology & Biotechnology: International Research in Process, Environmental & Clean Technology, 79*(5), 431–454.
Masoumi, Z. (2022). Flood susceptibility assessment for ungauged sites in urban areas using spatial modeling. *Journal of Flood Risk Management, 15*(1), e12767.
McGinnis, M. V. (1999). Making the watershed connection. *Policy Studies Journal, 27*(3), 497.
Motagh, M., Walter, T. R., Sharifi, M. A., Fielding, E., Schenk, A., Anderssohn, J., & Zschau, J. (2008). Land subsidence in Iran caused by widespread water reservoir overexploitation. *Geophysical Research Letters, 35*(16).
Mukherjee, A., Duttagupta, S., Chattopadhyay, S., Bhanja, S. N., Bhattacharya, A., Chakraborty, S., Sarkar, S., Ghosh, T., Bhattacharya, J., & Sahu, S. (2019). Impact of sanitation and socio-economy on groundwater fecal pollution and human health towards achieving sustainable development goals across India from ground-observations and satellite-derived nightlight. *Scientific Reports, 9*(1), 15193.
Napieralski, J. A., & Welsh, E. S. (2016). A century of stream burial in Michigan (USA) cities. *Journal of Maps, 12*(sup1), 300–303.
Nawrot, N., Wojciechowska, E., Rezania, S., Walkusz-Miotk, J., & Pazdro, K. (2020). The effects of urban vehicle traffic on heavy metal contamination in road sweeping waste and bottom sediments of retention tanks. *Science of the Total Environment, 749*, 141511.
Neary, D. G. (1998). Changing perceptions of watershed management from a retrospective viewpoint. *Proceedings RMRS., 13*, 167.
Nedeau, E. J., Merritt, R. W., & Kaufman, M. G. (2003). The effect of an industrial effluent on an urban stream benthic community: Water quality vs. habitat quality. *Environmental Pollution, 123*(1), 1–13.
Odeh, T., Mohammad, A. H., Hussein, H., Ismail, M., & Almomani, T. (2019). Over-pumping of groundwater in Irbid governorate, northern Jordan: A conceptual model to analyze the effects of urbanization and agricultural activities on groundwater levels and salinity. *Environmental Earth Sciences, 78*, 1–12.
Ouyang, W., Wang, Y., Lin, C., He, M., Hao, F., Liu, H., & Zhu, W. (2018). Heavy metal loss from agricultural watershed to aquatic system: A scientometrics review. *Science of the Total Environment, 637*, 208–220.
Patwardhan, A. D. (2017). *Industrial wastewater treatment*. PHI Learning Pvt. Ltd.
Pauchard, A., Aguayo, M., Peña, E., & Urrutia, R. (2006). Multiple effects of urbanization on the biodiversity of developing countries: The case of a fast-growing metropolitan area (Concepción, Chile). *Biological Conservation, 127*(3), 272–281.
Peng, J., Xu, Y., Cai, Y., & Xiao, H. (2011). Climatic and anthropogenic drivers of land use/cover change in fragile karst areas of Southwest China since the early 1970s: A case study on the Maotiaohe watershed. *Environmental Earth Sciences, 64*, 2107–2118.
Pickett, S. T., Cadenasso, M. L., Grove, J. M., Nilon, C. H., Pouyat, R. V., Zipperer, W. C., & Costanza, R. (2001). Urban ecological systems: Linking terrestrial ecological, physical, and socioeconomic components of metropolitan areas. *Annual Review of Ecology and Systematics, 32*(1), 127–157.

Platt, R. H., Beatley, T., Michaels, S., Goucher, N., & Fenstermacher, B. (2008). Urban stream restoration: Recovering ecological services in degraded watersheds. In *Growing greener cities: Urban sustainability in the twenty-first century* (pp. 127–151).

Potter, J. D., McDowell, W. H., Helton, A. M., & Daley, M. L. (2014). Incorporating urban infrastructure into biogeochemical assessment of urban tropical streams in Puerto Rico. *Biogeochemistry, 121*, 271–286.

Powell, J. W. (1879). *Report on the lands of the arid region of the United States: With a more detailed account of the lands of Utah. With maps*. US Government Printing Office.

Qi, S., Sun, G., Wang, Y., McNulty, S. G., & Myers, J. M. (2009). Streamflow response to climate and landuse changes in a coastal watershed in North Carolina. *Transactions of the ASABE, 52*(3), 739–749.

Rai, P. K., Mohan, K., Mishra, S., Ahmad, A., & Mishra, V. N. (2017). A GIS-based approach in drainage morphometric analysis of Kanhar River Basin, India. *Applied Water Science, 7*, 217–232.

Raphael, A., Ngaga, Y., & Lalika, M. (2023). Watershed degradation and water provision in Morogoro Municipality, Tanzania. *Ecohydrology & Hydrobiology, 23*(2), 272–279.

Richter, B. D., Mathews, R., Harrison, D. L., & Wigington, R. (2003). Ecologically sustainable water management: Managing river flows for ecological integrity. *Ecological Applications, 13*(1), 206–224.

Sadeghi, S. H., Hazbavi, Z., & Gholamalifard, M. (2019). Interactive impacts of climatic, hydrologic and anthropogenic activities on watershed health. *Science of the Total Environment, 648*, 880–893.

Samhouri, J. F., Shelton, A. O., Williams, G. D., Feist, B. E., Hennessey, S. M., Bartz, K., Kelly, R. P., O'Donnell, J. L., Sheer, M., Stier, A. C., & Levin, P. S. (2022). How much city is too much city? Biodiversity and ecosystem functioning along an urban gradient at the interface of land and sea. *Frontiers in Marine Science, 9*, 931319.

Sánchez, E., Colmenarejo, M. F., Vicente, J., Rubio, A., García, M. G., Travieso, L., & Borja, R. (2007). Use of the water quality index and dissolved oxygen deficit as simple indicators of watersheds pollution. *Ecological Indicators, 7*(2), 315–328.

Schueler, T. R. (1996). Crafting better urban watershed protection plans. *Watershed Protection Techniques, 2*(2), 329–337.

Schwarz, J., & Mathijs, E. (2017). Globalization and the sustainable exploitation of scarce groundwater in coastal Peru. *Journal of Cleaner Production, 147*, 231–241.

Sen, S., Shah, A., & Kumar, A. (2007). Watershed development programmes in Madhya Pradesh: Present scenario and issues for convergence. In *Forum for watershed research and policy dialogue*. Institute of Development Research.

Sewnet, A., & Abebe, G. (2018). Land use and land cover change and implication to watershed degradation by using GIS and remote sensing in the Koga watershed, North Western Ethiopia. *Earth Science Informatics, 11*, 99–108.

Seyam, M., Alagha, J. S., Abunama, T., Mogheir, Y., Affam, A. C., Heydari, M., & Ramlawi, K. (2020). Investigation of the influence of excess pumping on groundwater salinity in the Gaza Coastal Aquifer (Palestine) using three predicted future scenarios. *Water, 12*(8), 2218.

Shields, C. A., Band, L. E., Law, N., Groffman, P. M., Kaushal, S. S., Savvas, K., Fisher, G. T., & Belt, K. T. (2008). Streamflow distribution of non–point source nitrogen export from urban-rural catchments in the Chesapeake Bay watershed. *Water Resources Research, 44*(9).

Sincero, A. P., & Sincero, G. A. (2002). *Physical-chemical treatment of water and wastewater*. CRC Press.

Singh, A. K. (2017). Participatory watershed management: Some Indian experiences. *Regional Development: Theory and Practice, 4*, 1–32.

Sitterson, J., Knightes, C., Parmar, R., Wolfe, K., Avant, B. and Muche, M., 2018. An overview of rainfall-runoff model types.

Sobota, D. J., Harrison, J. A., & Dahlgren, R. A. (2009). Influences of climate, hydrology, and land use on input and export of nitrogen in California watersheds. *Biogeochemistry, 94*, 43–62.

Sonune, A., & Ghate, R. (2004). Developments in wastewater treatment methods. *Desalination, 167*, 55–63.

Tabari, H. (2020). Climate change impact on flood and extreme precipitation increases with water availability. *Scientific Reports, 10*, 13768. https://doi.org/10.1038/s41598-020-70816-2

Tabios, G. Q., III. (2023). Urban dimensions of floodings and holistic flood risk management: Case of Pasig-Marikina River basin in metro Manila. In *A better metro Manila? Towards responsible local governance, decentralization and equitable development* (pp. 347–370). Springer.

Tang, J., Liu, D., Shang, C., & Niu, J. (2024). Impacts of land use change on surface infiltration capacity and urban flood risk in a representative karst mountain city over the last two decades. *Journal of Cleaner Production, 454*, 142196.

Tournebize, J., Chaumont, C., & Mander, Ü. (2017). Implications for constructed wetlands to mitigate nitrate and pesticide pollution in agricultural drained watersheds. *Ecological Engineering, 103*, 415–425.

United Nations reports (2010). https://www.un.org/development/desa/pd/file/1942. Accessed on 03 Aug 2023.

Veale, B. & Cooke, S., (2018). Implementing integrated water management: illustrations from the Grand River watershed. In Integrated Water Management in Canada (pp. 28–45). Routledge.

Waikar, M. L., & Nilawar, A. P. (2014). Morphometric analysis of a drainage basin using geographical information system: A case study. *International Journal of Multidisciplinary and Current Research, 2*(2014), 179–184.

Wang, G., Mang, S., Cai, H., Liu, S., Zhang, Z., Wang, L., & Innes, J. L. (2016). Integrated watershed management: Evolution, development and emerging trends. *Journal of Forestry Research, 27*, 967–994.

Wang, L., Liu, M., Tao, W., Zhang, W., Wang, L., Shi, X., Lu, X., & Li, X. (2018). Pollution characteristics and health risk assessment of phthalate esters in urban soil in the typical semi-arid city of Xi'an, Northwest China. *Chemosphere, 191*, 467–476.

Wang, X., Yu, S., & Huang, G. H. (2004). Land allocation based on integrated GIS-optimization modeling at a watershed level. *Landscape and Urban Planning, 66*(2), 61–74.

Wolka, K., Tadesse, H., Garedew, E., & Yimer, F. (2015). Soil erosion risk assessment in the Chaleleka wetland watershed, Central Rift Valley of Ethiopia. *Environmental Systems Research, 4*(1), 1–12.

Worster, D. (2009). A river running west: Reflections on John Wesley Powell. *Journal of Cultural Geography, 26*(2), 113–126.

Yao, L., Chen, L., Wei, W., & Sun, R. (2015). Potential reduction in urban runoff by green spaces in Beijing: A scenario analysis. *Urban Forestry & Urban Greening, 14*(2), 300–308.

Yi, H., Güneralp, B., Kreuter, U. P., Güneralp, İ., & Filippi, A. M. (2018). Spatial and temporal changes in biodiversity and ecosystem services in the San Antonio River Basin, Texas, from 1984 to 2010. *Science of the Total Environment, 619*, 1259–1271.

Zalidis, G., Stamatiadis, S., Takavakoglou, V., Eskridge, K., & Misopolinos, N. (2002). Impacts of agricultural practices on soil and water quality in the Mediterranean region and proposed assessment methodology. *Agriculture, Ecosystems & Environment, 88*(2), 137–146.

Zhang, Y., Ya, X., Wang, R., Zou, Y., & Dong, X. (2022). Assessing and mapping human well-being for sustainable development amid drought and flood hazards: Dadu River Basin of China. *Environmental Science and Pollution Research, 29*(60), 90719–90737.

Zhou, L., Shen, G., Woodfin, T., Chen, T., & Song, K. (2018). Ecological and economic impacts of green roofs and permeable pavements at the city level: The case of Corvallis, Oregon. *Journal of Environmental Planning and Management, 61*(3), 430–450.

Zimmerman, C. L., & Civco, D. L. (2020). Impervious surface area: Effects. In *Fresh water and watersheds* (pp. 93–99). CRC Press.

Urban Heat Islands: A Manifestation of the Urban Thermal Environment

Richa Sharma and M. M. Anees

Abstract Urban heat island (UHI) is a well-documented phenomenon characterized by higher temperatures in urban areas compared to surrounding nonurban spaces. This discrepancy arises from various urban characteristics, such as altered landscape morphology and materials, resulting in changes in surface reflective and thermal properties. These alterations affect the energy budget of cities and contribute to atmospheric thermal changes. With rapid urbanization globally, UHI has become a significant concern due to its adverse impacts on human health, energy demands, urban flora and fauna, precipitation patterns, air quality, and climate change effects. Developing countries, experiencing rapid urban population growth without proportional infrastructural development, face heightened challenges from UHI effects. Despite a general scarcity of UHI studies in India, there has been an increasing trend, reflecting growing awareness. However, most research focuses on European, American, and recent Chinese cities, necessitating more India-specific studies, given its urbanization pace. Understanding land surface temperature (LST) changes, driven by Land Use/Land Cover (LULC) transformations, is crucial for managing urban microclimates. Remote sensing and geographic information system (GIS) techniques are pivotal in monitoring and understanding LULC changes, which are influenced by complex urban–rural linkages and ecological footprints. A conceptual framework illustrates the intricate interactions between different land use stages during urbanization, emphasizing the importance of studying LULC dynamics for effective urban planning and resource management.

Keywords Land surface temperature (LST) · Urbanization impacts · Land Use/Land Cover (LULC) · Remote sensing · Urban microclimates · Ecological footprints

---

R. Sharma (✉)
The Energy and Resources Institute (TERI), New Delhi, India
e-mail: richa.sharma@teri.res.in

M. M. Anees
Urban Studies, School of Global Affairs, Dr. B.R. Ambedkar University Delhi, New Delhi, India

P. K. Joshi et al. (eds.), *Geospatial Science for Urban Ecosystems*, The Urban Book Series, https://doi.org/10.1007/978-3-032-07155-2_6

1 UHI: The Phenomenon and Its Impacts

Thermal environment in urban areas is typically characterized by the phenomenon of urban heat island (UHI). UHI refers to the condition where the air and/or surface temperatures of urban areas are higher in comparison to the surrounding nonurban spaces (Oke et al., 2017). This phenomenon is usually a result of one or many of the typical characteristics of urban centers. These include altered morphology of the landscape, urban materials (Stache et al., 2022), and the unique geometry of urban areas, to mention a few. These alterations, in turn, bring about changes in the reflective and thermal properties of the surfaces, resulting in higher heat capacity and/or lower albedo, for instance. These changes in thermal properties of the urban landscape further cause variations in the energy budget of the city. Thermal changes in urban areas are not restricted to surfaces, but cities also experience thermal changes in the atmosphere due to causes such as higher greenhouse gas (GHG) emissions, accumulation of air pollutants, and other factors, including urban geometry that influences wind movement and increased anthropogenic heat generation due to higher population density. The phenomenon of UHI is universal in nature, irrespective of the size of the city, with variations only in the strength or intensity of the UHI and its impacts on the metabolism of the city.

UHI is of increasing concern, given the fact that the world is rapidly urbanizing at a fast pace. This has twofold implications: *first*, that more and more areas will urbanize, creating more UHIs globally, and *second*, that a greater number of people will be exposed to the UHI effects. UHI can adversely impact human health (Chaston et al., 2022), increase energy demands for cooling, impact the life cycle of urban flora and fauna, influence the precipitation patterns over the city, worsen the air pollution, and exacerbate the climate change impacts on the city population. Hence, there has been an increasing concern among the urban planners and practitioners to consider the UHI effect into development plans and while designing the climate change adaptation strategies. Like climate change, UHI also has a differential impact on the population. Thus, it becomes important to measure the UHI intensity and identify the vulnerable pockets and population in the city.

A comprehensive review of literature on UHI studies in India reveals a noticeable deficiency in research efforts. Nonetheless, when comparing publication statistics between India-specific studies and those encompassing international research, both show an upward trajectory. Despite this positive trend, there remains a substantial gap between the volume of publications from India and the total number of publications worldwide. The majority of UHI studies globally have predominantly focused on cities in Europe (Lauwaet et al., 2024) and the United States (Chakraborty et al., 2020; Kim et al., 2018). Cities like London (Benjamin et al., 2021; Jones & Lister, 2009; Kolokotroni et al., 2006), Paris (De Ridder et al., 2016; Tremeac et al., 2012), various locations in Spain (Barrao et al., 2022; Cuerdo-Vilches et al., 2023), and select German cities (Kabisch et al., 2023; Schwarz et al., 2012) have been extensively researched for their UHI effects, impacts, and mitigation strategies. Similarly, cities across the United States have also been extensively studied in this regard. However, UHI research in Asian cities, including those in China and India, has predominantly emerged as a more recent trend. Given the anticipated urban expansion in India, it becomes increasingly imperative to delve into UHI phenomena with a specific focus on understanding the unique characteristics and challenges posed by Indian urban environments.

2 Land Surface Temperature and Land Use/Land Cover

Land surface temperature (LST) refers to the average radiative temperature of the Earth's surface, representing its thermal state. It plays a pivotal role in determining the thermal characteristics of a specific area. Physical processes such as conduction and convection contribute to the development of LST, thereby influencing the overall thermal behavior of the land surface. Unlike air temperature, LST is measured remotely by satellites, capturing the temperature of the Earth's surface through the atmosphere. LST serves as a crucial indicator of surface hotness upon direct contact. In the realm of remote sensing, it holds significant importance as it reflects changes in land surface conditions related to Land Use/Land Cover (LULC) patterns. This temperature parameter directly impacts the surface energy budget (Bastiaanssen et al., 1998a, 1998b) and surface water hydrology by influencing the exchange of mass, energy, and momentum at the Earth's surface (Kerchove et al., 2013; Lu et al., 2012). Understanding LST dynamics is essential for assessing environmental changes and their implications on terrestrial ecosystems and human activities.

Cities are profoundly influenced by multiple local-scale factors, which are pivotal in shaping their environments due to extensive human activities and their consequential impacts (Mallick et al., 2013). Among these, the most significant are changes in land use (Drummond & Loveland, 2010) and land cover (Drohan et al., 2012), which occur globally (Lambin et al., 2001; Meyer & Turner, 1992; Turner et al., 1990). The alterations in LULC have direct and indirect consequences on global ecosystems. For instance, forest resources are fragmented and degraded as they are converted into urban or agricultural land (Freitas et al., 2007). These changes are increasingly recognized as fundamental drivers of climate change across various scales, from local to regional (Mahmood et al., 2010) and even global (Houghton, 1999; Turner et al., 1990). Understanding LULC dynamics is crucial for environmental monitoring, natural resource management, and climate change research. It provides essential inputs for science-based policymaking aimed at addressing these complex environmental challenges.

Monitoring, mapping, studying, and documenting changes in LULC are essential tasks for effectively managing natural resources (Loveland & DeFries, 2004) and addressing the evolving microclimate of urban areas (Carlson & Arthur, 2000; Vadakkuveettil & Grover, 2023). By studying LULC changes, we gain insights into both current conditions and post-change scenarios. This comprehensive understanding enriches our knowledge of the ongoing transformation of land within cities, elucidating the locations and mechanisms driving these changes. A detailed analysis of LULC changes not only provides snapshots of urban development but also enhances our comprehension of the underlying processes shaping urban landscapes. This knowledge is crucial for informed decision-making in urban planning, resource allocation, and environmental management strategies. A combination of remote sensing data, providing monitoring capability of changes over a long term, and GIS techniques, providing capabilities to classify the types of changes on the ground into interpretable processes, is most widely used in understanding the LULC changes in urban areas.

Urbanization as a form of land use not only directly impacts LULC changes but also influences them indirectly through intricate urban–rural linkages, which are directly linked to "ecological footprints." Figure 1 depicts the complex interactions among different land uses during the urbanization process. The transformation

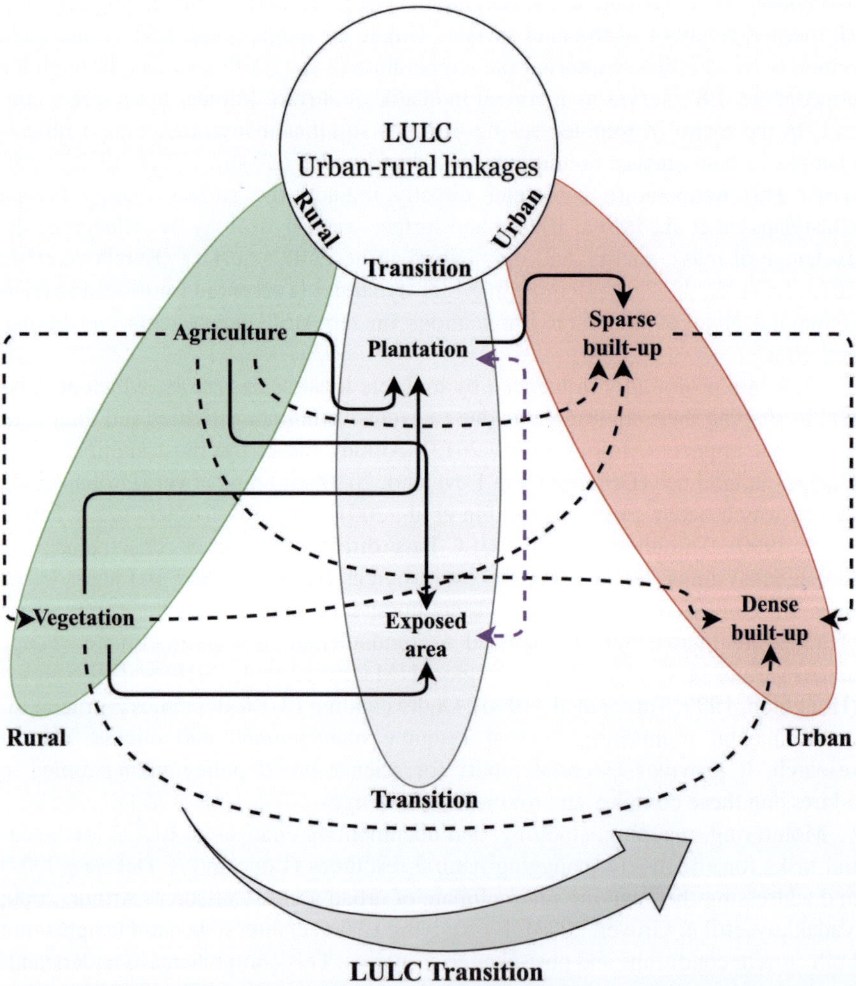

Fig. 1 Urbanization: the process and the dynamic implications for the overall LULC pattern

involves agriculture transitioning into vegetation such as scrub, plantation, and exposed areas as part of the initial phase. Simultaneously, agriculture directly contributes to the urban phase, characterized by built-up areas. In rural areas, vegetation like forests also undergoes transformation, shifting toward transitional phases like plantation and exposed areas, eventually leading to urban phases with sparse or dense built-up areas. This flow from agricultural to vegetation cover ultimately transitions into the urban landscape. The transitional phase illustrates further conversions, such as from plantation areas to built-up zones. Within each stage, transformations occur between exposed areas and plantation zones. The culmination of this process is observed in the urban stage, where sparse built-up areas expand, absorbing land from transitional phases (plantation and exposed areas) as well as rural phases (agriculture and vegetation). Among these categories, sparse built-up areas also contribute to the development of dense built-up zones. Dense built-up areas represent the final category where all transformations from other land uses converge, marking the completion of the urbanization process. Understanding these intricate dynamics is crucial for effectively managing urban growth, planning sustainable development, and mitigating the environmental impacts associated with LULC changes driven by urbanization.

3 Categorizing and Measuring UHIs

The urban heat island (UHI) phenomenon was first documented in 1818 by Luke Howard, who observed elevated temperatures in London compared to surrounding rural areas (Gartland, 2011). Rao (1972) pioneered the use of satellite data for studying UHI, laying the groundwork for subsequent research in this field. Since then, numerous researchers have investigated UHI effects in cities worldwide. However, the majority of these studies have predominantly focused on developed regions, such as North America (Bornstein, 1968; Bornstein & Lin, 2000; Buyantuyev & Wu, 2010; Childs, 2005; Gedzelman et al., 2003; Sarangi et al., 2021; Streutker, 2003; Weng et al., 2011) and Europe (Fabrizi et al., 2010; Florenzio et al., 2022; Kolokotroni et al., 2006; Pal et al., 2012; Tomlinson et al., 2011; Vez et al., 2000), highlighting a significant gap in research focusing on developing nations in Asia. The UHI effect often leads to various repercussions, including increased energy consumption for cooling purposes, particularly in tropical countries characterized by hot and humid climates. This heightened demand for cooling exacerbates energy consumption and greenhouse gas emissions. Furthermore, UHI contributes to discomfort among residents and poses health risks, particularly affecting vulnerable populations such as the elderly. The transformation of LULC exacerbates UHI effects by increasing the prevalence of impermeable surfaces and reducing green vegetation cover. This alteration disrupts the surface energy budget, leading to further temperature rises in urban areas (Senanayake et al., 2013). Additionally, anthropogenic heat sources from industrial and residential activities intensify this imbalance, exacerbating the heat island effect. Understanding and mitigating UHI effects are crucial for sustainable urban development and climate resilience strategies. Research efforts need to encompass a broader geographic scope, including comprehensive studies in developing nations, to address the varied impacts of UHI across different socioeconomic and environmental contexts.

UHIs are broadly categorized into two and studied as either surface UHI (SUHI) or atmospheric UHI (AUHI). AUHI is further categorized into boundary layer UHI (BLUHI) and canopy layer (CLUHI) (Arnfield, 2003; Oke, 1976) (Fig. 2). SUHI, as the name suggests, is the UHI dealing with the surfaces in the urban areas. BLUHI refers to the UHI effect in the layer above the average height of buildings, i.e., in the boundary layer of the city's atmosphere. BLUHI is a mesoscale phenomenon resulting from the combination of a number of factors pertaining to urban material and geometry, which include warming up due to urban pollutants, weakened wind movements due to the roughness of urban structures, and heat radiating from urban surfaces. CLUHI refers to the atmospheric UHI developing due to temperature differences in the canopy layer, the layer of atmosphere below rooftops and between the buildings, where humans thrive. The main reason for the development of this phenomenon is the obstructed sky view between different urban structures that trap the surface radiations (Giridharan et al., 2004). Anthropogenic heat sources also contribute to the development of CLUHI.

The UHI effect arises as natural surfaces are progressively replaced by built surfaces and materials that retain higher temperatures than natural counterparts. For example, SUHI persists in both the day and night, typically peaking in intensity during daytime and the summer months. In contrast, atmospheric UHIs tend to be weaker or absent during the day, reaching their peak intensity at night and during winter months. SUHIs demonstrate a greater variability in intensity between day and night compared to atmospheric UHIs. This variation exhibits the complex dynamics of heat distribution within urban areas, influenced by factors such as solar radiation absorption, surface materials, and urban morphology. Understanding these variations is essential for developing targeted strategies to mitigate UHI effects and enhance urban climate resilience across different seasons and times of day.

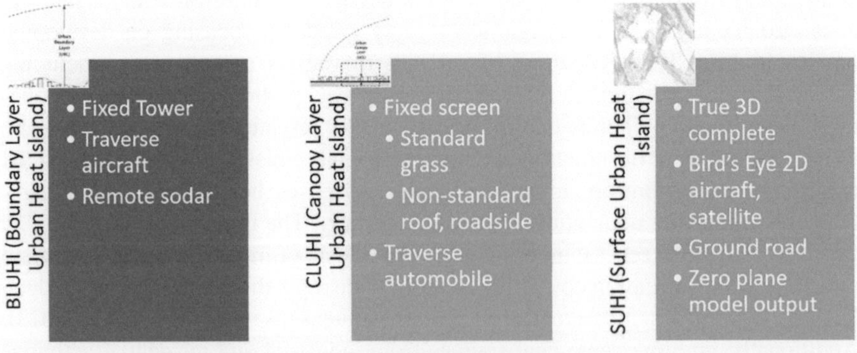

Fig. 2 Different types of UHIs and their measurements

The high concentration of surface temperatures in UHI zones is associated with the occurrence of heatwaves in urban areas (Chen et al., 2023; Chen et al., 2006). This results in abnormally hot weather periods that create uncomfortable conditions for the habitats, severely impacting the health of the citizens, resulting in increased mortality and morbidity (EPA, 2008). It is, therefore, important to measure the intensity of UHIs. BLUHI is assessed through the use of fixed towers, traverse aircraft, or remote sonars, providing valuable insights into the vertical extent and structure of the UHI, as well as the distribution of pollutants within it (Emeis, 2009). These methods enable researchers to examine how heat and pollutants vary vertically across the urban boundary layer, contributing to a deeper understanding of urban climate dynamics and air quality impacts. Light detection and ranging (LiDAR) data can also be used to analyze the depth of BLUHI (Pal et al., 2012). SUHI is measured using thermal imagery or infrared thermometers. These measurements could be done at various scales, ranging from broad city scales to street scale, using the thermal satellite data (Matson et al., 1978; Price, 1979; Voogt & Oke, 2003) or the infrared cameras. Thermal remote sensing employs contactless instruments to detect thermal radiation and estimate surface temperatures. However, this method is constrained by limited temporal coverage and susceptibility to cloud cover. Data can be collected via satellites, aircraft, or ground-based sensors. Monitoring CLUHI involves fixed screens placed in urban and rural areas, utilizing standard grass or nonstandard roof and roadside setups. Alternatively, mobile platforms like cars or bicycles are used for traverse measurements. In contrast, atmospheric UHI monitoring relies on in situ techniques that offer higher temporal resolution but suffer from lower spatial resolution. Satellite data plays a crucial role in comprehending both the response and physical mechanisms underlying SUHI formation. It provides a broader perspective on heat patterns across urban areas, supplementing ground-based measurements with valuable insights into urban climate dynamics and heat island development processes. Even though there are many ways of measuring different types of UHIs, what is most crucial is the selection of the urban and rural sites due to the relative nature of this phenomenon.

4 Environmental Parameters and UHI Measurement

This chapter delves into the dynamics of SUHI (SUHI=LST_{urban} - LST_{rural}), which is primarily assessed using LST. As mentioned earlier, LST represents the average radiative temperature of the Earth's surface, playing a pivotal role in determining its thermal characteristics. It governs how heat is distributed across the area of study, influenced by physical processes like conduction and convection. As heat is a critical factor shaping environmental conditions, LST serves as a key proxy for assessing thermal conditions. Environmental quality, a composite concept influenced by various factors, is intricately linked to LST. Beyond LST, the important indicators for monitoring environmental quality include metrics such as greenness, imperviousness (the proportion of impermeable surfaces), moisture intensity, and the extent of bare ground. These indicators collectively provide insights into the ecological health and resilience of an area, reflecting how urbanization and land use impact local climate and environmental conditions.

4.1 LST Estimation

LST serves as a gauge of the surface temperature of the Earth when physically contacted. In remote sensing, LST is observed by satellites passing through the atmosphere to capture ground-level temperatures. Unlike air temperature, LST provides crucial insights into changes in land surface conditions, particularly related to LULC patterns. It plays a pivotal role in regulating the surface energy budget, influencing how energy is exchanged and distributed across landscapes (Bastiaanssen et al., 1998a, 1998b). LST influences interactions involving mass, energy, and momentum at the Earth's surface (Kerchove et al., 2013; Lu et al., 2012). Besides assessing the indirect impacts of LULC changes on the environment, LST is crucial for studying phenomena that directly affect the thermal environment. These include monitoring UHI, assessing forest fires (Çolak & Sunar, 2023) and volcanic eruptions, and studying permafrost dynamics (Langer et al., 2010; Weng et al., 2004; Westermann et al., 2011). Beyond localized applications, LST plays a pivotal role in large-scale studies. It contributes to understanding how surface warming influences global weather patterns, the impact of greenhouse gases on Earth's thermal environment, and the response of glaciers to warming trends. Additionally, LST is essential for assessing global ecosystem dynamics, including vegetation patterns and wildlife distribution. Researchers also utilize LST data to calculate crop water requirements in different agricultural regions. Monitoring LST is critical due to its significant influence on living organisms, including human health. The retrieval of LST typically involves three main steps: (i) computing radiance, (ii) measuring brightness temperature, and (iii) converting this temperature into land surface temperature.

4.1.1 Calculation of Atmospheric Temperature or Thermal Radiance

To retrieve LST from satellite data, it is essential to first determine the reflectance of the thermal bands. Reflectance is calculated based on the spectral radiance of the thermal band, which represents the energy flux per solid angle emitted from a unit surface area in a specific direction. Spectral radiance refers to the brightness observed in a particular wavelength (or band) as detected by the sensor. This spectral radiance is quantified using specific equations tailored for Landsat datasets.

$$L_\lambda = \left[\frac{L_{max} - L_{min}}{QCal_{max} - QCal_{min}} \times \left(QCal - QCal_{min} \right) \right] + L_{min}$$

where L_λ is the spectral radiance at the sensor's aperture; L_{max} and L_{min} (W/(m^{2}srμm)) are the spectral radiances for the thermal band scaled to $QCal_{min}$ and $QCal_{max}$ values, respectively; and QCal represents the quantized calibrated pixel value in digital number (DN).

The equation is also expressed as follows, where M_L is the radiance multiplicative scaling factor for the band (RADIANCE_MULT_BAND_n from the metadata), Q_{cal} is the L1 pixel value in DN, and A_L is the radiance additive scaling factor for the band (RADIANCE_ADD_BAND_n from the metadata).

$$L_\lambda = M_L \times Q_{Cal} + A_L$$

4.1.2 At-Sensor Brightness Temperature Measurement

The subsequent step involves calculating the at-sensor brightness temperature from the radiance data. Brightness temperature is an estimation of the temperature assuming that the target behaves like a blackbody, which emits the radiance detected by the sensor. This temperature is often referred to as "blackbody temperature" because it assumes that the target has perfect emissivity (unity). The calculation involves applying the inverse of Planck's function to the measured radiation, expressed as follows:

$$T_B = \frac{K_2}{\ln\left(\frac{K_1}{L_\lambda} + 1\right)}$$

where T_B is the brightness temperature in Kelvin; K_1 and K_2 are prelaunch satellite calibration constants measured in W/(m^{2}srμm) and Kelvin, respectively; and L_λ (W/(m^{2}srμm)) is the radiance of the thermal band, as measured previously.

4.1.3 Calculating LST

There are two widely used methods for retrieving LST from thermal infrared bands: the mono-window algorithm (Qin et al., 2001) and the split-window algorithm (Sobrino et al., 2004). Qin's mono-window algorithm necessitates the prior knowledge of three variables: emissivity, transmittance, and effective mean atmospheric temperature. The computation of LST using the mono-window method is expressed by the following equation:

$$T_S = \{a(1-C-D) + [b(1-C-D) + C + D] \times T_B - D \times T_a\}C$$

where Ts is the LST, a and b are the model constants with fixed values for different datasets, T_B is the at-sensor brightness temperature computed previously, and Ta refers to the effective mean atmospheric temperature (Qin et al., 2001). The C and D parameters were derived from emissivity and transmittance (derived from water vapor) as follows:

$$C = \varepsilon \times \tau$$

$$D = (1-\tau)(1+(1-\varepsilon)) \times \tau$$

where ε is the measured emissivity, and τ is the transmittance computed from the water vapor content, as explained below. The third variable (Ta) is computed as follows using the near-surface air temperature (T_0 in K). This equation was developed by Qin et al., (2001) and is recommended for tropical countries.

$$Ta = 17.9769 + 0.91715T_0$$

Atmospheric transmittance is derived from the water vapor content, using profiles of high and low air temperatures (Table 1). The atmospheric temperature near the surface is set at 35 °C for the high temperature profile and 18 °C for the low temperature profile.

The water vapor content for this is computed using near-surface air temperatures and relative humidity (RH) in percentage:

$$w = 0.0981\left\{10 \times 0.6108 \times \exp\left[\frac{17.27(T_0 - 273.15)}{237.15 + (T_0 - 273.15)}\right] \times RH\right\} + 0.169$$

Emissivity Estimation

Brightness temperature provides a temperature estimate under the assumption that all objects behave like ideal blackbodies, which is not realistic. Since real objects have varying emissivity values, it is essential to correct for emissivity to accurately determine the LST. This correction is crucial for obtaining a true representation of thermal distribution in the image. The Normalized Difference Vegetation Index (NDVI) provides us with a globally acceptable biophysical parameter, which governs emissivity. The NDVI values calculated can be used to generate emissivity images through conditional statements. Emissivity values corresponding to NDVI, ranging from 0.157 to 0.727, can be calculated using the equation provided below (Zhang et al., 2006):

$$\varepsilon = 1.0094 + 0.047 \times \ln(\text{NDVI})$$

Table 1 Transmittance calculation

	τ equation	
Water vapor (w)	High air temperature profile (35 °C)	Low air temperature profile (18 °C)
0.4–1.6	0.974290–0.08007w	0.982007–0.09611w
1.6–3.0	1.031412–0.11536w	1.053710–0.14142w

Table 2 Emissivity estimation from NDVI

NDVI	Emissivity
Less than −0.18	0.985
−0.18–0.157	0.955
Greater than 0.727	0.99

For NDVI values out of this range, different emissivity values are assigned to three different ranges. These values are summarized in the table below (Table 2).

Split-Window Algorithm

The split-window algorithm takes advantage of the differential atmospheric absorption in two adjacent thermal infrared (TIR) channels to retrieve LST. Split-window algorithm could be linear or nonlinear.

This algorithm uses atmospheric transmittance and emissivity as inputs.

The linear algorithm is expressed as follows:

$$LST = A_0 + A_1 T_i + A_2 \left(T_i - T_j\right),$$

while the form of the nonlinear algorithm is expressed as follows:

$$LST = A_0 + A_1 T_i + A_2 \left(T_i - T_j\right) + A_3 \left(T_i - T_j\right)^2$$

where A_k (k = 0, 1, 2, 3) represent the coefficients that depend mainly on the land surface emissivity of the two TIR channels, atmospheric water vapor, and viewing zenith angle for a unique instrument.

LST Zones

Six LST zones have been mapped using the mean (μ) and standard deviation (SD) of LST values: <μ-2SD, <μ-SD AND >μ-2SD, <μ AND >μ SD, <μ+SD AND >μ, <μ+2SD AND >μ+SD and >μ+2SD. The zones with LST higher than the average are the representatives of SUHI. Thus, areas under zones IV, V, and VI are SUHI-affected areas with increasing influence of SUHI moving to higher zones.

4.2 Data Available

The infrared segment of the electromagnetic spectrum typically spans from 0.7 to 1000 μm. In terrestrial remote sensing, the range of 3–35 μm is specifically referred to as the thermal infrared region. Data collection focuses on atmospheric windows, which are spectral bands with minimal absorption, ensuring optimal transmission of infrared radiation. Several notable thermal infrared sensors include the Advanced Very High-Resolution Radiometer (AVHRR) aboard the National Oceanic and Atmospheric Administration/Television Infrared Observation Satellite (NOAA/TIROS) program's operational meteorological satellites, Moderate-Resolution Imaging Spectroradiometer (MODIS), Advanced Spaceborne Thermal Emission and Reflection Radiometer (ASTER), Landsat Thematic Mapper (TM), Enhanced Thematic Mapper (ETM+), Operational Land Imager (OLI), and Advanced Topographic Laser Altimeter System (ATLAS). These sensors play critical roles in various remote sensing applications, capturing thermal emissions from the Earth's surface for diverse scientific and operational purposes.

5 Case Studies

5.1 Delhi

The eight primary LULC classes—agriculture, forest, plantation, scrub, sparse built-up area, dense built-up area, exposed area, and water—mapped between 1998 and 2011 indicate a decline in agriculture from 1998 to 2003, followed by a slight recovery from 2003 to 2011, with significant dynamics observed within agricultural classes (Fig. 3). Losses in Kharif crops are predominantly due to their conversion into sparse built-up areas (28 km^2), while Zaid crops saw losses primarily to sparse built-up areas (10 km^2). The trends observed during 2003–2011 are generally consistent with previous years, albeit with a few variations. This comprehensive approach provides insights into the evolving urban landscape of Delhi and underscores the importance of monitoring LULC changes for sustainable urban planning and environmental management strategies.

The NDVI analysis reveals a decline in green cover, particularly noticeable along the periphery (Fig. 4(a)). Areas like the southwestern parts of Najafgarh and Dwarka, northwestern suburbs of Rohini, and industrial zones in Bawana and Narela exhibit reduced vegetation due to expanding residential and industrial land uses. Normalized Difference Water Index (NDWI) images depict a decrease in surface moisture intensity across the city (Fig. 4(b)). Agricultural lands in the periphery seem to maintain better moisture levels compared to the central urban areas. The central and southern ridge areas show slight improvements in moisture, contrasting with significant declines observed in new industrial zones in the north and minor decreases in

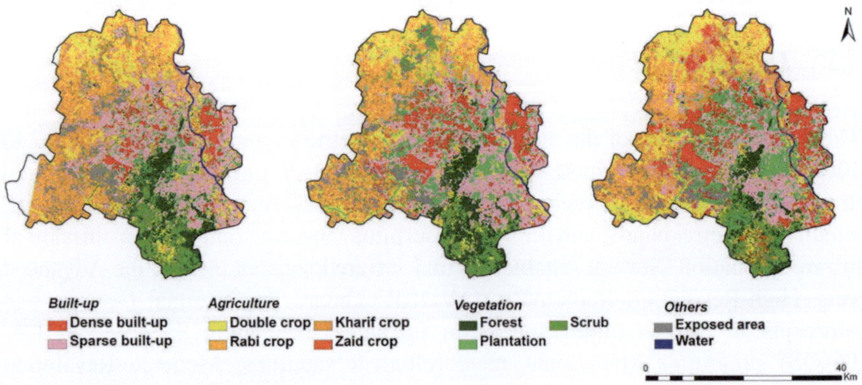

Fig. 3 Land Use/Land Cover results for 1998, 2003, and 2011

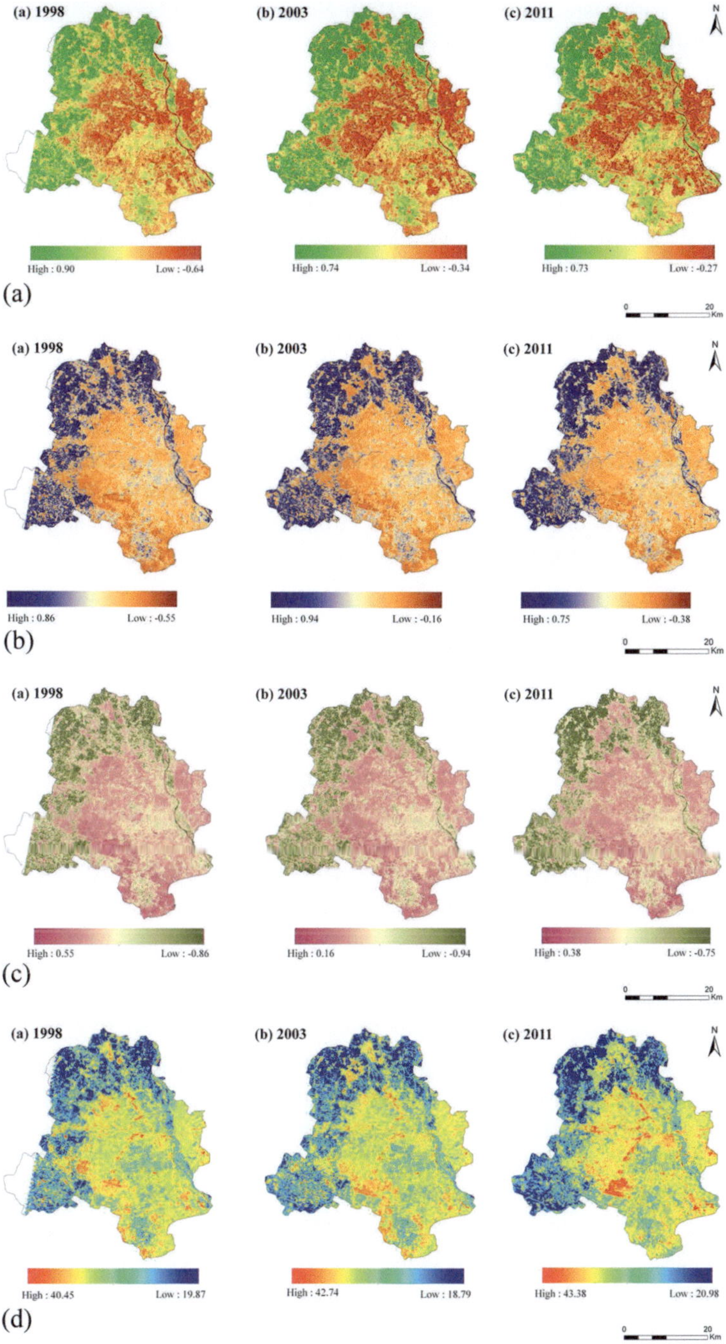

Fig. 4 (**a**) NDVI, (**b**) NDWI, (**c**) NDBI, and (**d**) LST maps for 1998, 2003, and 2011

expanding southwestern urban areas. High Normalized Difference Built-up Index (NDBI) values indicate extensive built-up areas, prominently seen in the southern ridge and expanding over time across the city (Fig. 4(c)). The trend in NDBI highlights increasing urban sprawl and built-up surfaces throughout the study period. This analysis provides an insight into how urbanization and industrialization have impacted vegetation cover and moisture levels in Delhi, demonstrating the utility of remote sensing–derived indices in understanding urban environmental dynamics.

The distribution of LST over the 3 years indicates a gradual spread of high LST values, extending outward from the city centers (Fig. 4(d)). In 2011, notably elevated LST values are observed in major industrial zones and particularly at the Indira Gandhi International (IGI) airport, as well as in central areas, including Old Delhi. Newly developed built-up areas, such as Bawana and Narela in the north and Dwarka and Najafgarh in the southwest, show a progressive increase in LST due to expanding urbanization in these regions over the years. Specifically, temperature rises of approximately 5–15 °C are recorded in Bawana and Narela, while Dwarka and Najafgarh experienced an increase of about 2 °C. This analysis underscores the spatial dynamics of urban heat islands and highlights how rapid urban growth contributes to elevated surface temperatures in different parts of Delhi.

5.2 Kolkata

Over the years, the delicate wetland ecosystem surrounding Kolkata has undergone significant transformation due to urban development pressures. Based on the processing of Landsat satellite imagery spanning 1989, 2006, and 2010, LST and associated surface characteristics are analyzed (Fig. 5). The LULC maps reveal a substantial urban sprawl of 108.94 km^2 between 1989 and 2010. These urban changes have significantly elevated surface temperatures across the study region. Satellite-derived temperature estimates indicate a clear warming trend, with mean surface temperatures increasing from 27.36 °C in 1989 to 30.025 °C in 2006 and further to 33.023 °C in 2010. These changes in LST align closely with the observed urbanization patterns in Kolkata and its surroundings. Analysis of biophysical parameters revealed noteworthy correlations: LST exhibits a positive relationship with NDBI, indicating higher temperatures in built-up areas; conversely, LST shows a negative correlation with NDVI, suggesting cooler temperatures in vegetated regions. Furthermore, NDBI and NDWI demonstrate a perfectly negative correlation, highlighting the contrast between built-up areas and water bodies. This comprehensive investigation underscores the profound impact of urban expansion on local climate and environmental conditions in Kolkata, emphasizing the critical need for sustainable urban planning strategies to mitigate these effects on the fragile wetland ecosystem.

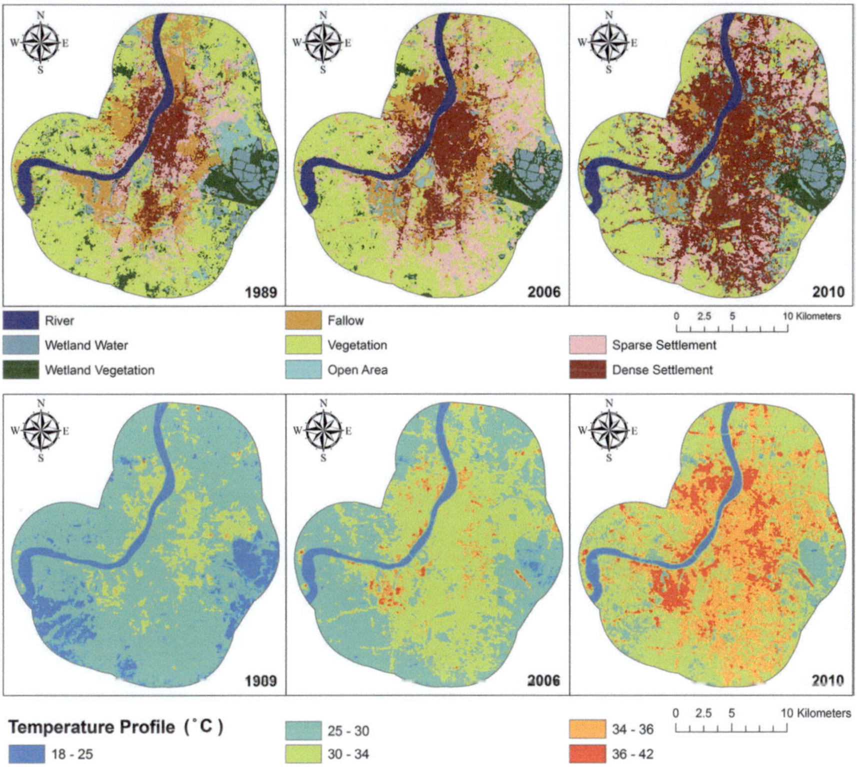

Fig. 5 LULC maps of Kolkata and the corresponding LST for the years 1989, 2006, and 2010

5.3 Chennai

Over the years (between 1991 and 2014), in the city of Chennai, bare land and marshland categories have been found to decrease consistently, while the built-up area is observed to show a consistent increase (Fig. 6). Water and vegetation first exhibited a fall and, later, registered an increase. Agriculture first increased slightly, by 15 km^2, between 1991 and 2000, and then came down to 37 km^2 from 115 km^2 in 2014. Amid all the land use dynamics, it is important to note that marshland, an ecologically critical LULC class, is observed to be steadily declining from 68 km^2 in 1991 to 44 km^2 in 2000 and finally to 23 km^2 in 2014. The built-up class, which was initially more confined to the northern parts of the study domain, spread over the entire city in the next two decades.

A qualitative analysis of the LST shows an expansion in the high-LST area (Fig. 7). A horizontal examination of the images at all three times exhibits an increase in LST, moving from west to east, i.e., on approaching the city from rural

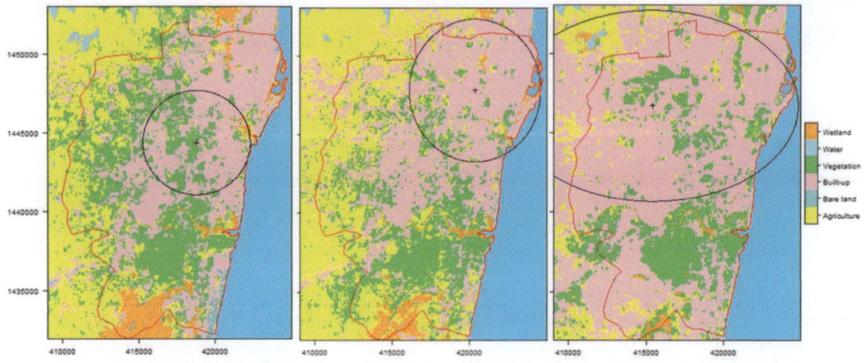

Fig. 6 UHI extent overlaid on LULC for the years 1991, 2000, and 2014

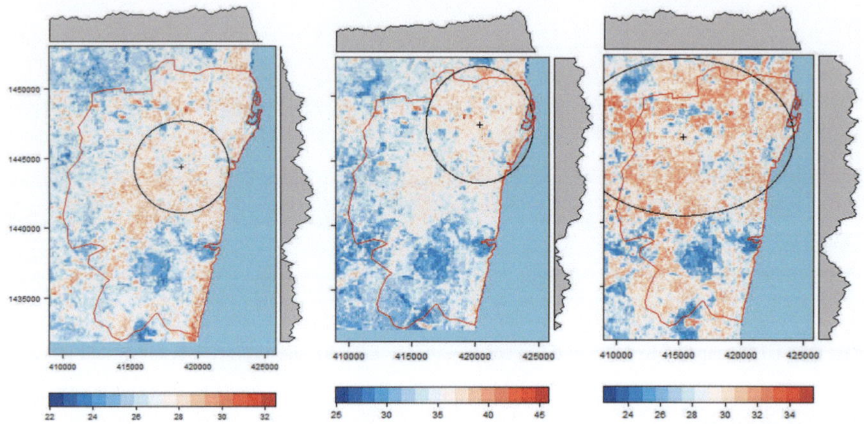

Fig. 7 SUHI extent displayed on LST for September 1991, 2000, and 2014 (the red line is the city boundary, the black point is the central location of the UHI, and the black circle/ellipse shows the extent of SUHI influence)

agricultural lands. LST, however, varies in the vertical direction in the three images moving from north to south. In 1991, LST increases initially on entering the city from the northern agricultural cropland, and it remains high and drops steeply at Guindy National Park, and beyond this, it again rises slightly. Such a typical UHI pattern of LST is observed in 2000 and 2014 as well, when the temperature in the northernmost agricultural fields is around 22–25 °C, which then goes on to increase by 10–15 °C as it approaches the city's built-up areas. However, the temperature falls by 10–15 °C on encountering the national park. In 2014, however, the LST beyond Guindy National Park also increases and is not just moderate as in the previous years.

The UHI intensity was observed to be 3.53, 2.19, and 1.4 for the years 2000, 2014, and 1991, respectively. The LST for October 2000 varies between 25 and 45 °C, while for September 1991 and 2014, it varies between 22 °C (minimum for 1991) and 34 °C (maximum for 2014), which explains the reason for the higher UHI intensity in 2000. For this reason, there is no trend in the UHI intensity over the years. The mean LST for SUHI extent each year is 27.6 °C, 35.8 °C, and 29.7 °C.

An important observation is that the area under the influence of SUHI steadily increases from 1991 to 2000 and from 2000 to 2014. The extent of UHI in 1991 is 36.8 km^2, which almost doubles to 60.2 km^2 in 2000 and further increases by almost 200% by 2014, covering 177.6 km^2. This indicates an expanding trend in UHI. However, the influence zone initially covered mainly the central to slightly northern parts of the city in 1991, including Thyagaraya Nagar (the congested shopping center), Nungambakkam (one of the oldest parts of Chennai), Teynampet (one of the busiest commercial localities), and Gopalapuram (a prime residential and commercial locality). In 2000, largely, the northern portion of the city is observed to be under the influence of SUHI, which includes new locations such as the residential area of Pattalam and the locality of Perambur (a mix of commercial and residential establishments), which was once a bamboo forest. The UHI extent by 2014 embraces more locations in the north and west. These include Valasaravakkam, Anna Nagar (a newly developed township), Arumbakkam (a residential locality), Kilpauk (a semi-residential area), Korattur (a twentieth-century, calm village, which is now an urban township), Kattupakkam (the city suburb), Kumananchavadi (an upcoming residential suburb), and the industrial area of Ambattur. Thus, by 2014, the whole of the city except the south had been distressed by UHI. The presence of Guindy National Park in the southern part of the study region averts the influence of UHI in this region. Thus, throughout the years, with the presence of thick and healthy vegetation cover in the south, this part of the city remains free from UHI owing to evapotranspiration by vegetation and thus provides a cooling effect.

For the year 1991, it is observed that the UHI extent mainly covers the built-up areas in the central parts of the city. The UHI influence zone is dominated by built-up and vegetation classes. In 2000, an increase in built-up area is observed in the northern parts, accompanied by an expansion of the UHI ellipse and its movement to the north. A drastic change is observed in LULC in 2014, when almost the whole of the study region has been converted to built-up areas, except Guindy National Park and some other smaller patches of vegetation and agriculture. This is associated with a larger UHI influence ellipse, mainly covering the upper half of the study region. Marshlands are observed to have disappeared both from the north and the south. This explains a tremendous increase in LST beyond the national park in longitudinal analysis.

The UHI influence zone in 1991 was 1.3 km^2 of agricultural land, 25.4 km^2 of built-up areas, 0.4 km^2 of marshland, and 9.7 km^2 of vegetation. The agricultural land under UHI increased to 5.1 km^2 by 2000 and further to 17.3 km^2 by 2014. A tremendous rise in the built-up area under the UHI influence is observed from 25.4 km^2 in 1991 to 46.2 km^2 in 2000 and finally 119.5 km^2 in 2014. Vegetation areas under UHI first registered a fall from 9.7 km^2 to 4.8 km^2, and in the second half, it increased to 14.2 km^2. Changes for other classes are negligible.

5.4 *Mumbai*

Six LULC classes were mapped for 2 years, 1998 and 2009. The maps (Fig. 8(a)) show a decrease in forest from 1998 to 2009 along the peripheral areas of the forest patch in the north. However, some improvement in forestland is observed in the southern patch. Mangroves show a huge loss in the northwestern and northern parts of the city. Vegetation class exhibits an improvement at the cost of the forest of Sanjay Gandhi National Park (the forest in the northern parts of Mumbai). However, vegetation has received a blow in the central parts of the city, where built-up area has intensified during the study period. The open area has claimed land from the mangrove cover. The built-up area has increased during the study period and has also been intensified in most of the locations. The increase in built-up area is at the cost of vegetation and open area. Water is a class that appears to have remained more or less unchanged.

The LULC statistics and their change analysis reveal a dynamic relationship between forest, vegetation, and open area. Here, each class converts to the other, and a part of the latter class reverts to the former one. For instance, 9.63 km^2 was converted to the vegetation class, and at the same time, 7.05 km^2 of vegetation has been transformed into forest. The most important transformations in LULC statistics include the conversion of vegetation (22.86 km^2) into built-up areas. Mangroves, the highly critical ecosystems, have lost a land area of 11 km^2 to open area. Forest and vegetation have lost 4.85 km^2 and 16.73 km^2 from 1998 to 2009. Another important observation is the increase of 30 km^2 (14%) in built-up area in 2009, as compared to 1998.

The LST ranged from 30 °C to 55 °C between the years. The maps (Fig. 8(b)) show increasing LST from 1998 to 2009 in the central parts of the city. These are the locations where an intensification of built-up area has taken place. The water class and the forest show the lowest LST (<34 °C) in both years. LST for mangroves is higher than these but lower than other land use classes. LST is the highest (>40 °C) for open area and built-up categories of LULC in both years. LST for vegetation lies in a range of 34 °C–38 °C. The 2009 LST map indicates elevated temperatures, particularly for the built-up class.

The degradation in forest peripheries is well-reflected by the increasing LST in that area. Land under mangroves (in 1998) shows an increase in LST (in 2009), as it loses its mangrove cover, particularly in the north of Kandivali West along the Malad and Manori creeks. High increases in LST are observed in areas where vegetation has been lost and/or areas where classes like open area and built-up area have increased.

Due to increasing encroachment (as shown by the LULC map), an activity that is engulfing the land along the peripheries of the Sanjay Gandhi National Park, the temperature pattern along the park fringes has altered. Areas of nonurban land that were converted to built-up areas during this period got warmer by an average increase of 4.45 °C. The construction activities along the eastern coastal areas did not increase much beyond 4–5 °C, which could be attributed to their proximity to the sea. Shivaji Nagar, Nirankar Nagar, New Gautam Nagar, Vaibhav Nagar, and Ramabai Ambedkar Nagar in the southeast; Jogeshwari East near Sanjay Gandhi National Park; and Asalfa, Mairwadi, Mohili, Lokmanya Tilak Nagar, and Dharavi in the central parts of the city have become potential UHI spots.

The zone-wise analysis (Fig. 8(c)) shows that the coolest zone of LST, Zone I, was dominated by forest (81%) in 1998 and by water (92%) in 2009. Zone II, the

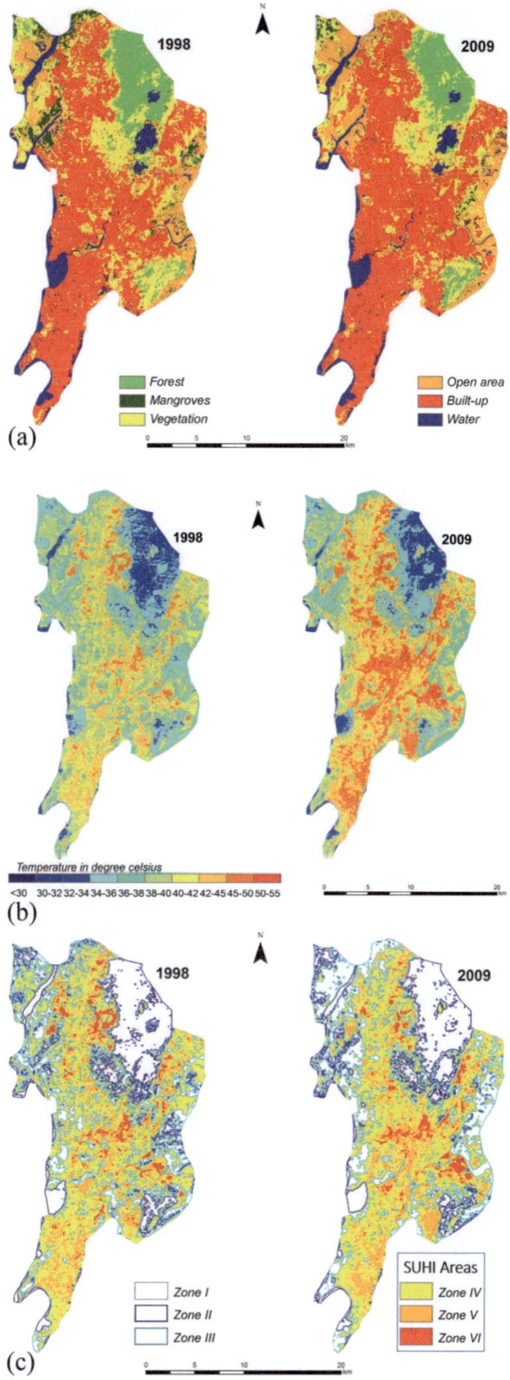

Fig. 8 (**a**) LULC maps of Mumbai for the years 1998 and 2009, (**b**) LST maps for 1998 and 2009, and (**c**) LST zones mapped for 1998 and 2009 (representing SUHI)

second coolest LST zone, comprised mainly forest (43%), vegetation (26%), and water (23%) in both years. The zone also had 5% mangrove in 1998, which reduced to 3% in 2009, and had 2% open area in 1998, which increased to 4% in 2009. Zone III comprised 6–7% water and 1.3% forest in both years. Mangroves constituted 7% of Zone III in 1998, which decreased to 2% in 2009. The contribution of vegetation to this class decreased from 48% to 36%, while the share of built-up area and open area increased from 17% to 23% and from 21% to 33%, respectively. The hotter LST zones, i.e., IV, V, and VI, had a relatively higher percentage composition of built-up class. Zone IV, for instance, had approximately 15% of land under vegetation during both years. In 1998, this zone had 10% open area, which decreased to 2% in 2009. The land lost by open area was the gain in built-up area for this zone. The hottest zones, V and VI, have an almost 98–100% built-up coverage.

5.5 *Surat*

In addition to megacities, numerous smaller urban clusters have experienced significant growth in recent years. Surat exemplifies one such city that has undergone rapid urbanization over the past decades. A comprehensive examination of the spatial and temporal dimensions of urban growth in Surat was conducted to analyze the interrelationships among various environmental factors using remote sensing datasets (Sharma et al., 2012). The study utilized Landsat Thematic Mapper satellite data, alongside geospatial techniques, to investigate urbanization trends and correlations among multiple satellite-derived biophysical parameters, such as Normalized Difference Vegetation Index (NDVI), Normalized Difference Built-up Index (NDBI), Normalized Difference Water Index (NDWI), Normalized Difference Bareness Index (NDBaI), Modified Normalized Difference Water Index (MNDWI), and land surface temperature (LST). Land use and land cover changes were assessed using hierarchical decision tree classification, achieving accuracies of 90.4% for 1990 and 85% for 2009. The findings revealed that Surat expanded by 42.75 km^2 over a decade, resulting in notable increases in surface temperatures (Fig. 9). For instance, transformations from vegetation to built-up areas corresponded to an average LST rise of 5.5 ± 2.6 °C, while transitions from vegetation to fallow land saw an increase of 6.7 ± 3 °C in LST. Conversions from fallow to built-up areas led to a temperature increase of 3.5 ± 2.9 °C, and shifts from built-up to dense built-up areas resulted in a 5.3 ± 2.8 °C rise in LST.

Directional profiling of LST highlighted the emergence of two new temperature peaks in 2009 along the north–south and northeast–southwest profiles (Fig. 10). This detailed analysis underscores the profound impacts of urbanization on local climate and land surface temperatures in Surat, emphasizing the need for sustainable urban planning and environmental management strategies to mitigate these effects.

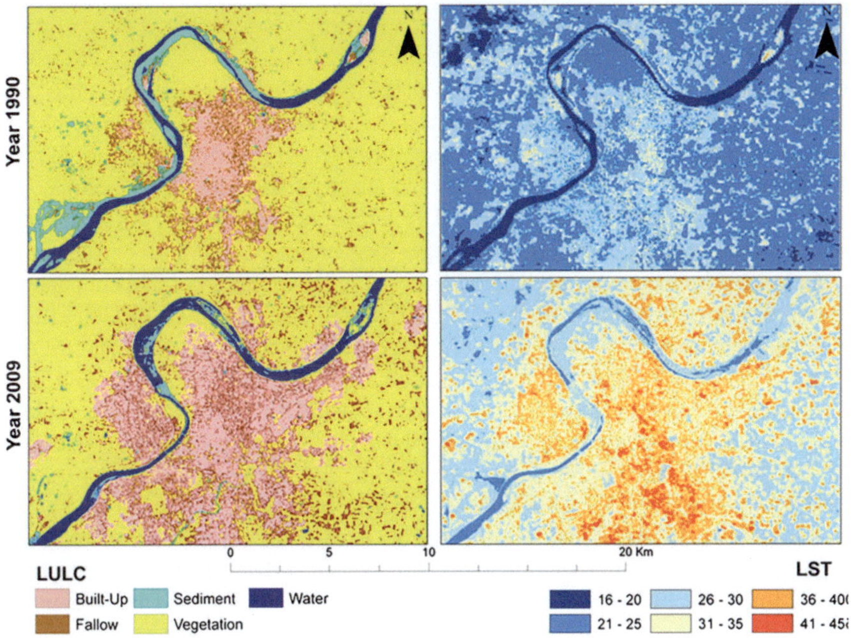

Fig. 9 LULC and LST for 1990 and 2009

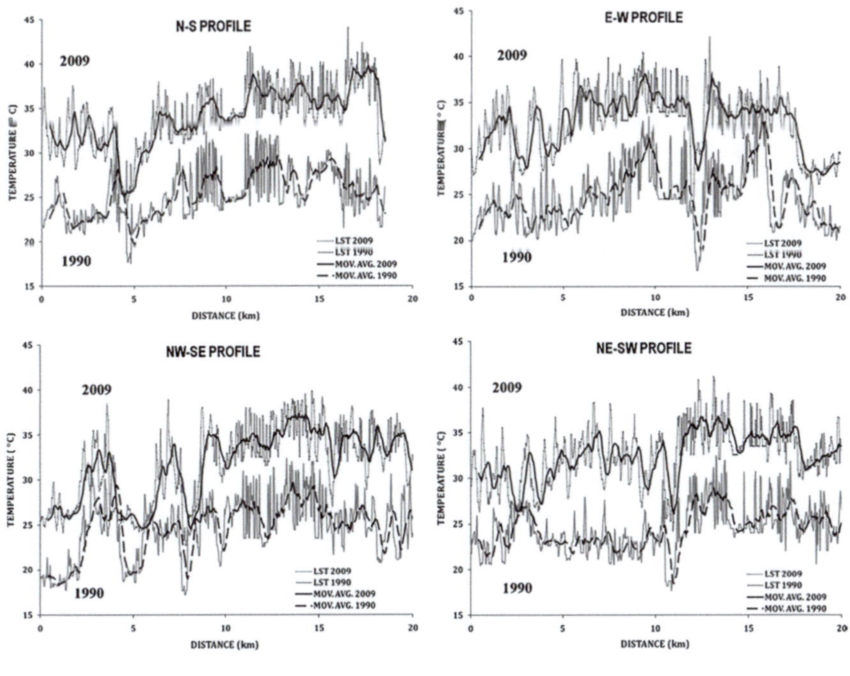

Fig. 10 Directional profiling of LST distribution in 1990 and 2009

6 Conclusion

With the additive effects of UHI and climate change–induced warming, cities are facing greater risks of extreme heat events and associated impacts, including health and economic impacts. This holds true for cities located in different geographies of the world, from developing to developed countries; cities with varying climatologies, for instance, arid, semiarid, tropical, or temperate; and cities of different sizes, be it smaller urban agglomerations or larger megacities.

In light of an increasing population living in urban areas, it becomes imperative to find solutions to mitigate the heat in urban areas. The mitigation of UHI effects will also help reduce the damage of climate change–induced extreme heat events. Some of these solutions include cool roofs, cool pavements, green roofs, and increasing the vegetation cover. Cities across the world are increasingly experimenting with different solutions for mitigating UHI effects, and nature-based solutions have become the need of the hour (Hayes et al., 2022). However, in order to efficiently implement any of these solutions, it is important to understand the spatial and temporal patterns of UHI in the city. This is what the current chapter has been about, i.e., ways to spatiotemporally map and measure urban heat island intensity in cities. Such mapping facilitates the identification of hotspots and high-intensity times of UHI, as well as improves the understanding of the causes contributing to UHI development in that particular area. Such findings can help guide urban planners and city managers in making policies and plans that diminish the detriments of UHI in the city.

References

Arnfield, A. J. (2003). Two decades of urban climate research: A review of turbulence, exchanges of energy and water, and the urban Heat Island. *International Journal of Climatology, 23*, 1–26.

Barrao, S., Serrano-Notivoli, R., Cuadrat, J. M., Tejedor, E., & Sánchez, M. A. S. (2022). Characterization of the UHI in Zaragoza (Spain) using a quality-controlled hourly sensor-based urban climate network. *Urban Climate, 44*, 101207.

Bastiaanssen, W., Menenti, M., Feddes, R., & Holtslag, A. (1998a). A remote sensing surface energy balance algorithm for land (SEBAL). 1. Formulation. *Journal of Hydrology, 212*, 198–212.

Bastiaanssen, W., Pelgrum, H., Wang, J., Ma, Y., Moreno, J., Roerink, G., & Van der Wal, T. (1998b). A remote sensing surface energy balance algorithm for land (SEBAL).: Part 2: Validation. *Journal of Hydrology, 212*, 213–229.

Benjamin, K., Luo, Z., & Wang, X. (2021). Crowdsourcing urban air temperature data for estimating urban heat island and building heating/cooling load in London. *Energies, 14*(16), 5208.

Bornstein, R. D. (1968). Observations of the urban heat island effect in New York City. *Journal of Applied Meteorology, 7*(4), 575–582.

Bornstein, R., & Lin, Q. (2000). Urban heat islands and summertime convective thunderstorms in Atlanta: Three case studies. *Atmospheric Environment, 34*(3), 507–516.

Buyantuyev, A., & Wu, J. (2010). Urban heat islands and landscape heterogeneity: Linking spatiotemporal variations in surface temperatures to land-cover and socioeconomic patterns. *Landscape Ecology, 25*(1), 17–33.

Carlson, T. N., & Arthur, S. T. (2000). The impact of land use—Land cover changes due to urbanization on surface microclimate and hydrology: A satellite perspective. *Global and Planetary Change, 25*(1–2), 49–65.

Chakraborty, T., Hsu, A., Manya, D., & Sheriff, G. (2020). A spatially explicit surface urban heat island database for the United States: Characterization, uncertainties, and possible applications. *ISPRS Journal of Photogrammetry and Remote Sensing, 168*, 74–88.

Chaston, T. B., Broome, R. A., Cooper, N., Duck, G., Geromboux, C., Guo, Y., Ji, F., Perkins-Kirkpatrick, S., Zhang, Y., & Dissanayake, G. S. (2022). Mortality burden of heatwaves in Sydney, Australia is exacerbated by the urban Heat Island and climate change: Can tree cover help mitigate the health impacts? *Atmosphere, 13*(5), 714.

Chen, X.-L., Zhao, H.-M., Li, P.-X., & Yin, Z.-Y. (2006). Remote sensing image-based analysis of the relationship between urban heat island and land use/cover changes. *Remote Sensing of Environment, 104*(2), 133–146.

Chen, K., Boomsma, J., & Holmes, H. A. (2023). A multiscale analysis of heatwaves and urban heat islands in the western US during the summer of 2021. *Scientific Reports, 13*(1), 9570.

Childs, P. P. (2005). Observations and numerical simulations of urban heat island and sea breeze circulations over New York City. *Pure and Applied Geophysics, 162*, 1955–1980.

Çolak, E., & Sunar, F. (2023). Investigating the usefulness of satellite-retrieved land surface temperature (LST) in pre-and post-fire spatial analysis. *Earth Science Informatics, 16*(1), 945–963.

Cuerdo-Vilches, T., Díaz, J., López-Bueno, J. A., Luna, M., Navas, M., Mirón, I., & Linares, C. (2023). Impact of urban heat islands on morbidity and mortality in heat waves: Observational time series analysis of Spain's five cities. *Science of the Total Environment, 890*, 164412.

De Ridder, K., Maiheu, B., Lauwaet, D., Daglis, I. A., Keramitsoglou, I., Kourtidis, K., Manunta, P., & Paganini, M. (2016). Urban heat island intensification during hot spells—The case of Paris during the summer of 2003. *Urban Science, 1*(1), 3.

Drohan, P. J., Brittingham, M., Bishop, J., & Yoder, K. (2012). Early trends in landcover change and forest fragmentation due to shale-gas development in Pennsylvania: A potential outcome for the Northcentral Appalachians. *Environmental Management., 49*(5), 1061–1075. https://doi.org/10.1007/s00267-012-9841-6. Epub 2012 Mar 25. PMID: 22447181.

Drummond, M. A., & Loveland, T. R. (2010). Land-use pressure and a transition to forest-cover loss in the Eastern United States. *BioScience, 60*(4), 286–298.

Emeis, S. (2009). Derivation of vertical wind and turbulence profiles, the mixing-layer height, and the vertical turbulent exchange coefficient from SODAR and ceilometer soundings in urban measurement campaigns. In *Meteorological and air quality models for urban areas* (pp. 133–141). Springer.

Fabrizi, R., Bonafoni, S., & Biondi, R. (2010). Satellite and ground-based sensors for the urban heat island analysis in the city of Rome. *Remote Sensing, 2*, 1400–1415.

Florenzio, N., Guastella, G., Magni, F., Pareglio, S., & Musco, F. (2022). The role of urban planning in climate adaptation: An empirical analysis of UHI in European cities. *Journal of Environmental Planning and Management, 66*, 1–19.

Freitas, E. D., Rozoff, C. M., Cotton, W. R., & Dias, P. L. S. (2007). Interactions of an urban heat island and sea-breeze circulations during winter over the metropolitan area of São Paulo, Brazil. *Boundary-Layer Meteorology, 122*(1), 43–65.

Gartland, L. (2011). *Heat islands: Understanding and mitigating heat in urban areas*. Routledge.

Gedzelman, S. D., Austin, S., Cermak, R., Stefano, N., Partridge, S., Quesenberry, S., & Robinson, D. A. (2003). Mesoscale aspects of the urban heat island around New York City. *Theoretical and Applied Climatology, 75*, 29–42.

Giridharan, R., Ganesan, S., & Lau, S. (2004). Daytime urban heat island effect in high-rise and high-density residential developments in Hong Kong. *Energy and Buildings, 36*(6), 525–534.

Hayes, A. T., Jandaghian, Z., Lacasse, M. A., Gaur, A., Lu, H., Laouadi, A., Ge, H., & Wang, L. (2022). Nature-based solutions (NBSs) to mitigate urban Heat Island (UHI) effects in Canadian cities. *Buildings, 12*(7), 925.

Houghton, R. A. (1999). The annual net flux of carbon to the atmosphere from changes in land use 1850–1990. *Tellus B, 51*(2), 298–313.

Jones, P. D., & Lister, D. H. (2009). The urban heat island in Central London and urban-related warming trends in Central London since 1900. *Weather, 64*(12), 323–327.

Kabisch, N., Remahne, F., Ilsemann, C., & Fricke, L. (2023). The urban heat island under extreme heat conditions: A case study of Hannover, Germany. *Scientific Reports, 13*(1), 23017.

Kerchove, R. V. D., Lhermitte, S., Veraverbeke, S., & Goossens, R. (2013). Spatio-temporal variability in remotely sensed land surface temperature, and its relationship with physiographic variables in the Russian Altay Mountains. *International Journal of Applied Earth Observation and Geoinformation, 20*, 4–19.

Kim, H., Gu, D., & Kim, H. Y. (2018). Effects of Urban Heat Island mitigation in various climate zones in the United States. *Sustainable Cities and Society, 41*, 841–852.

Kolokotroni, M., Giannitsaris, I., & Watkins, R. (2006). The effect of the London urban heat island on building summer cooling demand and night ventilation strategies. *Solar Energy, 80*(4), 383–392.

Lambin, E. F., Turner, B. L., Geist, H. J., Agbola, S. B., Angelsen, A., Bruce, J. W., Coomes, O. T., Dirzo, R., Fischer, G., Folke, C., George, P. S., Homewood, K., Imbernon, J., Leemans, R., Li, X., Moran, E. F., Mortimore, M., Ramakrishnan, P. S., Richards, J. F., & Xu, J. (2001). The causes of land-use and land-cover change: Moving beyond the myths. *Global Environmental Change, 11*(4), 261–269.

Langer, M., Westermann, S., & Boike, J. (2010). Spatial and temporal variations of summer surface temperatures of wet polygonal tundra in Siberia—Implications for MODIS LST-based permafrost monitoring. *Remote Sensing of Environment, 114*(9), 2059–2069.

Lauwaet, D., Berckmans, J., Hooyberghs, H., Wouters, H., Driesen, G., Lefebre, F., & De Ridder, K. (2024). High resolution modelling of the urban heat island of 100 European cities. *Urban Climate, 54*, 101850.

Loveland, T. R., & DeFries, R. S. (2004). Global land-cover characteristics and data sets. In G. Gutman et al. (Eds.), *Land change science: Observing, monitoring and understanding trajectories of change in Earth's surface* (pp. 231–256). Springer.

Lu, H., Koike, T., Yang, K., Hu, Z., Xu, X., Rasmy, M., Kuria, D., & Tamagawa, K. (2012). Improving land surface soil moisture and energy flux simulations over the Tibetan Plateau by the assimilation of the microwave remote sensing data and the GCM output into a land surface model. *International Journal of Applied Earth Observation and Geoinformation, 17*, 43–54.

Mahmood, R., Pielke, R. A., Sr., Hubbard, K. G., Niyogi, D., Bonan, G., Lawrence, P., et al. (2010). Impacts of land use/land cover change on climate and future research priorities. *Bulletin of the American Meteorological Society, 91*(1), 37–46.

Mallick, J., Rahman, A., & Singh, C. K. (2013). Modeling urban heat islands in heterogeneous land surface and its correlation with impervious surface area by using night-time ASTER satellite data in highly urbanizing city, Delhi-India. *Advances in Space Research, 52*(4), 639–655. https://doi.org/10.1016/j.asr.2013.04.025

Matson, M., Mcclain, E. P., McGinnis, D. F., Jr., & Pritchard, J. A. (1978). Satellite detection of urban heat islands. *Monthly Weather Review, 106*(12), 1725–1734.

Meyer, W. B., & Turner, B. L. (1992). Human population growth and global land-use/cover change. *Annual Review of Ecology and Systematics, 23*, 39–61.

Oke, T. (1976). The distinction between canopy and boundary-layer urban heat islands. *Atmosphere, 14*(4), 268–277.

Oke, T. R., Mills, G., & Voogt, J. (2017). *Urban climates*. Cambridge University Press.

Pal, S., Xueref-Remy, I., Ammoura, L., Chazette, P., Gibert, F., Royer, P., Dieudonfné, E., Dupont, J., Haeffelin, M., & Lac, C. (2012). Spatio-temporal variability of the atmospheric boundary layer depth over the Paris agglomeration: An assessment of the impact of the urban heat island intensity. *Atmospheric Environment, 63*, 261.

Price, J. C. (1979). Assessment of the urban heat island effect through the use of satellite data. *Monthly Weather Review, 107*(11), 1554–1557.

Qin, Z., Karnieli, A., & Berliner, P. (2001). A mono-window algorithm for retrieving land surface temperature from Landsat TM data and its application to the Israel-Egypt border region. *International Journal of Remote Sensing, 22*(18), 3719–3746.

Rao, P. (1972). Remote sensing of urban heat islands from an environmental satellite. *Bulletin of the American Meteorological Society, 53*(7), 647–648.

Sarangi, C., Qian, Y., Li, J., Leung, L. R., Chakraborty, T., & Liu, Y. (2021). Urbanization amplifies nighttime heat stress on warmer days over the US. *Geophysical Research Letters, 48*(24), e2021GL095678.

Senanayake, I. P., Welivitiya, W. D. D. P., & Nadeeka, P. M. (2013). Remote sensing–based analysis of urban heat islands with vegetation cover in Colombo City, Sri Lanka using Landsat-7 ETM+ data. *Urban Climate, 5*, 19–35.

Schwarz, N., Schlink, U., Franck, U., & Großmann, K. (2012). Relationship of land surface and air temperatures and its implications for quantifying urban heat island indicators—An application for the city of Leipzig (Germany). *Ecological Indicators, 18*, 693–704.

Sharma, R., Ghosh, A., & Joshi, P. (2012). Analysing spatio-temporal footprints of urbanization on environment of Surat city using satellite-derived bio-physical parameters. *Geocarto International, 28*, 1–19.

Sobrino, J. A., Jiménez-Muñoz, J. C., & Paolini, L. (2004). Land surface temperature retrieval from LANDSAT TM 5. *Remote Sensing of Environment, 90*(4), 434–440.

Stache, E. E., Schilperoort, B. B., Ottelé, M. M., & Jonkers, H. H. (2022). Comparative analysis in thermal behaviour of common urban building materials and vegetation and consequences for urban heat island effect. *Building and Environment, 213*, 108489.

Streutker, D. R. (2003). Satellite-measured growth of the urban heat island of Houston. *Texas. Remote Sensing of Environment, 85*(3), 282–289.

Tomlinson, C. J., Chapman, L., Thornes, J. E., & Baker, C. J. (2011). Including the urban heat island in spatial heat health risk assessment strategies: A case study for Birmingham, UK. *International Journal of Health Geographics, 10*(1), 42.

Tremeac, B., Bousquet, P., de Munck, C., Pigeon, G., Masson, V., Marchadier, C., Merchat, M., Poeuf, P., & Meunier, F. (2012). Influence of air conditioning management on heat island in Paris air street temperatures. *Applied Energy, 95*, 102–110.

Turner, B., Clark, W. C., Kates, R. W., Richards, J. F., Mathews, J. T., & Meyer, W. B. (1990). The Earth as transformed by human action: Global change and regional changes in the biosphere over the past 300 years. In *The earth as transformed by human action: Global change and regional changes in the biosphere over the past 300 years*. Cambridge University Press, with Clark University.

Vadakkuveettil, A., & Grover, A. (2023). Bi-temporal characterization of terrestrial temperature in relation to urban land use land cover dynamics and policies in Kozhikode Urban Area, India. *Land Use Policy, 132*, 106782.

Vez, J. P. M., Rodríguez, A., & Jiménez, J. I. (2000). A study of the urban heat island of Granada. *International Journal of Climatology, 20*, 899–911.

Voogt, J. A., & Oke, T. R. (2003). Thermal remote sensing of urban climates. *Remote Sensing of Environment, 86*(3), 370–384.

Weng, Q., Lu, D. and Schubring, J. (2004). Estimation of land surface temperature–vegetation abundance relationship for urban heat island studies. *Remote Sensing of Environment, 89*(4), 467–483.

Weng, Q., Rajasekar, U., & Hu, X. (2011). Modeling urban heat islands and their relationship with impervious surface and vegetation abundance by using ASTER images. *IEEE Transactions on Geoscience and Remote Sensing, 49*(10), 4080–4089.

Westermann, S., Langer, M., & Boike, J. (2011). Spatial and temporal variations of summer surface temperatures of high-Arctic tundra on Svalbard—Implications for MODIS LST-based permafrost monitoring. *Remote Sensing of Environment, 115*(3), 908–922.

Zhang, J., Wang, Y., & Li, Y. (2006). A C++ program for retrieving land surface temperature from the data of Landsat TM/ETM+ band 6. *Computers & Geosciences, 32*(10), 1796–1805.

Urban Ecosystem Services: Human Health and Well-Being

Shalini Dhyani, Dhanya Bhaskar, Harini Santhanam, Y. S. C. Khuman, Rakesh Kadaverugu, and Samudyatha Ramananda

Abstract Amid fast-growing urban sprawls, the urban population is projected to reach 6.5 billion by 2050. This necessitates a conservation-oriented outlook for the future of urban ecosystems, with reference to global warming and human health concerns. Natural ecosystems in urban areas have significant climate adaptation capacities that are crucial to developing socio-ecological resilience to ensure the good health and well-being of urban residents. Nature-based Solutions (NbS) for urban areas provide compelling evidence for supporting the development and implementation of a comprehensive urban resilience framework. This framework ensures urban sustainability by enhancing sociocultural security, generating economic benefits, and promoting the uninterrupted flow of ecosystem services (ES). There is a pressing need to develop integrated ecosystem management frameworks that assess urban risks, ecosystem health, and the impacts of the demand–supply gap in green spaces on human health and well-being. This chapter provides a comprehensive overview of human health linkages with urban green spaces (UGS), ecosystem services, urban socio-ecological system dynamics, and the relevance of urban–rural gradients. It explores key scientific concepts of environmental neuroscience (ENS), biophilia, environmental affordances, and environmental constraints and their applicability in order to mainstream potential NbS options and opportuni-

S. Dhyani (✉) · R. Kadaverugu
CSIR National Environmental Engineering Research Institute (CSIR-NEERI), Nagpur, India
e-mail: s_dhyani@neeri.res.in

D. Bhaskar · S. Ramananda
Indian Institute of Forest Management (IIFM), Bhopal, India

H. Santhanam
Manipal Academy of Higher Education (MAHE), Bengaluru, India

Y. S. C. Khuman
School of Inter-disciplinary and Transdisciplinary Studies, Indira Gandhi National Open University Maidan Garhi, New Delhi, India

P. K. Joshi et al. (eds.), *Geospatial Science for Urban Ecosystems*, The Urban Book Series, https://doi.org/10.1007/978-3-032-07155-2_7

ties in urban planning for the conservation of fast-depleting urban natural ecosystems in the Global South. With evidences and experiences from different countries across the world, the present chapter endorses and stresses the necessity of scaling up NbS for addressing urban stress and psychological benefits, along with diverse health risks, to develop resilient cities in the warming and fast-urbanizing world.

Keywords Urban ecosystems · Biodiversity · Health · Well-being · Blue-green infrastructure · Nature-based Solutions (NbS)

1 Introduction

Urban sprawl and expansion are profoundly reshaping global environments, as highlighted by Chen et al. (2020). Currently, over half of the world's population resides in rapidly expanding urban areas—a trend projected to escalate to 67% by 2050. This demographic shift underscores not only the escalating impact of urbanization but also its far-reaching and often unanticipated consequences on ecosystems worldwide. As per the UN-Habitat report of 2013, this massive rural–urban demographic shift has brought unprecedented global urban expansion, with hotspots concentrated in the developing nations of Asia and Africa (Zhang & Su, 2016). Urban growth is a crucial factor in fast-developing economies, predominantly in the context of accelerated commercialization and rising urban population (Dhanaraj & Angadi, 2021). Although urbanization has emerged as a pivotal factor influencing global extinction risks, our comprehension of urban ecology remains rudimentary, limiting the foundation upon which conservation-minded planners can make informed decisions (Douglas & Douglas, 2021). While infrastructure serves as the foundation for urban growth, its mere existence does not necessarily indicate the well-being of urban residents (Udnoor et al., 2020). Cities are continually expanding their boundaries and populations, leading to the observation that *"human history is characterized as the history of urbanisation from a climatological standpoint"* (Santamouris et al., 2001). While urbanization has bolstered economies, the encroachment and destruction of urban green spaces (UGS) pose significant threats to healthy living. This trend could potentially increase the burden of noncommunicable diseases (NCDs) and lead to substantial economic losses (Adhikari et al., 2019). Cities at the forefront of global climate change responses are increasingly tasked with mitigating the effects of climate change and adapting to it. They are significant sources of greenhouse gas emissions and are home to large populations that are highly vulnerable to the effects and risks associated with a changing climate (Frantzeskaki et al., 2019). The interplay between local and global climatic processes exposes urban ecosystems to high temperatures, extreme heat episodes, and persistent hot weather. Cities capitalize on their unique position and capability to implement both local and global climate solutions, aiming to mitigate the impacts and hazards of climate change (Frantzeskaki et al., 2019). In the Anthropocene, several climate change–induced drivers are accelerating the frequency and intensity of extreme weather events, such as heat waves, urban heat island intensity, and urban floods. This is especially true in the emerging coastal centers of the Global South (e.g., Kundu and Santhanam (2021c)). The urban heat island (UHI) effect is one of the most significant environmental consequences of urbanization, directly harming human health and well-being due to excess heat while also contributing to the deterioration of environmental quality (Marando et al., 2019). The double phenomena of UHI and heat waves that are exacerbated due to heat islands result in thermal discomfort of city dwellers, leading to various health issues, which have been frequently observed across different urban areas (Borah, 2020). Additionally, urban ecosystems are threatened by several anthropogenic actions, including heavy metal contamination, dry and wet deposition of harmful air pollutants, water contamination, and soil contamination. Among others, the impacts from anthropogenic noise, vibration, and light are also gradually interfering with the circadian rhythm of urban life.

There is a growing concern that the planet's health is dependent on the peaceful cohabitation of rapidly expanding cities and the natural world (Mansur et al., 2022). Urbanization has historically been related to human progress and advancement, but there are clear indications that urban settings can result in substantial discrimination and health issues (Kuddus et al., 2020). Studies reveal that our health is more conditioned by our surroundings and the manner of living rather than by the public health system. Only 11% of the so-called health determinants are related to our health systems, while 50% are related to the cities we live in, how we travel, what we eat, and the quality of the air we breathe (Ballester et al., 2020). Current estimates of rapid urbanization brings both basic issues as well as opportunities for designing communities that are more livable, healthy, and resilient (e.g., adaptation to climate change effects) (Elmqvist et al., 2015).

Expanding or enhancing UGS can enhance health, social, and environmental outcomes for all urban residents, particularly benefiting those with a lower socioeconomic status. Green areas are considered effective for reducing the negative effects of urbanization on human health. "Biophilia," defined as the need to be in touch with nature in order to have a physically and emotionally healthy existence, is a concept that medical specialists and urban planners alike are increasingly promoting (Ballester et al., 2020). Nevertheless, there is an increasing demand for improved monitoring of local green space management and its associated health and equity implications, along with better incorporation of health and equity outcomes in studies evaluating green space interventions (WHO, 2017). The rampant loss of green spaces resulting from urban sprawl has emerged as a major driver of some of the explicit connections between urbanization and human health damages (Adhikari et al., 2019). According to studies on land use transitions, rapid urbanization has resulted in the loss of green and blue areas in several Indian cities. Between 1973 and 2013, Bengaluru's built-up area increased by 925%, with green areas shrinking from 68% to 14% and wetlands declining from 3% to <1%. Likewise, from 1977 to 2017, Mumbai experienced a 60% reduction in vegetation and a 65% decline in water bodies. A technical assessment of land use and land cover for Greater Mumbai during the period from 1999 to 2019, released in 2020, revealed a further 2.5%

decrease in vegetation cover and a 1.4% reduction in water bodies. Between 2010 and 2030, Ahmedabad is expected to lose around 50% of its greenery while smaller tier II cities, like Nagpur, Lucknow, Faridabad, and Kanpur, are facing the same fate of exponential loss of green spaces and encroachment and concretization of wetlands and water bodies (Dhyani et al., 2018, 2021; Udas-Mankikar & Driver, 2021). Exploring how nature promotes physical exercise for mental and physical health benefits is crucial yet challenging, especially in densely populated cities with limited and diminishing access to natural environments (Remme et al., 2021). The successful implementation of the Sustainable Development Goals (SDGs) and the New Urban Agenda largely depends on cities (Ramirez-Rubio et al., 2019). To achieve inclusive, safe, resilient, and sustainable cities and human settlements, societies must take a leading role in the management and design of urban areas, as outlined in the UN SDG 11 of 2015.

2 Urban Environmental and Human Health: Teasing Out the Interlinkages

Though the basic science to understand the complexity of urban linkages with some of these challenges is not sufficiently developed yet, rapid global urbanization has brought a broader recognition to understand the interactions of the changing urban environment with human health and well-being (Bai et al., 2012). Urbanization, while contributing to countries' increased economic prosperity, also results in negative consequences, which include the emergence of a new set of human health issues, known as "city ailments" (Jha, 2020). Chronic stress, insufficient physical activity, and exposure to human-made environmental hazards are intrinsic aspects of contemporary urban living (Braubach et al., 2017). These conditions arise from the unique environment of urban settlements and the quality of life they foster. Urban roads are constantly congested, leading to disorderly scenarios, precarious to human physiological as well as psychological health. Enormous vehicles on roads, honking by drivers, noisy crowded markets, loudspeakers, engineering and welding workshops, and many others are the foremost reasons for unacceptable noise levels in urban areas (Das & Basak, 2020). Substandard housing, characterized by overcrowding and inadequate infrastructure, along with challenges such as air pollution, insufficient access to clean and safe drinking water, poor sanitation systems, and improper solid waste disposal services, contributes to a range of urban health hazards and concerns. Additionally, the prevalence of vector-borne diseases, industrial waste contamination, and the escalating volume of motor vehicle traffic further compound these issues, posing significant threats to public health in urban environments (Moore et al., 2003) (Fig. 1). Air pollution and water pollution, coupled with inadequate sanitation, have significantly exacerbated public health issues in rapidly expanding urban areas in developing and underdeveloped countries of the Global South (Pradhan et al., 2020). NCDs, such as diabetes, hypertension, and cardiovascular diseases, have recently surpassed cancer as the leading causes of death. Concurrently, there is an increasing incidence of cardiovascular illnesses, diabetes,

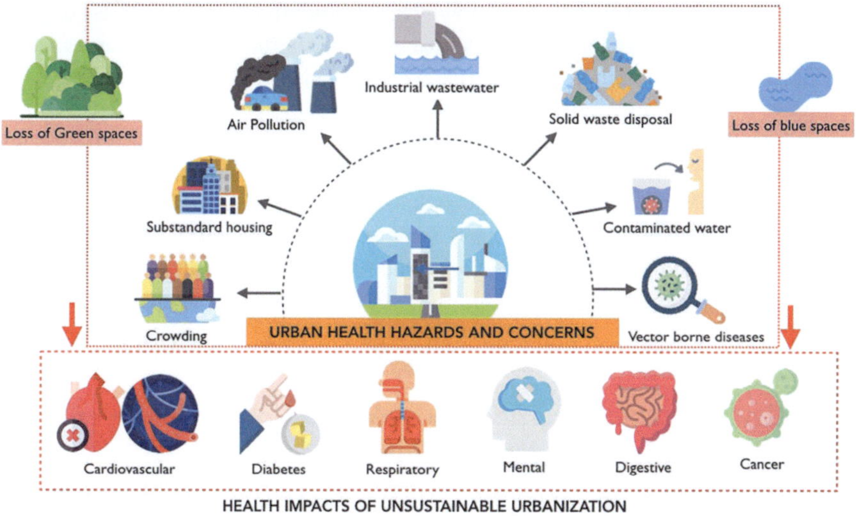

Fig. 1 Urban health issues due to fast-expanding urban sprawls and unplanned, unsustainable urbanization

respiratory diseases, neuro-mental illnesses, digestive disorders, and various types of cancers (Fig. 1). These NCDs are often referred to as "lifestyle diseases" or the "urban health penalty," primarily impacting individuals who are disproportionately affected by social inequalities (Ballester et al., 2020).

The importance of access to natural spaces for our mental and psychological health benefits is being acknowledged now, which has ramifications for how modern urban areas are designed or how they sprawl and expand without any planning in the Global South (Douglas & Douglas, 2021). According to Kondo et al. (2018), there is an inverse correlation between exposure to UGS and mortality rates, incidents of elevated heart rate, and violence. Additionally, there is a positive association between exposure to green spaces and attention, mood enhancement, and physical activity. In investigations of connections of UGS exposure with general health, weight, depression, and stress, the results were mixed, and no association could be established with UGS. However, the amount of research carried out so far on the impact of UGS on human blood pressure, heart rate, diabetes, cancer, respiratory symptoms, and birth defects is insufficient to make the broader connections.

3 Rural–Urban Interfaces: A Cauldron of Environmental and Human Health Issues

While the close connections between ecosystem services (ES) and human health and well-being are better understood in urban spaces, the socio-ecological transformations in urban fringe areas, often with severe repercussions to the environmental and human health, have traditionally received scarce attention in research and policymaking. Urban fringes, also known as rural–urban interfaces (RUIs) or peri-urban regions/interfaces, are transitional zones between rural and urban areas, characterized by their ambiguous boundaries with respect to urbanization. RUIs supply resources to the urban core and function as sinks for the city's wastes. These areas are characterized by rapid and drastic land use changes, livelihood heterogeneity, marginalization in terms of resource access and distribution, and most importantly, overlapping institutional structures (Marshall & Randhawa, 2017). The hydrological, ecological, geomorphological, and socioeconomic dynamics of RUIs are frequently overlooked by both rural and urban administrations, resulting in inadequate policy implementation.

3.1 Environmental Degradation in Rural–Urban Interfaces (RUIs) and Its Impacts on Human Health

Global assessments by the Peri-Urban Environmental Change Project (PU-ECH) of the International Council for Science (ICSU) highlighted various challenges, ranging from agricultural land conversion to accommodate peri-urban housing, immigration due to wars and civil conflicts, and industrial pollution to air, water, and soil contamination and their associated health risks in the RUIs of the Global South cities, such as Shanghai in China, Kumasi in Ghana, Maila Saba in Nairobi, and Karachi in Pakistan (Douglas, 2005). The diversion of productive resources, including water, land, and labor, to urban centers and the channeling of urban pollution and waste to RUIs lead to environmental degradation and the consequent marginalization of the poorer sections, such as farmers and livestock rearers, creating "degenerated peripheries" around many cities (Subramanian et al., 2015). The relocation of polluting industries and waste-dumping sites to RUIs has forced farmers to use contaminated water around the urban centers of Delhi and Kolkata. There have also been observations of unsustainable groundwater extraction and untreated sewage disposal. The health hazards continue through the food chain when degradation of the agroecosystem leads to low crop yields, poor food safety standards, and compromised human health (Marshall & Randhawa, 2017). Traditional activities like livestock grazing and fishing in lakes that are important for food and livelihood security of the urban poor have also been impacted due to the privatization of lakes, following urban expansion (Mundoli et al., 2015). Reduced access and unfair distribution of water to peri-urban and rural communities, on account of the privatization of water resources and urban pollution, have been reported from the cities of Cochabamba in Bolivia and Delhi in India (Mehta et al., 2014). The inequity in water distribution, providing greater access to the rich and forcing the poor to depend on diminishing groundwater and poorly chlorinated or contaminated informal supplies (Mehta et al., 2014), is perpetuated by the poor governance of the RUIs. Water insecurity in peri-urban areas, exacerbated by urbanization-induced stressors, such as changes in water flow patterns from rural to urban regions, competition for groundwater, and the use of peri-urban water sources for industrial waste disposal, has been documented in Khulna (Bangladesh), Kathmandu (Nepal), and Gurgaon and Hyderabad (India) (Narain et al., 2013). In coastal cities such as Mumbai and Dhaka, large-scale environmental degradation of peri-urban mangroves and floodplains is observed owing to changes in the use of rivers and waste disposal. An increased inflow of sewage and effluents contributes to high health risks, in addition to reduced resilience of the mangrove ecosystems (Marshall et al., 2009). The high incidence of coastal flooding in the RUIs owing to high coastal vulnerabilities, as well as the stress of the pandemic scenarios, can further impair the resilience of the marine and coastal ecosystems (e.g., Kundu and Santhanam (2021a, 2021c)). Climate change exacerbates the risks of emerging infectious diseases (EIDs) in RUIs by displacing people and other species into peri-urban areas, thereby increasing opportunities for the transfer of zoonotic diseases. RUIs are particularly vulnerable to escalating climate impacts, such as severe weather events, due to their limited resources and services to mitigate these events. This creates socio-ecological conditions that facilitate the outbreaks of EIDs (Matthew et al., 2022).

3.2 *Peri-Urban Agriculture and Urban Human Health*

Peri-urban agriculture serves as a critical bridge between urban human health and the outskirts of cities, contributing to urban food security. It has been recognized as a safety net for alleviating urban poverty, ensuring livelihood security, and enhancing nutrition availability in the Global South (Pribadi & Pauleit, 2016). Despite its potential as a "green infrastructure" that provides valuable ES, peri-urban agriculture has received poor recognition in urban policy and planning (Shackleton et al., 2021). The informal nature, low prioritization compared to urban developmental needs, and exclusion from agricultural support schemes arising from overlapping jurisdictional boundaries have pushed peri-urban agriculture toward vulnerability (Marshall et al., 2009). In a global meta-analysis, the replacement of peri-urban agriculture, triggered by use changes and growth of built-up areas, was found to be reported in 88% of reports. Around 42% of them indicated crop intensification, driven by a changing demand for food crops and resulting in increased inputs of inorganic fertilizers and pesticides. In Asia and Africa, agricultural land replacement was documented in 86%–95% of the studies, and crop intensification was reported in 41%–55% of studies (Follmann et al., 2021). Urban expansion pressures resulted in large-scale land acquisition and conversion and loss of farm livelihoods and may also create a deficit in the urban food market supply (Marshall & Randhawa, 2017; Coulibaly & Li, 2020). In China, urban expansion and developmental activities, like housing, roads, industries, power stations, and treatment plants, have displaced peri-urban agriculture. While proximity to large urban centers has increased the demand for crop and livestock produce, loss of cultivable land is also on the increase across peri-urban China. The pressure to move polluting industrial units away from the wealthy urban centers results in peri-urban water, soil, and land degradation, leading to reduced yields and lower production of fresh and safe produce for urban centers (Simon, 2008). Water scarcity, increased dependency on contaminated urban wastewater, an increase in inorganic inputs, and labor issues are common features of peri-urban farming across the Global South (Hofmann, 2013). In sub-Saharan African countries like Zimbabwe, Ghana, and Burkina Faso, the high demand in urban areas has driven water- and chemical-intensive monoculture food production to the peripheries as a means of sustaining livelihoods (Cofie et al., 2003). To cope with rising urban food demand, peri-urban areas frequently resort to using urban wastewater as a source of nutrients. Sewage irrigation leads to microbial and trace metal contamination of agricultural produce, especially of leafy vegetables (Nabulo et al., 2012). The levels of organic micro-contaminants and trace elements were notably higher in vegetables grown in peri-urban areas compared to those cultivated in rural areas (Margenat et al., 2019). Research from Kano, Nigeria; Kumasi, Ghana; Hubli-Dharwad and Kolkata, India, highlights that low-income families rely on wastewater and biodegradable components from urban waste for farming. Urban markets, waste disposal sites, sewage treatment plants, and nearby streams and rivers serve as readily available and cost-effective sources of biodegradable waste for agriculture and aquaculture in peri-urban areas. Women and children in the RUI areas are the most affected by their high exposure to wastes contaminated with chemicals and fecal matter (Hofmann, 2013).

4 Frameworks and Tools to Explore Urban Environment–Human Health Nexus

4.1 Urban Ecological Risk Assessment

Ecological risk is calculated by several methods; for instance, Kang et al. (2018) linked ecosystem health with ecosystem services to calculate the ecological risk. Committee et al. (2019) argue for a harmonized approach in the calculation of ecological risk by including the combined exposure to multiple harmful chemicals in relevant areas, including human health, animal health, and ecological health. At present, there is no clear working method for the calculation of the ecological risk in urban systems as a whole. However, ecological risk has been calculated by considering contamination by or exposure to toxins in the environment. According to Perrodin et al. (2011), most of the ecological risk assessment methods follow the steps in the sequence of (a) defining the problem, (b) quantifying the exposure to harmful elements, (c) characterizing the effects on the system, and last, (d) characterizing the risk. The risk to human health in the urban environment is from multiple spheres, including the atmosphere, lithosphere, and hydrosphere. Potential ecological risk index (RI), cancer risk (CR), and hazard quotient (HR) are commonly calculated from the measured pollutant data from urban environments to quantify the level of risk to human health (Urrutia-Goyes et al., 2018). Due to rapid economic development and haphazard urbanization, the soil in the urban areas is found with higher concentrations of heavy metals than natural soil. Kumar et al. (2019) observed that heavy metal concentrations in Indian natural soil are above the limits, and anthropogenic activities mainly influence this change. The study shows that contamination by Cd and As is predominant in India. Higher heavy metals in soil can directly enter the food web and can lead to biomagnification. Huang et al. (2018) calculated the soil health risk in peri-urban areas in China and found that contamination by As, Cd, and Pb is predominant. Urrutia-Goyes et al. (2018) calculated the ecological and human health risk due to street dust in the urban environment of Monterrey, Mexico. The study reported a high concentration of heavy metals due to the pollution from the wear of tires and brake pads and industrial emissions. Similarly, an increase in mass loads of antibiotics, persistent organic pollutants, and trace elements is reported in urban wastewater streams (Mirzaei et al., 2019). Adimalla (2020) shows that the geo-accumulation of As, Cr, Cu, Zn, Ni, and Pb is in moderate-to-extreme levels in almost all cities of India, posing potential health risk to adults and children. Dar et al. (2021) identified that the heavy metal contamination of urban wetlands is mainly from anthropogenic sources. Studies indicate that exposure to fine and ultrafine particulate matter (PM) is associated with cardiovascular diseases, hypertension, type 2 diabetes, obesity, cancer, and increased mortality (Alemayehu et al., 2020). The study also identifies a connection between prolonged exposure to air pollution and rising hospital admissions for conditions such as heart failure, hypertension, diabetes, and cancer across multiple countries. Moreover, there is a significant positive correlation between exposure to PM and mortality, as evidenced by epidemiological research (Tagliabue et al., 2016) (Tagliabue et al., 2016) and biomarker (Jia et al., 2017) studies reported across the world. There are very few studies relating the health damages in animals, birds, and trees in an urban setup. Noninvasive avian sample analysis is widely adopted to study the stress levels in urban birds. Oxidative indicators in house sparrows can be a bioindicator of pollution in urban centers (Herrera-Dueñas et al., 2014). Studies on mutations in animals, air pollution impacts on avian species, cancer risk in urban wildlife, modification of animal behavior due to urban environment (Somers et al., 2004; Sanderfoot & Holloway, 2017), etc. also reveal the relationship between urbanization and animal health. Nature plays a crucial role in human well-being, especially regarding health. With increasing urbanization, there is a growing need to study the connection between urban ecosystems and human health. To mitigate health risks, it is essential to explore urban ecology through the lens of urban design tools (Zabelskyte & Matijosaitiene, 2020).

4.2 Cities as Socio-Ecological Systems

Cities are a critical nexus in the human–nature connection, serving as massive demand centers for ES while simultaneously resulting in massive environmental consequences (Elmqvist et al., 2015). The flow of energy and materials happens through dynamic nature–society–economy interlinkages, with the human society controlling this flow of energy and materials via the socio-ecological system. ES determine the culture and society of a place and also place-based attachment. In short, ES determine the identity of a society, and in turn, the dynamic social characteristics affect the quality of ES. The roles of society and nature are mutual in managing the socio-ecological system. Solutions to the complex sustainability challenges need to address the reciprocal linkages between social and ecological systems (Frank et al., 2007; Barnard & Elliott, 2015; Berrouet et al., 2018; Egerer et al., 2020; Everard, 2020; Enssle & Kabisch, 2020). Human interactions with ecosystems are reflected through diverse cultural practices that are closely tied to human well-being. Cultural services denote the nonmaterial benefits derived from ecosystems that form the foundation of human culture, society, knowledge, philosophy, and heritage (de Groot et al., 2002). The concept of Nature's Contribution to People (NCP) expands the conceptual framework to enhance participation and deepen the understanding of the interrelationships between humans and nature (Kadykalo et al., 2019). Urban areas are centers of high productivity in terms of economic growth across the world while encompassing highly dynamic socio-ecological systems. Spatiotemporal perspectives play an important role in determining the functioning of the socio-ecological system in the same geographical region (e.g., Dhyani et al. (2021) and Santhanam (2021)). Neglecting the socio-ecological values can lead to ecological degradation of different urban ecosystems, as documented in the case of forests and UGS (Zepp et al., 2018; Afriyanie et al., 2020; Bonilla-Bedoya et al., 2020; Klobucar et al., 2021; Rasmi et al., 2021), rivers (Zingraff-Hamed et al., 2021), estuarine fisheries (Taylor & Suthers, 2021), and water (Niva et al., 2019). Different authors have suggested a holistic socio-ecological systems approach (Frank et al., 2017; Tawil & Olszewska, 2020; Afriyanie et al., 2020; Mak et al., 2021) to study urban dynamics, sustainability, and resource efficiency. It is necessary to generate a spatiotemporal understanding of the socio-ecological system at the local level to recognize adequate management strategies for urban centers and public spaces to improve public health and reduce inequality (Bonilla-Bedoya et al., 2020; Dhyani et al., 2021), as well as aid in a better assessment of the ecosystem service loss and the construction of better indicators (Atif et al., 2018; Pan et al., 2019). "Urban consonance," a novel paradigm in urban planning, highlights multifunctional landscapes as the cornerstone of urban design and emphasizes the importance of biodiversity and ES for enhancing human well-being and the resilience of cities. This approach embodies fundamental characteristics and principles aligned with a holistic socio-ecological perspective (Heymans et al., 2019).

4.3 Healthy Urban Ecosystems from the Perspective of Environmental Neuroscience (ENS)

Concepts such as "environmental affordances"[1] and "environmental constraints"[2] have been helpful in identifying both *"opportunities and inflexible barriers,"* which can influence and/or *"promote particular psychological and behavioral responses to stress"* (Environmental Affordance Model; Mezuk et al., 2013). From these studies, it is clear that the "socioeconomic positions" in a dwelling area can influence several health-relevant decision-making processes through their convergence with both the physical and social environments (Link & Phelan, 1995). The success of mainstreaming Nature-based Solutions (NbS), therefore, lies in balancing the internal and external sustainability components pertaining to the urban blue-green infrastructure (BGI) (Dhyani et al., 2021). From the perspective of design, development, and utilization of blue-green infrastructure (BGI), new fields such as "environmental neuroscience" (ENS) can be used to explore the relationships between the human-urban spaces. Environmental neuroscience (ENS) is defined as a type of integrative study "*......to understand the qualitative and quantitative relationships between the external environment, neurobiology, psychology and behavior.*" (Berman et al., 2019). The ENS methods can quantify the spatial and temporal dynamics of interactions between biological elements at various scales within built environments, including urban green spaces and rural or urban landscapes. Figure 2 explains the different ENS-based factors that influence the design and placement of BGI in urban environments. For example, considering the creation of healthy spaces as socio-ecological initiatives, under the ambit of ENS, it is possible to explore whether genetic profiles (e.g., polymorphisms) could affect how an individual benefits from his/her interactions with greenspace or blue infrastructure.

Another aspect that can be explored as regards the influences of different enriched environments on the structural and functional brain networks in human and other nonhuman species, which, in turn, impact the overall life quality in an urban space. It is evident that thoughtful planning of the "design," "placements," and "engagement" of BGI in urban centers, with a focus on long-term perspectives, is essential to create healthy urban environments that foster human well-being (Santhanam, 2021). Table 1 shows the list of BGI that can be matched with positive ENS impacts over a long term.

---

[1] In Environmental Psychology, environmental affordances are how we interpret settings as the means to meet our requirements. These requirements can range from shelter, nutrition, parking, safe strolling, seating, and psychological benefits, among others. An ecological niche exists when an environment fits practically all of an individual's demands. This would be the most suitable atmosphere for that individual.

[2] Environmental constraint is any factor in a person's life or environment that hinders skill development, freedom, psychological adjustment, or adaptive conduct or prevents the exhibition of already learned skills.

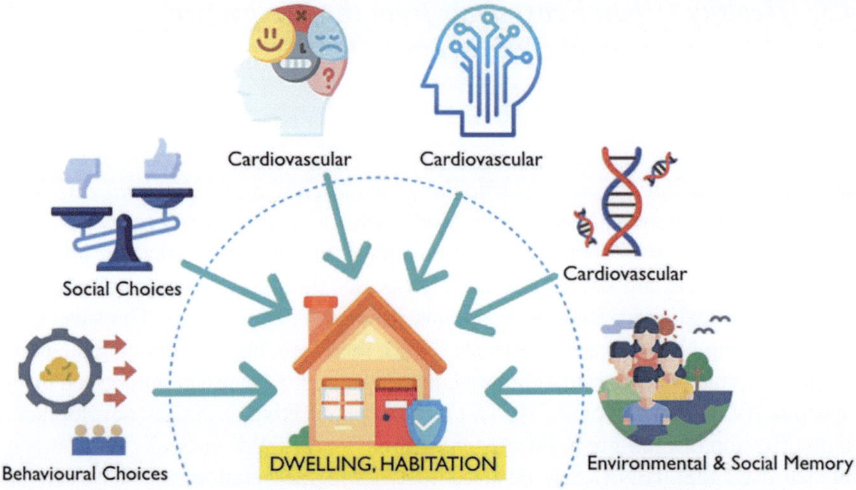

Fig. 2 The concept of dwelling or habitation as an important socio-ecological factor impacting the BGI in urban areas under the perspective of environmental neuroscience

Table 1 Correlation of the planning of NbS and the placement of urban blue-green infrastructure with the benefits of environmental neuroscience (ENS) from Santhanam (2021)

BGI/NbS	Activities	ENS benefits
Neighborhood lake area	Integrated shore-based facilities for physical training, physiotherapy, yoga, and cultural activities	Direct tangible benefits: Health and psychological Well-being
Road/street tree networks	Integrated green spaces for cycling, walking, and relaxing; bird-watching; and tree identification	Direct tangible benefits: Lung spaces for relaxing, psychological unburdening, ES from "locavory"
Converted brownfield sites into eco-friendly landscapes	Creation of eco-friendly learning, living spaces with NbS	Positive ENS with enhanced environmental and social memories and value addition to ES

5 Sustaining Human Health in an Urbanizing World: The Way Forward

The SDGs offer a comprehensive framework for addressing global urbanization, providing both global directives and local action opportunities, with a specific focus on reducing health inequalities. Health and well-being are explicitly prioritized in SDG 3, while they are also foundational to achieving SDG 11's objective of fostering inclusive, safe, resilient, and sustainable cities (Ramirez-Rubio et al., 2019). There is an increasing imperative to establish pathways that integrate urban blue-green spaces and their ES into urban planning, emphasizing their positive impacts on health and well-being. This approach seeks to harness the benefits of these natural environments to enhance urban livability, mitigate environmental challenges, and promote community health.

5.1 Valorizing Urban Ecosystem Services from Nature-Based Solutions

The perception of fast-expanding urban sprawls as not merely “places of work” but also as economic and cultural hubs by the inhabitants of urban regions indicates the needs of the urban population to connect to the conservation of biodiversity. Consequently, more and more individuals are choosing to reside in suburban or rural areas with abundant greenery. However, this, in turn, leads to greater automotive traffic, giving rise to traffic jams, accidents, stress, and increased environmental harm. The resulting traffic congestion and UHI effects are not only bad for the environment, but they also have social and economic consequences. Biodiverse habitats contribute to human health by offering a range of ES, including food and medicinal resources, and by purifying air and water (Flies et al., 2020). They also produce shade, release oxygen, absorb carbon dioxide, and reduce energy use and pollution emissions (Gherraz et al., 2020). The ES such as climate regulation, air quality maintenance, aesthetics, and cultural enrichment contribute to promoting human health and well-being (Salmond et al., 2016). The production and consumption of urban ES occur within the framework of independent but interconnected social, ecological, economic, and technical systems. Comprehending the dynamics of urban ES is crucial for efficient planning, management, and governance of urban green infrastructure (UGI) (Kremer et al., 2016). Embracing the concept of urban ES can profoundly reconnect cities with the biosphere, diminish the ecological footprint and debt of urban areas, and simultaneously bolster resilience, health, and quality of life of the residents (Gómez-Baggethun et al., 2013). Promoting biodiversity and well-being by linking them to land use planning, infrastructure, architecture, and buildings provide opportunities to realize the full potential of optimizing ecological services in these urban structures.[3] UGS play a crucial role in delivering a diverse array of ES, essential for maintaining a healthy urban environment. Across urban areas, the proportion of green and developed infrastructure varies between city centers and larger metropolitan regions. The green infrastructure outside the city core constitutes a substantial portion of the overall urban greenery, which is vital for supplying ES. It is imperative to consider these areas when planning urban growth and climate adaptation strategies (Kourdounouli & Jönsson, 2020).

[3] https://www.swecourbaninsight.com/climate-action/building-in-biodiversity-for-climate-for-health/

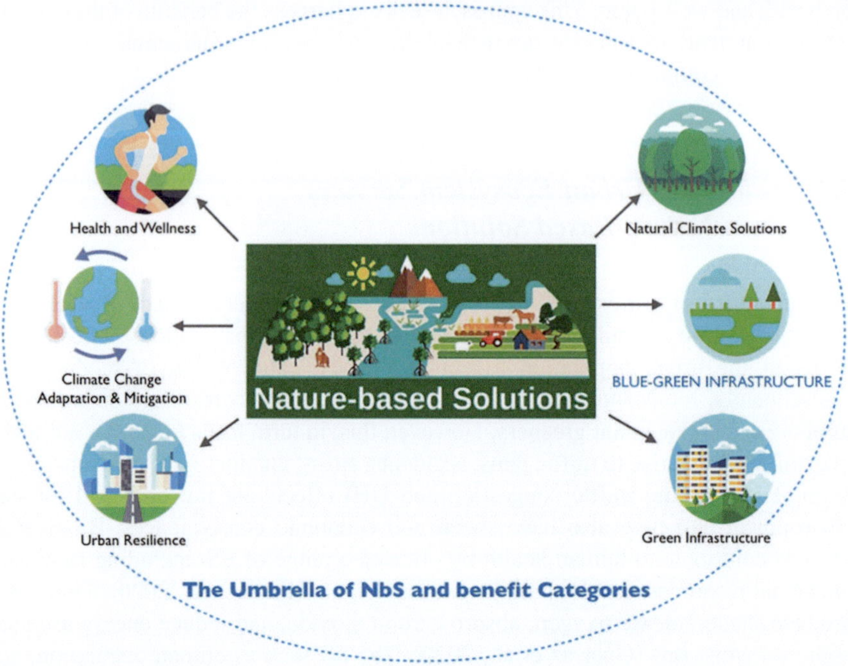

Fig. 3 The umbrella of NbS for improving urban resilience, health, wellness, climate change adaptation, and mitigation

Blue-green infrastructure (BGI) or green-blue infrastructure falls under the category of urban *Nature-based Solutions (NbS)*, focusing primarily on three key pillars: climate change mitigation and adaptation, promotion of public health, and enhancement of urban resilience (Dhyani et al., 2020, 2021, 2022; Dhyani & Kadaverugu, 2020; Dhyani, 2021). BGI serves as a crucial Nature-based Solution (NbS) for urban areas. It comprises a strategically planned network of natural and seminatural areas, alongside other environmental features, designed and managed to provide diverse ES. Recently, BGI has also come to encompass elements of biodiversity and organized systems that contribute to the natural capital of urban areas, integrating technological devices dependent on biodiversity into the built environment (Fig. 3).

The Delhi Development Authority is currently formulating a "blue-green masterplan" for Delhi, India, which delineates between blue and green infrastructure within the framework of urban planning. This initiative aims to enhance urban living conditions. Additionally, India's national flagship projects, the Smart Cities Mission and the Atal Mission for Rejuvenation and Urban Transformation (AMRUT), are integrating elements of blue and green infrastructure to further improve urban environments (Udas-Mankikar & Driver, 2021).

5.2 *Nature-Based Solutions for Developing Urban Resilience to Climate Change*

Renaturing cities and promoting green infrastructure are two of the most effective instruments we have for solving two of today's most pressing urban issues: climate change and city habitability (Ballester et al., 2020). According to the World Health Organization (WHO), urban residents should be able to access green space within 300 m of their home with a ratio of 10–15 m^2/capita of green space (WHO, 2017). However, there is a huge disparity in the availability of per capita green space to urban residents across the world. Many rapidly growing urban areas in the developing and underdeveloped countries of the Global South face significant demand–supply gaps in UGS due to inadequate access. Amid escalating concerns over climate change, numerous cities worldwide are shifting their urban planning and design strategies to integrate NbS. This approach contrasts with traditional infrastructure practices by blending blue elements (like lakes, wetlands, rivers, and seas) with green elements (such as trees, gardens, parks, urban forests, and playgrounds) (Udas-Mankikar & Driver, 2021). The primary NbS-based needs for the urban population include the requirement for green lung spaces and the spaces that aid in flood control as well as provide heat resilience (e.g., Santhanam and Majumdar (2020) and Dhyani et al. (2021)). Evidence-based design and NbS have been demonstrated to be useful for improving human well-being (Andreucci et al., 2019). Green spaces within cities exert a beneficial influence on the climate and microclimate by providing shade and promoting evapotranspiration, thus mitigating the UHI effect. Enhancing and strategically placing urban vegetation have been recognized as effective measures to mitigate UHI, improving human thermal comfort in urban environments. For instance, a study modeling the cooling potential of UGS in Lisbon revealed that to achieve a 1 °C reduction in air temperature relative to a reference station, an additional 50 square meters of vegetated area is required. These findings are applicable to other urban areas, allowing for the quantification of the cooling impact of plants to enhance urban thermal conditions and human well-being, thereby mitigating some of the anticipated impacts of climate change in the future (Reis & Lopes, 2019). NO_2 reduction, recreation opportunities, and a reduced UHI effect were all determined to be positive consequences of UGS (Kourdounouli & Jönsson, 2020). Any park positioned within an urban structure, regardless of scale, may effectively control heat in suburban neighborhoods. The type of vegetation, its distribution throughout the city, size, and other factors all play crucial roles (Baris et al., 2009). Preserving and enhancing UGS in densely populated cities are vital to ensuring a high quality of life and realizing social and economic benefits (Heidt & Neef, 2008). Parks can filter up to 80% of airborne pollutants, while trees along streets can filter up to 70%. Even in winter, when trees lose their leaves, they maintain 60% of their pollutant-filtering efficiency, reduce noise levels by up to 12 dB, and provide oxygen during calm weather conditions (Bernatzky, 1982). Although there are numerous information gaps in the social, ecological, and biophysical dynamics of employing urban vegetation to lessen the climate-induced impacts, it is accepted that it provides a cooling ecosystem function (Jenerette et al., 2011). Urban tree planting initiatives are actively promoted as a planning tool to help urban areas adapt to and mitigate climate change effects, enhance urban sustainability, and improve human health and well-being (Salmond et al., 2016). Climate change is projected to increase the average afternoon physiological equivalent temperature (PET) values by 2.4 K; however, integrating UGI could significantly reduce this impact. The UGS cooling effect refers to the UGI's capacity not only to cool its immediate surroundings but also to influence the broader urban environment. Planting trees had the greatest benefits, reducing PET by 13% on average when compared to existing vegetation. Trees give shade and evapotranspiration cooling by shading open spaces. Green facades, which have a moderating impact of 5%–10%, are another useful adaptation alternative. Green roofs, on the other hand, had negligible effects, indicating the relative benefits of NbS. It has been noted that simply increasing the percentage of green cover does not necessarily lead to an immediate reduction in potential evapotranspiration. Therefore, strategically placing vegetation in heat-exposed areas is more prudent than solely aiming for a higher overall green cover percentage (Zölch et al., 2016). Various studies have shown that the shape and kind of living and building elements in UGS have a significant impact on their performance (Graça et al., 2022). Some argue that large urban gardens exceeding 10 ha exert the most significant cooling impact in terms of distance and intensity. Nevertheless, the cooling effect of UGS is significantly influenced by their natural characteristics, quality, and prevailing climatic factors (Aram et al., 2019). Limited data suggests that land use not only influences the composition and configuration of UGS but also their cooling effects. Areas with simpler shapes, less fragmentation, and greater interconnectivity of UGS tend to exhibit lower land surface temperatures and produce a more pronounced cooling effect (Masoudi et al., 2021).

5.3 *Repurposing Urban Waste to Mitigate Peri-Urban Health Impacts*

The health impacts of peri-urbanization could be mitigated by reorienting urban metabolism to reduce the urban export of water and raw materials while treating the urban waste and sewage locally to avoid the impacts of waste discharge on peri-urban areas. The reuse of nutrients and water from treated wastewater in urban agriculture is often considered a feasible solution addressing food, energy, water, and health challenges (Miller-Robbie et al., 2017). For urban farmers in search of a reliable source of irrigation water, the perennial availability of nutrient-rich sewage lowers fertilizer costs, boosting the profits from the ever-increasing urban demand. In Bengaluru, for instance, peri-urban farmlands have shifted from traditional food grains, such as millets, to grow vegetables, fruits, and fodder grass that are more amenable to wastewater irrigation and are constantly demanded from the city. The incidence of skin disorders, gastrointestinal ailments, and vector-borne diseases in humans and cattle, on the other hand, indicates that sewage irrigation and the resultant contaminated drinking water and soil have ecological, biological, and societal consequences. Implications on human health are intense, given the fact that heavy metals were detected in the vegetables and milk from these locations (Patil et al., 2018). The Minor Irrigation Department of Karnataka launched a new initiative aimed at enhancing the value of wastewater. This project involves pumping 300 million liters of secondary-treated wastewater daily from the city to peri-urban lakes, aiming to replenish aquifers and support agricultural activities. This is one of the first projects in India to use secondary processed wastewater in such large quantities to fill the peri-urban tank ecosystem and offer water for agricultural use (Paliath, 2019). Converting urban sewage into a resource for water-scarce agricultural peripheries suggests the potential for bridging the water and nutrient divide between the peri-urban production landscapes and urban consumption centers; however, it could be risky to both producers and consumers without effective water pollution control measures (Patil et al., 2018). This calls for stringent regulations to ensure appropriate wastewater treatment for irrigation needs, securing farm livelihoods and the city's food supply. Existing regulatory frameworks, such as the Prime Minister's Irrigation Policy, 2015, and the Revised Municipal Solid Waste Management Rules, 2016, encourage the use of treated wastewater for irrigation and composted solid waste as fertilizer in peri-urban farmlands. Leveraging these frameworks could foster the closure of nutrient and water gaps between rural and urban areas, thereby promoting beneficial outcomes for human health and the environment in urban peripheries (Patil et al., 2018).

5.4 Leveraging "Biophilia" Supporting Positive Human Health Outcomes for Urban Residents

Urban BGI integrates various ES that significantly contribute to the mental health and well-being of urban residents, promoting healthy aging and addressing the challenges posed by climate change (Andreucci et al., 2019). BGI not only enhances the physical environment but also delivers multiple co-benefits. *First*, BGI plays a crucial role in creating a healthier urban environment by mitigating noise pollution, reducing heat stress, and enhancing air quality. These improvements are essential for fostering a conducive environment for mental and physical health. *Second*, BGI possesses aesthetic and sensory qualities that can have therapeutic effects on individuals, even if they do not directly interact with the green spaces. For instance, a small pocket park nestled within private land can provide visual relief and contribute to a sense of tranquility and well-being within the surrounding community. *Last*, physically accessible BGI areas facilitate physical activity and social interaction.

Fig. 4 Socio-ecological, economic, and health benefits from healthy and rich urban green spaces or blue-green infrastructure. (Adapted from USEPA)

By accommodating individuals with varying mobility needs, these spaces encourage diverse groups to engage in recreational activities, promoting fitness and fostering social connections (Sharp et al., 2020) (Fig. 4).

Monetising Human Health Benefits from Healthy UGS

The value of UGS for mental, social, and physical health in England has been valued at EUR 2–3 billion in avoided health expenses. All of the aforementioned health risks are linked to unemployment and sick leaves. Greening neighborhoods might save the Netherlands hundreds of millions of euros per year. Even a 1% reduction in sick leaves has been estimated to save billions of euros.[4] A study cited three important BGI characteristics that have health-related environmental consequences. According to the Scottish Natural Health Service, engaging in a brisk 30-min walk in a natural environment 5 days a week can significantly reduce health risks: It lowers the risk of heart attack and stroke by 20–30%, diabetes by 30–40%, hip fractures by 36–68%, bowel cancer by 30%, and breast cancer by 20% (Ballester et al., 2020). Kardan

[4] https://ec.europa.eu/environment/nature/ecosystems/pdf/Green%20Infrastructure/GI_health.pdf

et al. (2015) observed that adding approximately 10 more trees to a city block can improve the perception of human health to an extent comparable to the well-being associated with a $10,000 increase in annual income, moving into a community with a $10,000 higher median income, or experiencing the health benefits of being 7 years younger. The study also reported that the addition of 11 more trees in a city block reduced cardio-metabolic diseases attributed to the equivalent benefits of a $20,000 increase in annual individual income, relocating to a $20,000 higher median income neighborhood, or having the benefit of being 1.4 years younger than the actual age. According to a recent study, the presence of a green area nearby reduced melancholy and asthma by 25–30% and also increased the amount of time people spend exercising by a factor of three. Physical inactivity is a recognized factor impacting healthy life expectancy and is estimated to cause 600,000 deaths annually in Europe (Ballester et al., 2020).

The advantages of UGS to city dwellers have also been well-recognized, but understanding their emotional relationship with UGS is an ongoing and fascinating subject for researchers (Dasgupta et al., 2022). UGS, such as parks, playgrounds, and residential greenery, can significantly promote mental and physical health and reduce morbidity and mortality among urban residents. These spaces provide psychological relaxation and stress relief, foster social cohesion, support physical activity, and reduce exposure to air pollutants, noise, and excessive heat (Fig. 4). By understanding and utilizing existing evidence of the specific positive impacts of UGS on mental health, cardiovascular disease risk, obesity, diabetes, and mortality, we can enhance and ensure human well-being in rapidly expanding urban areas (Braubach et al., 2017). There is strong evidence indicating that green spaces enhance various cognitive functions, including attention, memory, and creativity, in individuals both with and without depression (Douglas & Douglas, 2021). This finding comes as no surprise to the growing number of psychologists and ecologists who are studying the effects of nature on mental health and well-being. Their research reveals complex relationships that are not yet fully understood. The COVID-19 pandemic has brought these issues into sharper focus, highlighting the diminishing access to nature in our increasingly urbanized society. This lack of access disproportionately affects socioeconomically disadvantaged and marginalized populations, who are often underserved in terms of green space availability (Douglas & Douglas, 2021; Kundu & Santhanam, 2021a).

Australia's "Healthy Parks, Healthy People" Initiative

The greenbelt of the city of Vitoria-Gasteiz in Spain and the Barcelona region's "Healthy Paths in Periurban Natural Areas" policy are two of the successful initiatives and policies that promote peri-urban parks.[5,6,7] The city of Vitoria-Gasteiz is widely recognized as a model of sustainable urban development, as evidenced by the numerous national and international awards it has received for its environmental quality initiatives. Notably, its greenbelt—a network of natural parks encircling the city—was named one of the top 100 initiatives at the United Nations' Third International Competition of Good Practices (Aguado-Moralejo et al., 2013). The city of Mollet del Vallès, Spain, with a clear strategy, went from 3.7 to 16 m^2 of UGS/inhabitant in 25 years, with 100% of the population now living within 300 m of any green space (Ballester et al., 2020).

In the case of coastal urban cities, the mainstreaming of NbS using indigenous and traditional knowledge systems can provide positive livelihood enhancements (Kundu & Santhanam, 2021b), as well as positive ENS reinforcements, aiding the long-term usage of ENS resources for a healthy environment. The co-benefits of the use of NbS are also further significant toward the preservation of the urban BGI (Santhanam, 2021). Thus, the co-benefits of incorporating the concepts of ENS in the planning and design of BGI can include a positive influence on the types and compositions of neighborhood ENS projects, improved chances for long-term survival of ENS schemes in the form of UGS, and a boost to local, state, and central governments and administrative bodies to plan for the regulation of ES as positive ENS and environmental affordance models for developing resilient cities in the Global South. This means adding a significant responsibility on urban planners to include reflective, robust, flexible, resourceful, inclusive, and integrated approaches for developing resilient future cities, which will be successful models of internalizing the principles and visions of SDG 11 (Fig. 5).

6 Way Forward

It is very clear that there exists a significant gap between the theory and practice of green space design in the Global South despite so many small cases of success. In addition, there is also a paucity of knowledge about how to use UGS for climate

[5] https://www.awe.gov.au/parks-heritage/national-parks/associated-programs/healthy-parks-healthy-people

[6] https://www.europarc.org/case-studies/healthy-paths-in-periurban-natural-areas/

[7] https://www.europarc.org/case-studies/vitoria-gasteiz-green-belt-actions-conservation-biodiversity/

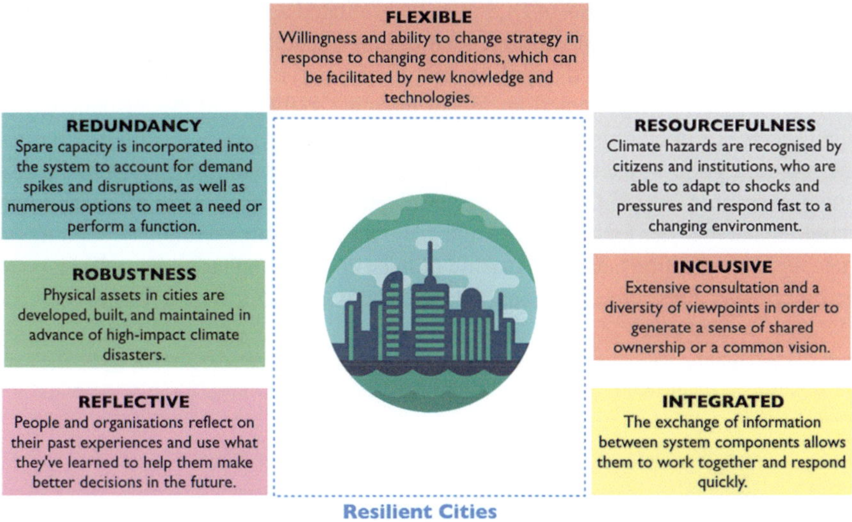

Fig. 5 Criteria and indicators to develop resilient cities in the Global South to ensure human health and well-being in the Anthropocene (https://prairieclimatecentre.ca/wp-content/uploads/2017/04/pcc-brief-climate-resilient-city-urban-ecosystems.pdf)

change mitigation and adaptation (Govindarajulu, 2014). It is crucial to recognize that, while UGS with multiple functions are highly sought after, balancing the diverse needs of human populations with the imperative of biodiversity conservation within constrained urban areas remains a significant challenge. To address this issue, the implementation of BGI through the comprehensive framework of the "One Health" approach can be proposed. This approach emphasizes the interconnectedness of human health, animal health, and environmental health, aiming to create synergies that benefit all three domains simultaneously (Felappi et al., 2020). Investing in people's health, notably through health-promotion programs (e.g., urban cycling paths), creating a network of green spaces in metropolitan areas to promote everyday physical exercise, investing wisely but not necessarily more in long-term health systems, and green infrastructure solutions can be more cost-effective than conventional approaches.[8] The concept of "*BGI*" has strong global recognition but is still considered new in Asia and Africa. The definition of this term in developing regions will be shaped by how it is integrated into policies at the national, state, and subregional levels. Landscape architects are getting interested and are looking at using the potential of green elements in urban landscape design to regulate air temperature, air quality, water and drainage, and noise issues.[9] The

[8] https://ec.europa.eu/environment/nature/ecosystems/pdf/Green%20Infrastructure/GI_health.pdf

[9] http://www.adaptivecircularcities.com/wp-content/uploads/2016/07/T02-ACC-WP3-Green-Blue-infrastructure-for-Healthy-Urban-Living-Final-report-160701.pdf

social and economic benefits of BGI must be evaluated against the expenses, including "lost opportunity" costs. Unintended effects of BGI must also be considered, such as the health risk posed by biting arthropods (e.g., ticks and mosquitoes) transmitting diseases; allergic reactions to, for example, pollen; and the introduction of alien species or infections. When such "ecological disservices" are weighed against the reported good impacts, BGI and/or GI appear to provide various benefits at a lower economic cost.[10] Climate change adaptation policies in the Global South should make it easier for cities to design adaptation measures without depending on state legislative actions, as UGS can aid climate change adaptation by reducing risk and serving as a resource for urban regeneration (García Sánchez et al., 2018).

Acknowledgments The authors thank the Knowledge Resource Center (KRC) of CSIR National Environmental Engineering Research Institute (NEERI) for checking for plagiarism of the chapter KRC.

References

Adhikari, B., Pokharel, S., & Mishra, S. R. (2019). Shrinking urban greenspace and the rise in non-communicable diseases in South Asia: An urgent need for an advocacy. Frontiers in sustainable. *Cities, 1*.

Adimalla, N. (2020). Heavy metals pollution assessment and its associated human health risk evaluation of urban soils from Indian cities: A review. *Environmental Geochemistry and Health, 42*, 173–190. https://doi.org/10.1007/s10653-019-00324-4

Afriyanie, D., Julian, M. M., Riqqi, A., et al. (2020). Re-framing urban green spaces planning for flood protection through socio-ecological resilience in Bandung City, Indonesia. *Cities, 101*, 102710. https://doi.org/10.1016/j.cities.2020.102710

Aguado-Moralejo, I., Echebarria, C., & Barrutia, J. (2013). The Green Belt of Vitoria-Gasteiz. A successful practice for sustainable urban planning. *Boletín de la Asociación de Geógrafos Españoles*, 181–193.

Alemayehu, Y. A., Asfaw, S. L., & Terfie, T. A. (2020). Exposure to urban particulate matter and its association with human health risks. *Environmental Science and Pollution Research, 27*, 27491–27506. https://doi.org/10.1007/s11356-020-09132-1

Andreucci, M. B., Russo, A., & Olszewska-Guizzo, A. (2019). Designing urban green blue infrastructure for mental health and elderly wellbeing. *Sustainability, 11*, 6425. https://doi.org/10.3390/su11226425

Aram, F., Higueras García, E., Solgi, E., & Mansournia, S. (2019). Urban green space cooling effect in cities. *Heliyon, 5*, e01339. https://doi.org/10.1016/j.heliyon.2019.e01339

Atif, S. B., Saqib, Z., Ali, A., et al. (2018). Identification of key-trends and evaluation of contemporary research regarding urban ecosystem services: A path towards socio-ecological sustainability of urban areas. *Applied Ecology and Environmental Research, 16*, 3545–3581.

Bai, X., Nath, I., Capon, A., et al. (2012). Health and wellbeing in the changing urban environment: Complex challenges, scientific responses, and the way forward. *Current Opinion in Environmental Sustainability, 4*, 465–472. https://doi.org/10.1016/j.cosust.2012.09.009

[10] https://ec.europa.eu/environment/nature/ecosystems/pdf/Green%20Infrastructure/GI_health.pdf

Ballester EJ, et al. (2020). *Urban green spaces are our Most effective tool to improve urban life*. Urbanet. https://www.urbanet.info/urban-green-spaces-improve-urban-life/. Accessed 16 Feb 2022.

Baris, M. E., Suuml, S. K., & Yazgan, M. E. (2009). The contribution of trees and green spaces to the urban climate: The case of Ankara. *AJAR, 4*, 791–800. https://doi.org/10.5897/AJAR.9000683

Barnard, S., & Elliott, M. (2015). The 10-tenets of adaptive management and sustainability: An holistic framework for understanding and managing the socio-ecological system. *Environmental Science & Policy C*, 181–191. https://doi.org/10.1016/j.envsci.2015.04.008

Berman, M. G., Stier, A. J., & Akcelik, G. N. (2019). Environmental neuroscience. *American Psychologist, 74*, 1039–1052. https://doi.org/10.1037/amp0000583

Bernatzky, A. (1982). The contribution of tress and green spaces to a town climate. *Energy and Buildings, 5*, 1–10. https://doi.org/10.1016/0378-7788(82)90022-6

Berrouet, L. M., Machado, J., & Villegas-Palacio, C. (2018). Vulnerability of socio—Ecological systems: A conceptual framework. *Ecological Indicators, 84*, 632–647. https://doi.org/10.1016/j.ecolind.2017.07.051

Bonilla-Bedoya, S., Estrella, A., Santos, F., & Herrera, M. Á. (2020). Forests and urban green areas as tools to address the challenges of sustainability in Latin American urban socio-ecological systems. *Applied Geography, 125*, 102343. https://doi.org/10.1016/j.apgeog.2020.102343

Borah, R. (2020). Development of Urban Heat Island and its relation to heat waves. In R. B. Singh, B. Srinagesh, & S. Anand (Eds.), *Urban health risk and resilience in Asian cities* (pp. 127–139). Springer.

Braubach, M., Egorov, A., Mudu, P., et al. (2017). Effects of urban green space on environmental health, equity and resilience. In N. Kabisch, H. Korn, J. Stadler, & A. Bonn (Eds.), *Nature-based solutions to climate change adaptation in urban areas: Linkages between science, policy and practice* (pp. 187–205). Springer International Publishing.

Chen, G., Li, X., Liu, X., et al. (2020). Global projections of future urban land expansion under shared socioeconomic pathways. *Nature Communications, 11*, 537. https://doi.org/10.1038/s41467-020-14386-x

Cofie, O., Veenhuizen, R., & Drechsel, P. (2003). *Contribution of urban and peri-urban agriculture to food security in Sub-Saharan Africa*. undefined.

Committee, E. S., More, S. J., Bampidis, V., et al. (2019). Guidance on harmonised methodologies for human health, animal health and ecological risk assessment of combined exposure to multiple chemicals. *EFSA Journal, 17*, e05634. https://doi.org/10.2903/j.efsa.2019.5634

Coulibaly, B., & Li, S. (2020). Impact of agricultural land loss on rural livelihoods in Peri-urban areas: Empirical evidence from Sebougou, Mali. *Land, 9*, 470. https://doi.org/10.3390/land9120470

Dar, S. A., Hamid, A., Rashid, I., & Bhat, S. U. (2021). Identification of anthropogenic contribution to wetland degradation: Insights from the environmetric techniques. *Stochastic Environmental Research and Risk Assessment*. https://doi.org/10.1007/s00477-021-02121-x

Das, T. K., & Basak, R. (2020). Noise pollution and its consequences on urban health in Sylhet City. In R. B. Singh, B. Srinagesh, & S. Anand (Eds.), *Urban health risk and resilience in Asian cities* (pp. 231–252). Springer.

Dasgupta, R., Basu, M., Hashimoto, S., et al. (2022). Residents' place attachment to urban green spaces in Greater Tokyo region: An empirical assessment of dimensionality and influencing socio-demographic factors. *Urban Forestry & Urban Greening, 67*, 127438. https://doi.org/10.1016/j.ufug.2021.127438

de Groot, R. S., Wilson, M. A., & Boumans, R. M. J. (2002). A typology for the classification, description and valuation of ecosystem functions, goods and services. *Ecological Economics, 41*, 393–408.

Dhanaraj, K., & Angadi, D. P. (2021). Urban expansion quantification from remote sensing data for sustainable land-use planning in Mangaluru, India. *Remote Sensing Applications: Society and Environment, 23*, 100602. https://doi.org/10.1016/j.rsase.2021.100602

Dhyani, S. (2021). *Nature based solutions for reducing disaster risks and resilience building: Moving ahead from theory to practices*. Focus Global Reporter. http://www.focusglobalreporter.org/nature-based-solutions-for-reducing-disaster-risks-and-resilience-building-moving-ahead-from-theory-to-practices/. Accessed 7 Feb 2022.

Dhyani, S., & Kadaverugu, R. (2020). Food security and cultural benefits from urban green spaces: Exploring urban foraging as a silently growing global movement. *Climate Change and Environmental Sustainability, 8*. https://doi.org/10.5958/2320-642X.2020.00022.8

Dhyani, S., Lahoti, S., Khare, S., et al. (2018). Ecosystem based Disaster Risk Reduction approaches (EbDRR) as a prerequisite for inclusive urban transformation of Nagpur City, India. *International Journal of Disaster Risk Reduction, 32*, 95–105. https://doi.org/10.1016/j.ijdrr.2018.01.018

Dhyani, S., Karki, M., & Gupta, A. K. (2020). Opportunities and advances to mainstream nature-based solutions in disaster risk management and climate strategy. In S. Dhyani, A. K. Gupta, & M. Karki (Eds.), *Nature-based solutions for resilient ecosystems and societies* (pp. 1–26). Springer.

Dhyani, S., Majumdar, R., & Santhanam, H. (2021). Scaling-up nature-based solutions for mainstreaming resilience in Indian cities. In M. Mukherjee & R. Shaw (Eds.), *Ecosystem-based disaster and climate resilience: Integration of blue-green infrastructure in sustainable development* (pp. 279–306). Springer.

Dhyani, S., Basu, M., Santhanam, H., & Rajarshi, D. (Eds.). (2022). *Blue-green infrastructure across Asian countries* (1st ed.).

Douglas, I. (2005). Peri-urban ecosystems and societies: Transitional zones and contrasting values. In *The Peri-urban Interface*. Routledge.

Douglas, K., & Douglas, J. (2021). *Green spaces aren't just for nature—They boost our mental health too*. New Scientist. https://www.newscientist.com/article/mg24933270-800-green-spaces-arent-just-for-nature-they-boost-our-mental-health-too/. Accessed 16 Feb 2022.

Egerer, M., Fouch, N., Anderson, E. C., & Clarke, M. (2020). Socio-ecological connectivity differs in magnitude and direction across urban landscapes. *Scientific Reports, 10*, 4252. https://doi.org/10.1038/s41598-020-61230-9

Elmqvist, T., Setälä, H., Handel, S., et al. (2015). Benefits of restoring ecosystem services in urban areas. *Current Opinion in Environmental Sustainability, 14*, 101–108. https://doi.org/10.1016/j.cosust.2015.05.001

Enssle, F., & Kabisch, N. (2020). Urban green spaces for the social interaction, health and well-being of older people—An integrated view of urban ecosystem services and socio-environmental justice. *Environmental Science & Policy, 109*, 36–44. https://doi.org/10.1016/j.envsci.2020.04.008

Everard, M. (2020). Managing socio-ecological systems: Who, what and how much? The case of the Banas river, Rajasthan, India. *Current Opinion in Environmental Sustainability, 44*, 16–25. https://doi.org/10.1016/j.cosust.2020.03.004

Felappi, J. F., Sommer, J. H., Falkenberg, T., et al. (2020). Green infrastructure through the lens of "One Health": A systematic review and integrative framework uncovering synergies and trade-offs between mental health and wildlife support in cities. *Science of the Total Environment, 748*, 141589. https://doi.org/10.1016/j.scitotenv.2020.141589

Flies, E. J., Jones, P., Buettel, J. C., & Brook, B. W. (2020). Compromised ecosystem services from urban aerial microbiomes: A review of impacts on human immune function. *Frontiers in Ecology and Evolution, 8*.

Follmann, A., Willkomm, M., & Dannenberg, P. (2021). As the city grows, what do farmers do? A systematic review of urban and peri-urban agriculture under rapid urban growth across the Global South. *Landscape and Urban Planning, 215*, 104186. https://doi.org/10.1016/j.landurbplan.2021.104186

Frank, D. J., Longhofer, W., & Schofer, E. (2007). World society, NGOs and environmental policy reform in Asia. *International Journal of Comparative Sociology, 48*, 275–295. https://doi.org/10.1177/0020715207079530

Frank, B., Delano, D., & Caniglia, B. S. (2017). Urban systems: a socio–ecological system perspective. *Sociology International Journal, 1*(1). https://doi.org/10.15406/sij.2017.01.00001
Frantzeskaki, N., McPhearson, T., Collier, M. J., et al. (2019). Nature-based solutions for urban climate change adaptation: Linking science, policy, and practice communities for evidence-based decision-making. *Bioscience, 69*, 455–466. https://doi.org/10.1093/biosci/biz042
García Sánchez, F., Solecki, W. D., & Ribalaygua Batalla, C. (2018). Climate change adaptation in Europe and the United States: A comparative approach to urban green spaces in Bilbao and New York City. *Land Use Policy, 79*, 164–173. https://doi.org/10.1016/j.landusepol.2018.08.010
Gherraz, H., Guechi, I., & Alkama, D. (2020). Quantifying the effects of spatial patterns of green spaces on urban climate and urban heat island in a semi-arid climate. *Bulletin de la Société Royale des Sciences de Liège*. https://doi.org/10.25518/0037-9565.9821
Gómez-Baggethun, E., Gren, Å., Barton, D. N., et al. (2013). Urban ecosystem services. In T. Elmqvist, M. Fragkias, J. Goodness, et al. (Eds.), *Urbanization, biodiversity and ecosystem services: Challenges and opportunities: A global assessment* (pp. 175–251). Springer.
Govindarajulu, D. (2014). Urban green space planning for climate adaptation in Indian cities. *Urban Climate, 10*, 35–41. https://doi.org/10.1016/j.uclim.2014.09.006
Graça, M., Cruz, S., Monteiro, A., & Neset, T.-S. (2022). Designing urban green spaces for climate adaptation: A critical review of research outputs. *Urban Climate, 42*, 101126. https://doi.org/10.1016/j.uclim.2022.101126
Heidt, V., & Neef, M. (2008). Benefits of urban green space for improving urban climate. In M. M. Carreiro, Y.-C. Song, & J. Wu (Eds.), *Ecology, planning, and Management of Urban Forests: International perspectives* (pp. 84–96). Springer.
Herrera-Dueñas, A., Pineda, J., Antonio, M. T., & Aguirre, J. I. (2014). Oxidative stress of House Sparrow as bioindicator of urban pollution. *Ecological Indicators, 42*, 6–9. https://doi.org/10.1016/j.ecolind.2013.08.014
Heymans, A., Breadsell, J., Morrison, G. M., et al. (2019). Ecological urban planning and design: A systematic literature review. *Sustainability, 11*, 3723. https://doi.org/10.3390/su11133723
Hofmann, P. (2013). Wasted waste-disappearing reuse at the peri-urban interface. *Environmental Science & Policy, 31*, 13–22. https://doi.org/10.1016/j.envsci.2013.03.011
Huang, Y., Chen, Q., Deng, M., et al. (2018). Heavy metal pollution and health risk assessment of agricultural soils in a typical peri-urban area in southeast China. *Journal of Environmental Management, 207*, 159–168. https://doi.org/10.1016/j.jenvman.2017.10.072
Jenerette, G. D., Harlan, S. L., Stefanov, W. L., & Martin, C. A. (2011). Ecosystem services and urban heat riskscape moderation: Water, green spaces, and social inequality in Phoenix, USA. *Ecological Applications, 21*, 2637–2651.
Jha, R. (2020). *The impact of urbanisation on health*. ORF. https://www.orfonline.org/expert-speak/impact-urbanisation-health-67644/. Accessed 15 Feb 2022.
Jia, Y.-Y., Wang, Q., & Liu, T. (2017). Toxicity research of PM2.5 compositions in vitro. *International Journal of Environmental Research and Public Health, 14*, 232. https://doi.org/10.3390/ijerph14030232
Kadykalo, A. N., López-Rodriguez, M. D., Ainscough, J., et al. (2019). Disentangling 'ecosystem services' and 'nature's contributions to people'. *Ecosystems and People, 15*, 269–287. https://doi.org/10.1080/26395916.2019.1669713
Kang, P., Chen, W., Hou, Y., & Li, Y. (2018). Linking ecosystem services and ecosystem health to ecological risk assessment: A case study of the Beijing-Tianjin-Hebei urban agglomeration. *Science of the Total Environment, 636*, 1442–1454. https://doi.org/10.1016/j.scitotenv.2018.04.427
Kardan, O., Gozdyra, P., Misic, B., et al. (2015). Neighborhood greenspace and health in a large urban center. *Scientific Reports, 5*, 11610. https://doi.org/10.1038/srep11610
Klobucar, B., Östberg, J., Wiström, B., & Jansson, M. (2021). Residential urban trees—Socio-ecological factors affecting tree and shrub abundance in the city of Malmö, Sweden. *Urban Forestry & Urban Greening, 62*, 127118. https://doi.org/10.1016/j.ufug.2021.127118

Kondo, M. C., Fluehr, J. M., McKeon, T., & Branas, C. C. (2018). Urban green space and its impact on human health. *International Journal of Environmental Research and Public Health, 15*, 445. https://doi.org/10.3390/ijerph15030445

Kourdounouli, C., & Jönsson, A. M. (2020). Urban ecosystem conditions and ecosystem services—A comparison between large urban zones and city cores in the EU. *Journal of Environmental Planning and Management, 63*, 798–817. https://doi.org/10.1080/09640568.2019.1613966

Kremer, P., Hamstead, Z., Haase, D., et al. (2016). Key insights for the future of urban ecosystem services research. *Ecology and Society, 21*. https://doi.org/10.5751/ES-08445-210229

Kuddus, M. A., Tynan, E., & McBryde, E. (2020). Urbanization: A problem for the rich and the poor? *Public Health Reviews, 41*, 1. https://doi.org/10.1186/s40985-019-0116-0

Kumar, V., Sharma, A., Kaur, P., et al. (2019). Pollution assessment of heavy metals in soils of India and ecological risk assessment: A state-of-the-art. *Chemosphere, 216*, 449–462. https://doi.org/10.1016/j.chemosphere.2018.10.066

Kundu, S. K., & Santhanam, H. (2021a). All pain and no gain: Factors impacting local and regional sustainability due to COVID-19 pandemic with respect to the Indian marine fisheries. *Current Research in Environmental Sustainability, 3*. https://doi.org/10.1016/j.crsust.2021.100086

Kundu, S. K., & Santhanam, H. (2021b). *Indigenous knowledge as 'socio-economic strongboxes' in emerging coastal cities - Technology versus ecosystem services from selected marine fish landing centres in Odisha, India*. Procs. 3rd ESP Asia conference. Nagasaki, Japan, Dec. 14-17, 2021. Session S9 - Ecosystem services and Indigenous and local communities well-being. https://www.espconference.org/asia2021/wiki/623728/book-of-abstracts

Kundu, S. K., & Santhanam, H. (2021c). A report on the impacts of cyclone Yaas over a fish landing centre vulnerable to cyclonic storms and natural hazards—Talsari, Northern Odisha. *Local Environment, 26*(9), 1043–1050. https://doi.org/10.1080/13549839.2021.1964457

Link, B. G., & Phelan, J. (1995). Social conditions as fundamental causes of disease. *Journal of Health and Social Behavior*, 80–94.

Mak, B., Francis, R. A., & Chadwick, M. A. (2021). Living in the concrete jungle: A review and socio-ecological perspective of urban raptor habitat quality in Europe. *Urban Ecosystem, 24*, 1179–1199. https://doi.org/10.1007/s11252-021-01106-6

Mansur, A. V., McDonald, R. I., Güneralp, B., et al. (2022). Nature futures for the urban century: Integrating multiple values into urban management. *Environmental Science & Policy, 131*, 46–56. https://doi.org/10.1016/j.envsci.2022.01.013

Marando, F., Salvatori, E., Sebastiani, A., et al. (2019). Regulating ecosystem services and green infrastructure: Assessment of Urban Heat Island effect mitigation in the municipality of Rome, Italy. *Ecological Modelling, 392*, 92–102. https://doi.org/10.1016/j.ecolmodel.2018.11.011

Margenat, A., Matamoros, V., Díez, S., et al. (2019). Occurrence and human health implications of chemical contaminants in vegetables grown in peri-urban agriculture. *Environment International, 124*, 49–57. https://doi.org/10.1016/j.envint.2018.12.013

Marshall, F., & Randhawa, P. (2017). *India's Peri-urban frontier: Rural-urban transformations and food security*. International Institute of Environment and Development (IIED).

Marshall F, Waldman L, Macgregor H, et al (2009) On the edge of sustainability: Perspectives on peri-urban dynamics.

Masoudi, M., Tan, P. Y., & Fadaei, M. (2021). The effects of land use on spatial pattern of urban green spaces and their cooling ability. *Urban Climate, 35*, 100743. https://doi.org/10.1016/j.uclim.2020.100743

Matthew, R., Chiotha, S., Orbinski, J., & Talukder, B. (2022). Climate change, peri-urban space and emerging infectious disease. *Landscape and Urban Planning, 218*, 104298.

Mehta, L., Allouche, J., Nicol, A., & Walnycki, A. (2014). Global environmental justice and the right to water: The case of peri-urban Cochabamba and Delhi. *Geoforum, 54*, 158–166. https://doi.org/10.1016/j.geoforum.2013.05.014

Mezuk, B., Abdou, C. M., Hudson, D., et al. (2013). "White box" epidemiology and the social neuroscience of health behaviors: The environmental affordances model. *Society and Mental Health, 3*, 79–95. https://doi.org/10.1177/2156869313480892

Miller-Robbie, L., Ramaswami, A., & Amerasinghe, P. (2017). Wastewater treatment and reuse in urban agriculture: Exploring the food, energy, water, and health nexus in Hyderabad, India. *Environmental Research Letters, 12*, 075005. https://doi.org/10.1088/1748-9326/aa6bfe

Mirzaei, R., Mesdaghinia, A., Hoseini, S. S., & Yunesian, M. (2019). Antibiotics in urban wastewater and rivers of Tehran, Iran: Consumption, mass load, occurrence, and ecological risk. *Chemosphere, 221*, 55–66. https://doi.org/10.1016/j.chemosphere.2018.12.187

Moore, M., Gould, P., & Keary, B. S. (2003). Global urbanization and impact on health. *International Journal of Hygiene and Environmental Health, 206*, 269–278. https://doi.org/10.1078/1438-4639-00223

Mundoli, S., Manjunath, B., & Nagendra, H. (2015). Effects of urbanisation on the use of lakes as commons in the peri-urban interface of Bengaluru, India. *International Journal of Urban Sustainable Development, 7*, 89–108. https://doi.org/10.1080/19463138.2014.982124

Nabulo, G., Black, C. R., Craigon, J., & Young, S. D. (2012). Does consumption of leafy vegetables grown in peri-urban agriculture pose a risk to human health? *Environmental Pollution, 162*, 389–398. https://doi.org/10.1016/j.envpol.2011.11.040

Narain, V., Khan, M. S. A., Sada, R., et al. (2013). Urbanization, peri-urban water (in)security and human well-being: A perspective from four South Asian cities. *Water International, 38*, 930–940. https://doi.org/10.1080/02508060.2013.851930

Niva, V., Taka, M., & Varis, O. (2019). Rural-urban migration and the growth of informal settlements: A socio-ecological system conceptualization with insights through a "water lens". *Sustainability, 11*, 3487. https://doi.org/10.3390/su11123487

Paliath, S. (2019). *Why domestic wastewater from Bengaluru is being sent to Kolar and Chikkaballapur*. In: Scroll.in. https://scroll.in/article/944539/why-domestic-wastewater-from-bengaluru-is-being-sent-to-kolar-and-chikkaballapur. Accessed 3 Mar 2022.

Pan, H., Zhang, L., Cong, C., et al. (2019). A dynamic and spatially explicit modeling approach to identify the ecosystem service implications of complex urban systems interactions. *Ecological Indicators, 102*, 426–436. https://doi.org/10.1016/j.ecolind.2019.02.059

Patil, S., Bhaskar, D., Vanjari, R., & Purushothaman, S. (2018). Urbanisation and new Agroecologies the story of Bengaluru's peripheries. *Economic and Political Weekly, 53*, 71–77.

Perrodin, Y., Boillot, C., Angerville, R., et al. (2011). Ecological risk assessment of urban and industrial systems: A review. *Science of the Total Environment, 409*, 5162–5176. https://doi.org/10.1016/j.scitotenv.2011.08.053

Pradhan, B., Sharma, P., & Pradhan, P. K. (2020). Urban growth and environment and health hazards in Kathmandu Valley, Nepal. In R. B. Singh, B. Srinagesh, & S. Anand (Eds.), *Urban health risk and resilience in Asian cities* (pp. 293–324). Springer.

Pribadi, D. O., & Pauleit, S. (2016). Peri-urban agriculture in Jabodetabek Metropolitan Area and its relationship with the urban socioeconomic system. *Land Use Policy, 55*, 265–274. https://doi.org/10.1016/j.landusepol.2016.04.008

Ramirez-Rubio, O., Daher, C., Fanjul, G., et al. (2019). Urban health: An example of a "health in all policies" approach in the context of SDGs implementation. *Globalization and Health, 15*, 87. https://doi.org/10.1186/s12992-019-0529-z

Rasmi, D. M., Zayed, M. A., Dewidar, K. M., & Gabr, H. S. (2021). Towards forming a socio-ecological action model for urban open spaces' design in New Cairo, Egypt. *Journal of Engineering and Applied Science, 68*, 3. https://doi.org/10.1186/s44147-021-00005-z

Reis, C., & Lopes, A. (2019). Evaluating the cooling potential of urban green spaces to tackle urban climate change in Lisbon. *Sustainability, 11*, 2480. https://doi.org/10.3390/su11092480

Remme, R. P., Frumkin, H., Guerry, A. D., et al. (2021). An ecosystem service perspective on urban nature, physical activity, and health. *PNAS, 118*. https://doi.org/10.1073/pnas.2018472118

Salmond, J. A., Tadaki, M., Vardoulakis, S., et al. (2016). Health and climate related ecosystem services provided by street trees in the urban environment. *Environmental Health, 15*, S36. https://doi.org/10.1186/s12940-016-0103-6

Sanderfoot, O. V., & Holloway, T. (2017). Air pollution impacts on avian species via inhalation exposure and associated outcomes. *Environmental Research Letters, 12*, 083002. https://doi.org/10.1088/1748-9326/aa8051

Santamouris, M., Papanikolaou, N., Livada, I., et al. (2001). On the impact of urban climate on the energy consumption of buildings. *Solar Energy, 70*, 201–216. https://doi.org/10.1016/S0038-092X(00)00095-5

Santhanam, H. (2021). *Greening the grass, blueing the water: aligning environmental neuroscience with blue-green infrastructure planning to synergise the delivery of Nature-based Solutions*. Procs. 3rd ESP Asia conference. Nagasaki, Japan, Dec. 14–17, 2021. Session B10 - Sustainable urban planning and policy in Global South: endorsing and mainstreaming ecosystem services from blue-green infrastructure. https://www.espconference.org/asia2021/wiki/623728/book-of-abstracts

Santhanam, H., & Majumdar, R. (2020). Permeable pavements as sustainable nature-based solutions for the management of urban lake ecosystems. In S. Dhyani, A. K. Gupta, & M. Karki (Eds.), *Nature-based solutions for resilient ecosystems and societies* (pp. 329–345). Springer.

Shackleton CM, Cilliers SS, Davoren E, du Toit MJ (2021) Urban ecology in the Global South.

Sharp, L., Kenyon, A., & Choe, E. Y. (2020). *Designing Blue Green Infrastructure (BGI) for water management, human health, and wellbeing: Summary of evidence and principles for design*. The University of Sheffield.

Simon, D. (2008). Urban environments: Issues on the Peri-urban fringe. *Annual Review of Environment and Resources, 33*, 167–185. https://doi.org/10.1146/annurev.environ.33.021407.093240

Somers, C. M., McCarry, B. E., Malek, F., & Quinn, J. S. (2004). Reduction of particulate air pollution lowers the risk of heritable mutations in mice. *Science, 304*, 1008–1010. https://doi.org/10.1126/science.1095815

Subramanian, A., Kundu, A., & Pradhan, B. (2015). Dichotomy or continuum. *Economic and Political Weekly*, 7–8.

Tagliabue, G., Borgini, A., Tittarelli, A., et al. (2016). Atmospheric fine particulate matter and breast cancer mortality: A population-based cohort study. *BMJ Open, 6*, e012580. https://doi.org/10.1136/bmjopen-2016-012580

Tawil, N., & Olszewska, N. (2020). *Fast urban growth, human psycho-socio-ecological needs and "enactivism": The future of cities*.

Taylor, M. D., & Suthers, I. (2021). The socio-ecological system of urban fisheries in estuaries. *Estuaries and Coasts, 44*, 1–8. https://doi.org/10.1007/s12237-021-00916-3

Udas-Mankikar, S., & Driver, B. (2021). *Blue-green infrastructure: An opportunity for Indian cities*. ORF. https://www.orfonline.org/research/blue-green-infrastructure-an-opportunity-for-indian-cities/. Accessed 17 Feb 2022.

Udnoor, K., Narayanan, P., & Anguluri, R. (2020). Urban health infrastructure in small cities: Is it availability or accessibility? In R. B. Singh, B. Srinagesh, & S. Anand (Eds.), *Urban health risk and resilience in Asian cities* (pp. 389–407). Springer.

Urrutia-Goyes, R., Hernandez, N., Carrillo-Gamboa, O., et al. (2018). Street dust from a heavily-populated and industrialized city: Evaluation of spatial distribution, origins, pollution, ecological risks and human health repercussions. *Ecotoxicology and Environmental Safety, 159*, 198–204. https://doi.org/10.1016/j.ecoenv.2018.04.054

WHO. (2017). *Urban green space interventions and health: A review of impacts and effectiveness*. Full report (2017). https://www.euro.who.int/en/health-topics/environment-and-health/urban-health/publications/2017/urban-green-space-interventions-and-health-a-review-of-impacts-and-effectiveness.-full-report-2017. Accessed 16 Feb 2022.

Zabelskyte, G., & Matijosaitiene, I. (2020). *Relationship between urban ecosystem services and human health risks: Systematic review* (pp. 275–284). WIT Press.

Zepp, H., Inostroza, L., Sutcliffe, R., et al. (2018). Neighbourhood environmental contribution and health. A novel indicator integrating urban form and urban green. *Change and Adaptation in Socio-Ecological Systems, 4*, 46–51. https://doi.org/10.1515/cass-2018-0005

Zhang, Q., & Su, S. (2016). Determinants of urban expansion and their relative importance: A comparative analysis of 30 major metropolitans in China. *Habitat International, 58*, 89–107. https://doi.org/10.1016/j.habitatint.2016.10.003

Zingraff-Hamed, A., Hüesker, F., Albert, C., et al. (2021). Governance models for nature-based solutions: Seventeen cases from Germany. *Ambio, 50*, 1610–1627. https://doi.org/10.1007/s13280-020-01412-x

Zölch, T., Maderspacher, J., Wamsler, C., & Pauleit, S. (2016). Using green infrastructure for urban climate-proofing: An evaluation of heat mitigation measures at the micro-scale. *Urban Forestry & Urban Greening, 20*, 305–316. https://doi.org/10.1016/j.ufug.2016.09.011

Urban Vulnerability and Climate Change

Roopam Shukla, M. M. Anees, and Deepika Mann

Abstract Urban ecosystems—hosting the highest density of the world's population—are being recognized to be increasingly vulnerable to climate-related hazards, such as floods and heat waves. Moreover, the vulnerability within urban systems is not only influenced by frequent hazards but also by the changing socioeconomic, political, and governance processes. Given the higher concentration of assets and larger population at risk, climate change vulnerability in urban ecosystems demands prompt attention. This chapter provides a detailed understanding of climate-related hazards and vulnerability in urban ecosystems. First, the basic theories and multiple frameworks of vulnerability are explained. Second, vulnerability to climatic and social hazards is documented to highlight the critical factors for urban vulnerability. At last, the chapter summarizes the proposed risk reduction approaches of nature-based solutions (NbS) and urban sustainability transitions for resilient urban ecosystems.

Keywords Climate-related hazards · Urban vulnerability · Socioeconomic factors · Nature-based solutions · Urban sustainability transitions

R. Shukla
Centre of Excellence in Disaster Mitigation and Management, Indian Institute of Technology Roorkee, Roorkee, Uttarakhand, India

M. M. Anees (✉)
Urban Studies, School of Global Affairs, Dr. B.R. Ambedkar University Delhi, New Delhi, India
e-mail: mmanees@aud.ac.in

D. Mann
Department of Aerospace Engineering, University of the Bundeswehr, Munich, Neubiberg, Bavaria, Germany

P. K. Joshi et al. (eds.), *Geospatial Science for Urban Ecosystems*, The Urban Book Series, https://doi.org/10.1007/978-3-032-07155-2_8

1 Defining Urbanization and Understanding Global Trends

Urbanization refers to the increasing concentration of population in urban areas, as opposed to rural areas. This phenomenon is often accompanied by the growth and expansion of cities, both in terms of physical size and population. Urbanization typically involves the migration of people from rural areas to urban centers, where there is a greater availability of opportunities, including jobs, education, healthcare, and various forms of entertainment and amenities. It also encompasses the transformation of societies from agrarian-based lifestyles to ones centered on industry and services. Additionally, urbanization often brings about changes in social structures, cultural dynamics, and economic developments. It is a complex process influenced by a range of factors, including economic advancements, technological innovations, and policy decisions, and has significant implications on the environment, infrastructure, and the overall quality of life in the regions affected.

As we navigate the twenty-first century, one of the central narratives shaping our era is the unprecedented and accelerated growth of urban areas across the globe. According to the United Nations, the global population stood at 7.6 billion in 2018, with a significant proportion, 4.2 billion, residing in urban centers. As we look toward the mid-century, it is anticipated that 68% of a vastly expanded global population of 9.7 billion will be urban dwellers, tallying to approximately 6.6 billion individuals. Urban sustainable development is at the heart of shaping a future that is in harmony with the environmental and socioeconomic demands of such a massive urban populace. However, the trajectory and nature of urbanization are highly fragmented and vary greatly between nations due to differing levels of socioeconomic development. In developed nations, urban growth is generally characterized by a well-planned expansion that caters efficiently to population surges. These cities are marked by robust public services and infrastructure, encompassing sectors like water and energy supplies, sanitation facilities, educational establishments, and green spaces. In stark contrast, many developing countries grapple with urban centers that are expanding at a pace that outstrips their economic growth and development. These cities are frequently marred by a deficit in basic infrastructure, overcrowding, and an escalating climate crisis. Climate-induced extreme events are rising in the cities, with intensive urban floods and urban heat islands, among others.

2 Identifying Vulnerabilities

2.1 Overview of the Vulnerability Concept

The vulnerability of any system is its susceptibility to any harm due to external or internal disturbance. Some of the well-known concepts like marginalization, fragility, and risk are often compared and related to it (Liverman, 1990; Cutter et al., 2003; Ionescu et al., 2005). The Intergovernmental Panel on Climate Change

(IPCC), in its Fifth Assessment Report (AR5), defines vulnerability as the "propensity or predisposition to be adversely affected." However, the causes and determinants of this predisposition to stress or shock are debatable (Ribot, 2014; Räsänen et al., 2016) and are areas of interest for social scientists. The vulnerability of a system is highly context-specific and driven by a variety of determinants. The determinants vary depending on the type of system (urban, natural, and economic), the scale of analysis (global, national, subnational, and local), and the attribute under consideration (such as livelihoods, biodiversity, and human health) (Adger, 2006; Füssel, 2007). Under the current conditions of climate change, it is imperative to understand how various systems are vulnerable to it and how it impacts the population and economies (Fussel & Klein, 2006; Ford et al., 2013). Studies on vulnerability have spanned multiple disciplines over the decades and have led to frameworks that bring in clarity as well as an overlap of concepts. Additionally, researchers have focused on utilizing these foundational concepts to devise assessment methods that are handy to assess on-ground conditions of vulnerability (Preston et al., 2011; Tate, 2012). Over the decades, this multidisciplinary approach has led to a deeper understanding of factors that contribute to vulnerability. This has resulted in the development of effective strategies that contribute to enhanced resilience against climate-related stresses and shocks.

2.2 *Theoretical Frameworks of Vulnerability*

Vulnerability is understood broadly using two main approaches: starting point and endpoint vulnerability (Kelly & Adger, 2000). These are also identified as contextual and outcome vulnerability, respectively (O'Brien et al., 2007). The outcome vulnerability, rooted in the risk–hazard model, views vulnerability as the result of an external hazard (Eriksen & Kelly, 2007). In this approach, the focus is on the impacts of climate change on the system, assuming an apolitical and technical perspective (Nagoda, 2015). The current approach to managing disasters in urban areas is dominated by technical interventions aimed at predicting hazards or mitigating their impacts. In contrast, contextual vulnerability considers a broader range of factors, including social, economic, and political conditions, that influence a system's susceptibility to harm. This approach emphasizes the underlying conditions and processes that shape vulnerability, recognizing that these factors are often deeply interconnected and influenced by power dynamics and social inequities. Contextual vulnerability considers vulnerability as an intrinsic property of a system that is influenced by sociocultural, economic, and ecological factors (Adger, 2006; Ribot, 2014). This is based on the social constrictive and entitlement theory. An evolution in the IPCC's definition of vulnerability is noted with regard to the centrality of climate change or variability in determining vulnerability (Crane et al., 2017), as shown in Fig. 1.

The distinction between contextual and outcome vulnerability approaches highlights the complexity of vulnerability and underscores the need for diverse

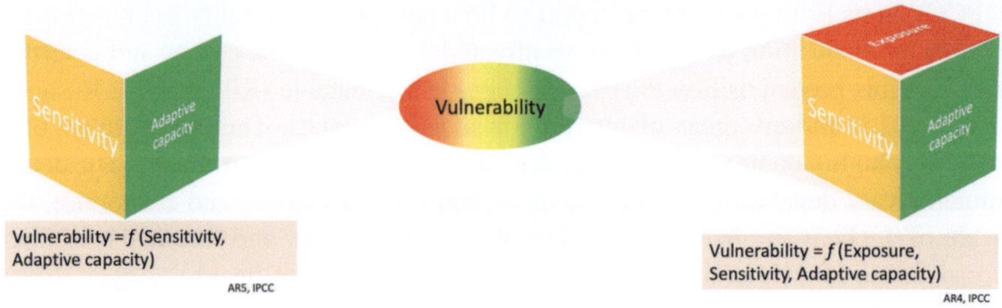

Fig. 1 Evolution of vulnerability framework from IPCC Assessment Report (AR) 4 to AR5

methodologies to address both immediate risks and long-term structural factors. This comprehensive understanding is essential for developing effective strategies for climate change adaptation and resilience building.

An observable shift toward contextual vulnerability emphasizes the inherent characteristics of a system and addresses the processes and multidimensional aspects of climate–society interactions (O'Brien et al., 2007). This approach focuses on the complex interplay between various factors, highlighting how climate change exacerbates existing inequalities and marginalization (Nagoda, 2015). By adopting a contextual vulnerability perspective, adaptation strategies aim to address these inherent complexities, recognizing that vulnerability is not uniform but varies across different social and spatial contexts. Studies utilizing this approach reveal both the spatial and social dynamics of vulnerability, illustrating the differentiated nature of vulnerability among various communities and regions. This comprehensive understanding is crucial for designing more effective and equitable climate adaptation measures. For example, in a study conducted in a rapidly growing urban area (Hamza & Zetter, 1998), researchers employed the contextual vulnerability approach to understand the vulnerability of different neighborhoods to the impacts of climate change. The study analyzed not only the physical exposure to climate-related hazards but also the underlying socioeconomic conditions and social dynamics within each neighborhood.

2.3 *Types of Vulnerabilities in Urban Systems*

Urban systems are complex and have multi-linear interactions, leading to both positive and negative feedback loops. These feedbacks often lead to increased vulnerability, and thus, it is important to understand the types and interconnections among various vulnerabilities in any urban system. Typically, there are five major types of vulnerabilities that exist in any urban system, as presented in Table 1. The study of different types of urban vulnerability is pivotal for creating resilient urban landscapes. By dissecting the contributing factors that give rise to vulnerability and identifying their primary determinants and indicators, we can gain a holistic

Table 1 Urban vulnerability types, definition, contributing factors, and recommended strategies for mitigation

Type of vulnerability	Definition of vulnerability	Main contributing factors	Recommended strategies
Human–social vulnerability	Factors that directly or indirectly harm the well-being of urban inhabitants and factors related to the occupants that might impact their safety during events	Population density, socioeconomic conditions, health services, infrastructure, ratio of adults to pupils, number of disabled pupils, type of school, working hours of the school	Implement community-led resilience programs and mental health initiatives, conduct regular evacuation drills, and ensure accessible escape routes for the disabled
Physical vulnerability	Damage to the built environment and infrastructure, including structural resilience to withstand seismic forces	Buildings, roads, bridges, utility systems, age of buildings, construction materials, number of stories, retrofitting, number of exits, number of windows	Adopt smart city technologies and real-time monitoring systems for infrastructure, health, promote earthquake-resistant construction techniques, and conduct regular structural audits
Economic vulnerability	Factors affecting the financial health and productivity of the urban system	Businesses, workforce, markets, investments	Diversify economic streams and promote local businesses to reduce dependency on singular industries
Cultural–environmental vulnerability	Aspects concerning ecological and cultural heritage degradation	Parks, cultural sites, biodiversity zones, water bodies	Develop green corridors and promote eco-tourism; preserve cultural hubs through digital archiving

understanding of urban fragility. Leveraging each of these urban vulnerability types, informed by key determinants and real-world data, offers a blueprint of the challenges faced by specific urban segments. Understanding these types is not just an academic exercise; it is a tool for policymakers, urban planners, and community leaders to understand which strategies to prioritize for vulnerability and risk mitigation.

3 Climate-Related Hazards in Urban Systems

3.1 Flooding

Urban areas are especially vulnerable to flooding due to impervious surfaces, inadequate drainage systems, and changes in precipitation patterns induced by climate change. Flooding can result in severe economic, social, and environmental impacts, exacerbating urban vulnerability. Flooding can harm critical infrastructure, disrupt essential services, and result in significant economic losses. Moreover, it poses risks to human health and safety, including the potential for injuries, waterborne diseases, and displacement of populations. Vulnerable communities, such as those in low-lying areas or informal settlements, often bear the brunt of flooding impacts due to their limited resources and lack of adequate infrastructure.

To address the vulnerability to flooding, urban areas need to adopt a multifaceted approach. This involves implementing effective urban planning strategies, such as land-use regulations and floodplain zoning, to prevent further urban expansion into flood-prone areas. Developing and maintaining robust stormwater management systems, including green infrastructure solutions, can help reduce flood risks and enhance urban resilience. Additionally, enhancing early warning systems and improving emergency response capabilities are essential for timely evacuation and effective flood management.

3.2 Urban Drought

Urban areas also face the challenge of drought, which can have far-reaching consequences on water availability, food security, and socioeconomic systems. Climate change–induced shifts in precipitation patterns can exacerbate the frequency and severity of droughts in certain regions, further magnifying urban vulnerability.

Droughts can lead to water scarcity, impacting various sectors, such as agriculture, industry, and domestic water supply. This, in turn, affects food production, economic productivity, and public health. Vulnerable populations, particularly those in marginalized communities, may experience disproportionate impacts due to limited access to water resources and social inequalities.

Mitigating the vulnerability to drought in urban areas requires a multipronged approach. Water demand management strategies, such as water conservation measures and efficient irrigation practices, can help reduce water consumption and enhance resilience. Diversifying water sources through investments in alternative water supply systems, such as rainwater harvesting or wastewater reuse, can enhance water security. Furthermore, promoting sustainable land and water management practices, including water-sensitive urban design, can contribute to overall drought resilience in urban areas.

4 Social Challenges in Urban Systems

Disasters or hazards, in conventional terms, are caused by exceptional or unusual events, usually “natural” disasters. This has, however, been questioned in the past three decades, as disasters are now seen or understood not only as natural or unusual events but as failures of development. Social hazards can be defined as events originating within society that have the potential to create significant ripple effects on human activity systems, leading to societal transformations aimed at overcoming the resulting negative impacts. Cities can indeed be viewed as hotspots for hazards that, without effective management, can create extreme vulnerability and risk for large populations. The accumulation of risk in urban centers, such as the growing number of people at risk from floods or accidental fires, is not simply a by-product of urban development. Rather, it results from the complex and dynamic interactions between human-induced and natural hazards, combined with extreme conditions of vulnerability. Urban populations' vulnerability to disasters is not a natural occurrence but is constructed and intensified by economic, social, and political systems. Poor urban planning, inadequate infrastructure, and socioeconomic inequalities contribute significantly to this heightened vulnerability. Addressing these issues requires comprehensive strategies that incorporate better management practices, resilient infrastructure, and policies aimed at reducing socioeconomic disparities. By understanding and mitigating these risk factors, cities can become more resilient to hazards and better protect their populations. Some of the most prominent social challenges in urban systems are discussed below.

(a) ***Population Density***: Urban areas are associated with concentrated or high-density populations due to concentrated job opportunities for nonagricultural activities. This increases the number of lives at risk due to hazards in urban systems. Due to a significant portion of the world's urban population residing in low-latitude, low-income cities, it is probable that the most impoverished urban dwellers will experience adverse effects from interacting with various elements of nature. As urban populations continue to grow, so will the spectrum of hazards they face. These include the rapid spread of infectious and parasitic diseases within densely populated areas, heightened by close proximity among residents. Furthermore, the presence of potentially hazardous industrial processes and the accumulation of industrial waste pose additional risks. The density of motorized vehicles in urban areas adds another layer of concern due to their close interaction with pedestrians and residents. As cities expand and urbanize further, these challenges are expected to intensify, necessitating proactive measures to mitigate risks and safeguard public health and safety.

(b) ***Inequality:*** Large sections of populations are living in informal settlements where no or little attention is paid to health and safety standards, such as solid and liquid waste disposal systems. These section are often called the urban poor. In urban areas, the urban poor are usually bound to be in the most vulnerable locations for any disaster because the land available for housing is unaffordable for low-income groups. This makes the urban poor more vulnerable to urban disasters. The ongoing degradation of green spaces and water sources in slums within and around urban areas will persist for generations unless substantial efforts are undertaken to ensure universal access to safe water and sanitation. Social and individual risk factors exacerbate these challenges, encompassing elevated levels of crime, violence, and social marginalization, as well as insufficient access to education and vocational training among urban populations living in poverty. Social unrest among the populace is escalating, driven by a growing trend of inequality affecting middle- and lower-income groups. Consequently, there is an erosion of social cohesion and trust in governmental institutions. Over the past decade, civil unrest has doubled, fueled by economic hardships and political instability.

(c) ***Environmental Risks:*** As cities grow larger, their urban heat islands become more pronounced, heightening the risk of increased mortality during heat waves. Additionally, urban systems must address environmental challenges, such as ambient air quality, noise pollution levels, soil and water contamination, and the management of solid waste disposal. These factors collectively impact urban environments and necessitate proactive management strategies. Land subsidence risks are further identified to be growing in urban systems. Land subsidence can arise from various causes, including both natural processes and human activities, such as underground resource extraction. Anthropogenic factors like excessive groundwater pumping, mining activities, and geological processes significantly contribute to land subsidence. Moreover, the rapid growth of urban populations over recent decades has driven up water demand, leading to substantial increases in both water supply infrastructure and wastewater production in many developing regions worldwide. These trends underscore the critical need for sustainable water management practices to mitigate the impacts of land subsidence and ensure long-term environmental stability. Finally, increasing demand for clean water and decreasing precipitation in the ever-growing cities are going to lead to water shortages in already water-stressed cities, for instance, Delhi and Mexico city.

(d) ***Other Risks:*** Economic risks encompass a range of challenges faced by low-income households, including inadequate access to affordable housing, food, and clean water, as well as persistent unemployment and disparities in accessing essential resources and services, such as affordable healthcare. These factors contribute significantly to socioeconomic inequalities within communities. In addition to economic risks, technological hazards also pose substantial threats. These include traffic accidents, industrial and chemical disasters, and contamination stemming from mass-produced foods and synthetic goods. These technological risks underscore the importance of robust safety measures and regulatory frameworks to safeguard public health and environmental integrity in urban and industrial settings alike.

5 Social Vulnerability Assessment Approaches

The magnitude and extent of climatic stresses are known to vary across different geographic regions. Social vulnerability is identified as an important concept that focuses on the social components that bring threshold crossing, resulting in varied losses within the same region (Brooks et al., 2005; Kelkar et al., 2011). The approach of emphasizing the social components of the vulnerable population is representative of a “people-centric” approach, as it is vital to highlight the inequalities within communities. Inequalities are especially more pronounced in urban areas as resources are unevenly distributed and are more pronounced by preexisting and evolving factors, such as socioeconomic conditions, population density, surface features, mobility and accessibility, and the availability of green and open spaces, among others (Krellenberg & Welz, 2017). Thus, it becomes imperative to understand the various assessment approaches that highlight the factors leading to disproportionate access to resources and the resultant vulnerability.

5.1 Inherent Vulnerability

The inherent vulnerability approach evolves from the contextual vulnerability construct (as also discussed in the Introduction) and identifies vulnerability as a state rather than an endpoint (Adger, 2006; Cutter & Finch, 2008). This approach is focused on investigating and highlighting which inherent conditions (such as socioeconomic conditions and prevailing biophysical conditions) within a community lead to differential vulnerability (Lankao & Qin, 2011). The inherent vulnerability approach deviates from an approach of vulnerability as an impact and brings in perspectives that can answer the following questions: *Why are some neighborhoods or communities affected more than others? Do their social and economic conditions make them more vulnerable than others?* If so, *which specific attributes should be focused on for assistance?* Such a perspective on quantifying vulnerability, in turn, also opens the following debate: *What is the role of service providers in the city in reducing vulnerability*, and *what is the role of communities in reducing their vulnerability?*

Much of the research on inherent vulnerability has focused on the differential or disproportionate access to resources, assets, and prevailing ecological and biophysical conditions (Anees et al., 2020; Kelkar et al., 2011). Methodological advancements in determining the inherent vulnerability of urban populations have mainly focused on two aspects: (i) indicators to quantify vulnerability and (ii) methods to integrate these indicators.

5.2 Indicators to Quantify Vulnerability

Due to the unobservable nature of vulnerability, researchers rely on measurable objective indicators, which, as a sum, attempt to explain most of the variables that may be contributing to an increase or decrease in vulnerability. However, due to a variety of indicators available for different communities across the world and a subjective understanding of the vulnerability concept, a strong consensus on methods is missing. Despite this, a few indicators are common among different studies, such as those that indicate income, employment, gender, age, household assets, and education level (Frigerio et al., 2016; Frigerio & de Amicis, 2016; Krellenberg & Welz, 2017; Salvati, 2014). These indicators are mostly chosen with the premise that they capture multidimensional factors that may influence the outcomes of exposure to disaster (Cutter, 1996). Social science researchers have identified many more specific indicators that are aggravated, especially in light of disaster response and recovery. These include indicators such as special needs population (differently abled population), immigrants, homeless people, and renters (Cutter et al., 2003). Other than indicators relevant to the population and their socioeconomic constraints, the spatial construct of urban areas adds an additional layer of variation among communities. This includes differences in urbanization levels, disadvantaged proximity to rivers and streams, availability or unavailability of green spaces, and other ecological parameters (Frigerio & de Amicis, 2016; Salvati, 2014).

Due to the nested and interconnected nature of different indicators in urban areas, the selection of indicators is often one of the challenging steps of inherent vulnerability assessment. Most of the studies rely on previous literature to select a set of indicators, with some form of context-specific reasoning. It is also common to remove these selected indicators based on their statistical measures, such as mutual dependence on other indicators or low-standard deviation (indicating low variance among the population). The selected indicators are often categorized into IPCC's three contributing factors to vulnerability: exposure, sensitivity, and adaptive capacity (Hahn et al., 2009). It is also common to categorize indicators as different components, such as the sociodemographic profile, health, infrastructure, resource and energy, physical accessibility, technology, finance, livelihood, social networks, sanitation, and environmental parameters, among others (Anees et al., 2020; Gerlitz et al., 2017; Hahn et al., 2009; Panthi et al., 2016).

5.3 Methods to Integrate Indicators

A critical requirement of vulnerability research is its capability to communicate findings among larger communities and policymakers, especially when employing a "people-centric" approach, such as the inherent vulnerability approach. One of the most used vulnerability assessment methods is indicator-based (il Choi, 2019; Nasiri et al., 2016), in which aggregation and weightage are two important steps (preceded by the selection of indicators, as discussed above). This method provides a simple framework to deconstruct complex relationships leading to vulnerability. However, this can also lead to misinterpretations when techniques are used without context. Different methods have been developed to minimize shortcomings (such as introducing sources of uncertainty and nonuniformity in vulnerability studies (OECD, 2008)) and utilize the statistical advantages provided by them. Weighing techniques, such as analytical hierarchy process (AHP), principal component analysis (PCA), and equal weightage, are the most commonly employed techniques (Diaz-Sarachaga & Jato-Espino, 2020). These methods represent three different approaches to the weighing of indictors: (i) AHP: employing expert judgment, when there is credible information on the impact of different indicators, and a consensus is needed among different experts; (ii) PCA: utilizing statistical construct of the indicators to derive the most important/varied information from among all the indicators; and (iii) equal weightage: when there is an absence of credible information or consensus for varied weights to indicators. Aggregation methods that are commonly used include additive or multiplicative methods, with many different forms (il Choi, 2019).

The selection of indicators and methods are also subjectively selected, depending on the type of vulnerability. The major focus of research has been on separate disasters, such as heat (Reischl et al., 2016; Romero-Lankao et al., 2012), flooding (Mavhura et al., 2017; Nasiri et al., 2016; Rasch, 2016), coastal flooding (Hossain, 2015; Rani et al., 2015; Yoo et al., 2011), and earthquakes (Frigerio et al., 2016). Fewer studies focus on multi-hazard scenarios in urban systems, such as (Tapia et al., 2017) heat waves, drought, and flooding and (Krellenberg & Welz, 2017) flood and heat waves. Some studies have focused on the vulnerability of the population to nonspecific disasters (Anees et al., 2020).

6 Solutions for Urban Resilience: An Integrated Approach

As urban centers across the globe continue to grow and evolve, the imperative to understand and address urban vulnerabilities has never been greater. The rapid pace of urbanization, coupled with global climatic and social shifts, presents multifaceted challenges to the sustainability and resilience of cities. This chapter has traversed the landscape of urban vulnerabilities, offering insights into their origins, manifestations, and potential solutions. The increasing convergence of climate-related hazards, such as flooding and urban heating, with social vulnerabilities in the urban systems underscores the complex interplay of factors contributing to urban fragility. At the heart of these vulnerabilities lies the tension between rapid urban growth and the lagging pace of adaptive urban planning.

Figure 2 presents a compelling visual representation of the various solutions that can be adopted by urban areas to address sustainability challenges. The three overlapping circles depict the three main solution sets: nature-based solutions (NbS), techno-infrastructure solutions, and socio-governance solutions. The interplay between these solutions is showcased by their intersections, emphasizing the importance of an integrated approach. The figure underscores the importance of a multifaceted and integrated approach to urban sustainability. While each city may have its unique challenges and strengths, the confluence of nature-based, technological, and socio-governance strategies emerges as a consistent theme in the quest for urban resilience and sustainability. Figure 2B shows Dehradun city, nestled in the foothills of the Himalayas, which emphasizes a balance between all three solutions, with a greater focus on techno-infrastructure solutions, given the fragile geography of the city. Figure 2 (C) shows Prayagraj city, situated at the confluence of three rivers, and the image of Prayagraj shows its expansive river systems and green belts.

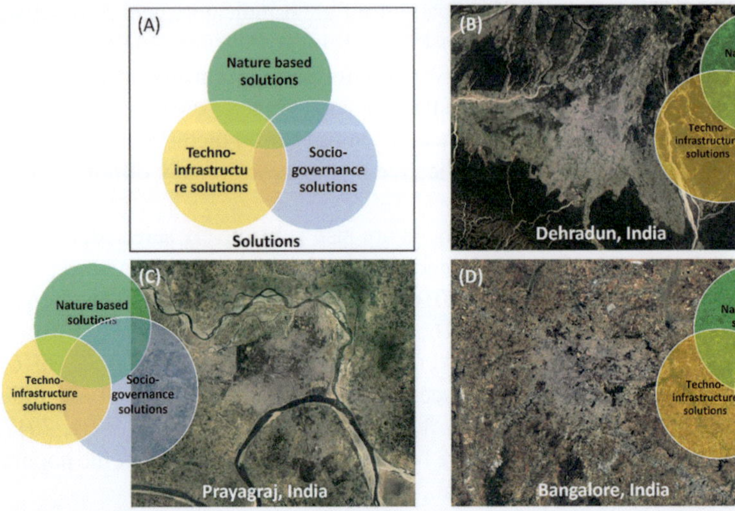

Fig. 2 Integrated approaches to urban sustainability and vulnerability reduction across Indian urban areas. Panel A illustrates the general model of an integrated solution set. Nature-based solutions, depicted in green, emphasize the utilization of natural processes and ecosystems to enhance urban resilience. Techno-infrastructure solutions, shown in yellow, represent the harnessing of technological advancements and infrastructure development to address urban challenges. The blue circle, socio-governance solutions, underscores the role of policies, governance mechanisms, and community engagement in fostering urban sustainability

The emphasis here leans toward NbS and socio-governance, reflecting the city's reliance on its rivers and the sociocultural significance they hold. Figure 2 (D) shows Bangalore city's satellite image, which reveals its densely populated urban areas. The circles suggest a substantial emphasis on techno-infrastructure solutions, given its urban nature. Yet the presence of green and blue highlights the city's efforts to incorporate nature-based measures and socio-governance strategies.

In conclusion, the urban vulnerabilities of today demand solutions rooted in integration, innovation, and inclusivity. As urban planners, policymakers, and citizens chart the future course of cities, the emphasis must remain on creating urban landscapes that are not only resilient to challenges but also vibrant, inclusive, and sustainable for generations to come.

6.1 Nature-Based Solutions (NbS)

Among the three solutions discussed in the last section, techno-infrastructure and socio-governance solutions have been the default choice of planners and policymakers. This choice is a result of technological advancements that provide quick and effective solutions and our developed understanding of the role of communities in making any solution successful on the ground. A recent but very promising addition to the solution is the cost-effective, multifunctional, and sustainable NbS. The International Union for Conservation of Nature (IUCN) defines NbS as "actions to protect, sustainably manage, and restore natural or modified ecosystems that address societal challenges effectively and adaptively, simultaneously providing human well-being and biodiversity benefits."

While there is a lot of emphasis on integrating NbS with techno-infrastructure solutions, thus blending the gray–green infrastructure for optimum results, NbS is also being recognized for reducing people's vulnerabilities to climate change, especially in the Global South (Woroniecki et al., 2023). This gives more opportunities to integrate NbS with socio-governance solutions, as people's experiences with vulnerabilities and the possible solutions provided by NbS bring a higher acceptance, both among communities and authorities in charge of applying such solutions. A synergy of these two solutions is also necessary for the urban settings, as the isolation of either solution would create a misfit, both in the expectations of people and the long-term management of such infrastructure that requires governance and community participation. Thus, to be successful in fulfilling the goal of reducing vulnerability and increasing resilience, cities need to inculcate an environment that brings around a change in people's perception and community involvement, governance, and financing mechanisms and ensures training and adaptation of the practitioners.

NbS stands out not only as a cost-saving solution to many of the urban problems but also as the one that is most in sync with nature and capable of addressing multiple challenges at a time. Being in sync with nature is of critical advantage in urban ecosystems as it reduces the risk of creating complex, unforeseen additional problems that are commonly associated with contrasting techno-infrastructure solutions. At the same time, NbS offers the capability of solving multiple problems at no additional cost and, in the long term, reducing the maintenance cost to a minimum. The co-benefits of NbS, such as for public health, climate change mitigation, biodiversity, and social well-being, have played an important role in viewing NbS as a no-harm solution in contested urban spaces, where solutions are regularly linked to public perception and effective results. However, the longer return value and fear of negative public perception due to lower short-term effectiveness demote NbS as an option in many urban areas.

The relationship of NbS to vulnerability and resilience is a dual-scope opportunity with the capability to remedy both local and global problems. These include issues like air and water quality, rising temperatures/heat waves, coastal and urban flooding, erosion, and water scarcity, among many others. Ecosystem-based disaster risk reduction (Eco-DRR) has become a popular term that involves the use of NbS to address both short-term and long-term impacts from disasters. The multiple benefits of NbS, for example, as an Eco-DRR solution for reducing vulnerabilities from any disaster and as an ecosystem service provider, such as for improving the social well-being of residents, provide an attractive package solution to policymakers. A permeable surface is capable of both lowering the risk of flooding and improving water availability in the area. Evidence from across the world has highlighted the capability of NbS to address the vulnerability of urban residents (Panno et al., 2017; Zölch et al., 2017). Woroniecki et al. (2023) reported that NbS leads to a reduction in vulnerability, mainly by lowering ecosystem sensitivity to climate impacts and reducing social sensitivity. However, NbS is still in its nascent stages as it requires novel finance mechanisms and the development of an evidence base for policymakers to adopt them with confidence and, in turn, advocate their benefits to the public. The mainstreaming of NbS is the future, and to put policy into practice, there is a need for a long-term strategic vision, including developing policies for blending an optimum level of gray and green infrastructure. One of the key players who can take up this challenge and integrate it into their existing provisions and regulations is the urban planning division (Bush & Doyon, 2019).

7 Conclusion

Today's urban vulnerabilities require solutions that are integrated, innovative, and inclusive. Urban planners, policymakers, and citizens must focus on creating urban systems that are resilient and sustainable for future generations. It is critical to realize that the process of urbanization brings significant changes in social structures, cultural dynamics, and economic developments, influenced by economic advancements, technological innovations, and policy decisions. Therefore, understanding the urban system vulnerability, which determines a system's susceptibility to harm, is crucial. We emphasize this critical need to understand and assess the urban vulnerability and social challenges and strongly advocates for a balanced and strategic approach to urban development and planning by integrating nature-based solutions in enhancing urban resilience and sustainability.

References

Adger, W. N. (2006). Vulnerability. *Global Environmental Change, 16*(3), 268–281. https://doi.org/10.1016/j.gloenvcha.2006.02.006

Anees, M. M., Shukla, R., Punia, M., & Joshi, P. K. (2020). Assessment and visualization of inherent vulnerability of urban population in India to natural disasters. *Climate and Development, 12*(6), 532–546. https://doi.org/10.1080/17565529.2019.1646629

Brooks, N., Adger, W. N., & Kelly, P. M. (2005). The determinants of vulnerability and adaptive capacity at the national level and the implications for adaptation. *Global Environmental Change, 15*(2), 151–163. https://doi.org/10.1016/j.gloenvcha.2004.12.006

Bush, J., & Doyon, A. (2019). Building urban resilience with nature-based solutions: How can urban planning contribute? *Cities, 95*. https://doi.org/10.1016/j.cities.2019.102483

Crane, T. A., Delaney, A., Tamás, P. A., Chesterman, S., & Ericksen, P. (2017). A systematic review of local vulnerability to climate change in developing-country agriculture: In search of transparency, coherence and comparability. *Wiley Interdisciplinary Reviews: Climate Change, 8*(4), e464.

Cutter, S. L. (1996). Vulnerability to environmental hazards. *Progress in Human Geography, 20*(4), 529–539. https://doi.org/10.1177/030913259602000407

Cutter, S. L., & Finch, C. (2008). Temporal and spatial changes in social vulnerability to natural hazards. *Proceedings of the National Academy of Sciences of the United States of America, 105*(7), 2301–2306. https://doi.org/10.1073/pnas.0710375105
Cutter, S. L., Boruff, B. J., & Shirley, W. L. (2003). Social vulnerability to environmental hazards. *Social Science Quarterly, 84*(2), 242–261. https://doi.org/10.1111/1540-6237.8402002
Diaz-Sarachaga, J. M., & Jato-Espino, D. (2020). Analysis of vulnerability assessment frameworks and methodologies in urban areas. *Natural Hazards, 100*(1), 437–457. https://doi.org/10.1007/s11069-019-03805-y
Eriksen, S. H., & Kelly, P. M. (2007). Developing credible vulnerability indicators for climate adaptation policy assessment. *Mitigation and Adaptation Strategies for Global Change, 12*(4), 495–524. https://doi.org/10.1007/s11027-006-3460-6
Ford, J. D., McDowell, G., & Pearce, T. (2013). The dynamic multiscale nature of climate change vulnerability: an Inuit harvesting example. *Annals of the Association of American Geographers, 103*(5), 1193–1211. https://doi.org/10.1080/00045608.2013.776880
Frigerio, I., & de Amicis, M. (2016). Mapping social vulnerability to natural hazards in Italy: A suitable tool for risk mitigation strategies. *Environmental Science and Policy, 63*, 187–196. https://doi.org/10.1016/j.envsci.2016.06.001
Frigerio, I., Ventura, S., Strigaro, D., Mattavelli, M., de Amicis, M., Mugnano, S., & Boffi, M. (2016). A GIS-based approach to identify the spatial variability of social vulnerability to seismic hazard in Italy. *Applied Geography, 74*, 12–22. https://doi.org/10.1016/j.apgeog.2016.06.014
Füssel, H.-M. (2007). Vulnerability: A generally applicable conceptual framework for climate change research. *Global Environmental Change, 17*(2), 155–167. https://doi.org/10.1016/j.gloenvcha.2006.11.006
Füssel, H.-M., & Klein, R. J. T. (2006). Climate change vulnerability assessments: An evolution of conceptual thinking. *Climatic Change, 75*(3), 301–329. https://doi.org/10.1007/s10584-006-0329-3
Gerlitz, J. Y., Macchi, M., Brooks, N., Pandey, R., Banerjee, S., & Jha, S. K. (2017). The multidimensional livelihood vulnerability index–an instrument to measure livelihood vulnerability to change in the Hindu Kush Himalayas. *Climate and Development, 9*(2), 124–140. https://doi.org/10.1080/17565529.2016.1145099
Hahn, M. B., Riederer, A. M., & Foster, S. O. (2009). The livelihood vulnerability index: A pragmatic approach to assessing risks from climate variability and change-a case study in Mozambique. *Global Environmental Change, 19*(1), 74–88. https://doi.org/10.1016/j.gloenvcha.2008.11.002
Hamza, M., & Zetter, R. (1998). Structural adjustment, urban systems and disaster vulnerability in developing countries. *Cities, 15*(4), 291–299. https://doi.org/10.1016/S0264-2751(98)00020-1
Hossain, M. N. (2015). Analysis of human vulnerability to cyclones and storm surges based on influencing physical and socioeconomic factors: Evidences from coastal Bangladesh. *International Journal of Disaster Risk Reduction, 13*, 66–75. https://doi.org/10.1016/j.ijdrr.2015.04.003
il Choi, H. (2019). Assessment of aggregation frameworks for composite indicators in measuring flood vulnerability to climate change. *Scientific Reports, 9*(1), 1–15. https://doi.org/10.1038/s41598-019-55994-y
Ionescu, C., Klein, R. J. T., Hinkel, J., & KaviKumar, K. S. (2005). Towards a formal framework of vulnerability to climate change (NeWater Working Paper No. 2). Potsdam Institute for Climate Impact Research.
Kelly, P. M., & Adger, W. N. (2000). Theory and practice in assessing vulnerability to climate change and facilitating adaptation. *Climatic Change, 47*(4), 325–352. https://doi.org/10.1023/A:1005627828199
Kelkar, U., Balachandra, P., & Gurtoo, A. (2011). Assessing Indian cities for vulnerability to climate change. In *2nd international conference on environmental science and development* (Vol. 4, pp. 246–250).
Krellenberg, K., & Welz, J. (2017). Assessing urban vulnerability in the context of flood and heat Hazard: Pathways and challenges for indicator-based analysis. *Social Indicators Research, 132*(2), 709–731. https://doi.org/10.1007/s11205-016-1324-3

Lankao, P. R., & Qin, H. (2011). Conceptualizing urban vulnerability to global climate and environmental change. *Current Opinion in Environmental Sustainability, 3*(3), 142–149. https://doi.org/10.1016/j.cosust.2010.12.016

Liverman, D. M. (1990). Vulnerability to global environmental change. In R. E. Kasperson, K. Dow, D. Golding & J. X. Kasperson (Eds.), Understanding Global Environmental Change: The Contributions of Risk Analysis and Management (pp. 27–44).

Mavhura, E., Manyena, B., & Collins, A. E. (2017). An approach for measuring social vulnerability in context: The case of flood hazards in Muzarabani district, Zimbabwe. *Geoforum, 86*(December 2016), 103–117. https://doi.org/10.1016/j.geoforum.2017.09.008

Nagoda, S. (2015). New discourses but same old development approaches? Climate change adaptation policies, chronic food insecurity and development interventions in northwestern Nepal. *Global Environmental Change, 35*, 570–579.

Nasiri, H., Mohd Yusof, M. J., & Mohammad Ali, T. A. (2016). An overview to flood vulnerability assessment methods. *Sustainable Water Resources Management, 2*(3), 331–336. https://doi.org/10.1007/s40899-016-0051-x

O'Brien, K., Eriksen, S., Nygaard, L. P., & Schjolden, A. (2007). Why different interpretations of vulnerability matter in climate change discourses. *Climate Policy, 7*(1), 73–88. https://doi.org/10.1080/14693062.2007.9685639

OECD. (2008). Handbook on constructing composite indicators. In *Methodology and UserGuide* (p. 162). https://doi.org/10.1787/9789264043466-en

Panno, A., Carrus, G., Lafortezza, R., Mariani, L., & Sanesi, G. (2017). Nature-based solutions to promote human resilience and wellbeing in cities during increasingly hot summers. *Environmental Research, 159*, 249–256. https://doi.org/10.1016/j.envres.2017.08.016

Panthi, J., Aryal, S., Dahal, P., Bhandari, P., Krakauer, N. Y., & Pandey, V. P. (2016). Livelihood vulnerability approach to assessing climate change impacts on mixed agro-livestock smallholders around the Gandaki River Basin in Nepal. *Regional Environmental Change, 16*(4), 1121–1132. https://doi.org/10.1007/s10113-015-0833-y

Preston, B. L., Yuen, E. J., & Westaway, R. M. (2011). Putting vulnerability to climate change on the map: a review of approaches, benefits, and risks. *Sustainability Science, 6*(2), 177–202. https://doi.org/10.1007/s11625-011-0129-1

Rani, N. N. V. S., Satyanarayana, A. N., & v, & Bhaskaran, P. K. (2015). Coastal vulnerability assessment studies over India: A review. *Natural Hazards, 77*(1), 405–428. https://doi.org/10.1007/s11069-015-1597-x

Rasch, R. J. (2016). Assessing urban vulnerability to flood hazard in Brazilian municipalities. *Environment and Urbanization, 28*(1), 145–168. https://doi.org/10.1177/0956247815620961

Räsänen, A., Juhola, S., Nygren, A., Käkönen, M., Kallio, M., Monge Monge, A., & Kanninen, M. (2016). Climate change, multiple stressors and human vulnerability: a systematic review. *Regional Environmental Change, 16*(8), 2291–2302. https://doi.org/10.1007/s10113-016-0974-7

Reischl, C., Rauter, R., & Posch, A. (2016). Urban vulnerability and adaptation to heatwaves: A case study of Graz (Austria). *Climate Policy, 18*(1), 1–13. https://doi.org/10.1080/14693062.2016.1227953

Ribot, J. (2014). Cause and response: vulnerability and climate in the Anthropocene. *The Journal of Peasant Studies, 41*(5), 667–705. https://doi.org/10.1080/03066150.2014.894911

Romero-Lankao, P., Qin, H., & Dickinson, K. (2012). Urban vulnerability to temperature-related hazards: A meta-analysis and meta-knowledge approach. *Global Environmental Change, 22*(3), 670–683. https://doi.org/10.1016/j.gloenvcha.2012.04.002

Salvati, L. (2014). A socioeconomic profile of vulnerable land to desertification in Italy. *Science of the Total Environment, 466–467*, 287–299. https://doi.org/10.1016/j.scitotenv.2013.06.091

Tapia, C., Abajo, B., Feliu, E., Mendizabal, M., Martinez, J. A., Fernández, J. G., Laburu, T., & Lejarazu, A. (2017). Profiling urban vulnerabilities to climate change: An indicator-based vulnerability assessment for European cities. *Ecological Indicators, 78*, 142–155. https://doi.org/10.1016/j.ecolind.2017.02.040

Tate, E. (2012). Social vulnerability indices: A comparative assessment using uncertainty and sensitivity analysis. *Natural Hazards, 63*(2), 325–347. https://doi.org/10.1007/s11069-012-0152-2

Woroniecki, S., et al. (2023). Contributions of nature-based solutions to reducing people's vulnerabilities to climate change across the rural Global South. *Climate and Development, 15*(7), 590–607.

Yoo, G., Hwang, J. H., & Choi, C. (2011). Development and application of a methodology for vulnerability assessment of climate change in coastal cities. *Ocean and Coastal Management, 54*(7), 524–534. https://doi.org/10.1016/j.ocecoaman.2011.04.001

Zölch, T., Henze, L., Keilholz, P., & Pauleit, S. (2017). Regulating urban surface runoff through nature-based solutions—An assessment at the micro-scale. *Environmental Research, 157*, 135–144. https://doi.org/10.1016/j.envres.2017.05.023

Urbanization with Remote Sensing and Predictive Models

Swapan Talukdar, Shahfahad, Mohd Waseem Naikoo, Tanmoy Das, and Atiqur Rahman

Abstract Urban sprawl and expansion present significant challenges in the process of sustainable urban management. Urban growth and the escalating complexity of urban challenges have exacerbated environmental impacts, creating pronounced disparities between urban centers and their surrounding regions. During the past few years, these complexities have increased significantly; thus, addressing these challenges has become significantly important for better urban planning and management. Modern urban geography studies increasingly rely on remote sensing data and artificial intelligence to track the pace of urban expansion and spatial patterns. The multi-temporal land use land cover (LULC) maps of urban areas derived from remote sensing data predictive models provide planners and policy makers with valuable insights into the development trajectories of cities, encompassing their systems, functions, and structures. Remote sensing and GIS-based analytical approaches and forecasting models are used for mapping, monitoring, measuring, and analyzing the urban expansion. At a global level, more and more interest groups are getting involved in critical urban environmental issues to support sustainable urban develop-

S. Talukdar
Department of Geography, Asutosh College, University of Calcutta, Kolkata, West Bengal, India

Shahfahad
Department of Geography, Faculty of Sciences, Jamia Millia Islamia, New Delhi, India

Department of Geography, Institute of Science, Banaras Hindu University, Varanasi, Uttar Pradesh, India

M. W. Naikoo
Department of Geography, Faculty of Sciences, Jamia Millia Islamia, New Delhi, India

Department of Geography and Disaster Management, University of Kashmir, Srinagar, Jammu & Kashmir, India

T. Das · A. Rahman (✉)
Department of Geography, Faculty of Sciences, Jamia Millia Islamia, New Delhi, India
e-mail: arahman2@jmi.ac.in

P. K. Joshi et al. (eds.), *Geospatial Science for Urban Ecosystems*, The Urban Book Series, https://doi.org/10.1007/978-3-032-07155-2_9

ment. These groups use remote sensing and GIS tools to address the issues related to urban planning and policy making. Despite all these, there is still a gap between the urban remote sensing research and the broader theoretical understanding of urban development. This study aims to address and bridge this gap and provide policymakers, local urban authorities, and planners with a better understanding of urbanization through the integration of theoretical insights and research-based results.

Keywords Urban growth · Remote sensing · GIS · Artificial intelligence · Agent-based model

1 Introduction

The transformation of rural natural land to urban areas is referred to as "urban expansion," which leads to an increase in the area of a city (Seto et al., 2011). This phenomenon is common worldwide due to swift population growth in cities and the progress of socio-economic factors (Wu, 2014; Angel et al., 2016). Due to consistent urbanization in developing nations, the global urban population crossed 50% of the total population in 2008 (UN-Habitat, 2008). Urbanization has significant impacts on local and global ecosystems, even if it is not the predominant form of land use. This influence stems from activities such as waste disposal, industrial accumulation, and intensive use of energy and resources, leading to a range of environmental challenges such as pollution, habitat degradation, and resource depletion. As a result, scientific studies aimed at creating ecological answers to these issues are receiving more attention. Addressing the challenges associated with the degradation of freshwater resources and the movement of materials through ecosystems is becoming increasingly important. Such research is critical to the adoption of sustainable urban practices and the mitigation of the negative environmental impacts of urbanization (Palmer et al., 2004).

Global urbanization continues to accelerate, drastically altering landscapes and influencing the functioning of ecosystems in and around cities (Riley et al., 2007; Shahfahad et al., 2024). The resulting ecological changes due to urbanization have a direct influence on the urban and global environment as well as on human health (Musacchio, 2009). Alberti (2005) investigated the consequences of urban growth on the functionality of ecosystems and provided insights for future research aimed at reducing the negative impacts of urbanization on ecosystems. Further, Berling-Wolff and Wu (2004) analyzed the past development of urban growth models, examined early models related to transportation and land use planning, and assessed the potential for integrating different approaches to urban modeling. Jusuf et al. (2007) presented an innovative urban planning framework for a specific area that combines landscape ecology principles with traditional feng shui concepts to develop strategies for urban planning. Fang et al. (2005) conducted spatial simulations to explore the interactions within urban expansion dynamics, while Pauleit et al. (2005) modeled the environmental effects of urban land use and land cover change (LULC) in Merseyside, UK. These studies highlight the diversity of methods and the advancement of models for urban growth and planning. The global urban expansion has led to various environmental challenges, including increased carbon emissions, water scarcity, biodiversity loss, regional warming, and pollution. This urban sprawl disrupts natural habitats, exacerbates climate change, strains water resources, increases urban temperatures, and degrades air and water quality. The more cities grow, the more pronounced these problems become, and they represent a significant obstacle to sustainable development and the health of ecosystems. Tackling these challenges demands thorough planning and creative approaches to minimize the adverse effects of urbanization on the urban environment (Luo et al., 2021; Li et al., 2022). Accurate data on global urban growth is vital for assessing the influence of urban expansion (Seto et al., 2011; Liu et al., 2018). Therefore, assessing the impact of urbanization has become a central topic of urban environment research (Wu, 2014; Rahman et al., 2023; Naikoo et al., 2024).

Comprehensive data on the Earth's surface are necessary for efficient planning and monitoring of urban change. As highlighted by Araya and Cabral (2010), geospatial technology is central to the collection of this data. In the field of modeling, there are two main categories: data- and process-based models, each playing a different role in the analysis and management of urban areas. Cellular automata (CA), widely used in urban development simulation, model urban growth by representing spatial processes as discrete, dynamic systems that evolve over time and space (Alkheder & Shan, 2005). In addition to CA, techniques such as logistic regression and fuzzy logic are used to objectively estimate event probabilities, which are newer methods in the field (Fang et al., 2005). These approaches greatly improve our understanding and prediction of urban expansion patterns.

2 Urban Growth and Expansion

Cities or urban areas are permanent human settlements governed by structured administrative systems or local governments (Goodal, 1987). The centralization of activities in cities encourages interaction between residents, businesses, and various services, supported by complex networks of facilities and amenities that embody modern civilization (Banister, 2012). These community resources are the main drivers of urban growth and distinguish them from cities, rural areas, and villages through their concentration of infrastructure and services. From an ecological perspective, urbanization means the transformation of natural landscapes into residential, commercial, and industrial areas (Tan et al., 2005). This process reflects the increasing demand for land over time, as noted by Mubareka et al. (2011). Urbanization is a key human endeavor focused on enhancing the quality of life, livability, and safety within communities. The scale and complexity of these characteristics are crucial indicators of the level of urbanization in metropolitan areas. This process drives not only industrialization but also economic growth (Kuddus et al., 2020; Zhang et al., 2022; Naikoo et al., 2024). The term "urban growth" denotes the rate of expansion of population, land area, or effective land use within urban regions. This phenomenon, which is synonymous with the growth of metropolitan areas or cities, dates back to 3000–2000 BC, when people began to move from the countryside to cities—places characterized by dense population, economic strength, and developed infrastructure.

The nineteenth century marked a pivotal moment in the evolution of modern cities, transforming them into hubs of cultural, economic, and social activity.

Since 1800, the pace of urbanization has increased dramatically and reached an unprecedented scale in human history. The growth of cities is closely linked to urbanization or urbanism—a term used to describe the increasing concentration of people in these urban areas (Liang & Yang, 2019). This relationship highlights the significant influence of urbanization on social development and spatial transformation over time. Urban growth serves not only as a marker of a country's economic strength but also as a reflection of its overall level of development. The growth of urban areas is influenced by various factors, including the availability of natural resources, infrastructure development, commercial activities, educational institutions, and mining. Together, these factors drive urban expansion and shape the physical and socio-economic landscapes of these regions. As cities grow, they become centers of innovation, trade, and cultural exchange, strengthening their role as engines of national and regional progress. Since the late 1980s, globalization has triggered a new wave of urbanization (Sassen, 2017). Urban growth and urban sprawl are seen as distinct phenomena that characterize physical urban development processes in both developing and developed countries. The urban sprawl has been identified as an important aspect of urbanization and suburbanization in the United States, distinguished by low-density, single-purpose land use, and widespread development (Ewing, 1997). In contrast, urban growth in the majority of developing countries does not exhibit sudden development and low density and is referred to as urban expansion (Li et al., 2021). The process of urban growth and urban expansion is illustrated in the schematic diagram (Fig. 1).

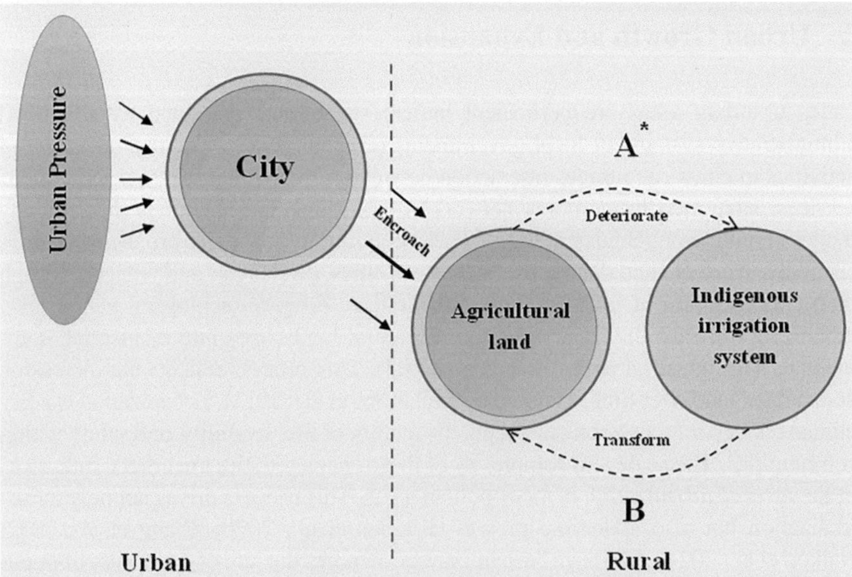

Fig. 1 Schematic diagram of the process of urban expansion

Several terms have been developed to describe the interrelated phenomena of urban growth and urban sprawl, with each term emphasizing different geographic and spatial patterns and the mechanisms that drive these changes. Researchers have developed a variety of methods, often using GIS techniques, to classify and analyze the spatial pattern of urban growth and sprawl. Important indicators used in these studies include the time cost of transportation, land use mix, distribution of population centers, development density, and compactness. These metrics help both researchers and urban planners understand the way cities expand and sprawl over time, offering crucial insights about the dynamics of urban development and its subsequent environmental and social impact.

3 Urban Area Mapping and Monitoring

The use of geospatial techniques is essential for understanding urban expansion and associated human activities (Chai & Seto, 2019; Shahfahad et al., 2022). The rapidly changing nature of the urban environment requires advanced monitoring systems that can keep pace with these changes. Remote sensing is central to mapping, monitoring, and analyzing these evolving urban landscapes (Shahfahad et al., 2023). Geospatial techniques such as GIS and remote sensing have significant importance for quantifying and studying urban growth patterns (Wentz et al., 2014). GIS, in particular, is crucial for tracking and analyzing LULC changes and urban growth dynamics (Bhatta et al., 2010). By combining spatial data with analytical tools, GIS allows researchers and planners to gain deeper insights into the complex interactions between urbanization, environmental change, and socio-economic factors. This comprehensive method aids informed decision-making for sustainable urban development and resource management in fast-growing urban areas.

In recent years, various satellite systems like Landsat, SPOT, ASTER, and MODIS have been instrumental in mapping, assessing, and forecasting changes in LULC dynamics. These satellites provide important data that serve a variety of research and planning purposes, offering detailed insights into environmental changes, urban expansion, agricultural patterns, and natural resource management. Recently, the use of hyperspectral satellite sensors in scientific research and planning applications has increased significantly. These advanced sensors capture detailed spectral data across the electromagnetic spectrum, enabling more precise differentiation between vegetation types, soil compositions, urban infrastructure, and other landscape features than conventional sensors. The integration of these satellite technologies into geospatial analysis systems has significantly improved the understanding and management of landscapes around the world. By integrating high-resolution imagery, spectral data, and spatial analysis, scientists and planners can more effectively track LULC changes over time, evaluate the effects of human activities and climate change, and create strategies for sustainable land use planning and natural resource conservation.

There is increasing interest in applying machine learning algorithms to map LULC changes using remote sensing imagery (Talukdar et al., 2020). For this, machine learning techniques are generally divided into two categories: supervised and unsupervised methods (Alloghani et al., 2020). Supervised machine learning includes support vector machines (SVMs), random forests (RFs), spectral angle mappers (SAMs), artificial neural networks (ANNs), radial basis functions (RBFs), decision trees (DTs), multilayer perceptrons (MLPs), naive Bayes (NBs), maximum likelihood classifiers (MLCs), fuzzy logic, etc. (Kafy et al., 2021; Chachondhia et al., 2021). On the other hand, unsupervised algorithms include clustering algorithms, K-means, fuzzy C-means, and ISODATA (Georganos et al., 2018). Table 1 shows the effectiveness of machine learning algorithms in high-precision classification of urban areas and highlights their efficiency in accurately categorizing urban land use. For example, in a study conducted in the city of English Bazar, the ANN algorithm was used for the classification of urban LULC.

The classification using the default model can predict a low accuracy result; therefore, tuning the hyperparameters is required. In this study, the hyperparameter tuning for three models has been presented in Fig. 2. Fig. 2a shows the process of hyperparameter tuning to predict LULC of 2001, followed by LULC of 2011 (Fig. 2b) and LULC of 2021 (Fig. 2c). Figure 2a shows that the model reached its peak accuracy of 0.94 with a decay rate of 0.1 and 10 neurons in the hidden layers. On the other hand, for building the LULC of 2011, the best parameters, like a decay of 0.1 and 15 hidden-layer neurons, have been obtained using the grid search technique. For the 2021 LULC data, the optimal hyperparameters were identified as a decay rate of 0.1 and 11 hidden-layer neurons, resulting in an accuracy of 0.94. The decay of 0.001 and 0.01 has not provided good accuracy for LULC prediction. Therefore, based on the optimal parameters, the LULC of three periods has been built in this research.

As per the research objectives and the existing LULC patterns, five distinct LULC classes were identified. Following the LULC classification scheme (level 1) of the National Remote Sensing Center (NRSC), the study identified five LULC classes in the study area, i.e., water bodies, open land, built-up areas, vegetation,

Table 1 Application of machine learning algorithms in urban land use mapping and achieving high accuracy

Authors	City	Models used for mapping	Accuracy
Mostafa et al. (2021)	Gharbia governorate, Egypt	SVM	0.943
Frimpong and Molkenthin (2021)	Kumasi Metropolis, Ghana	RF	0.97
Jamali (2021)	Shiraz, Iran	SVM	0.98
Zeng et al. (2020)	South Africa	RF	0.97
Din and Mak (2021)	Hyderabad, Pakistan	SVM	0.98
Hossain et al. (2021)	Gopalganj, Bangladesh	Ann	0.894
Nyamekye et al. (2020)	New Juaben municipality, Ghana	RF	0.963
Poghosyan (2018)	Soviet cities	SVM	0.94
Khan and Sudheer (2022)	Islamabad, Pakistan	RF	0.98

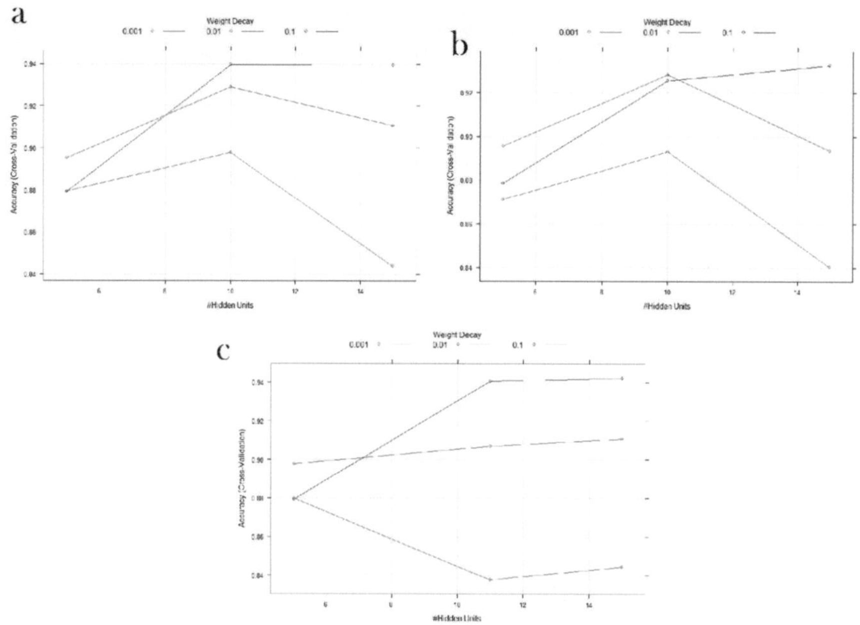

Fig. 2 Application of ANN with hyperparameter tuning using grid search for LULC modeling for (**a**) 2001, (**b**) 2011, and (**c**) 2021

and cropland (Fig. 3). While additional LULC categories could be created, their representation is minimal, making it difficult to clearly depict them. Therefore, these fuzzy pixels were included in the adjacent classes of the system. The final classification results for the years 2001 to 2021 offer a detailed outline of the main LULC components in the urban area of English Bazar and its surroundings. The study shows that the LULC pattern has changed significantly over the last two decades. In 2001, arable land was the predominant land use category with about 44% of the area of English Bazar, followed by built-up area (42%), open land with 8%, vegetation with 4%, and water bodies with 2%. By 2011, however, notable changes were evident, especially in the transformation of agricultural and open land into built-up surfaces. The built-up surfaces expanded rapidly across the study area, with 12% of arable land and 10% of open land being converted to built-up zones. Additionally, 13% of open land was converted into arable land, with a slight shift of vegetation to open space.

Between 2001 and 2011, there was a modest increase in vegetation cover, particularly in mango plantations, a trend that continued to grow between 2011 and 2021. The most significant change in the latter period was the substantial increase in built-up area, which replaced much of the cultivated areas and open land. The built-up area, which covered 42% of the total area in 2001, rose to over 78% by 2021, an increase of 36% in 20 years. Another notable change has been in the extent of water bodies, particularly in the southern areas of the English Bazar, which has increased because of severe flooding in 2017, which inundated low-lying areas. The

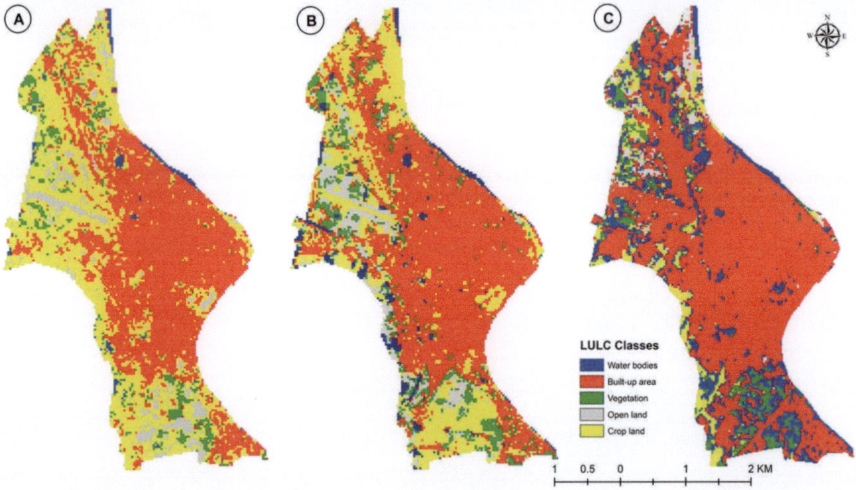

Fig. 3 Urban area mapping for English Bazar city, West Bengal, using an ANN model

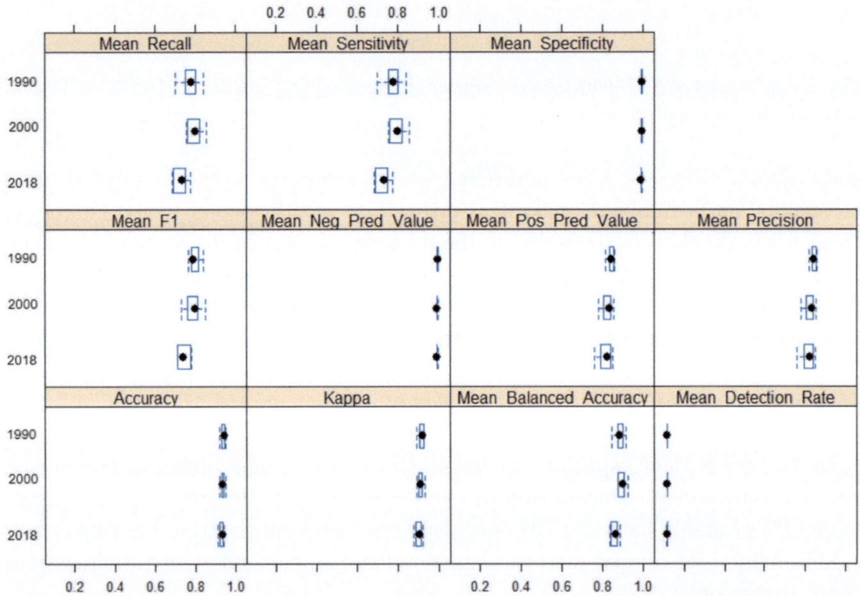

Fig. 4 LULC validation using a confusion matrix

rate of urbanization has also accelerated, increasing by 2.4% between 2001 and 2011 and by a further 3.6% between 2011 and 2021.

The study evaluated performance and validation of the LULC models by calculating several metrics derived from the confusion matrix, including average recall, sensitivity, specificity, positive predictive value, negative predictive value, kappa coefficient, overall accuracy, F1 score, detection rate, precision, and balanced accuracy over 2001–21 (Fig. 4). All metrics were measured on a scale from 0 to 1. The value close to 1 indicates higher performance of the model to detect land use objects, while close to 0 indicates that the object detection ability of the model is poor. Figure 5 shows that the overall accuracy for all models is above 0.85, which signifies that the ANN model performs better for LULC modeling. On the other hand, the kappa coefficient values for LULC of 2001, 2011, and 2021 were 0.94, 0.92, and

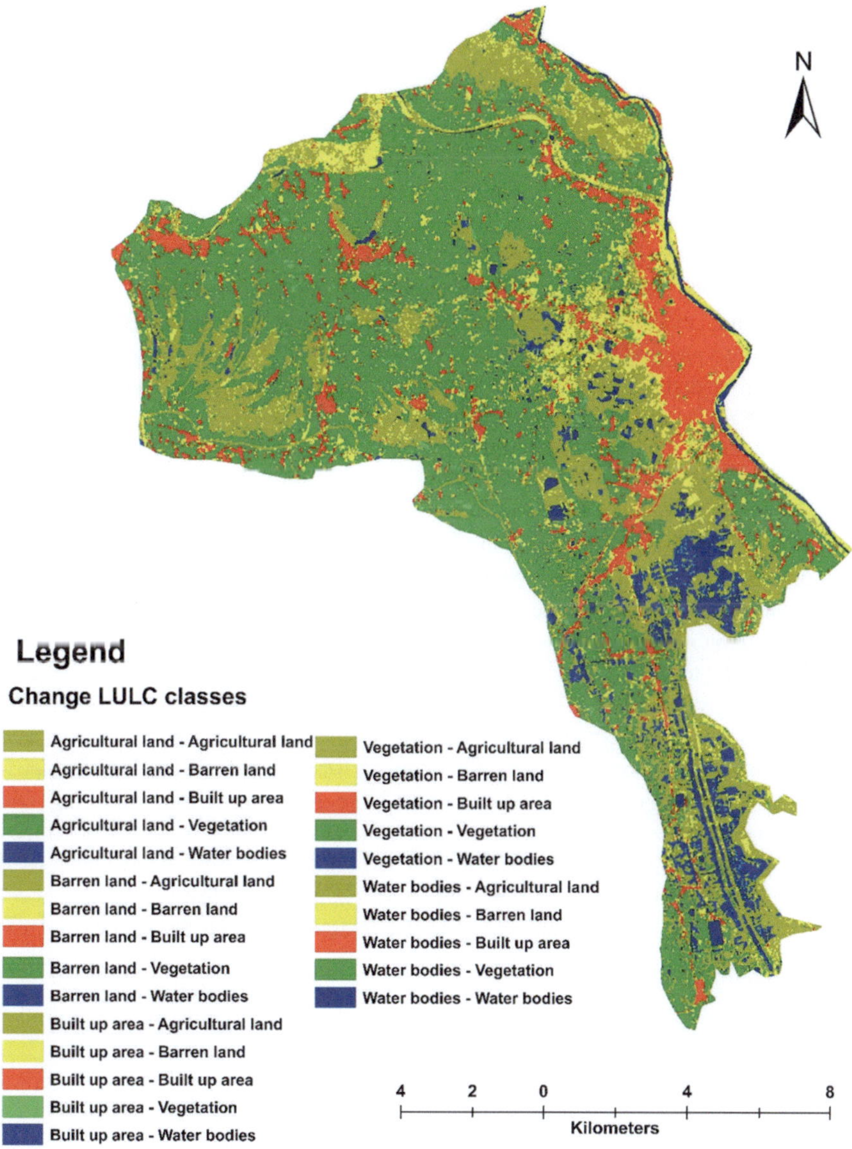

Fig. 5 Change detection analyses between the LULC of 2001 and 2021 for English Bazar municipality

0.937, indicating very good agreement of predicted LULC with ground truth. A high value of precision indicates that the detection of relevant classes was retrieved and predicted correctly. In the present study, we obtained high-precision values for LULC of the 2001, 2011, and 2021 LULC models. However, it can be stated that the accuracy may be enhanced because of the application of a hyper-tuned ANN model instead of a default ANN model and conventional classifiers.

4 Change Detection

The change detection process is defined as the analysis of a characteristic or phenomenon throughout its unique time window. Satellite mapping, environmental monitoring, and the study of land use shifts and cover all benefit significantly from change detection methods. Using remote sensing satellites, which capture satellite images at varying resolutions, can detect changes. Whether on a global, national, or regional scale, environmental concerns can be monitored and analyzed using remote sensing techniques (Afaq & Manocha, 2021). The change detection process examines the shifts that occur in each region over time by comparing images taken at various intervals. Landsat images comprise a Red, Green, and Blue (RGB) band, which can be used with multifeature-based fusion approaches to spot changes over time (Wang et al., 2020).

Monitoring environmental conditions, such as the impacts of natural disasters and urban expansion, detecting changes in vegetation cover, evaluating desertification processes, and identifying specific alterations in urban or natural environments are common applications of the change detection method (Salem et al., 2020). One of the fundamental characteristics of remotely sensed data is the ability to detect changes owing to the frequent coverage at short intervals and consistent image quality provided by satellites in Earth's orbit (Thakur et al., 2020). There are several methods for detecting changes, including LULC, deforestation, urbanization, and changes related to natural disasters (Shi et al., 2020; Yan et al., 2019). Remote sensing (RS) technology has become indispensable for long-term monitoring and assessment. For example, to detect the land use transformation of highly urbanized NCR regions, the Landsat images with 4–5 TM and 8 OLI sensors have been used to classify the LULC image for 1990 and 2020. Then, we employed a change detection method to examine the change of LULC classes between the LULC of 1990 and 2020 (Fig. 5).

5 Urban Sprawl Assessment

In the field of spatial planning, the phrase "urban sprawl" has become common parlance. Although the phrase has been widely used and its location identified, a consensus on what it means has yet to be reached (Torrens, 2008). Various research institutes, colleges, and organizations have tried numerous definitions of "urban sprawl" since the first attempts began in the 1970s (Jaeger et al., 2010). According to Hess, "urban sprawl" initially served as an adjective to characterize a particular style of urban development, as a verb to depict the process of such development, and as a noun to denote a distinct urban spatial configuration (Hess, et al., 2001). The term "sprawl" has been defined in a number of different ways, most of which originate from European and North American literature published in the late twentieth and early twenty-first centuries, when it first emerged as a major trend in urban growth. Nonetheless, there has not yet been a broadly agreed description of the phenomena, despite its acknowledgment and the significance it has for sustainable urban development (Sahana et al., 2018; Gao et al., 2016; Zhang et al., 2021; Shah et al., 2022;). There is also a lack of a universally applicable explanation of urban sprawl; instead, definitions tend to focus on the specific geographical qualities, physiognomy of the research region, and data at hand. The phenomenon of "urban sprawl" manifests itself in various ways in different cities, mostly owing to commercial, geographic, and regulatory variables, making it difficult to draw broad conclusions and agree on a uniform definition (Lopez, 2014; Catalán et al., 2008; Paulsen, 2014).

However, urban sprawl is a major concern for many cities throughout the globe (Kumar et al., 2018). Therefore, many people understand urban sprawl to be the unplanned and uncontrolled expansion of an urban core to its perimeter, which puts a strain on the development in the area immediately next to the city's borders and has a variety of other negative effects (Borana et al., 2020; Pan et al., 2020; Patra et al., 2021). Sprawl in an urban area is closely linked to city growth. In addition to urbanization, Bhatta et al. (2010) identify urban expansion as one of the processes of urban development because of the universal nature of its occurrence and consequent far-reaching effects. While urbanization refers to changes in a city's communal dimensions, urban expansion refers to changes in population within a given economy and society. As such, urban growth may be seen as a multifaceted process of physical and geographical transformation that has far-reaching effects on several sectors, including society, the economy, population, politics, and the environment. Barnes et al. (2001) note that sprawl occurs because of a complex interplay of economic, social, political, and physical pressures, and that this is true regardless of the criteria used to define and assess urban development, growth, or urbanization. Municipal fragmentation, infrastructure investment patterns, and city subsidies are all factors. Urban sprawl experts agree that studying the phenomenon in its many forms is essential for mitigating its negative consequences (Hatab et al., 2019; Baqa et al., 2021; Xia et al., 2020).

However, there are several methods available for computing urban sprawl (Rahman et al., 2010; Langa et al., 2021). This research employed the frequency method to demonstrate the urban sprawl of the English bazaar block, and the results were encouraging. The probability of an event happening at the upper threshold of the relative frequency that it happens in repetition under substantially comparable circumstances is a central tenet of frequency theory. Now that we have this model, it is easy to get a handle on the freshly formed, more expansive territory. This model stacks the urbanized areas of three distinct time periods to produce a map that shows the same time period three times, twice, and once. Therefore, we were able to do a rough calculation of the cumulative growth over time. The sky-blue coloration (Fig. 6) indicates that these regions were heavily populated in 2001, 2011, and 2021, with the number three denoting a permanent or stable built-up area. The frequency approach, represented by the number two, shows that the area has been developed twice in those areas (in 2001 and 2011). One incidence of recently constructed urban space is represented by the value of red, suggesting that this area is now undergoing urbanization. Based on the results, value one indicates that fresh and

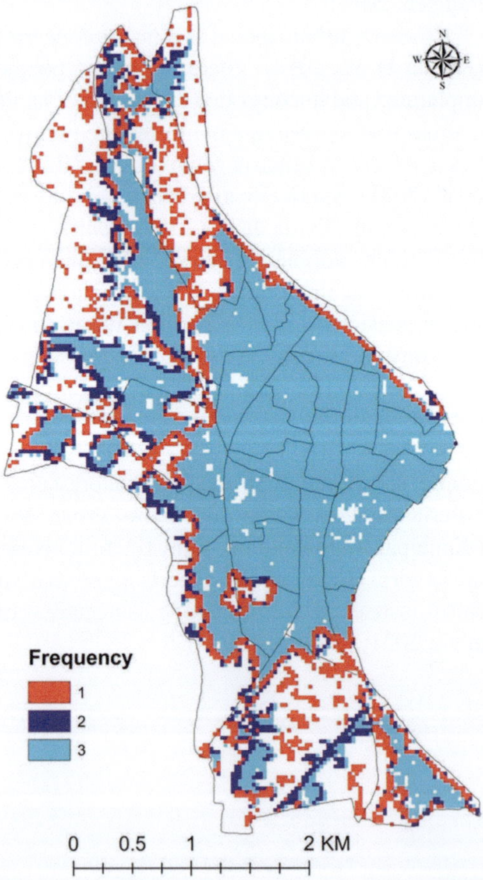

Fig. 6 Urban sprawl modeling using the frequency approach in English Bazar municipality

isolated built-up nodes spanned 2.8 km^2, and value two indicates that the built-up transition area covered 2.35 km^2 (Fig. 6). The transition zone is the most precarious because it is constantly shifting as it gains territory from newly developed and isolated built-up regions and loses territory to established and permanent urban cores. New and previously isolated built-up nodes have been transformed into urban areas with the help of medium and small enterprises, as well as health and educational institutions, in recent years.

6 Urban Compactness Assessment

The concept of compactness encompasses the form and arrangement of urban elements such as spatial planning, land use types, and transportation infrastructure (Xu et al., 2019; Abdullahi et al., 2018; Wang, 2022). Traditionally, this term has been used to describe the physical characteristics of an urban system. However, it is important to recognize that human behavior and the lifestyle choices of urban residents also contribute to the overall density of metropolitan areas. The extent to which a city can operate independently without external support serves as a measure of its compactness (Burton, 2002). In addition, compactness can be seen as a reflection of residents' travel patterns to meet daily needs such as work, shopping, and leisure activities. Assessing compactness is critical to effective land use planning, evaluating site capacity, and guiding the development of individual sites. Assessing the compactness of a city, considering its various elements and its link to urban sustainability, is complex (Tan & Lu, 2019; Naikoo et al., 2023). Consequently, there is no universally accepted methodology for this assessment, and each study tends to analyze compactness based on its specific objectives and available data. For example, urban density and land use diversity are often measured using census tracts, which vary in size and resolution. Because results may vary by zoning method, cell size, and type of input data, these assessments cannot be considered comprehensive or consistently reliable. Furthermore, assessments are typically conducted at the zone or district level, with most previous research using a zonal approach to analyze the compactness of cities. However, such assessments based on predefined zones are often neither thorough nor reliable enough. Density metrics such as population, building, and housing density are very sensitive to changes in zonal area, which can be influenced by changes in zonal boundaries (Koziatek & Dragićević, 2019). These boundary changes also affect the ownership and availability of community facilities and services within each zone, leading to different outcomes depending on the zoning technique used. Li and Yeh (2004) proposed to assess urban compactness through methods that include entropy and compactness indices. At the national level, the concept of cellular structures and the concentration of built-up cells within specific areas is often used when analyzing urban compactness. Mubareka et al. (2011) developed a composite measure for analyzing urban growth trends that defines the degree of compactness in European cities. This composite index was created using principal component and cluster analysis. Thinh et al. (2002) introduced a method for assessing compactness through GIS grid analysis and a gravity approach applied to built-up grid areas of 500 m × 500 m within the study area. Urban compactness, which includes both the physical density of built-up areas and functional aspects, provides valuable insights into the conditions of cities (Lan et al., 2021). Therefore, compactness assessment without preconceived assumptions or local rules provides a flexible and universally applicable approach.

In this study, SAGA GIS was used to build dominance, diversity, and connectivity models to analyze the compactness of built-up areas (Fig. 7). The research focused on assessing the current urban landscape, including built form and structure. Diversity in cities promotes cosmopolitanism and economic productivity by supporting a wide range of economic activities and opportunities. This diversity, along with urban density, has become a central principle of urban design and

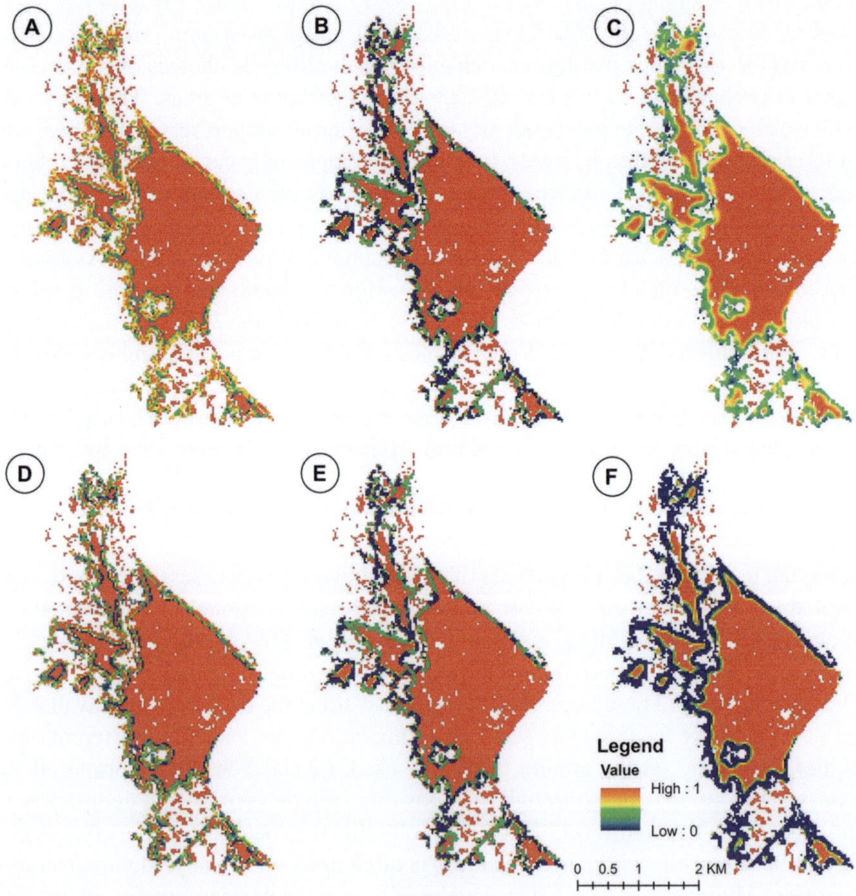

Fig. 7 Physical compactness assessment of English Bazar city in terms of (**a**, **d**) dominance, (**b**, **e**) diversity, and (**c**, **f**) connectivity for 2011 and 2021

development. Communities within the urban hierarchy play different roles that attract a large population. Connectivity, i.e., the ease with which people and goods can move between different points, either directly or via intermediate routes, is a key indicator of overall urban development. Each of these three models is scored on a scale of 0 to 1, with 0 indicating low concentration and 1 indicating high concentration. The study has identified significant concentration zones within the city and its surroundings, especially along the NH-34 corridor and in other areas. Rathbari serves as an important hub for travelers between the Malda and North Bengal regions. As a result, development has extended to the outskirts of the urban area and beyond the existing road network.

7 Sprawl Versus Compact Development

During the second part of the twentieth century, a new phenomenon emerged: the sprawling city. It is defined by a variety of indications, such as a lack of housing and commercial building density, an abundance of privately owned autos, an absence of public transportation, congestion, environmental deterioration, and a lack of community among local inhabitants (Squires, 2002; Lang, 2000). The "natural development" of cities, the "flight from blight (Alonso, 1964)," and the "mono-centric city model" (Mieszkowski & Mills, 1993) are only some of the hypotheses that have been put out to try to understand and explain the reasons for urban sprawl. "Sprawl" refers to the process by which cities develop by occupying an ever-larger land area, whereas "densification" refers to the process by which people live closer together in the same space (i.e., densification). Sprawl is facilitated by a number of factors, including the growth of peripheral economies, the accessibility of cheap land, cultural preferences for low-density living, and the development of transportation networks to lower commute costs. In this article, we showed that the emergence of new buildings along the periphery of English Bazar municipality and even beyond the administrative boundary of the municipality is considered urban sprawl, denoted by frequency 1 (highlighted in red) (Fig. 6). Also, Fig. 8 shows that marginal expansion and enclave expansion of the old urban patch can be considered as urban sprawl, where new economic activities can be observed, like a new nursing home, hotel, and car showroom in the periphery of English Bazar city.

Conversely, factors conducive to compact development include increases in energy costs, which elevate the expense of long commutes; the growing significance of time management in modern competitive environments; urban planning strategies such as the implementation of greenbelts to constrain spatial expansion; and advancements in efficient and affordable public transportation systems (The World Bank, 2011). In this article, it can be seen in Fig. 6 that the connectivity is much higher than in the outside part of the English Bazar city. Also, the temporal model for 2011–2021 shows that the urban sprawl is only noticeable in the southeastern part of the English Bazar city, while other parts do not observe urban sprawl. But the main city has observed high-rise buildings, new green belts, and existing vacant

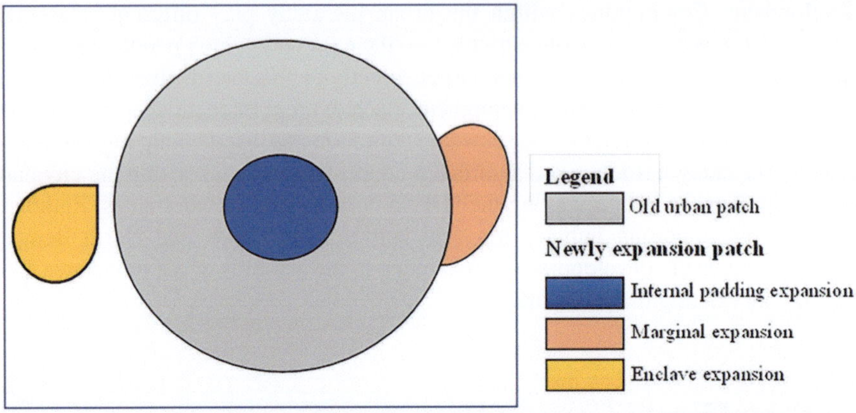

Fig. 8 Explanation of urban sprawl and urban compactness

space is being converted into high-rise buildings, which has caused compactness. Figure 8 shows internal padding expansion within an old urban patch is an example of compactness.

8 Spatial Land Use Change and Landscape Pattern Analysis

Unique patterns within the landscape play a crucial role in supporting ecological processes. As this pattern can be viewed from different angles, research increasingly focuses on developing effective tools to analyze and evaluate the spatial configurations of different landscape types (Jun et al., 2022; Lin et al., 2018; Gustafson, 2019). Landscapes are composed of both natural and man-made elements, with a rapid expansion of cities in India and globally becoming increasingly common in recent decades. This accelerated urbanization has raised significant concerns about the impact of landscape fragmentation on ecosystems and biodiversity. As a result, scientists are increasingly interested in the ecological impacts of large-scale urban growth and its effects on landscape structure. The quantification of landscape pattern is typically done through the use of language, statistical analysis, visual representations, and landscape metrics. In the past decades, landscape metrics has become a focal point in landscape ecology, with a wide range of metrics developed and applied in various disciplines, including urban ecology, landscape planning, landscape change monitoring, forest dynamics, and ecological network planning (Dadashpoor et al., 2019; Duarte et al., 2018; Guo et al., 2020; Hoffman et al., 2020). Key indicators such as patch number, mean patch area, patch density, and the Shannon diversity index are widely recognized as fundamental indicators for landscape analysis (Akın & Erdoğan, 2020; Liu et al., 2018; Sun et al., 2020; Zhang et al., 2018). Through a systematic and refined approach, researchers can effectively analyze horizontal landscape patterns by using a set of commonly accepted metrics and guidelines for selecting appropriate indicators.

In this study, a GIS-based landscape fragmentation analysis was conducted to assess the landscape pattern of the English Bazar community. Six fragmentation indices—large core, medium core, small core, perforated, edge, and patch—were used to examine the landscape pattern of the built-up area over time (Fig. 9). The large core, covering over 500 hectares, represents the most stable and durable built-up area. In contrast, the small core, comprising less than 250 acres, reflects newly developed built-up areas that have expanded outward from the primary core. The results show that the urban center, which occupied 2.92 km^2 in 2001, grew to 3.8 km^2 and further to 7.42 km^2. This expansion indicates that the size of the core area identified in 2001 has tripled over time. Between 2001 and 2011, small and medium-sized core areas merged with larger core areas, and by 2021, the medium-sized core areas had merged into large core areas. This pattern closely mirrors the ongoing process within the study area, where the dense urban center is expanding. The small and medium core areas have also increased in size since 2001, growing from 0.78 km^2 and 0.64 km^2 to 0.92 km^2 and 1.32 km^2, respectively. This considerable expansion of the small and medium-sized core areas is due to the conversion of new, isolated built-up nodes. The core areas in the northern and southwestern parts of the study area, especially along SH-10 and NH-34, have experienced the fastest growth. Neighborhoods such as Deshbandhu Para, Ramchandrapur, Netaji Park, Mokdompur, Malanchya Pally, and Barrack Colony form the small- and medium-sized nodes in the study region. New patches were identified in scattered locations in the west, south, and north of English Bazar municipality, while fringe and perforated areas were more prevalent in the eastern and southern parts of the area. In 2001, the

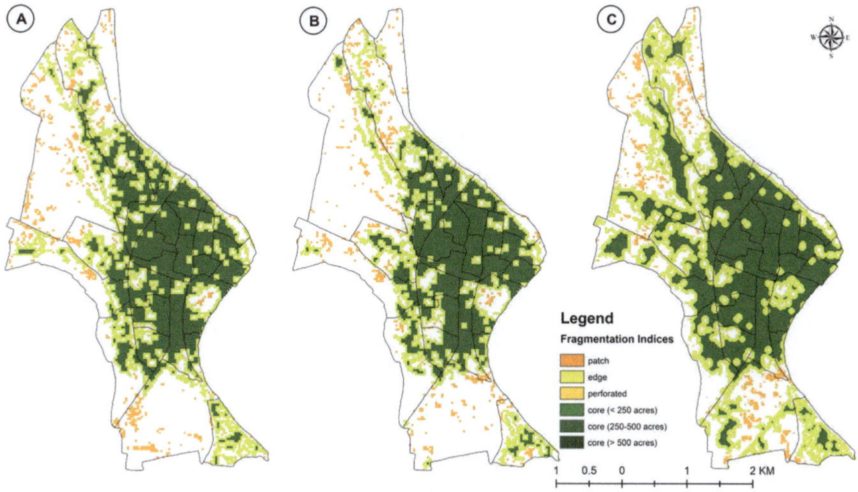

Fig. 9 Landscape pattern analysis using the landscape fragmentation tool for English Bazar Municipality for the period of (**a**) 2001, (**b**) 2011, and (**c**) 2021

"perforated" and "patch" categories accounted for 0.62 km^2 and 0.68 km^2, respectively; by 2010, these areas had grown to 1.26 km^2 and 1.05 km^2 (Fig. 8). These formerly rural areas have experienced accelerated urbanization due to economic and infrastructural development, which has led to the emergence of new, separate built-up centers. Over time, some perforated and patched areas have merged with smaller and medium-sized cores. This trend shows a strong tendency toward expansion of all core sizes—small, medium, and large. The study also revealed a significant growth of the road and rail network in the region.

9 Spatial Urban Modeling and Prediction

Predicting future urban expansion and associated problems is an investigation of probable environmental changes. The discovery and study of pre-patterns of urban expansion and their issues aid in the optimal design of infrastructure in urban settings. Understanding urban sprawl behavior and its issues might also help with resource management and smart environmental planning. In this regard, several modeling tools have been developed to comprehend the dynamics and patterns of urban expansion and its related issues. These models have been used in the identification of development patterns, the description and evaluation of future development impacts, and the investigation of the possible effects of various policies, as well as in the prediction of future changes. Yet, there have been no flawless models developed to anticipate the challenges that would accompany the expansion of metropolitan areas. However, each model has its benefits and disadvantages, and each model is seen to be restricted.

Researchers in the field of urban studies have utilized predictive models to learn things like the rate of urban expansion in the future, the nature of urbanization currently, and the geographical pattern of issues related to urbanization (Table 2). Past urbanization patterns, as well as anticipated development and LULC shift patterns, might be better comprehended with the aid of urban growth simulations. Thanks to recent developments in computing, urban growth models (UGMs) may be used to foresee how LULC can evolve in the future. Multiple UGMs have been developed in recent years to simulate and predict future urban growth. These include CA (Clarke et al., 2007), Markov chain or Markov (Sivakumar & Ghosh, 2016), Artificial Neural Network (ANN) (Schneider & Pontius Jr, 2001), CA-Markov (Subedi et al., 2013), ANN-CA (Yang et al., 2016), and conversion of land use (Waddell, 2002). The most well-known ones are SLEUTH, CLUE, Markov, and ANN for predicting LULC (Kim & Newman, 2020). Using data on land use behavior change, the SLEUTH model projects how LULC may shift in the future (Kuo & Tsou, 2018). CLUE is a model that projects how land use and other factors will interact in the future (Veldkamp & Fresco, 1996). The Markov model gives a stochastic estimate of the likelihood of LULC shifts in light of long-term trends, but it does so without evaluating the potential shifts' geographic foci (Moghadam & Helbich, 2013). Therefore, UGMs like CA, which almost exclusively focus on sites

Table 2 Different types of predictive models for analyzing different urban issues

Predictive models	Urban areas	Purpose	References
Cellular automata	Tehran	Urban growth	Shafizadeh-Moghadam et al. (2017)
Markov chain	Kaduna, Nigeria	Urban expansion	Bununu (2017)
Multilayer perceptron-Markov chain	Jashore city, Bangladesh	Ecosystem service value modeling	Morshed et al. (2021)
Artificial neural networking	Cumilla, Bangladesh	Seasonal urban thermal field variance	Kafy et al. (2021)
Random forest	Tehran	Green space mapping	Sharifi and Hosseingholizadeh (2019)
Random forest	Jodhpur, India	Carbon sequestration quantification	Uniyal et al. (2022)
Markov chain–cellular automata	Chongqing, China	Land use policies and plans	Xu et al. (2019)
Random forest	Fukushima prefecture, Japan	Land price estimation	Derdouri and Murayama 2020)
Support vector machine	Shenzhen, China	Monitoring housing rental prices	Hu et al. (2019)
Support vector machine	Ouagadougou, Burkina Faso	Mapping pattern of urban development	Schug et al. (2018)

of change, may forecast both the future location and amount of LULC change by combining a Markov model with such an approach (Arsanjani et al., 2012). To simulate potential LULC changes, the ANN-Markov model is among the most reliable models available (Pahlavani et al., 2017).

Anthropogenic variables have a significant impact on the current state and future development of the urban surface ecological status (USES) (Xu ct al., 2019; Hu & Xu, 2018). To achieve sustainable development goals such as sustainable cities and communities, it is important to analyze and model the USES to support urban environmental management and planning and to advise decision makers and the community on ecosystem services (Estoque, 2020).

The use of remote sensing has now advanced to the point where it can accurately assess the state of the ecosystem in a given area (Weng, 2009; Zhou et al., 2014). However, quantifying SES with a single spectral indicator is challenging due to the complex interaction between SES and biophysical and ecological parameters (Van Kamp et al., 2003). Several ecological indicators have been developed to assess the health of ecosystems. For example, the Normalized Difference Vegetation Index (NDVI) and leaf area index have been used to track environmental changes (Stefanov and Netzband, 2005; Eckert et al., 2015; Brehaut & Danby, 2018); land surface temperature (LST) has been used to assess the impact of urbanization (Weng, 2009; Meng et al., 2018); an index-based built-up index (IBI) (Estoque and Murayama, 2015) and the normalized difference impervious surface index (NDMI) (Qi et al., 2022) were used to quantify the extent of urbanized and impervious surfaces. The complexity and diversity of influencing variables make it insufficient to use only one or two ecological indicators to measure the state of an ecosystem.

A composite index developed to assess urban environmental quality integrates remote sensing data for environmental and ecological parameters as well as socio-economic indices (Yue et al., 2019). This type of index can be considered robust and accurate for ecological research. Consequently, the synthetic index can be constructed using a variety of frameworks. For environmental policy, the OECD created the Pressure-State-Response (PSR) using a weighting approach (such as the analytic hierarchy process (AHP) (Jin et al., 2016), the analytic network process (ANP) (Nekhay et al., 2009), or the Delphi method (Lee et al., 2013)), which can summarize multiple types of remote sensing data and statistical data into a single index. Indicators of pressure (from human activities), the state of the environment, and the response of society are the three types of indicators that can be used in the identification and evaluation processes of the PSR model (Hu & Xu, 2018). Subsequently, it has been used to assess a variety of ecosystems, including forests (Santibáñez-Andrade et al., 2015), soils (Zhou et al., 2014), wetlands (Jin et al., 2016), farms (Grecchi et al., 2014), rivers (Novoa et al., 2018), cities (Han et al., 2015; Xu, 2013), etc. A large amount of remote sensing and socio-economic data was used to build the PSR framework, and subjective experiences could change the weighting of indicators. Therefore, using the PSR as a framework, we developed an ecological index (RSEI) by combining four indicators (greenness, wetness, dryness, and heat) using principal component analysis (PCA), AHP, fuzzy AHP, and fuzzy gamma models to assess the ecological situation in Asir city in Saudi Arabia.

In this chapter, we used a fuzzy logic model to determine the probability of urban growth using a fuzzy logic approach. We used the ArcGIS Fuzzy Membership Tool to integrate all parameters unidirectionally to create a fuzzy model of dominance, diversity, and connectivity ranging from 0 to 1. This approach led to consistent results, as we had previously observed. Next, we used the fuzzy overlay tool along with operators such as AND, OR, and gamma to integrate the three fuzzified models into a final expansion model. The resulting map shows the predicted probability of future development in the English Bazar block, represented by values between 0 and 1 (Fig. 10).

The result of the fuzzy logic model ranged from 0 to 1, which means a high to low probability of expansion in the stretch format, depending on the factors (dominance, diversity, and connectedness). The stretch expansion probability index was then categorized into three classes using the natural discontinuity algorithm: high, medium, and low expansion probability zones. The results show that the zones with high probability of development account for 85% of the area, while the zones with moderate (12%) and low probability of development account for the remaining 3% (Fig. 10). All parameters had a positive effect on building development, indicating that where parameter values are higher, the probability of building development is also higher, as our results show. As the eastern part of the study region had higher values for the factors' dominance, diversity, and connectivity, the probability of building expansion was higher there. A higher concentration of all factors indicates that the location has a higher accessibility rate, which will lead to higher future

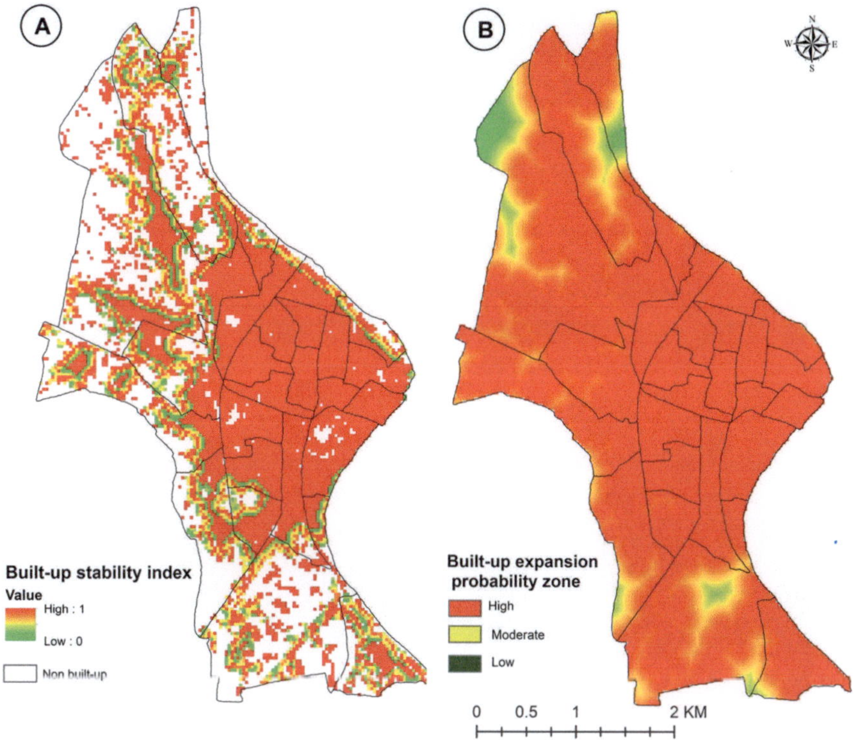

Fig. 10 Built-up expansion probability model for English Bazar city using fuzzy logic model

development expansion. It appears that the surrounding rural areas of the community, particularly the area adjacent to NH-34, will have a greater chance of becoming urbanized, so expansion of the study area will be concentrated to the southwest and north. In addition, the western portion of the city has a significant probability of population growth despite its status as a rural residential neighborhood. The areas with the least potential for expansion were discovered in the northern part of the study area due to the presence of mango forests or water bodies and poor connectivity. Although the municipality of English Bazar Block is a Type I town with a population of 274,627 and a population density of 1100/km^2, the municipality has a high chance or probability of expanding its built-up area much faster in the future, as shown in our study. The convenient location and the fertile soil, which have favored agricultural activities, especially mango cultivation and some food processing businesses, as well as the increasing commercial efficiency of the city, contribute to its rapid expansion.

In this research, we used four models, such as PCA, fuzzy logic, FAHP, and integration of all models with fuzzy logic for integrating four parameters under the PSR framework to build RUSEI models for the Asir region, Saudi Arabia (Fig. 11). The area coverage under five categories of RUSEI zones using four models has been

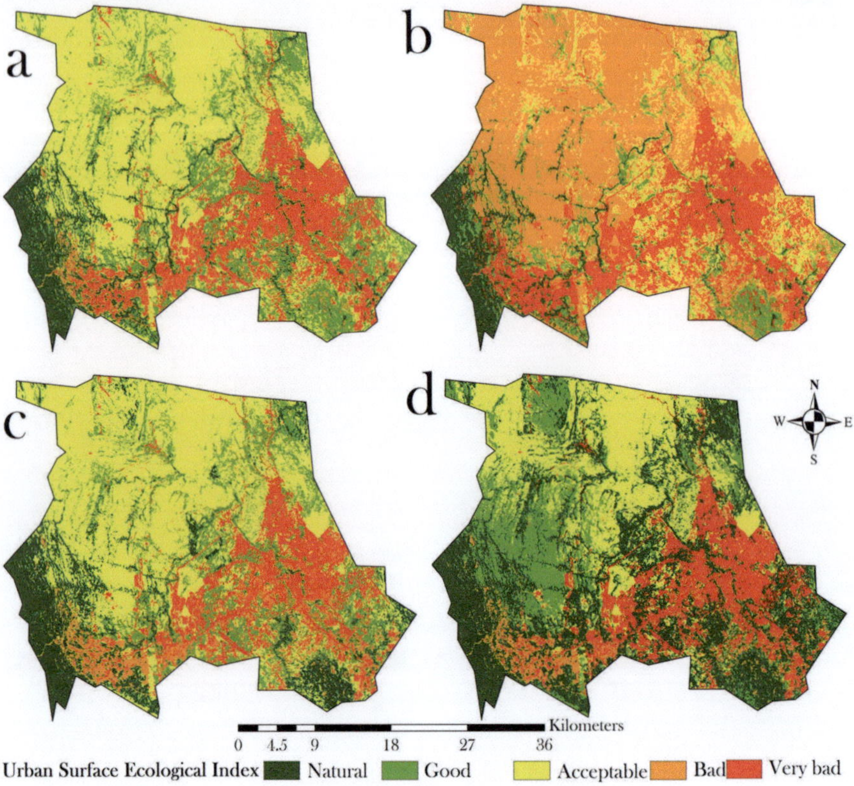

Fig. 11 RUSEI model for Asir city using different analytical techniques: (**a**) PCA, (**b**) fuzzy logic, (**c**) FAHP, and (**d**) integrated model

Table 3 Computation of area coverage under different RUSEI zones using various models (unit km^2)

RUSEI zones	PCA	Fuzzy logic	FAHP	Integrated model
Natural	124.20	91.50	176.75	434.40
Good	310.85	157.90	257.50	224.15
Acceptable	570.35	185.00	571.00	347.00
Bad	66.70	571.00	111.20	82.85
Very bad	204.60	271.30	160.00	188.40

computed (Table 3). The PCA model has predicted 124.16 km^2 areas as natural, followed by 310.83 km^2 areas as a good RUSEI zone. On the other hand, 204.61 and 66.68 km^2 areas were computed as bad and very bad RUSEI zones (Table 3). The FAHP model has predicted 160.05 km^2 and 111.23 km^2 areas as bad and very bad RUSEI zones. While the integrated model computed 188.39 km^2 and 82.86 km^2 areas as bad and very bad RUSEI zones. Therefore, based on the area computation, we can state that all models predicted RUSEI zones differently. However, FAHP and integrated models have similarities for the zones of good, acceptable, and very bad RUSEI. Therefore, it is very difficult to state which model performs perfectly based on area computation. In addition, these areas are conceptual and imaginary models; therefore, the ground truth data is not available. Hence, to claim the best model, the sensitivity and uncertainty analysis should be conducted.

10 Measures for Planned Urban Growth

10.1 Prioritization of Sector-Specific Activities

Maintain the sector-wise activities that mean a particular area should be mentioned for a particular purpose, like a specific area for the market, an area for all the offices, an area for schools and colleges, and an area for agricultural product storage. Especially the health sector and the showroom (four/two-wheeler) must be shifted. Most of them are situated along the NH-34, and these sectors attract more people, and they settle surrounding them; as a result, unplanned growth will happen. So, if we centralized them into a specific place separately, it could help with the planned built-up growth. Establishing a health center a little away from NH is better than beside the NH. Because of traffic congestion, excessive noise is creating very harmful problems for the patient.

10.2 Reconnecting the Urban Area with the Surrounding Rural Area

Urban and rural regions both depend on each other. Mainly, urban regions do not operate in isolation. Urban areas depend heavily on their surrounding rural areas for various essential resources and services such as food, labor, water supply, and disposal of food waste. And in rural areas, people depend on urban areas in terms of the health sector, institutional purposes, and administrative work. So, if we developed better connectivity between rural and urban areas, the disparity should be minimized.

As well as improving the health sector, road network, and administrative services, the improvement of the lifestyle and social well-being of the rural people can reduce the pressure on the urban areas. As there is a trend to migrate from rural to the nearby urban area on a regional scale, for better living standards. These facilities act as a major pull factor for the rural population, and hence, the population has risen in the urban area. So, if we provide better facilities in rural areas, this pressure might be reduced, and through better road connectivity, they can easily access the urban regions for their needs.

10.3 Urban Agriculture Promotion

Agriculture is often associated with rural areas, but around 800 million people worldwide practice urban agriculture and community gardening (FAO Framework for the Urban Food Agenda). Preserving agricultural land in cities can significantly shorten supply chains and reduce carbon emissions associated with transporting food from rural areas to urban centers. By promoting the cultivation and sale of fresh produce in urban areas, the environmental impact of food distribution can be reduced. This strategy also promotes the development of inclusive local supply chains and improves access to healthy food through initiatives such as farmers' markets. In addition, farmland and community gardens serve as important green spaces in the urban environment and help reduce air pollution by absorbing pollutants and improving air quality. Integrating agriculture into urban spaces not only promotes sustainable food production but also supports healthier communities, reduces the ecological footprint, and strengthens the resilience of cities.

10.4 Expanding Green Areas to Enhance Environmental Health and Quality of Life

The more cities grow, the more the number of green spaces and vegetation decreases (Shahfahad et al., 2024). Apart from their visual appeal, trees and green spaces are crucial for improving air quality, mitigating urban heat islands, promoting physical activity, and improving public health (Islam et al., 2024; Yang et al., 2023). The decline in green spaces contributes to increased air pollution and higher local temperatures, which in turn increases the risk of heart and lung disease, obesity, and vulnerability to new health threats. To address these issues, it is crucial to harmonize food systems with the natural environment. This alignment helps to reduce pollution and promote healthier lifestyles through better nutrition and increased physical activity, contributing to sustainable urban development and more resilient communities.

A study by Pal & Ziaul (2017) points to the UHI effect in the community of English Bazar, a common problem in growing cities. Expanding green spaces or creating green belts can help mitigate the UHI effect, reduce particulate matter, and provide outdoor thermal comfort for residents. The expansion of built-up areas is a dynamic process, especially in countries like ours, where it is difficult to define fixed boundaries. It is therefore important to accompany and control this expansion process through careful planning. To ensure planned urban growth, several important steps should be taken: First, it is important to assess the city's land requirements for the next 30 to 40 years, which forms the basis for strategic planning. Second, it is crucial to obtain planning permission for a sufficiently large expansion area to meet future land needs. Third, the protection of ecologically sensitive areas within the expansion zone is necessary to maintain the ecological balance. Fourth, for efficient traffic flow and mobility in the city, it is essential to plan main roads through the expansion area, ideally at a distance of one kilometer. Finally, securing rights of way for these roads and protecting sensitive environmental sites from development is crucial for long-term infrastructure planning and environmental protection. Together, these steps contribute to sustainable and well-managed urban growth.

References

Abdullahi, S., Pradhan, B., & Mojaddadi, H. (2018). City compactness: Assessing the influence of the growth of residential land use. *Journal of Urban Technology, 25*(1), 21–46.

Afaq, Y., & Manocha, A. (2021). Analysis on change detection techniques for remote sensing applications: A review. *Ecological Informatics, 63*, 101310.

Akın, A., & Erdoğan, M. A. (2020). Analysing temporal and spatial urban sprawl change of Bursa city using landscape metrics and remote sensing. *Modeling Earth Systems and Environment, 6*(3), 1331–1343.

Alberti, M. (2005). The effects of urban patterns on ecosystem function. *International Regional Science Review, 28*(2), 168–192.

Alkheder, S., & Shan, J. (2005). Cellular Automata urban growth simulation and evaluation-A case study of Indianapolis. In *Proceedings of the 8th international conference on geocomputation* (Vol. 31).

Alloghani, M., Al-Jumeily, D., Mustafina, J., Hussain, A., & Aljaaf, A. J. (2020). A systematic review on supervised and unsupervised machine learning algorithms for data science. In *Supervised and unsupervised learning for data science* (pp. 3–21). Springer.

Alonso, W. (1964). *Location and land use: Toward a general theory of land rent*. Harvard University Press.

Angel, S., Blei, A. M., Lamson-Hall, P., Galarza, N., Gopalan, P., Kellergis, A., Civco, D. L., Kumar, S., Madrid, M., Shingade, S., & Hurd, J. D. (2016). Atlas of urban expansion—The 2016 edition. New York University, The Lincoln Institute of Land Policy & UN-Habitat.

Araya, Y. H., & Cabral, P. (2010). Analysis and modeling of urban land cover change in Setúbal and Sesimbra, Portugal. *Remote Sensing, 2*(6), 1549–1563.

Arsanjani, J. J., Helbich, M., Kainz, W., & Boloorani, A. D. (2012). Integration of logistic regression, Markov chain and cellular automata models to simulate urban expansion. *International Journal of Applied Earth Observation and Geoinformation, 21*, 265–275. https://doi.org/10.1016/j.jag.2011.12.014

Banister, D. (2012). Assessing the reality—Transport and land use planning to achieve sustainability. *Journal of Transport and Land Use, 5*(3), 1–14.

Baqa, M. F., Chen, F., Lu, L., Qureshi, S., Tariq, A., Wang, S., Jing, L., Hamza, S., & Li, Q. (2021). Monitoring and modeling the patterns and trends of urban growth using urban sprawl matrix and CA-Markov model: A case study of Karachi, Pakistan. *Land, 10*(7), 700.

Barnes, K. B., Morgan, J. M., III, Roberge, M. C., & Lowe, S. (2001). *Sprawl development: its patterns, consequences, and measurement* (Vol. 1, p. 24). Towson University.

Berling-Wolff, S., & Wu, J. (2004). Modeling urban landscape dynamics: A case study in Phoenix, USA. *Urban Ecosystems, 7*(3), 215–240.

Bhatta, B., Saraswati, S., & Bandyopadhyay, D. (2010). Quantifying the degree-of-freedom, degree-of-sprawl, and degree-of-goodness of urban growth from remote sensing data. *Applied Geography, 30*(1), 96–111.

Borana, S. L., Vaishnav, A., Yadav, S. K., & Parihar, S. K. (2020). Urban growth assessment using remote sensing, GIS and Shannon's entropy model: A case study of Bhilwara city, Rajasthan. In *In 2020 3rd international conference on emerging technologies in computer engineering: Machine learning and internet of things (ICETCE)* (pp. 1–6). IEEE.

Brehaut, L., & Danby, R. K. (2018). Inconsistent relationships between annual tree ring-widths and satellite-measured NDVI in a mountainous subarctic environment. *Ecological Indicators, 91*, 698–711.

Bununu, Y. A. (2017). Integration of Markov chain analysis and similarity-weighted instance-based machine learning algorithm (SimWeight) to simulate urban expansion. *International Journal of Urban Sciences, 21*(2), 217–237.

Burton, E. (2002). Measuring urban compactness in UK towns and cities. *Environment and Planning B: Planning and Design, 29*(2), 219–250.

Catalán, B., Saurí, D., & Serra, P. (2008). Urban sprawl in the Mediterranean?: Patterns of growth and change in the Barcelona metropolitan region 1993–2000. *Landscape and Urban Planning, 85*(3–4), 174–184.

Chachondhia, P., Shakya, A., & Kumar, G. (2021). Performance evaluation of machine learning algorithms using optical and microwave data for LULC classification. *Remote Sensing Applications: Society and Environment, 23*, 100599.

Chai, B., & Seto, K. C. (2019). Conceptualizing and characterizing micro-urbanization: A new perspective applied to Africa. *Landscape and Urban Planning, 190*, 103595.

Clarke, K. C., Gazulis, N., Dietzel, C., & Goldstein, N. C. (2007). A decade of SLEUTHing: Lessons learned from applications of a cellular automaton land use change model. In *Classics in IJGIS: twenty years of the international journal of geographical information science and systems* (pp. 413–427).

Dadashpoor, H., Azizi, P., & Moghadasi, M. (2019). Land use change, urbanization, and change in landscape pattern in a metropolitan area. *Science of the Total Environment, 655*, 707–719.

Derdouri, A., & Murayama, Y. (2020). A comparative study of land price estimation and mapping using regression kriging and machine learning algorithms across Fukushima prefecture, Japan. *Journal of Geographical Sciences, 30*(5), 794–822.

Din, U. S., & Mak, H. W. L. (2021). Retrieval of land-use/land cover change (LUCC) maps and urban expansion dynamics of Hyderabad, Pakistan via Landsat datasets and support vector machine framework. *Remote Sensing, 13*(16), 3337.

Duarte, G. T., Santos, P. M., Cornelissen, T. G., Ribeiro, M. C., & Paglia, A. P. (2018). The effects of landscape patterns on ecosystem services: Meta-analyses of landscape services. *Landscape Ecology, 33*(8), 1247–1257.

Eckert, S., Hüsler, F., Liniger, H., & Hodel, E. (2015). Trend analysis of MODIS NDVI time series for detecting land degradation and regeneration in Mongolia. *Journal of Arid Environments, 113*, 16–28.

Estoque, R. C. (2020). A review of the sustainability concept and the state of SDG monitoring using remote sensing. *Remote Sensing, 12*(11), 1770.

Estoque, R. C., & Murayama, Y. (2015). Classification and change detection of built-up lands from Landsat-7 ETM+ and Landsat-8 OLI/TIRS imageries: A comparative assessment of various spectral indices. *Ecological Indicators, 56*, 205–217.

Ewing, R. (1997). Is Los Angeles-style sprawl desirable? *Journal of the American Planning Association, 63*(1), 107–126.

Fang, S., Gertner, G. Z., Sun, Z., & Anderson, A. A. (2005). The impact of interactions in spatial simulation of the dynamics of urban sprawl. *Landscape and Urban Planning, 73*(4), 294–306.

Frimpong, B. F., & Molkenthin, F. (2021). Tracking urban expansion using random forests for the classification of landsat imagery (1986–2015) and predicting urban/built-up areas for 2025: A study of the Kumasi Metropolis, Ghana. *Land, 10*(1), 44.

Gao, B., Huang, Q., He, C., Sun, Z., & Zhang, D. (2016). How does sprawl differ across cities in China? A multi-scale investigation using nighttime light and census data. *Landscape and Urban Planning, 148*, 89–98.

Georganos, S., Grippa, T., Lennert, M., Vanhuysse, S., Johnson, B. A., & Wolff, E. (2018). Scale matters: Spatially partitioned unsupervised segmentation parameter optimization for large and heterogeneous satellite images. *Remote Sensing, 10*(9), 1440.

Goodal, B. (1987). *The penguin dictionary of human geography*. Penguin Books.

Grecchi, R. C., Gwyn, Q. H. J., Bénié, G. B., Formaggio, A. R., & Fahl, F. C. (2014). Land use and land cover changes in the Brazilian Cerrado: A multidisciplinary approach to assess the impacts of agricultural expansion. *Applied Geography, 55*, 300–312.

Guo, A., Yang, J., Xiao, X., Xia, J., Jin, C., & Li, X. (2020). Influences of urban spatial form on urban heat island effects at the community level in China. *Sustainable Cities and Society, 53*, 101972.

Gustafson, E. J. (2019). How has the state-of-the-art for quantification of landscape pattern advanced in the twenty-first century? *Landscape Ecology, 34*(9), 2065–2072.

Han, B., Liu, H., & Wang, R. (2015). Urban ecological security assessment for cities in the Beijing–Tianjin–Hebei metropolitan region based on fuzzy and entropy methods. *Ecological Modelling, 318*, 217–225.

Hatab, A. A., Cavinato, M. E. R., Lindemer, A., & Lagerkvist, C. J. (2019). Urban sprawl, food security and agricultural systems in developing countries: A systematic review of the literature. *Cities, 94*, 129–142.

Hess, G. R., Daley, S. S., Dennison, B. K., Lubkin, S. R., McGuinn, R. P., Morin, V. Z., Potter, K. M., Savage, R. E., Shelton, W. G., Snow, C. M., & Wrege, B. M. (2001). Just what is sprawl, anyway. *Carolina Planning, 26*(2), 11–26.

Hoffman, J. S., Shandas, V., & Pendleton, N. (2020). The effects of historical housing policies on resident exposure to intra-urban heat: A study of 108 US urban areas. *Climate, 8*(1), 12.

Hossain, M. T., Zarin, T., Sahriar, M. R., & Haque, M. N. (2021). Machine learning based modeling for future prospects of land use land cover change in Gopalganj District, Bangladesh. *Physics and Chemistry of the Earth, Parts A/B/C*, 103022.

Hu, X., & Xu, H. (2018). A new remote sensing index for assessing the spatial heterogeneity in urban ecological quality: A case from Fuzhou City, China. *Ecological Indicators, 89*, 11–21.

Hu, L., He, S., Han, Z., Xiao, H., Su, S., Weng, M., & Cai, Z. (2019). Monitoring housing rental prices based on social media: An integrated approach of machine learning algorithms and hedonic modeling to inform equitable housing policies. *Land Use Policy, 82*, 657–673.

Islam, M. R., Talukdar, S., Rihan, M., & Rahman, A. (2024). Evaluating cooling effect of blue-green infrastructure on urban thermal environment in a metropolitan city: Using geospatial and machine learning techniques. *Sustainable Cities and Society, 113*, 105666.

Jaeger, J. A., Bertiller, R., Schwick, C., & Kienast, F. (2010). Suitability criteria for measures of urban sprawl. *Ecological Indicators, 10*(2), 397–406.

Jamali, A. (2021). Land use land cover modeling using optimized machine learning classifiers: A case study of Shiraz, Iran. *Modeling Earth Systems and Environment, 7*(3), 1539–1550.

Jin, Y., Yang, W., Sun, T., Yang, Z., & Li, M. (2016). Effects of seashore reclamation activities on the health of wetland ecosystems: A case study in the Yellow River Delta, China. *Ocean & Coastal Management, 123*, 44–52.

Jun, Y. E., Siqi, K. A. N. G., Genshen, F. U., Haiyan, L. Ü., Wenqi, Q. I. A. N., & Xuehai, T. A. N. G. (2022). Research progress of urban green space landscape pattern change based on satellite remote sensing. *Bulletin of Surveying and Mapping, 3*, 23.

Jusuf, S. K., Wong, N. H., Hagen, E., Anggoro, R., & Hong, Y. (2007). The influence of land use on the urban heat island in Singapore. *Habitat International, 31*(2), 232–242.

Kafy, A. A., Rahman, M. S., Islam, M., Al Rakib, A., Islam, M. A., Khan, M. H. H., et al. (2021). Prediction of seasonal urban thermal field variance index using machine learning algorithms in Cumilla, Bangladesh. *Sustainable Cities and Society, 64*, 102542.

Khan, A., & Sudheer, M. (2022). Machine learning-based monitoring and modeling for spatio-temporal urban growth of Islamabad. *The Egyptian Journal of Remote Sensing and Space Science, 25*(2), 541–550.

Kim, Y., & Newman, G. (2020). Advancing scenario planning through integrating urban growth prediction with future flood risk models. *Computers, Environment and Urban Systems, 82*, 101498. https://doi.org/10.1016/j.compenvurbsys.2020.101498

Koziatek, O., & Dragićević, S. (2019). A local and regional spatial index for measuring three-dimensional urban compactness growth. *Environment and Planning B: Urban Analytics and City Science, 46*(1), 143–164.

Kuddus, M. A., Tynan, E., & McBryde, E. (2020). Urbanization: A problem for the rich and the poor? *Public Health Reviews, 41*(1), 1–4.
Kumar, M., Singh, R. B., Pravesh, R., Kumar, P., Tripathi, D. K., & Sahu, N. (2018). Urban growth dynamics and modelling using remote sensing data and multivariate statistical techniques. *Current Science, 114*, 2080–2091.
Kuo, H.-F., & Tsou, K.-W. (2018). Modeling and simulation of the future impacts of urban land use change on the natural environment by SLEUTH and cluster analysis. *Sustainability, 10*, 72.
Lan, T., Shao, G., Xu, Z., Tang, L., & Sun, L. (2021). Measuring urban compactness based on functional characterization and human activity intensity by integrating multiple geospatial data sources. *Ecological Indicators, 121*, 107177.
Lang, R. E. (2000). *Office sprawl: The evolving geography of business*. Brookings Institution, Center on Urban and Metropolitan Policy.
Langa, C., Hara, J., Wang, J., Nakamura, K., Watanabe, N., & Komai, T. (2021). Dynamic evaluation method for planning sustainable landfills using GIS and multi-criteria in areas of urban sprawl with land-use conflicts. *PLoS One, 16*(8), e0254441.
Lee, G., Jun, K. S., & Chung, E. S. (2013). Integrated multi-criteria flood vulnerability approach using fuzzy TOPSIS and Delphi technique. *Natural Hazards and Earth System Sciences, 13*(5), 1293–1312.
Li, R., Zhang, K., Li, Q., Yang, L., Wang, S., Liu, Z., Zhang, X., Chen, H., Yi, Y., Feng, J., & Wang, Q. (2022). Characteristics and degradation of organic aerosols from cooking sources based on hourly observation of organic molecular markers in urban environment. *Atmospheric Chemistry and Physics Discussions*, 1–24.
Li, X., & Yeh, A. G. O. (2004). Analyzing spatial restructuring of land use patterns in a fast growing region using remote sensing and GIS. *Landscape and Urban Planning, 69*(4), 335–354.
Li, X., Zhou, Y., Hejazi, M., Wise, M., Vernon, C., Iyer, G., & Chen, W. (2021). Global urban growth between 1870 and 2100 from integrated high resolution mapped data and urban dynamic modeling. *Communications Earth & Environment, 2*(1), 201.
Liang, W., & Yang, M. (2019). Urbanization, economic growth and environmental pollution: Evidence from China. *Sustainable Computing: Informatics and Systems, 21*, 1–9.
Lin, C. H. U., Xin-ran, Z. H. A. N. G., Tian-wei, W. A. N. G., Zhao-xia, L. I., & Chong-fa, C. A. I. (2018). Spatial-temporal evolution and prediction of urban landscape pattern and habitat quality based on CA-Markov and InVEST model. *Yingyong Shengtai Xuebao, 29*(12).
Liu, Y., Peng, J., & Wang, Y. (2018). Efficiency of landscape metrics characterizing urban land surface temperature. *Landscape and Urban Planning, 180*, 36–53.
Lopez, R. (2014). Urban sprawl in the United States: 1970-2010. *Cities and the Environment (CATE), 7*(1), 7.
Luo, Y., Sun, W., Yang, K., & Zhao, L. (2021). China urbanization process induced vegetation degradation and improvement in recent 20 years. *Cities, 114*, 103207.
Meng, Q., Zhang, L., Sun, Z., Meng, F., Wang, L., & Sun, Y. (2018). Characterizing spatial and temporal trends of surface urban heat island effect in an urban main built-up area: A 12-year case study in Beijing, China. *Remote Sensing of Environment, 204*, 826–837.
Mieszkowski, P., & Mills, E. S. (1993). The causes of metropolitan suburbanization. *The Journal of Economic Perspectives, 7*, 135–147.
Moghadam, S. H., & Helbich, M. (2013). Spatiotemporal urbanization processes in the megacity of Mumbai, India: A Markov chains-cellular automata urban growth model. *Applied Geography, 40*, 140–149. https://doi.org/10.1016/j.apgeog.2013.01.009
Morshed, S. R., Fattah, M. A., Haque, M. N., & Morshed, S. Y. (2021). Future ecosystem service value modeling with land cover dynamics by using machine learning based artificial neural network model for Jashore city, Bangladesh. *Physics and Chemistry of the Earth, Parts A/B/C*, 103021.
Mostafa, E., Li, X., Sadek, M., & Dossou, J. F. (2021). Monitoring and forecasting of urban expansion using machine learning-based techniques and remotely sensed data: A case study of Gharbia governorate, Egypt. *Remote Sensing, 13*(22), 4498.

Mubareka, S., Koomen, E., Estreguil, C., & Lavalle, C. (2011). Development of a composite index of urban compactness for land use modelling applications. *Landscape and Urban Planning, 103*(3–4), 303–317.

Musacchio, L. R. (2009). The scientific basis for the design of landscape sustainability: A conceptual framework for translational landscape research and practice of designed landscapes and the six Es of landscape sustainability. *Landscape Ecology, 24*(8), 993–1013.

Naikoo, M. W., Talukdar, S., Ishtiaq, M., & Rahman, A. (2023). Modelling built-up land expansion probability using the integrated fuzzy logic and coupling coordination degree model. *Journal of Environmental Management, 325*, 116441.

Naikoo, M. W., et al. (2024). A geospatial approach to mapping and monitoring real estate-induced urban expansion in the National Capital Region of Delhi. *PFG–Journal of Photogrammetry, Remote Sensing and Geoinformation Science, 92*(2), 177–200.

Nekhay, O., Arriaza, M., & Boerboom, L. (2009). Evaluation of soil erosion risk using analytic network process and GIS: A case study from Spanish mountain olive plantations. *Journal of Environmental Management, 90*(10), 3091–3104.

Novoa, J., Chokmani, K., & Lhissou, R. (2018). A novel index for assessment of riparian strip efficiency in agricultural landscapes using high spatial resolution satellite imagery. *Science of the Total Environment, 644*, 1439–1451.

Nyamekye, C., Kwofie, S., Ghansah, B., Agyapong, E., & Boamah, L. A. (2020). Assessing urban growth in Ghana using machine learning and intensity analysis: A case study of the new Juaben municipality. *Land Use Policy, 99*, 105057.

Pahlavani, P., Askarian Omran, H., & Bigdeli, B. (2017). A multiple land use change model based on artificial neural network, Markov chain, and multi objective land allocation. *Earth Observation and Geomatics Engineering, 1*, 82–99.

Palmer, M., Bernhardt, E., Chornesky, E., Collins, S., Dobson, A., Duke, C., et al. (2004). Ecology for a crowded planet. *Science, 304*, 1251–1252.

Pal, S., & Ziaul, S. K. (2017). Detection of land use and land cover change and land surface temperature in English Bazar urban centre. *The Egyptian Journal of Remote Sensing and Space Science, 20*(1), 125–145.

Pan, X., Wang, Y., Liu, Z., He, C., Liu, H., & Chen, Z. (2020). Understanding urban expansion on the Tibetan plateau over the past half century based on remote sensing: The case of Xining city, China. *Remote Sensing, 13*(1), 46.

Patra, P. K., Behera, D., Naik, S. P., & Goswami, S. (2021). Spatio-temporal variation of vegetation and urban sprawl using remote sensing and GIS: A case study of Cuttack City, Odisha, India. *Journal of Geosciences Research, 6*(2), 213–219.

Pauleit, S., Ennos, R., & Golding, Y. (2005). Modeling the environmental impacts of urban land use and land cover change—A study in Merseyside, UK. *Landscape and Urban Planning, 71*(2–4), 295–310.

Paulsen, K. (2014). Geography, policy or market? New evidence on the measurement and causes of sprawl (and infill) in US metropolitan regions. *Urban Studies, 51*(12), 2629–2645.

Poghosyan, A. (2018). Quantifying urban growth in 10 post-soviet cities using Landsat data and machine learning. *International Journal of Remote Sensing, 39*(23), 8688–8702.

Qi, S., Song, B., Liu, C., Gong, P., Luo, J., Zhang, M., & Xiong, T. (2022). Bamboo Forest mapping in China using the dense Landsat 8 image archive and Google earth engine. *Remote Sensing, 14*(3), 762.

Rahman, A., Aggarwal, S. P., Netzband, M., & Fazal, S. (2010). Monitoring urban sprawl using remote sensing and GIS techniques of a fast growing urban Centre, India. *IEEE Journal of Selected Topics in Applied Earth Observations and Remote Sensing, 4*(1), 56–64.

Rahman, A., Roy, S. S., Talukdar, S., & Shahfahad. (2023). Rethinking progress in approaches and techniques for the urban environmental studies. In *Advancements in urban environmental studies: Application of geospatial technology and artificial intelligence in urban studies* (pp. 471–478). Springer International Publishing.

Riley, S. P., Bromley, C., Poppenga, R. H., Uzal, F. A., Whited, L., & Sauvajot, R. M. (2007). Anticoagulant exposure and notoedric mange in bobcats and mountain lions in urban southern California. *The Journal of Wildlife Management, 71*(6), 1874–1884.

Sahana, M., Hong, H., & Sajjad, H. (2018). Analyzing urban spatial patterns and trend of urban growth using urban sprawl matrix: A study on Kolkata urban agglomeration, India. *Science of the Total Environment, 628*, 1557–1566.

Salem, M., Tsurusaki, N., & Divigalpitiya, P. (2020). Land use/land cover change detection and urban sprawl in the peri-urban area of greater Cairo since the Egyptian revolution of 2011. *Journal of Land Use Science, 15*(5), 592–606.

Santibáñez-Andrade, G., Castillo-Argüero, S., Vega-Peña, E. V., Lindig-Cisneros, R., & Zavala-Hurtado, J. A. (2015). Structural equation modeling as a tool to develop conservation strategies using environmental indicators: The case of the forests of the Magdalena river basin in Mexico City. *Ecological Indicators, 54*, 124–136.

Sassen, S. (2017). "Locating cities on global circuits": From environment and urbanization (2002). In *The globalizing cities reader* (pp. 25–30). Routledge.

Schneider, L. C., & Pontius, R. G., Jr. (2001). Modeling land-use change in the Ipswich watershed, Massachusetts, USA. *Agriculture, Ecosystems & Environment, 85*, 83–94.

Schug, F., Okujeni, A., Hauer, J., Hostert, P., Nielsen, J. Ø., & van der Linden, S. (2018). Mapping patterns of urban development in Ouagadougou, Burkina Faso, using machine learning regression modeling with bi-seasonal Landsat time series. *Remote Sensing of Environment, 210*, 217–228.

Seto, K. C., Fragkias, M., Güneralp, B., & Reilly, M. K. (2011). A meta-analysis of global urban land expansion. *PLoS One, 6*(8), e23777.

Shafizadeh-Moghadam, H., Asghari, A., Tayyebi, A., & Taleai, M. (2017). Coupling machine learning, tree-based and statistical models with cellular automata to simulate urban growth. *Computers, Environment and Urban Systems, 64*, 297–308.

Shah, A., Ali, K., & Nizami, S. M. (2022). Spatio-temporal analysis of urban sprawl in Islamabad, Pakistan during 1979–2019, using remote sensing. *GeoJournal, 87*(4), 2935–2948.

Shahfahad, Naikoo, M. W., Das, T., Talukdar, S., Asgher, M. S., Asif, & Rahman, A. (2022). Prediction of land use changes at a metropolitan city using integrated cellular automata: Past and future. *Geology, Ecology, and Landscapes*, 1–19.

Shahfahad, Talukdar, S., Naikoo, M. W., Rahman, A., Gagnon, A. S., Islam, A. R. M. T., & Mosavi, A. (2023). Comparative evaluation of operational land imager sensor on board landsat 8 and landsat 9 for land use land cover mapping over a heterogeneous landscape. *Geocarto International, 38*(1), 2152496.

Shahfahad, Talukdar, S., Naikoo, M. W., & Rahman, A. (2024). Urban expansion and vegetation dynamics: The role of protected areas in preventing vegetation loss in a growing mega city. *Habitat International, 150*, 103129.

Sharifi, A., & Hosseingholizadeh, M. (2019). The effect of rapid population growth on urban expansion and destruction of green space in Tehran from 1972 to 2017. *Journal of the Indian Society of Remote Sensing, 47*(6), 1063–1071.

Shi, S., Zhong, Y., Zhao, J., Lv, P., Liu, Y., & Zhang, L. (2020). Land-use/land-cover change detection based on class-prior object-oriented conditional random field framework for high spatial resolution remote sensing imagery. *IEEE Transactions on Geoscience and Remote Sensing, 60*, 1–16.

Sivakumar, R., & Ghosh, S. (2016). Spatiotemporal dynamic study of lakes and development of mathematical prediction model using geoinformatic techniques. *Arabian Journal of Geosciences, 9*(1), 1–14.

Squires, G. D. (2002). Urban sprawl and the uneven development of metropolitan America. In *Urban sprawl: Causes, consequences, and policy responses* (pp. 1–22). The Urban Insitute.

Stefanov, W. L., & Netzband, M. (2005). Assessment of ASTER land cover and MODIS NDVI data at multiple scales for ecological characterization of an arid urban center. *Remote Sensing of Environment, 99*(1–2), 31–43.

Subedi, P., Subedi, K., & Thapa, B. (2013). Application of a hybrid cellular automaton–Markov (CA-Markov) model in land-use change prediction: A case study of Saddle Creek Drainage Basin, Florida. *Applied Ecology and Environmental Sciences, 1*, 126–132.

Sun, X., Tan, X., Chen, K., Song, S., Zhu, X., & Hou, D. (2020). Quantifying landscape-metrics impacts on urban green-spaces and water-bodies cooling effect: The study of Nanjing, China. *Urban Forestry & Urban Greening, 55*, 126838.

Talukdar, S., Singha, P., Mahato, S., Pal, S., Liou, Y. A., & Rahman, A. (2020). Land-use land-cover classification by machine learning classifiers for satellite observations—A review. *Remote Sensing, 12*(7), 1135.

Tan, F., & Lu, Z. (2019). The impact of urban compactness on urban sustainable development in China: The case of Nanjing. *Sustainable Development, 27*(3), 270–280.

Tan, M., Li, X., Xie, H., & Lu, C. (2005). Urban land expansion and arable land loss in China—A case study of Beijing–Tianjin–Hebei region. *Land Use Policy, 22*(3), 187–196.

Thakur, T. K., Patel, D. K., Bijalwan, A., Dobriyal, M. J., Kumar, A., Thakur, A., Bohra, A., & Bhat, J. A. (2020). Land use land cover change detection through geospatial analysis in an Indian biosphere reserve. *Trees, Forests and People, 2*, 100018.

The World Bank. (2011). *Malaysia economic monitor: Smart cities*.

Thinh, N. X., Arlt, G., Heber, B., Hennersdorf, J., & Lehmann, I. (2002). Evaluation of urban land-use structures with a view to sustainable development. *Environmental Impact Assessment Review, 22*(5), 475–492.

Torrens, P. M. (2008). A toolkit for measuring sprawl. *Applied Spatial Analysis and Policy, 1*(1), 5–36.

UN-Habitat. (2008). *Global human settlements report. Harmonious cities*. Earthscan.

Uniyal, S., Purohit, S., Chaurasia, K., Rao, S. S., & Amminedu, E. (2022). Quantification of carbon sequestration by urban forest using Landsat 8 OLI and machine learning algorithms in Jodhpur, India. *Urban Forestry & Urban Greening, 67*, 127445.

Van Kamp, I., Leidelmeijer, K., Marsman, G., & De Hollander, A. (2003). Urban environmental quality and human Well-being: Towards a conceptual framework and demarcation of concepts; a literature study. *Landscape and Urban Planning, 65*(1–2), 5–18.

Veldkamp, A., & Fresco, L. O. (1996). CLUE: A conceptual model to study the conversion of land use and its effects. *Ecological Modelling, 85*, 253–270.

Waddell, P. (2002). UrbanSim: Modeling urban development for land use, transportation, and environmental planning. *Journal of the American Planning Association, 68*, 297–314.

Wang, H. C. (2022). Prioritizing compactness for a better quality of life: The case of US cities. *Cities, 123*, 103566.

Wang, S. W., Gebru, B. M., Lamchin, M., Kayastha, R. B., & Lee, W. K. (2020). Land use and land cover change detection and prediction in the Kathmandu district of Nepal using remote sensing and GIS. *Sustainability, 12*(9), 3925.

Weng, Q. (2009). Thermal infrared remote sensing for urban climate and environmental studies: Methods, applications, and trends. *ISPRS Journal of Photogrammetry and Remote Sensing, 64*(4), 335–344.

Wentz, E. A., Anderson, S., Fragkias, M., Netzband, M., Mesev, V., Myint, S. W., Quattrochi, D., Rahman, A., & Seto, K. C. (2014). Supporting global environmental change research: A review of trends and knowledge gaps in urban remote sensing. *Remote Sensing, 6*(5), 3879–3905.

Wu, J. (2014). Urban ecology and sustainability: The state-of-the-science and future directions. *Landscape and Urban Planning, 125*, 209–221.

Xia, C., Yeh, A. G. O., & Zhang, A. (2020). Analyzing spatial relationships between urban land use intensity and urban vitality at street block level: A case study of five Chinese megacities. *Landscape and Urban Planning, 193*, 103669.

Xu, H. Q. (2013). A remote sensing urban ecological index and its application. *Acta Ecol. Sin, 33*(24), 7853–7862.

Xu, T., Gao, J., & Li, Y. (2019). Machine learning-assisted evaluation of land use policies and plans in a rapidly urbanizing district in Chongqing, China. *Land Use Policy, 87*, 104030.

Yan, J., Wang, L., Song, W., Chen, Y., Chen, X., & Deng, Z. (2019). A time-series classification approach based on change detection for rapid land cover mapping. *ISPRS Journal of Photogrammetry and Remote Sensing, 158*, 249–262.

Yang, X., Chen, R., & Zheng, X. Q. (2016). Simulating land use change by integrating ANNCA model and landscape pattern indices. *Geomatics, Natural Hazards and Risk, 7*, 918–932.

Yang, Y., Xu, Y., Duan, Y., Zhang, S., Zhang, Y., & Xie, Y. (2023). How can trees protect us from air pollution and urban heat? Associations and pathways at the neighborhood scale. *Landscape and Urban Planning, 236*, 104779.

Yue, H., Liu, Y., Li, Y., & Lu, Y. (2019). Eco-environmental quality assessment in China's 35 major cities based on remote sensing ecological index. *IEEE Access, 7*, 51295–51311.

Zeng, H., Wu, B., Wang, S., Musakwa, W., Tian, F., Mashimbye, Z. E., et al. (2020). A synthesizing land-cover classification method based on Google earth engine: A case study in Nzhelele and Levhuvu Catchments, South Africa. *Chinese Geographical Science, 30*(3), 397–409.

Zhang, X., Liu, Y., & Zhou, L. (2018). Correlation analysis between landscape metrics and water quality under multiple scales. *International Journal of Environmental Research and Public Health, 15*(8), 1606.

Zhang, Y., Ning, G., Chen, S., & Yang, Y. (2021). Impact of rapid urban sprawl on the local meteorological observational environment based on remote sensing images and GIS technology. *Remote Sensing, 13*(13), 2624.

Zhang, X., Han, L., Wei, H., Tan, X., Zhou, W., Li, W., & Qian, Y. (2022). Linking urbanization and air quality together: A review and a perspective on the future sustainable urban development. *Journal of Cleaner Production, 346*, 130988.

Zhou, D., Zhao, S., Liu, S., Zhang, L., & Zhu, C. (2014). Surface urban heat island in China's 32 major cities: Spatial patterns and drivers. *Remote Sensing of Environment, 152*, 51–61.

Case Studies

Nondegradable Waste Incursion, Environment and Ecology: A Case of Densely Populated Urban Settings in India

Rudrodip Majumdar

Abstract India is one of the world's fastest-growing economies, and this rapid economic expansion has been accompanied by significant urbanization. Studies on the evolution of urban agglomerations across India over a period of more than a century (1870–2020) bring forth a strong relationship between economic growth and urbanization in the country. As the country aims for a highly aspirational target of a US$40 trillion economy by 2047 (the hundredth year since its independence), it is intuitive that there will be a rise in the number of densely populated urban settings. This would pose a great challenge to society at large since the rapidly expanding dense urban settings would put pressure on the natural resources due to heavy anthropogenic activities associated with the creation and development of highly dynamic urban habitats. Since this growth is envisaged to be fueled primarily by unabated collective aspiration and economic growth, at times, there may be a departure from sustainable practices. A crucial area requiring greater attention is the careful management of nondegradable solid waste produced in densely populated cities and their surrounding areas. The Municipal Waste Management Rules (2000), implemented by the Ministry of Environment and Forest in the Government of India at the time, required urban local bodies to collect organic (or food) waste, domestic hazardous waste, and recyclable waste separately to ensure the safe and scientific handling and disposal of municipal waste. However, the realization of this noble goal has not materialized to the desired extent. Plastic waste has been a menace to the cities in India. This chapter aims to focus on the trail of plastic waste in densely populated Indian cities, starting from generation to disposal after usage. It also discusses the implications of plastic and other nondegradable wastes on the urban environment and ecology. A pragmatic way forward has been mentioned, which could be used as the baseline toward formulating a strategy to deal with the issue. Evidence has been highlighted to indicate the possibilities of creating a "waste-to-wealth" value chain so that the initiatives can be justified from an investment point of view.

R. Majumdar (✉)
School of Natural Sciences and Engineering, National Institute of Advanced Studies, Bengaluru, India
e-mail: rudrodip@nias.res.in

P. K. Joshi et al. (eds.), *Geospatial Science for Urban Ecosystems*, The Urban Book Series, https://doi.org/10.1007/978-3-032-07155-2_10

Keywords Plastic waste · Ecological implications · Waste management · Recovery and recycling · Waste-to-wealth · Sustainability strategy

1 Introduction

India has witnessed rapid urbanization over the past few decades, since 1990, marked by the accelerated growth of various urban centers across the country in terms of economic and anthropogenic activities (Abhishek et al., 2017; Bhagat, 2017; NIUA, 2023). India's urban population grew by over 150 million between 1990 and 2014, and projections suggest that this number will grow further by another 500 million by 2050, indicating that Indians would comprise around 20% of the world's new urban dwellers between 2020 and 2050 (India Policy Brief, 2014).

According to open-source information based on the 2011 Census data, the country now has 8 Tier I cities, 97 Tier II cities, numerous Tier III cities, and over 5000 Tier IV towns. Table 1 shows the tier-wise classification of the urban centers based on population (as per the Census data) (Reserve Bank of India (RBI), 2023). Cities belonging to the categories Tier I, Tier II, and Tier III are more commonly called Tier X, Tier Y, and Tier Z cities as per the convention used for estimating the house rent allowances (HRA) by the Inland Revenue Services (DOE, 2015; DISPUR, 2008). The Department of Expenditure in the Ministry of Finance, Government of India, issued this reclassification to determine the house rent allowances payable to Central Government employees. However, this reclassification also provides an indication of the cost of living, which in turn reflects the population densities in these urban centers. This dual purpose highlights the financial considerations of the government as well as the demographic realities of India's cities (van Doorn et al., 2019). Another popular way of classification comprises the categories metropolitan center (population of 1 million or above), urban center (population ranging between 1,00,000 and 9,99,999), and semi-urban center (population ranging between 10,000 and 99,999) (Reserve Bank of India (RBI), 2023).

While rising living costs put substantial pressure on the budget of urban dwellers, it has a common correlation with income growth (De Pietro, 2017). Robust income growth frequently coincides with periodic rises in the cost of living within densely populated, competitive urban centers (Choe & Roberts, 2011). Figure 1 presents the income growth in India in terms of GDP per capita (measured in US$) and GDP per capita on a purchasing power parity (PPP) basis (measured in terms of

Table 1 Major classes of the urban centers (tier-wise) based on population

Class (Tier)	Population
Tier 1	More than 1,00,000
Tier 2	50,000–99,999
Tier 3	20,000–49,999
Tier 4	10,000–19,999

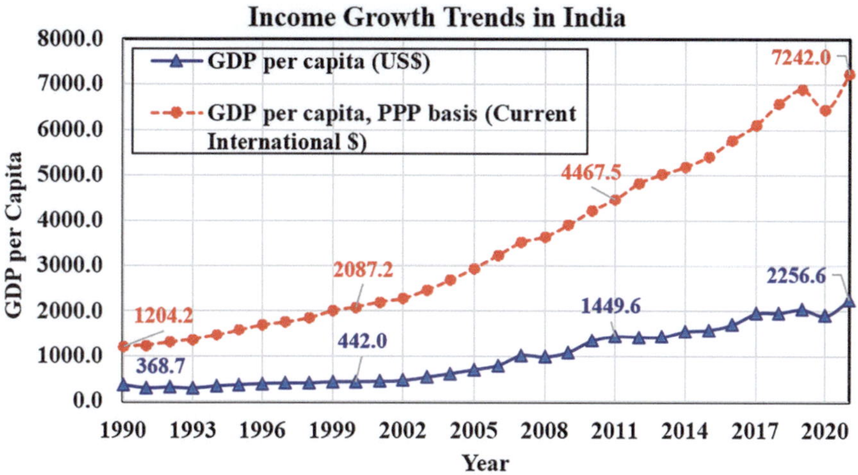

Fig. 1 Income growth trends in India (1990–2021) based on GDP per capita

current international dollars) between 1990 and 2021 based on World Bank data (The World Bank Data, 2023a, 2023b). The figure clearly shows that the income levels in India in terms of GDP per capita on a PPP basis have increased by about six times between 1990 and 2021. It is also noteworthy that India's GDP per capita on a PPP basis in 2021 stood at ~38.9% of the global average of 18604.9 current international dollars (The World Bank Data, 2023c). Looking at the recent trends, it would be reasonable to argue that India's economic growth is likely to continue in an undeterred manner in the foreseeable future until some external inhibiting factors (including geopolitical developments) curb the same abruptly.

As suggested by the 2011 Census data, urbanization in India has accelerated more rapidly than anticipated. Between 2001 and 2011, the urban population surged from 286 million to 377 million, marking an increase of 91 million people. This growth contrasts sharply with the rural population, which saw a rise of only 90.5 million over the same period. This unexpected rate of urban expansion highlights significant demographic shifts and underscores the increasing appeal and migration toward urban centers in the country (Bhagat, 2017). This trend has persisted, with the urban population in India comprising 35.39% of the total population by 2021 (The World Bank Data, 2023d). Since 1990, the percentage share of the urban population in India has been continuously increasing. Table 2 shows the year-wise urban population in India along with the Y-o-Y percentage increase in the same. The table suggests that, on average, there has been annual growth of 2.29% in the urban population in India between 2017 and 2021.

Increasing income levels, as well as rising urban population, indicate a further accelerated growth in the spatial expansion of major urban centers across the country, which is likely to be accompanied by large population densities in the growing and developing cities. Most of the Tier I and Tier II cities in India have undergone rapid urbanization over the past few decades (Sankhe et al., 2010). Expanding urban

Table 2 Year-wise urban population in India and Y-o-Y growth (%)

Year	Population	Increase (%) w.r.t to the previous year
2021	498,179,071	2.15%
2020	487,702,168	2.29%
2019	476,786,386	2.34%
2018	465,871,825	2.39%

areas are confronting ongoing challenges due to obsolete drainage systems, rapid alterations in land use patterns, and an increased frequency of flash floods. Additionally, these areas face significant threats from fatal heat stress and recurring drought incidents. These combined issues underscore the urgent need for modernized infrastructure and adaptive strategies to ensure the resilience and sustainability of urban environments (Liu et al., 2014; Majidi et al., 2019). Although novel interventions in terms of Nature-based Solutions (NbS) are being pursued and mainstreaming of these solutions is being considered for the urban settings in developing countries, including India, scaling up of such initiatives has remained a challenge and requires intensification of concerted efforts as well as more investments (Dhyani et al., 2021; Santhanam & Majumdar, 2020).

One major issue that continues to plague all the growing urban centers is the continuous accumulation of a large quantum of waste. Table 3 provides the spatial extent of the top nine metropolitan areas (urban agglomerates) in India with a population of more than 4 million as per the 2011 Census data (MMRDA, 2023; MCGM, 2020; Greater Hyderabad Municipal Corporation, 2023; Hyderabad Metropolitan Development Authority (HMDA), 2023; The Hindu, 2023; The Times of India, 2020, 2023a; GCC, 2023; Kolkata Metropolitan Development Authority (KMDA), 2023; Kolkata Municipal Corporation, 2017; Amdavad Municipal Corporation (AMC), 2023; Jadhav, 2021; Bangalore Metropolitan Region Development Authority (BMRDA), 2023; Bangalore Metropolitan Region Development Authority (BMRDA) New Map, 2019; Bengaluru Smart City, Government of Karnataka, 2023), as well as the areas currently under the command of the respective city municipal corporation bodies. The table also provides the population for both the metropolitan areas and the city municipal corporation areas (Census 2011 data) (Urban Agglomerations Census, 2011; CPCB, 2023a). Further, the table highlights the trends of daily solid waste generation in the respective city municipal corporation areas, as estimated by the Central Pollution Control Board (CPCB) (CPCB, 2023a).

From the total area of the proposed urban agglomerate and the current extent of the city municipal corporation area pertaining to each metropolitan city of interest, it is evident that huge city clusters are going to emerge in the next 10–15 years around all the metro cities. Although the city municipal corporation areas currently hold a large fraction of the total population, the other regions identified within the larger metropolitan area around each Tier I city are likely to witness a high influx of new urban population as the peripheral city clusters continue to develop and the areas under the city municipal corporation become saturated in terms of carrying

Table 3 Spatial extent and population of the nine major "Metropolitan areas" and the solid waste generation trends in the city municipal corporation areas

Metropolitan area/ Urban agglomerate	State	Metropolitan area spatial stretch (sq. km)	City municipal corporation area (sq. km)	Population (2011 Census data)		Waste generation (TPD) (city municipal corporation area)			
				Metropolitan area	City municipal corporation area	1999–2000	2004–05	2010–11	2015–16
Mumbai	Maharashtra	6640	603	1,83,94,912	1,24,42,373	5355	5320	6500	11,000
Delhi	India's capital territory	1483	700	1,63,49,831	1,10,34,555	400	5922	6800	8700
Kolkata	West Bengal	1887	206	1,40,35,959	44,96,694	3692	2653	3670	4000
Chennai	Tamil Nadu	5904	426	86,53,521	70,88,000	3124	3036	4500	5000
Bengaluru	Karnataka	8005	741	85,20,435	84,43,675	200	1669	3700	3700
Hyderabad	Telangana	7257	650	76,74,689	67,31,790	1566	2187	4200	4000
Ahmedabad	Gujarat	1866	505	63,61,084	55,77,940	1683	1302	2300	2500
Pune	Maharashtra	7257	5162	50,57,709	31,24,458	700	1175	1300	1600
Surat	Gujarat	4255	474	45,91,246	44,67,797	900	1000	1200	1680

*TPD: Tonnes Per Day

capacity. Table 3 clearly shows that the daily solid waste generation increased sharply right after the year 2000 in the case of Delhi, Bengaluru, Hyderabad, and Pune. In the case of Bengaluru, the monotonic increase in the quantum of solid waste generated can somewhat be linked to the rapid urban sprawl that took place within the first decade of the new millennium. Figure 2(a) indicates the representative area chosen around the Bruhat Bengaluru Mahanagara Palike (BBMP) area for the *land use land cover* (LULC) analysis (comprising eight land use classes) using the K-means clustering unsupervised classification algorithm. Figure 2(b) shows the sharp increase in the urban built-up area (marked by the red color on the LULC map) between 2003 and 2007.

From the land use pattern indicating an increase in the urbanization within the BBMP area (Bengaluru) with time (see Fig. 2(a)) and the increasing trend of solid waste generation shown for Bengaluru city in Table 3, one would expect that as the urban centers increasingly become the hub of economic activities, they would

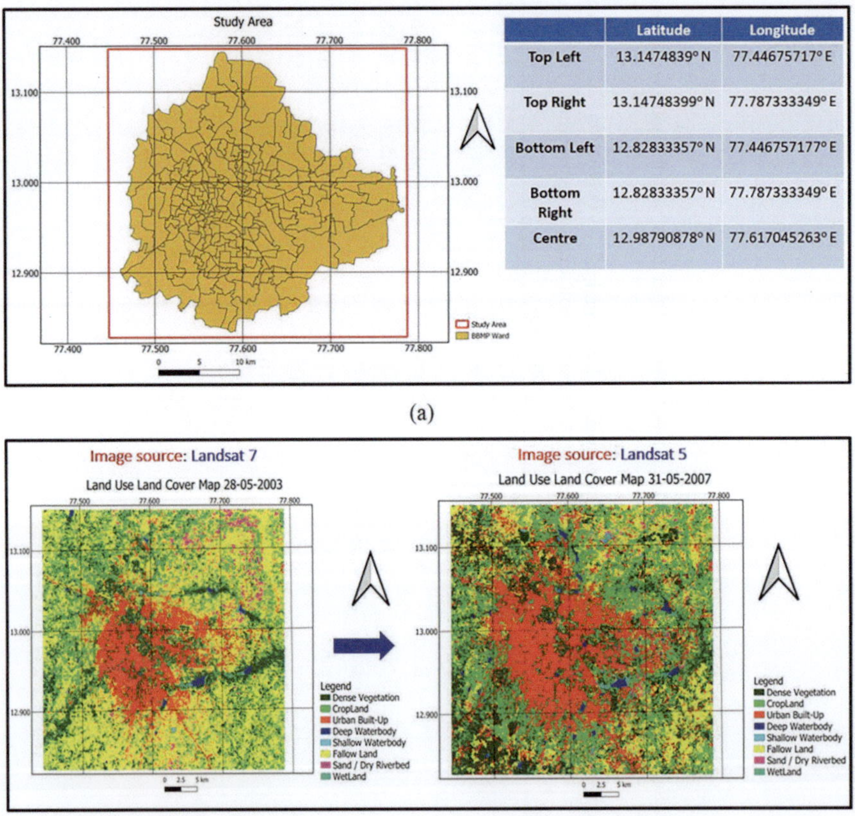

(a)

(b)

Fig. 2 (**a**) BBMP area and immediate surroundings, and (**b**) change in urban built-up area between 2003 and 2007 attract more people, which would ultimately lead to a higher quantum of daily waste generation. This conjecture is backed by the extent of daily solid waste generation in the top two metro cities, Mumbai and Delhi (see Table 3). Mumbai is India's financial capital, and Delhi is the national capital. Both these cities are high-profile in the context of economic activities in India. As of 2015–16, the quantity of solid waste generated daily in Mumbai is about 11,000 tonnes, whereas that for Delhi was estimated to be around 8700 tonnes (see Table 3). Incidentally, these two cities occupy the top positions in terms of the total population as well.

The foregoing discussion indicates that the cumulative quantity of municipal solid waste (MSW) generated in the cities and towns of India is already quite huge, and it is expected to increase to gigantic proportions in the foreseeable future (say, within the next two decades). Therefore, there is an urgent need for scientific segregation and prudent management of this waste. The international seminar titled "*Sustainable solid waste management for cities: opportunities in South Asian Association for Regional Cooperation (SAARC) countries*," held from 25 to 27 March 2015 and jointly organized by the Council of Scientific and Industrial Research-National Environmental Engineering Research Institute (CSIR-NEERI) and the Royal Society, offered significant insights into this critical area. The seminar explored strategies and opportunities for sustainable solid waste management in SAARC countries, highlighting key challenges and potential solutions aimed at enhancing environmental sustainability across urban centers in the region. During the deliberations, it was identified that the waste dumps are devoid of any environmental protection and can potentially lead to health hazards and urban environment degradation. Hence, Indian and UK solid waste management (SWM) experts agreed that the emphasis should be on source segregation of waste. They advocated for the use of specialized waste processing facilities to effectively separate recyclable materials, thereby reintegrating valuable resources into their respective commercial value chains (Kumar et al., 2017). In densely populated urban centers in India, the disposal of solid waste is a critical concern. Improper disposal can result in the release of hazardous gases and leachates through microbial decomposition, influenced by climatic conditions, waste characteristics, and landfill practices. This can contribute to pollution of the air, soil, and water in the surrounding areas (Singh, 2020).

The structural organization of this chapter is as follows: Section 2 of this chapter discusses the typical composition of municipal solid waste (MSW) in India. Section 3 brings forth insights from a recent visit to a peri-urban locality near Bengaluru. Section 4 discusses the impact of plastic waste and nondegradable discards on the local environment and ecology in urban settings. Section 5 explains the pathways of plastic intrusion into urban water bodies and their repercussions. Section 6 provides a broad view of the worldwide current status of plastic waste collection and recycling, with insights from the advanced nations, and the state of affairs in India is assessed. Section 7 discusses the way forward for prudent plastic waste management in Indian urban centers. Finally, conclusions are drawn in Sect. 8 based on the discussions presented in this chapter.

2 Characteristics of Municipal Solid Waste (MSW) in India

It is important to mention that although the term "municipal waste" mainly encompasses the waste resulting from municipal activities, it also includes street waste, dead animals, market waste, and abandoned vehicles in its ambit. For holistic urban solid waste management, the term is, however, applied functionally in an even wider sense, where it stands for residential, institutional, and commercial wastes as well (CPHEEO, 2023). It excludes the wastes generated from healthcare facilities and industries, as well as electrical and electronic wastes (CPCB, 2023b).

The amount of municipal solid waste (MSW) generated varies based on living standards, the scale and nature of commercial activities, urban residents' dietary preferences, and seasonal factors (Kaushal et al., 2012). In a metropolitan city in India, on average, about 0.45 to 0.65 kg of waste is generated per person per day (Sharma & Jain, 2018; Vij, 2012). As per a recent study, the estimated MSW generation in Bangalore City is about 0.5 kg per person per day (Naveen, 2021).

Solid wastes can be categorized into three main types: (A) Biodegradable waste or organic waste: This includes food scraps, kitchen waste, green waste (such as vegetables and fruits), and paper products. These materials can decompose naturally over time. (B) Inert and nonbiodegradable waste: This category encompasses materials that do not decompose easily, such as construction and demolition waste, dirt, and debris. These items generally require specialized handling and disposal methods. (C) Recyclable waste: This category consists of materials that can be recycled and reused, such as soft plastics (packets, wrappers), hard plastics (water bottles, trays made of polyethylene terephthalate (PET), milk cartons, high-density polyethylene (HDPE) bottles), cardboard (packing and shipping boxes), and glass containers. Recycling these materials helps conserve resources and reduce environmental impact. Each category poses unique challenges and opportunities for waste management practices aimed at reducing environmental pollution and promoting sustainability (Singh, 2020).

The typical composition of municipal solid waste (MSW) in India includes about 40–60% compostable materials, 30–50% inert substances, and 10–30% recyclable items (Sharma & Jain, 2018). Typically, dry waste comprises about 35% of the total MSW by weight in India. In recent times, the quantum of recyclable plastic in dry waste has been on the rise in several cities due to the higher extent of commercial activities and lifestyle changes.

According to a recent assessment published jointly by the Indian Institute of Science (IISc), Bangalore, and the Marico Innovation Foundation (MIF), India produces approximately 3.4 million tonnes of plastic waste (P-waste) annually. Sadly, only about 30% of this plastic waste is currently recycled, highlighting significant challenges and opportunities for improving recycling infrastructure and practices in the country (Innovation in Plastics - The Potential and Possibilities, 2023). The ever-growing number of retail chains and increased use of plastic packaging for groceries, food, cosmetics, and other consumer products have led to an increase in plastic consumption. Most of the plastic packaging is disposed of within a relatively short time. Typically, the discarded packaging accounts for about 46% of the plastic waste generated annually (Geyer, 2020). As per the MIF report, plastic consumption in the country has increased from 14 million tonnes in FY 2016–17 to 20 million tonnes in FY 2019–20 (Innovation in Plastics - The Potential and Possibilities, 2023). Consequently, the accumulation of plastic waste within the densely populated city premises has also increased. The report also highlights that the combined quantum of plastic waste generated annually in Maharashtra, Gujarat, and Tamil Nadu amounts to about 38% of the total plastic waste currently generated in India. Of late, the streams of plastic have transcended the administrative boundaries of the cities, and such spillovers tend to pollute and plague the suburban and peri-urban regions with a multitude of problems.

3 Insights from a Recent Visit to a Peri-Urban Locality Near Bengaluru

Very recently, in May 2023, the author visited Dasanapura locality, which is officially a village located in Bengaluru North Taluka of Bengaluru Rural district (i.e., an outskirt region with respect to the Bengaluru city municipality map (BBMP area)). The area is now slowly developing toward becoming a town, and in the future, it will become an integral part of the larger Bengaluru Metropolitan Agglomerate. As the author entered the Huskur Main Road from the Makali-Aluru Main Road, within a few hundred meters, the author observed a lot of plastic litter on both sides of the road. Big plastic bags full of either residential or commercial rejects were observed at several spots (or dump sites). This scene continued for about a 3.5 km stretch, after which the author reached his destination. The author interacted with five local individuals regarding this state of affairs. Two of them were residents of a gated residential complex (both were adult females), two were street vendors with hut-like roadside shops (both were adult males), and the remaining one was an auto rickshaw driver (an adult male). All of them unequivocally mentioned that they feel as if the "Swachh Bharat Mission is barred from entering this area." The author took an auto ride during his return and requested the auto driver to drive a bit slower. The author observed a few feral dogs and cows scavenging in the piles of litter, which were full of soft plastic waste (packets, wrappers, and polythene bags). Also, several household rejects, such as smashed plastic bottles and discarded slippers, were observed. One of the female respondents mentioned to the author that she had witnessed people coming in personal cars to dump plastic bags full of waste at one of the sites near the *Huskur Main Road.* This particular case is only an illustrative example. However, similar situations may be present in many other developing peri-urban areas across India.

This experience brought forth a few insights regarding the menace of urban waste dumps and their possible implications. The large footprint of the litter (heavily loaded with plastic litter) in the peri-urban areas gives a direct impression that the urban dwellers are not exhibiting responsible behavior, and it would not be prudent to rely on the behavioral transformation and civic sense of urban dwellers alone to change the current status. The urban local bodies (ULB) responsible for managing the suburban and the peri-urban areas should start investing resources in devising innovative mechanisms for the effective collection of residential and commercial wastes, preferably in a segregated manner to the extent possible at the source itself. This will help them keep track of the situation on the ground and would help manage the large quantum of different types of municipal solid waste in a more organized manner as these areas start developing rapidly toward becoming sizable urban centers.

4 Impact of Plastic Waste and Other Discards on the Local Environment and Ecology

The evaluation of the impact of plastic waste and other discards on the local environment and ecology is of paramount importance. Local-level case studies add value to the global database, as they provide interesting insights from the micro-level observations, which may prove to be useful while devising administrative strategies for intervention and management. In the Indian urban settings, the feral dogs that scavenge in the dump sites usually tear apart the polythene bin bags in search of food. Often this behavior is triggered by their acute sense of smell, which picks up the scent of leftovers, kitchen discards, and rejects from the meat shops and the fish market. In search of food, they tend to scatter a lot of degradable waste by destroying the garbage bin bags. Most of these bags are nowadays made of biodegradable plastic, and therefore, they are prone to damage in the presence of sharp objects, such as pointed and sharp household discards (e.g., blades, wires, etc.), and also the sharp teeth of feral dogs, rats, pigs, and cows. The meat scraps and fish rejects rotting in the open become a part of the landfill and create a foul smell, which makes the surrounding environment very unpleasant. Gases like phosphine and hydrogen sulfide are released from decaying substances. Phosphine is a highly toxic systemic poison and causes severe irritation in the respiratory tract (Lu et al., 2018). The release of hydrogen sulfide from MSW dumping sites has been associated with odor-related nuisances and adverse health effects for residents in nearby areas (Nunes et al., 2021). When used as landfills, decaying bio-wastes also produce "marsh gas" which contains methane gas as a major constituent (about 50–75% by volume). Methane is a greenhouse gas (GHG) with a large global warming potential. Although methane is short-lived (atmospheric lifetime of ~12 years) as compared to carbon dioxide, it possesses about 25 times higher global warming potential (C2ES, 2023). Marsh gas also produces a substantial volume of carbon dioxide gas, which is a major GHG as well, highly responsible for changing the climate toward a warming world.

Another important aspect is directly associated with the health and well-being of the nonhuman beings living close to the urban waste dumping sites. Reports suggest that on many occasions, urban feral cows consume polythene bags and thin plastic wrappers from the dumping sites owing to the dearth of natural food and grazing spaces. Such scavenging behavior often proves to be fatal for them since the consumed plastics tend to choke their gut. Nondigestible plastic foreign objects accumulate in the rumen, the first stomach of ruminant animals like cows, where they cause issues such as ruminal impaction, indigestion, recurrent tympany (excessive gas build-up in the gastrointestinal tract resulting in abdominal distension), and various other detrimental health effects. These conditions not only impair digestion but also pose serious threats to the overall health and welfare of the animals, necessitating urgent attention to prevent such occurrences and ensure the well-being of nonhumans (Priyanka & Dey, 2018). Cows regularly feeding on garbage and sanitary wastes may develop negative hormones and may contract diseases (e.g., intramammary infections or udder problems) due to prolonged resting on the pathogen-laden, untidy dumping grounds. Birds such as crows, pariah kites, and herons occasionally get entangled in plastic bags and may die due to suffocation (Mongabay-India, 2023). It is noteworthy that Ahmedabad-based *Jivdaya Charitable Trust* (JCT) handled more than a hundred birds (mostly black ibis, a species from the spoonbill family) in 2022 that were rescued in a gravely ill condition caused by the ingestion of plastics from water bodies. Many of these birds were brought in with massive swelling in their stomachs because of plastic getting stuck in their stomach lining. Some of them even choked as plastic beads got stuck in their food pipe (The Times of India, 2023b). The *Jivdaya Charitable Trust* also rescued several cows, dogs, and cats that had fallen sick due to plastic ingestion.

The author has first-hand experience regarding the serious implications of discarding residential rejects without prudent segregation in the locality of Subedarpalya near the Mathikere suburban region within the BBMP area. Due to the presence of sharp objects in the disposed garbage bin bag (a knife on one occasion and a discarded shaving blade on another instance), street dog puppies got injured upon tearing the bin bag apart. Discarded pencil compasses and similar other objects with pointed and sharp protrusions (including those made of hard plastics—such as toys) may prove to be dangerous for inquisitive as well as hungry feral nonhuman beings. Therefore, such household rejects should be disposed of safely and responsibly.

Looking at the increasing number of solid waste dump sites in the urban, suburban, and peri-urban areas in India, it would be reasonable to infer that plastic waste would continue to remain all-pervasive in the foreseeable future until drastic measures are put in place to curb this menace. Since the use of plastic is intertwined with the various economic activities in urban areas, it will not be an easy task to eradicate plastic from our lives. Although there is a worldwide effort toward adopting sustainable and eco-friendly alternatives, plastics continue to be used due to their low cost, durability, and ubiquity. Figure 3 schematically shows the different stages spread across the typical life cycle of plastics. The illustrative diagram draws inspiration from the process flow and the indicative numbers available in the previously reported studies (Geyer, 2020; Geyer et al., 2017; Tsakona et al., 2021). It

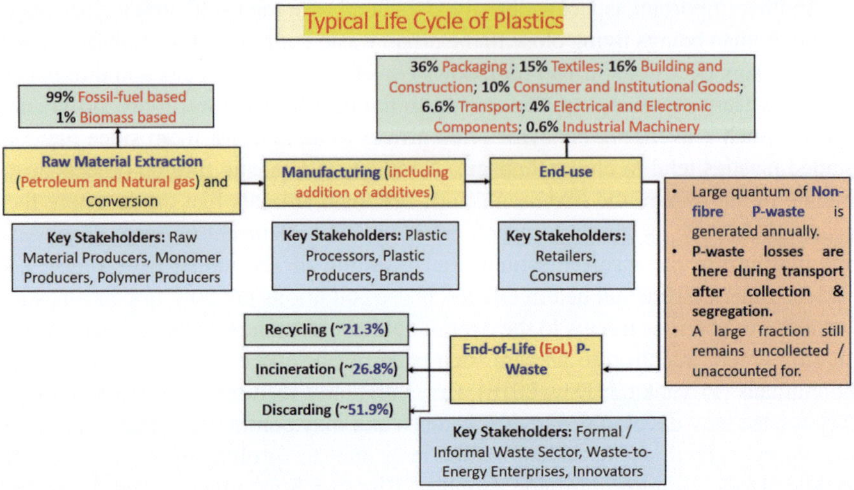

Fig. 3 An indicative illustration of the typical life cycle of plastics

shows that each part of the value chain involves multiple stakeholders. Therefore, it is necessary to consider the inputs from all the stakeholders while planning for a major transition away from plastics. An integrated view would help in minimizing the negative impact caused by the disruptive changes during the transition. Instead of an abrupt shift, a gradual change has to take place through a deeper understanding of the lasting hazardous effects (or long-term residual effects) of plastics. A detailed plan for occupational rehabilitation and financial compensation needs to be charted out to ensure livelihood security for all the workers employed across the plastic value chain.

5 Intrusion of Plastics into Urban Water Bodies and Its Repercussions

Plastics encompass a diverse range of materials classified into major categories such as acrylics, polyesters, silicones, polyurethanes, and halogenated plastics. Halogenated plastics, for example, include chlorine-based varieties like chlorinated polyethylene (CPE), chlorinated polyvinyl chloride (CPVC), and chlorosulfonated polyethylene (CSPE), each with unique properties and applications. These materials play critical roles in various industries due to their durability, versatility, and specific chemical characteristics, though their environmental impact and recycling challenges also require careful consideration and management strategies.

It is important to note that plastics may require anywhere between 20 and 500 years to degenerate, depending on the chemical structure of the polymer base and side chains and environmental factors such as sunlight exposure. For example, single-use plastic grocery bags and plastic coffee cups made of low-density polyethylene (LDPE) take about 2–3 decades to decompose, whereas plastic water bottles and soda can rings made of polyethylene terephthalate (PET) may take up to 450 years to fully break down (Chariot Energy, 2023). Once these materials are dumped into the environment (in the form of landfills as well as by other means), they can remain as they are across a few human generations (a 25-year temporal block can be considered as one generation for practical purposes).

Discarded plastic bags often choke the open drains as well as the sewers (or underground drainage networks, known as UGDs) in dense urban settings, leading to overflow of the drainage system (The Times of India, 2023c; The New Indian Express, 2023). The municipal sewage water flowing through the city drainage systems is loaded with pathogenic contents as well as a lot of toxic chemicals (Majumdar, 2021). Therefore, during times of more-than-average rainfall, the city dwellers would be vulnerable to various contagious and infectious diseases. If the imprudent disposal of plastic debris into the flowing natural waterbodies continues in an undeterred manner, it might cause a rise in the urban river water levels as well (Honingh et al., 2020), which would potentially inundate the areas located on the riverbanks. The evacuation from low-lying areas would be very challenging in such cases, and the emergency preparedness of the urban local bodies will be put to a stern test. Spillage of sewage water into urban water bodies (such as lakes and ponds), triggered by the plastic-choked drainage systems, would contaminate the water and may render it unfit for direct use. Evidently, discarded plastic bags can potentially pose significant problems for urban flood protection systems and stormwater drainage, as well as public health (Gupta, 2005).

As the accumulation of large bulks of plastic waste continues to take place in the natural environment, under the effect of the sunlight and wind, these plastics gradually break down into smaller pieces. The pieces that are smaller than 5 mm in size are classified as "*microplastics*." The tiny plastic particles with sizes in the range of 1 μm to 5 mm are grouped into this category. Plastic waste particles smaller than this specified range are termed "*nano-plastic*." In addition to the natural degeneration of P-waste, plastic pellets used in plastic manufacturing facilities and the plastic granules used in cosmetics and toothpaste may also serve as primary sources of microplastics. The wearing of vehicle tires is also a major source of microplastics in the urban environment. Stormwater has the potential to carry microplastics into the urban waterbodies from landfills as well as from the roads. The effluents from plastic package manufacturing facilities can potentially discharge a substantial quantity of microplastics into the drains as well as the water bodies. Freshwater fishes living in ponds and lakes may consume microplastics, which will then start accumulating in their tissues, leading to tissue damage, oxidative stress, and changes in immune-related gene expression in the fishes (Bhuyan, 2022).

Both human beings, as well as several aquatic/semi-aquatic bird species (e.g., cormorant, ibis, etc.), consume the fish growing in the urban water bodies. Therefore, microplastics can enter the food chain via the consumption of fish. Microplastics may further enter the tissues of the predatory species preying upon the wetland birds. In a nutshell, a typical urban setting loaded with P-waste may witness the propagation of microplastics through the food chain, through inhalation, and even via the skin contact route. Extended exposure to microplastics may lead urban residents to encounter oxidative stress, cytotoxicity, neurotoxicity, and immune system disruption. However, more evidence is needed before coming to definitive conclusions in this regard. One of the most important aspects will be to investigate the potential and likelihood of transfer of microplastics from one organ to the tissues of other organs within the human body, as the level of accumulation of microplastics increases. In a recent research investigation conducted in Italy, Raman microspectroscopy was carried out to analyze human breastmilk samples collected from 34 women, and the study indicated microplastic contamination in 26 out of 34 samples. The identified microparticles were categorized according to their shape, color, dimensions, and chemical composition. Among the various types of microplastics observed, polyethylene (PE), polyvinyl chloride (PVC), and polypropylene (PP) were notably more abundant, with sizes ranging from 2 to 12 μm. These findings underscore the prevalence of these specific plastics in the environment and highlight their potential impacts on ecosystems and human health, warranting further investigation into effective mitigation strategies (Ragusa et al., 2022). The emergence of such evidence raises significant concerns, particularly for the highly vulnerable population of infants, who represent future generations within the broader societal context. This underscores the importance of addressing these issues promptly to safeguard the health and well-being of young children, ensuring a sustainable and healthy environment for generations to come.

The microplastics can reach the groundwater aquifer by means of seepage through the pores and fractures of the soil layers. During the seepage, the interaction of microplastics with colloidal aggregates present in the subsoil layers can partially affect the microplastic dynamics, and such changes would make the detection of microplastics in the groundwater systems quite challenging (Viaroli et al., 2022). Recently, groundwater samples were collected from five sampling sites encompassing seven capped groundwater monitoring bores in Bacchus Marsh, Victoria, Australia. Analysis using an advanced Laser Direct Infra-Red (LDIR) imaging system revealed the presence of microplastics in all sampled locations. The sizes of the microplastic granules were found to be in the range of 18–491 μm, with the average size being 89 ± 55 μm (Samandra et al., 2022). The presence of microplastic contamination in an unconfined groundwater aquifer situated in an agricultural region suggests that these microplastics may exert wide-ranging impacts on both human habitats and ecosystems.

Since the microplastics can only degenerate but cannot decompose, they can potentially become hosts to organic matter while remaining within the water bodies, in the suspended form, or as a part of the bottom sediment. Such a kind of accumulation of organic matter may also be followed by the formation of pathogen colonies, as it happens in the case of sewage sludge (Majumdar, 2021), which can potentially lead to disease breakouts in case a community comes into direct contact with the pathogen-laden microplastics. In the context of disease spread, it is noteworthy that times of COVID-19 have shown serious repercussions faced by several densely populated localities in India due to improper disposal of single-use plastic

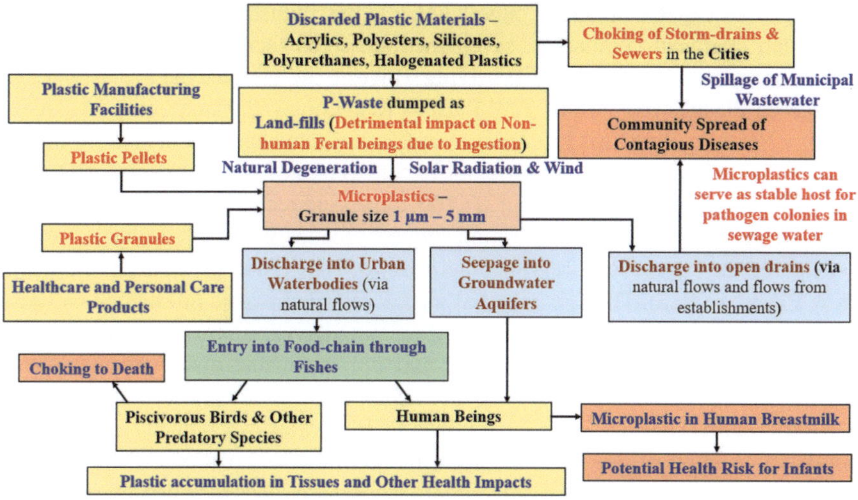

Fig. 4 Illustrative representation of microplastic incursion pathways and implications on the environment and ecology

items such as disposable face masks (de Sousa, 2021). Such evidence indicates the potential for community spread of contagious diseases via the route of P-wastes (Vaid et al., 2021).

This indicates the level of penetration the *P-waste* can have if it is allowed to remain in the natural ecosystem. In the bulk form as well as the degenerated form, the plastics have the potential to strangle an ecosystem as well as a habitat as a whole from multiple directions. Figure 4 provides a simplified schematic representation of the pathways of microplastic incursion into the natural environment and its implications on the environment as well as the ecology.

6 Current Status of Plastic Waste Collection and Recycling

6.1 Insights from the Advanced Nations

Recent reports indicate that out of the 40 million tonnes of plastic waste generated in the United States in 2021, only approximately 2 million tonnes, which accounts for about 5%, were recycled. About 85% of the generated *P-waste* went to landfills, and 10% was incinerated (Osborne, 2022). According to the United States Environmental Protection Agency (US EPA), the amount of plastic waste (P-waste) generated was approximately 35.41 million tonnes in 2017 and increased slightly to 35.68 million tonnes in 2018. In both years, around 8.5% of the reported total P-waste was recycled. The relative extent of plastic recycling has been reducing in the USA since then (Osborne, 2022; US EPA, 2023). As per the information shared by the American Chemistry Council (ACC), the industry trade association for American chemical companies, seven medium-sized facilities across the United States are already engaged in recycling P-waste to produce new plastic (e.g., Evergreen Plastics Inc. (Ohio), KW Plastics Recycling (Alabama), etc.) (APNews, 2023; ENF Recycling, 2023a). *Evergreen Plastics* recycles more than 11.6 billion PET bottles every year, producing about 73,500 tonnes of *rPET* (recycled PET) pellets. The enterprise collects about 87,100 tonnes of recyclable material annually (Evergreen, 2023). KW Plastics is an industry leader in recycled plastics and is considered the largest reprocessing enterprise for HDPE and PP resins in the world (KW Plastics, 2023). The company has a peak recycled plastic production capacity of 500,000 tonnes per year. KW Plastics Recycling Division specializes in manufacturing a diverse array of polyethylene and polypropylene post-consumer resins (PCR) tailored for various industries, including personal care products, automotive, and agriculture. They also offer customized PCR solutions specifically designed for construction applications such as pipes and coatings, recreational facilities, and flexible packaging. Their innovative approach ensures that recycled materials are utilized effectively across multiple sectors, promoting sustainability and environmental responsibility in manufacturing processes (KW Plastics, 2023). Several other companies are involved in transforming difficult-to-recycle plastics into alternative fuels used in aviation, marine, and automotive industries. In December 2022, ExxonMobil commenced operations at the state-of-the-art advanced plastic recycling facility in Baytown, Texas (ExxonMobil, 2023). The facility has a P-waste recycling capacity of 40,000 US tonnes per year. Other major corporations, including New Hope Energy, Nexus Circular, Eastman, and Encina, are also in the process of constructing large-scale plastic recycling facilities as part of their dedication to advancing the circular economy and promoting sustainability (APNews, 2023). The Alliance to End Plastic Waste (AEPW), a nongovernmental and nonprofit organization funded by industry stakeholders, was established in 2019. The Alliance to End Plastic Waste (AEPW) is composed of 30 companies spanning the entire plastic value chain. These include petrochemical giants like Shell and ExxonMobil, leading enterprises in the consumer goods segment such as Dow and Procter & Gamble (P&G), and recycling experts Veolia and Suez. AEPW has committed US$1.5 billion over the next 5 years to initiatives aimed at enhancing plastic recycling capabilities and supporting cleanup endeavors worldwide. This collaborative effort underscores their collective commitment to addressing the global challenge of plastic waste through innovative solutions and sustainable practices (Packaging Insights, 2023).

Before 2017, China once represented nearly 60% of the global trade in waste plastic. Between 1992 and 2006, the cumulative amount of waste plastic imported by China was estimated to be about 170 million tonnes. Following the release of the "Implementation Plan for Banning the Import of Wastes and Promoting the Reform of the Solid Waste Import Management System" in 2017 (GBF [2017] No. 70), China's import of waste plastics decreased significantly, becoming virtually negligible by 2019 (IPEN, 2022). Since 2019, China has no longer been a major importer of waste plastics. China generated 63 million tonnes of waste plastics in 2019. Out of these, about 32% went to landfills, 30% was recycled, 31% was incinerated in waste-to-energy plants for producing power, and 7% of the waste was discarded (IPEN, 2022). However, the plastic product recycling rates in China during the period 2017–20 remained on the lower side. Improvements have taken place in China's domestic plastic waste recycling industry ecosystem since 2018, and it is expected that the Chinese P-waste recycling industry will witness a 6.35% Y-o-Y growth in the near term. The recycled P-waste volume is expected to reach about 29 million tonnes by 2023 (SICAB, 2020). Dadong Plastic Granulation Plant, Huafei Plastic Recycling Plant, Baoding Kairui Plastic Products Manufacturing Co., Ltd., BioPack Co., Limited, and Chongqing Hongcan Environmental Protection Technology Co., Ltd. are some of the major enterprises with facilities to produce pellets, granules, and flakes from recycled plastic (Veolia, 2023; ENF Recycling, 2023b). Huafei Plastic Recycling Plant has a design capacity of producing 100,000 tonnes of recycled PET flakes, 100,000 tonnes of recycled PET pellets, and 50,000 tonnes of recycled polypropylene (PP) / high-density polyethylene (HDPE) / acrylonitrile butadiene styrene (ABS) particles per year (Veolia, 2023). The National Development and Reform Commission (NDRC), the top state planning agency in China, has planned to boost the plastic recycling and incineration capabilities in the country. In the "*five-year plan (2021-25)*" published in 2021, NDRC has clearly articulated the goals regarding the promotion of "green" plastic products and taking stringent action against the overuse of plastic in packaging and agriculture (Reuters, 2023). China has announced plans to ban the production of ultra-thin plastic bags and personal care products containing plastic microbeads, similar to existing bans in the United States and Europe (Reuters, 2023).

Japan ranks second globally in the Plastic Management Index (PMI) due to its advanced waste management system and active collaboration from local communities (Heinrich Böll, 2022). In recent years, Japan has seen an increase in plastic waste generation, reaching 4.1 million tonnes in 2019. Approximately 77.2% of this amount consisted of single-use containers and plastics used in packaging. Japan ranks second globally in per capita plastic packaging waste emissions (32.4 kg per capita per year), following the United States (40 kg per capita per year), making it the largest emitter in Asia. In terms of per capita PET bottle consumption, Japan's figures are among the highest (183 PET bottles per person per year). However, it also has a very high collection rate (93%) and recycling rate (~ 85.8%) attributable to the cutting-edge plastic waste management system, advanced infrastructure and technologies, as well as progressive policies for waste processing and recycling (Heinrich Böll, 2022). Annually, the Ministry of Environment of Japan reports on the generation and treatment status of general and industrial wastes (Kodera, 2012). Municipal bodies regularly monitor the plastic waste content in municipal waste. Depending on the segregation categories determined by municipalities, waste plastics are either incinerated as mixed burnable wastes or recycled, such as plastic containers and packaging (P-waste). Plastic materials from discarded toys and electronic devices typically end up in landfills. A composition survey of mixed municipal solid waste (burnable household waste) delivered to the Chuo incineration plant in Tokyo revealed that plastic waste accounted for 16% by weight (Kodera, 2012).

Far Eastern Ishizuka Green PET Corporation is one of the larger recycling plants in Japan, with a material processing capacity of 120,000 tonnes per year and a recycled PET (pellets/granules/flakes) producing capacity of 85,000 tonnes per year (ENF Recycling, 2023c). Kyoei J&T Recycling Corporation, with a recycled PET (pellets/granules/flakes) producing capacity of 50,000 tonnes per year, and Maruki Sangyo Co., Ltd., with a recycled plastic (granules/pellets) producing capacity of 54,000 tonnes per year, are some of the other major facilities (ENF Recycling, 2023c).

In 2022, the Industrial Technical Centre for Plastics and Composites (IPC)—France spearheaded a collaboration involving 19 European partners in the Buddie-Pack project. This project focuses on reusable plastic packaging (RPP) for widespread market adoption. The project is funded by the European Union. Partners countries include France, Germany, Ireland, the Netherlands, Spain, and the United Kingdom. Leading enterprises, innovation centers, and universities are collaborating in this consortium. The key target sectors for this project are the food and cosmetic/personal care sectors. The six specific objects of this project are mainly focused on developing technological solutions for the design and manufacture of packaging with appropriate functional properties required for reuse, understanding consumers' expectations regarding RPP, assessing drivers and barriers to behavioral changes, developing profitable business models to ensure success of RPP, reducing microplastic release in the environment and ensuring safety, carrying out large-scale demo trials to validate RPP across the value chains from a techno-economic and social point of view, and finally promoting reusable plastic packaging across the EU countries. Implementing large-scale reusable plastic packaging is anticipated to reduce the industry's energy and water consumption, thereby promoting sustainability through enhanced efficiency in energy and water use (Packaging Europe, 2023; Buddie Pack, 2023).

6.2 *Current State of Affairs in India*

Several enterprises are engaged in the collection, segregation, and recycling of plastic waste in India (ENF Recycling, 2023d). However, the current waste-handling capabilities and recycling capacities are small in comparison with the quantum of P-waste generated annually across Indian cities and towns. *Pashupati Group* is a prominent enterprise engaged in plastic recycling in India. The group has two companies operating in Uttarakhand, namely Pashupati Polytex Pvt. Ltd. (recycling capacity ~31,000 tonnes/year; produces recycled PET flakes) and Pashupati Extrusions Pvt. Ltd. (recycling capacity ~21,600 tonnes/year; produces recycled PET granules/pellets). Telangana-based Kundana Techno Tex Pvt. Ltd. (Pashupati group) has a PET recycling capacity of 54,000 tonnes/year. Rajasthan-based Saurabh Clean Tech Pvt. Ltd. (Pashupati Group) accepts a wide range of plastics as *P-waste* material, such as polyethylene (PE), polyvinyl chloride (PVC), high-density polyethylene (HDPE), and linear low-density polyethylene (LLDPE)—a copolymer of ethylene and another longer olefin—and produces recycled granules/pellets and flakes. The company has a recycling capacity of 7200 tonnes/year.

Through its different companies, Pashupati Group is engaged in the production of a wide range of recycled plastic products, such as Recycled Polyester Stable Fiber (RPSF) (used in thermal wears, work wears & uniforms, and also in carpets), Recycled PET (rPET) flakes (used in manufacturing PET films, sheet, monofilament, polyester synthetic fiber, plastic yarn etc.), Recycled PET chips (nonfood grade) (used in manufacturing monofilament, polyester synthetic fiber, polyester sheet and yarn), Recycled PET chips (food grade) (used in manufacturing bottles for water, carbonated soft drinks, juices; thermoformed containers for fruit, baked goods, meat and cheese; containers for vegetable oil and drinking cups), Recycled Polyolefin Granules (RPG) and Recycled Polyolefin Regrinds (RPR) (used in the extruders and molders), Polypropylene and HDPE Woven Fabric and Bags (used in packaging sugar, flour, food grains, cement, chemical, fertilizer, and textiles), rPET Black Master Batch (used in producing RPSF, poly oriented yarn, blow molding, sheet PET filament yarn, etc.), and Recycled Polyester yarn (Pashupati Group, 2023). Similar recycled plastic product profiles are available from other small and medium-sized facilities in the country.

Delhi-based Al Mehtab Industries (AMI) Pvt. Ltd. is a leading reprocessor of high-quality recyclable materials, equipped with a technically advanced plastic bottle reprocessing facility. It has a recycling capacity of 26,000 tonnes/year and produces recycled PET granules/pellets and flakes. Uttar Pradesh-based Ganesha Ecosphere Ltd. has been a leading PET bottle recycler, engaged in the recycling of discarded PET bottles to polyester spun yarn, rPET fiber, and rPET yarn (Ganesha Ecosphere, 2023). It processes about 135,000 tonnes of P-waste material per year and has a recycling capacity of about 140,000 tonnes/year (ENF Recycling, 2023d). Other noteworthy P-waste recycling enterprises are Alliance Fibres Ltd. (Gujarat) (recycling capacity: 36,000 tonnes/year; produces recycled PET flakes), AGL Polyfil Private Limited (West Bengal) (recycling capacity: 26,250 tonnes/year; produces PET flakes), Luckystar Group (Maharashtra) (processes 30,000 tonnes of P-waste material per year; produces recycled PP, HDPE, LDPE, LLDPE, HIPS granules/pellets), MTL New Initiatives Pvt. Ltd. (Karnataka) (recycling capacity: 7000 tonnes/year; produces HDPE granules/pellets), Shakti Corporation Pvt. Ltd. (Maharashtra) (recycling capacity: 8400 tonnes/year; produces PET, PP, PA, PS, HDPE, LDPE, ABS, HIPS granules/pellets), Mittal Polygrains LLP (Maharashtra) (recycling capacity: 8400 tonnes/year; produces BOPP, EVA, HDPE, LDPE, LLDPE granules/pellets) (ENF Recycling, 2023d).

Despite the recycling facilities in place, *P-waste* recycling has not been effective to the desired extent in the country, the primary reason being informal waste collection with no streamlined channeling. This reduces the collection efficiency in terms of recycling readiness. Further, single polymer packaging often mixes with food waste, and thereafter it becomes nonrecyclable. Lack of training among sanitation workers has been identified as a hindrance to effective plastic segregation since they often tend to mix the segregated wastes for the convenience of their usual (and uninformed) style of disposal. Also, economic incentives for preventing the leakage of low-value single-use plastics are not in place. Direct support from the urban local bodies, as well as the incentives and subsidies for the waste-based business ecosystem, are by and large absent in India, barring a few localized progressive examples (The Times of India, 2023d).

Advanced waste-to-energy (WTE) technologies, including bio-methanation, incineration, gasification, and pyrolysis, are established methods for converting a wide range of solid wastes into valuable forms of energy. These technologies can generate electricity and biogas, which primarily consists of methane, carbon dioxide, and small amounts of other gases. Additionally, some WTE processes produce syngas, composed mainly of carbon monoxide and hydrogen. These approaches play a crucial role in sustainable waste management by recovering energy from waste streams while minimizing environmental impact (MNRE, 2023).

Amongst the abovementioned technological options, incineration is a well-established technology and has been deployed at a commercial scale in India to generate electricity from the combustion of municipal and industrial solid wastes (MNRE, 2023). However, the success of such facilities has been limited in India, and a deeper technology assessment is required, considering the diverse nature of the solid waste produced in the country (Sambyal & Agarwal, 2017). Currently, there is no major capability of dedicated plastic incineration or pyrolysis in the country.

Incineration of plastic waste in the open field leads to the release of several toxic gases such as dioxins (a prominent class of persistent environmental pollutants, or POPs), furans, polychlorinated biphenyls (PCBs), polycyclic aromatic hydrocarbons (PAHs), and mercury into the atmosphere (Verma et al., 2016). Analyses of particulate soot emissions and residue ash emanating from highly controlled combustion (600–750 °C) of common types of plastic (e.g., PVC, LDPE, HDPE, PP, PS, PET) have indicated higher concentrations of low-molecular-weight PAHs in the airborne particulate soot as compared to the residue solid ash for all types of plastics. Residue ash from PVC combustion contained higher concentrations of higher-ringed PAHs as compared to that from the other types of plastics (Valavanidis et al., 2008). Polystyrene poses risks to the Central Nervous System (CNS), while hazardous brominated compounds are known carcinogens and mutagens. Dioxins can settle on crops and enter natural waterways, potentially entering the food chain and posing significant health risks (Verma et al., 2016). Therefore, plastic incineration needs to be performed in a very careful and prudent manner, even if it is aimed at reducing the quantum of waste and producing useful energy from the discards.

7 Way Forward for Prudent Plastic Waste Management in Indian Cities

NITI Aayog and UNDP jointly published a comprehensive handbook on sustainable management of urban plastic waste in 2021 (NITI Aayog, 2021). The handbook was aimed at facilitating the capacity building of the urban local body (ULB) officials and other relevant stakeholders at the city level for effective and efficient management of plastic waste. The core spirit of the handbook is aligned with a stricter implementation of the Reduce-Reuse-Recycle philosophy. In India, the plastic waste management (PWM) Rules (2016 & 2018) and the amendment introduced in 2021 focus primarily on single-use plastics. These regulations define various categories of plastics and suggest recycling methods tailored to each polymer type. In the NITI Aayog-UNDP Handbook, ample emphasis has been laid on the need for managing pre-consumer plastic waste generated by plastic manufacturing facilities as well as the packaging industry, in conjunction with post-consumer waste. Although *Extended Producer Responsibility* (EPR) and the *Buy Back Models* (BBM) have been discussed in the context of a reverse collection ecosystem, such initiatives would require incentivization and close monitoring during the nascent stage to ensure success.

One important aspect that needs special attention and strengthening of activities in the context of urban plastic waste (UPW) is the identification of the fate of *P-waste* using a hybrid approach comprising material flow analysis and on-ground monitoring (Rinasti et al., 2022). Both these approaches should be used such that they complement each other and reduce the duplication of effort. Attention should be paid to estimating the plastic leakages from the collection services, informal value chains, disposal facilities, and formal sorting facilities, as well as during the waste transportation. A detailed mapping of the leakages will help in hotspot identification, which would help the municipalities and allied administrative bodies in planning effective interventions. Attractive financial arrangements need to be devised for the different parts of plastic production, use, and waste value chains to retain materials in the system.

Conventional incineration of rubber and plastic wastes is usually carried out at relatively low temperatures (~1000 K), in which toxic gas formation, including dioxins and furans, is a significant concern. The air plasma incinerator developed by Bhabha Atomic Research Centre (BARC) provides us with an indigenous technological solution capable of creating a process bed with a temperature higher than 1500 °C (Ghorui et al., 2022; BARC, 2023). The syngas produced in this system comprises CH_4, CO, and H_2, and offers an energy density of more than 6.3 MJ/m^3. This indigenous equipment offers reduced device and installation costs (due to simple design and use of low power), low operational cost (since it uses atmospheric air as plasma gas), portability (due to low physical weight), and low maintenance cost (R & D News, 2023; The Times of India, 2023e). Pune Municipal Corporation (PMC) has opted to utilize the indigenous plasma gasification technology developed by BARC to convert municipal solid waste into hydrogen gas (Punekar News, 2023).

Two such systems have been installed in the BARC premises for the rapid processing of all types of incinerable wastes. Such systems can be installed in the centralized P-waste processing facilities set up jointly by the state government, urban local bodies, and private enterprises engaged in waste recovery and recycling. In case the P-waste becomes nonrecyclable due to mixing with other types of waste, plasma incinerators can be used effectively to reduce its volume. Since plastic waste mitigation would lead to the betterment of the sanitation systems within the city limits, there is an opportunity to attract CSR funds from large companies. Schedule VII in section 135 of the Companies Act, 2013 (18 of 2013), highlights "sanitation" as one of the key activities that can be undertaken by the companies as part of the Company Social Responsibility (CSR) Policies (Ministry of Corporate Affairs, 2014). A waste management fund can be created by the ULBs, where CSR and philanthropic donations can be turned into a sizable corpus for sustainable management of urban plastic waste.

Investment should be made to support the research and development (R&D) activities pertinent to the innovation of bio-benign plastic materials (e.g., polylactic acid (PLA) and polyhydroxyalkanoate (PHA)) to reduce the negative impact on the environment and ecology (Ellen MacArthur Foundation, 2023; NITI Aayog, 2022; Ecoideaz, 2023). Innovation is required in terms of delamination and/or selective dissolution techniques for separating the components of multilayered packaging waste to facilitate effective chemical recycling (Innovation in Plastics - The Potential and Possibilities, 2023). Investments are also required to upscale the existing recycling capabilities. Such progressive interventions would ensure that the existing value chains are not completely disrupted, and gradually, the complete value chain can be transformed into a "green" value chain with the right support and incentives.

Multiple Sustainable Development Goals can be achieved by creating an efficient plastic waste management (PWM) ecosystem in cities and towns across India (NITI Aayog, 2021). Through the means of supporting innovation and the creation of entities to fill the gaps in the P-waste value chain, SDG 9 can be achieved, which emphasizes industry, innovation, and infrastructure.

A reduction in plastic consumption would reduce the potential waste load from the source end. Since the manufacturing of recycled plastic products causes up to 87% reduction in energy consumption as compared to virgin plastics, SDG 12 (responsible consumption and production) can be realized by adopting plastic recycling (BJs Park, 2023). SDG 7 (access to clean, modern, and affordable energy) can be realized by generating electricity from nonrecyclable plastic waste via the environmentally benign high-temperature combustion route using an air plasma incinerator. Recycling of plastics would lead to a reduced extent of P-waste in landfills and result in the avoidance of toxic gas emissions attributable to the burning of plastics in the open field. A lesser amount of plastic waste in landfills would ensure environmental safety for nonhuman feral beings, and prudent handling would also ensure that the plastic waste does not end up choking the open drains and the sewers in dense urban settings. Chances of discharge of plastics (and microplastics) into the natural flows and urban water bodies would also diminish considerably. These would translate into the realization of SDG 3 (good health and well-being), SDG 6 (clean water and sanitation), SDG 11 (sustainable cities and communities), and SDG 15 (life on land). Finally, the creation of a robust and sustainable waste-to-wealth value chain would create employment opportunities in a socio-environmentally important segment, realizing SDG 8 (decent work and economic growth). By adopting P-waste recycling and by enforcing various allied sustainable measures, India can potentially unlock various international funding opportunities, such as Green Climate Fund (GCF) and Global Environment Facility (GEF) funding, which can be used for innovation, capacity building, infrastructure modernization, and training purposes in the areas that call for the *Reduce*, *Recycle,* and *Reuse* (3R) approach.

8 Conclusions

This chapter offers a comprehensive overview of the extensive impacts attributable to the nondegradable wastes in India's densely populated urban areas, specifically focusing on the urban plastic waste problem. From the elaborate discussions presented in the chapter, it is evident that the plastic problem is multifaceted. Therefore, a multipronged and multidisciplinary approach is required to address the issue. There is an immediate necessity to understand the different parts involved in the plastic life cycle in the country and to evaluate the leakages of plastics from various parts of the value chain. Taking the learnings from the European countries and Japan, India needs to bolster the community engagement channels to create awareness about the long-term detrimental effects of plastic waste on the environment and the ecology and to secure greater cooperation from local communities. Investments should be made toward innovation of alternative materials, capacity building, infrastructure modernization, and training purposes to ensure that the transition away from nondegradable plastics takes place in a sustainable and gradual manner. Finally, the urban plastic problem should be conceived as a wholesome, complex issue by policymakers, instead of conceptualizing it as a series of smaller targets aimed at short-term goals. Through the adoption of the right incentivization framework, by providing necessary logistical support, and by facilitating collaboration amongst various enterprises operating in different parts of the plastic value chain, the urban local bodies would become better equipped to deploy a multipronged work plan toward alleviating the plastic menace. Efficient collection and prudent segregation of municipal solid waste would form a core part of the strategy. With suitable financial and organizational arrangements, sustainable business-oriented models can be devised and customized for different urban settings in India to realize the overarching goal of "*Green Cities*."

Acknowledgments The author is highly grateful to Dr. Shailesh Nayak (Director, NIAS) for his invaluable insights and comments, which enhanced the quality of the chapter. The author also gratefully acknowledges the help provided by Mr. Souvik Sankar Mitra (Research Scientist, IIRS Dehradun) in preparing the LULC images.

References

Abhishek, N., Jenamani, M., & Mahanty, B. (2017). Urban growth in Indian cities: Are the driving forces really changing? *Habitat International, 69*, 48–57.

Amdavad Municipal Corporation (AMC). (Retrieved from: https://ahmedabadcity.gov.in/portal/jsp/Static_pages/introduction_of_amdavad.jsp) [Accessed on 10 May 2023].

APNews. *Advanced recycling: Plastic crisis solution or distraction?* (By Jennifer McDermott). (Retrieved from: https://apnews.com/article/science-united-states-providence-business-climate-and-environment-b9f202a703ea7fa4231053d544b3266e) [Accessed on 10 May 2023].

Bangalore Metropolitan Region Development Authority (BMRDA). (Retrieved from: https://bmrda.karnataka.gov.in/info-1/Introduction/en) [Accessed on 10 May 2023].

Bangalore Metropolitan Region Development Authority (BMRDA) New Map. (2019). (Retrieved from: https://citizenmatters.in/wp-content/uploads/sites/14/2019/06/BMRDA_Map_New.pdf) [Accessed on 10 May 2023].

BARC. Technical Specifications of a 60 kW Triple Torch Air Plasma Incinerator. (Retrieved from: https://www.barc.gov.in/technologies/air_plasma/air_plasma.pdf) [Accessed on 10 May 2023].

Bengaluru Smart City, Government of Karnataka. (Retrieved from: https://blrsmartcity.karnataka.gov.in/page/Explore+Bengaluru/Introduction/en) [Accessed on 10 May 2023].

Bhagat, R. B. (2017). Urbanisation in India: Trend, pattern and policy issues. In *Proceedings of 28th international population conference, Cape Town, 29th October - 4th November, 2017*. (Retrieved from: https://iussp.confex.com/iussp/ipc2017/mediafile/Presentation/Paper2674/Bhagat_IUSSP%202017_Cape%20Town_Urbanisation.pdf) [Accessed on 10 May 2023]

Bhuyan, M. S. (2022). Effects of microplastics on fish and in human health. *Frontiers in Environmental Science, 10*, 827289. https://doi.org/10.3389/fenvs.2022.827289

BJs Park. *Recycled plastic - Benefits of using recycled plastic in parks*. (Retrieved from: https://bjspark.com/recycled-plastic/) [Accessed on 10 May 2023].

Buddie Pack. *Homepage*. (Retrieved from: https://www.buddie-pack.com/) [Accessed on 10 May 2023].

C2ES. *Short-lived climate pollutants*, Center for Climate and Energy Solutions (C2ES). (Retrieved from: https://www.c2es.org/content/short-lived-climate-pollutants/) [Accessed on 10 May 2023].

Chariot Energy. *How long does it take for plastic to decompose?*. (Retrieved from: https://charriotenergy.com/blog/how-long-until-plastic-decomposes/) [Accessed on 10 May 2023].

Choe, K. A., & Roberts, B. (2011). *Competitive cities in the 21st century: Cluster-based local economic development* (Australian Government (AusAID), Urban Development Series). Asian Development Bank (ADB). (Retrieved from: https://www.adb.org/sites/default/files/publication/29242/cluster-based-local-economic-development.pdf) [Accessed on 10 May 2023]

CPCB. *Solid waste generation in 46 metrocities*, Central Pollution Control Board (CPCB). (Retrieved from: https://cpcb.nic.in/uploads/MSW/trend_46_cities_list.pdf) [Accessed on 10 May 2023a].

CPCB. *Government perspective on municipal solid wastes management*, Central Pollution Control Board (CPCB). (Retrieved from: https://cpcb.nic.in/zobengaluru/MSW_Perspectives%20.pdf) [Accessed on 10 May 2023b].

CPHEEO. *Composition and quantity of solid waste*. (Retrieved from: https://cpheeo.gov.in/upload/uploadfiles/files/chap3.pdf) [Accessed on 10 May 2023].

De Pietro, A. (2017, December 28). *Cost of living is surging in these major cities and what it could mean for 2018*. Forbes. (Retrieved from: https://www.forbes.com/sites/andrewdepietro/2017/12/28/cost-of-living-is-surging-in-these-major-cities-and-what-it-could-mean-for-2018/?sh=11c0adc771c6) [Accessed on 10 May 2023].

de Sousa, F. D. B. (2021). Plastic and its consequences during the COVID-19 pandemic. *Environmental Science and Pollution Research, 28*, 46067–46078. https://doi.org/10.1007/s11356-021-15425-w

Dhyani, S., Majumdar, R., & Santhanam, H. (2021). Scaling-up nature-based solutions for mainstreaming urban resilience in Indian cities. In M. Mukherjee & R. Shaw (Eds.), *Ecosystem-based disaster and climate resilience* (Disaster and risk research: GADRI book series) (pp. 279–306). Springer. https://doi.org/10.1007/978-981-16-4815-1_12

DISPUR. *Recommendations of the sixth central pay commission- decision of government relating to grant of dearness allowance to central government servants* (Office Memorandum no. 1(3)/2008-E-II(B) dated 29th August, 2008) Department of Expenditure, Ministry of Finance, Government of India. (Retrieved from: https://dispur.nic.in/sixthpay/sixth-pay-allowances.pdf) [Accessed on 10 May 2023].

DOE. *Re-classification/Upgradation of Cities/Towns on the basis of Census-2011 for the purpose of grant of House Rent Allowance (HRA) to Central Government Employees* (Office Memorandum no. 2/5/2014-E. II(B) dated 21st July, 2015), Department of Expenditure,

Ministry of Finance, Government of India. (Retrieved from: https://doe.gov.in/sites/default/files/21-07-2015.pdf) [Accessed on 10 May 2023].

Ecoideaz. *Pioneering bioplastic companies in india bring about a positive change*. (Retrieved from: https://www.ecoideaz.com/expert-corner/pioneering-bioplastic-companies-india-bring-positive-change) [Accessed on 10 May 2023].

Ellen MacArthur Foundation. *The new plastics economy - Rethinking the future of plastics*. (Retrieved from: https://www.mckinsey.com/~/media/McKinsey/dotcom/client_service/Sustainability/PDFs/The%20New%20Plastics%20Economy.ashx) [Accessed on 10 May 2023].

ENF Recycling. *Plastic recycling plants in United States*. (Retrieved from: https://www.enfrecycling.com/directory/plastic-plant/United-States) [Accessed on 10 May 2023a].

ENF Recycling. *Plastic recycling plants in China*. (Retrieved from: https://www.enfrecycling.com/directory/plastic-plant/China) [Last Accessed on 10 May 2023b].

ENF Recycling. *Plastic recycling plants in Japan*. (Retrieved from: https://www.enfrecycling.com/directory/plastic-plant/Japan) [Accessed on 10 May 2023c].

ENF Recycling. *Plastic recycling plants in India*. (Retrieved from: https://www.enfrecycling.com/directory/plastic-plant/India) [Accessed on 10 May 2023d].

Evergreen. *The Evergreen sustainability solution homepage*. (Retrieved from: https://www.evergreentogether.com/) [Last Accessed on 10 May 2023].

ExxonMobil. *ExxonMobil starts operations at large-scale advanced recycling facility*. (Retrieved from: https://corporate.exxonmobil.com/news/news-releases/2022/1214_exxonmobil-starts-operations-at-large-scale-advanced-recycling-facility) [Accessed on 10 May 2023].

Ganesha Ecosphere. *Homepage*. (Retrieved from: https://ganeshaecosphere.com/) [Accessed on 10 May 2023].

GCC. *Brief note on Greater Chennai Corporation (GCC)*. (Retrieved from: https://chennaicorporation.gov.in/gcc/about GCC/greater chennai corporation/brief-note-about-GCC/) [Accessed on 10 May 2023].

Geyer, R. (2020). Chapter 2: Production, use, and fate of synthetic polymers. In T. M. Letcher (Ed.), *Plastic waste and recycling: environmental impact, societal issues, prevention, and solutions* (pp. 13–32). Academic Press. https://doi.org/10.1016/B978-0-12-817880-5.00002-5

Geyer, R., Jambeck, J. R., & Law, K. L. (2017). Plastics-production, use, and fate of all plastics ever made. *Science Advances, 3*(7), e1700782. https://doi.org/10.1126/sciadv.1700782

Ghorui, S., Bhandari, S., Tiwari, N., Misra, V. C., & Chakravarthy, Y. (2022). *Air plasma incinerator in pursuit of 'Clean India Mission'*, Feature Article, BARC Newsletter (March–April 2022). (Retrieved from: https://www.barc.gov.in/barc_nl/2022/2022030415.pdf) [Accessed on 10 May 2023].

Greater Hyderabad Municipal Corporation. (Retrieved from: https://www.ghmc.gov.in/Menu_Files/GHMCRTI.pdf) [Accessed on 10 May 2023].

Gupta, K. (2005). The drainage systems of India's cities. *Waterlines, 23*(4), 22–24. Practical Action Publishing. (Retrieved from: https://www.jstor.org/stable/24684703) [Accessed on 10 May 2023]

Heinrich Böll. (2022). *Challenges and efforts in Japan: Japan's plastic waste management – Challenges and potential solutions, plastic atlas Asia*. (Retrieved from: https://hk.boell.org/sites/default/files/2022-05/PlasticAtlasAsia2022_en_WEB_0.pdf) [Accessed on 10 May 2023].

Honingh, D., et al. (2020). Urban river water level increase through plastic waste accumulation at a rack structure. *Frontiers in Earth Science, 8*, 28. https://doi.org/10.3389/feart.2020.00028

Hyderabad Metropolitan Development Authority (HMDA). (Retrieved from: https://www.hmda.gov.in/) [Accessed on 10 May 2023].

India Policy Brief. (2014, November). *OECD better policy series*, Organisation for Economic Co-operation and Development (OECD). (Retrieved from: https://www.oecd.org/policy-briefs/India-Challenges-of-Urbanisation.pdf) [Accessed on 10 May 2023].

Innovation in Plastics - The Potential and Possibilities. (12 January 2023), *Joint study by Indian Institute of Science & Praxis Global Alliance, supported by Marico Innovation Foundation.* (Retrieved from: https://packaging360.in/whitepapers/innovation-in-plastics-the-potential-and-possibilities/) [Accessed on 10 May 2023].

IPEN. (2022). *Plastic waste management and burden in China*, International Pollutants Elimination Network (IPEN). (Retrieved from: https://ipen.org/sites/default/files/documents/ipen-china-2021-epa_v1_2.pdf) [Accessed on 10 May 2023].

Jadhav, A. (2021, July 1). *Largest area under PMC, Pune officially becomes biggest city in Maharashtra.* (Retrieved from: https://indianexpress.com/article/cities/pune/largest-area-under-pmc-pune-officially-becomes-biggest-city-in-maharashtra-7383298/) [Accessed on 10 May 2023].

Kaushal, R. K., Varghese, G. K., & Chabukdhara, M. (2012). Municipal solid waste management in India - Current state and future challenges: a review. *International Journal of Engineering, Science and Technology, 4*, 1473–1489.

Kodera, Y. (2012). *Plastics recycling—Technology and business in Japan.* (Retrieved from: https://cdn.intechopen.com/pdfs/40492/InTech-Plastics_recycling_technology_and_business_in_japan.pdf) [Last Accessed on 10 May 2023].

Kolkata Metropolitan Development Authority (KMDA). (Retrieved from: https://kmda.wb.gov.in/page/cms/introducing_kmda_011393) [Last Accessed on 10 May 2023].

Kolkata Municipal Corporation. (2017, December). *Monthly e-newsletter by Kolkata Municipal Corporation*, 19(19). (Retrieved from: https://www.kmcgov.in/KMCPortal/downloads/EletterDecember2017_09_01_2018.pdf) [Accessed on 10 May 2023].

Kumar, S., et al. (2017). Challenges and opportunities associated with waste management in India. *Royal Society Open Science, 4*, 160764. https://doi.org/10.1098/rsos.160764

KW Plastics. *Homepage.* (Retrieved from: https://www.kwplastics.com/) [Accessed on 10 May 2023].

Liu, W., Chen, W., & Peng, C. (2014). Assessing the effectiveness of green infrastructures on urban flooding reduction: A community scale study. *Ecological Modelling, 291*, 6–14.

Lu, M., et al. (2018). Phosphine production in anaerobic wastewater treatment under tetracycline antibiotic pressure. *Journal of Environmental Sciences, 69*, 239–250.

Majidi, A. N., Vojinovic, Z., Alves, A., et al. (2019). Planning nature-based solutions for urban flood reduction and thermal comfort enhancement. *Sustainability, 11*, 6361.

Majumdar, R. (2021). Chapter 11: Sludge hygienisation—a novel technology for urban areas to deal with incursion of COVID-19 viral particles in wastewater. In C. Chakrabarty et al. (Eds.), *The impact of the COVID-19 pandemic on green societies: Environmental sustainability.* Springer. https://doi.org/10.1007/978-3-030-66490-9_11

MCGM. (2020, January). Draft Interim Report: Air Quality and Emission Source Apportionment Studies for Ten Cities of Maharashtra - MUMBAI CITY, Maharashtra Pollution Control Board, Joint Study by CSIR-NEERI and IIT Bombay. (Retrieved from: https://r3app.mcgm.gov.in/irj/go/km/docs/documents/MCGM%20Department%20List/Environment/Docs/04%20Source%20Apportionment%20Report-MCGM.pdf) [Accessed on 10 May 2023].

Ministry of Corporate Affairs, "*Companies Act, 2013 Notification: Schedule VII*" Gaz. India Extraordinary [Part II-Sec. 3(i)], vol. 4, no. D, pp. 1–2, 2014, [Online]. (Retrieved from: http://www.mca.gov.in/Ministry/pdf/CompaniesActNotification3_2014.pdf) [Accessed on 10 May 2023].

MMRDA. *Basic information on Greater Mumbai and Mumbai Metropolitan Region (MMR)* (as on 2020). (Retrieved from: https://mmrda.maharashtra.gov.in/documents/10180/12218/About%20MMR/) [Accessed on 10 May 2023].

MNRE. *Waste to energy (Current status in India).* Ministry of New and Renewable Energy (MNRE) Homepage. (Retrieved from: https://mnre.gov.in/waste-to-energy/current-status) [Accessed on 10 May 2023].

Mongabay-India. *Fantastic beasts and why plastic is bad for them.* (Retrieved from: https://india.mongabay.com/2022/03/fantastic-beasts-and-why-plastic-is-bad-for-them/) [Accessed on 10 May 2023].

Naveen, B. P. (2021). Scenarios of waste management nexus in Bangalore. *Energy Nexus, 1*, Article 100004. https://doi.org/10.1016/j.nexus.2021.100004

NITI Aayog. (2021). *UNDP handbook on sustainable urban plastic waste management.* (Retrieved from: https://www.niti.gov.in/sites/default/files/2021-10/Final_Handbook_PWM_10112021.pdf) [Accessed on 10 May 2023].

NITI Aayog. (2022, May). *Report on alternative products and technologies to plastics and their applications.* (Retrieved from: https://www.niti.gov.in/sites/default/files/2022-07/Plastics%20Alternative%20Study_Final_Report_compressed.pdf) [Accessed on 10 May 2023].

NIUA. *Urbanisation and urban sprawl: A perspective on India's urban growth.* (Retrieved from: https://www.niua.org/tod/todfisc/book.php?book=1&section=2) [Accessed on 10 May 2023].

Nunes, M. I., et al. (2021). Hydrogen sulfide levels in the ambient air of municipal solid waste management facilities: A case study in Portugal. *Case Studies in Chemical and Environmental Engineering, 4*, Article 100152. https://doi.org/10.1016/j.cscee.2021.100152

Osborne, M. (09 May 2022). *At least 85 percent of U.S. plastic waste went to landfills in 2021*, Smithsonian Magazine. (Retrieved from: https://www.smithsonianmag.com/smart-news/the-us-recycled-just-5-percent-of-its-plastic-in-2021-180980052/) [Accessed on 10 May 2023].

Packaging Europe. *European companies across plastics value chain collaborate on project to facilitate wider use of recycled plastic packaging.* (Retrieved from: https://packagingeurope.com/news/european-companies-across-plastics-value-chain-collaborate-on-project-to-facilitate-wider-use-of-recycled-plastic-packaging/8929.article) [Accessed on 10 May 2023].

Packaging Insights. *Plastic value chain giants pledge US$1.5bn investment in recycling infrastructure and plastic clean-up.* (Retrieved from: https://www.packaginginsights.com/news/plastic-value-chain-giants-pledge-us15bn-investment-in-recycling infrastructure and plastic clean-up.html) [Accessed on 10 May 2023].

Pashupati Group. *Homepage.* (Retrieved from: https://pashupatigrp.com/recycling) [Accessed on 10 May 2023].

Priyanka, M., & Dey, S. (2018). Ruminal impaction due to plastic materials - An increasing threat to ruminants and its impact on human health in developing countries. *Veterinary World, 11*(9), 1307–1315. https://doi.org/10.14202/vetworld.2018.1307-1315. PMID: 30410238; PMCID: PMC6200578.

Punekar News. *Pune: PMC to use plasma gasification technology to turn trash into hydrogen gas.* (Retrieved from: https://www.punekarnews.in/pune-pmc-to-use-plasma-gasification-technology-to-turn-trash-into-hydrogen-gas/) [Accessed on 10 May 2023].

R & D News. *BARC technology for air plasma torch.* (Retrieved from: https://rndnewsindia.com/barc-technology-for-air-plasma-torch/) [Accessed on 10 May 2023].

Ragusa, A., et al. (2022). Raman microspectroscopy detection and characterisation of microplastics in human breastmilk. *Polymers, 14*(13), 2700.

Reserve Bank of India (RBI). Details of tier-wise classification of centres based on population, (Retrieved from: https://rbidocs.rbi.org.in/rdocs/content/pdfs/NOTI306180517_A1.pdf) [Accessed on 10 May 2023].

Reuters. *China to ramp up recycling, incineration in new plastic pollution push.* (Retrieved from: https://www.reuters.com/world/china/china-ramp-up-recycling-incineration-new-plastic-pollution-push-2021-09-15/) [Accessed on 10 May 2023].

Rinasti, A. N., et al. (2022). Fate identification and management strategies of non-recyclable plastic waste through the integration of material flow analysis and leakage hotspot modeling. *Scientific Reports, 12*, 16298. https://doi.org/10.1038/s41598-022-20594-w

Samandra, S., et al. (2022). Microplastic contamination of an unconfined groundwater aquifer in Victoria, Australia. *Science of the Total Environment, 802*, Article 149727. https://doi.org/10.1016/j.scitotenv.2021.149727

Sambyal, S. S., & Agarwal, R. (15 October 2017). *Waste - Burn it all*. Down to Earth. (Retrieved from: https://www.downtoearth.org.in/news/waste/burn-it-all-58827) [Accessed on 10 May 2023].

Sankhe, S., et al. (2010, April 1). *India's urban awakening: Building inclusive cities, sustaining economic growth*. Report of McKinsey Global Institute. (Retrieved from: https://www.mckinsey.com/featured-insights/urbanization/urban-awakening-in-india) [Accessed on 10 May 2023].

Santhanam, H., & Majumdar, R. (2020). Permeable pavements as sustainable nature-based solutions for the management of urban lake ecosystems. In S. Dhyani, A. Gupta, & M. Karki (Eds.), *Nature-based solutions for resilient ecosystems and societies* (Disaster Resilience and Green Growth (DRGG)) (pp. 329–345). Springer. https://doi.org/10.1007/978-981-15-4712-6_19

Sharma, K. D., & Jain, S. (2018). Overview of municipal solid waste generation, composition, and management in India. *Journal of Environmental Engineering (ASCE), 145*(3). https://doi.org/10.1061/(ASCE)EE.1943-7870.0001490

SICAB. (2020). *China waste plastic recycling industry (Thematic report 04)*, Edited by China-Italy Chamber of Commerce, Sino-Italian Capacity Building for Environmental Protection. (Retrieved from: https://www.sicab.net/wp-content/uploads/2020/05/04.-China-Waste-Plastic-Recycling-Industry-Report.pdf) [Accessed on 10 May 2023].

Singh, S. (2020). *Solid waste management in urban India: Imperatives for improvement* (ORF Occasional Paper (Paper ID 283)). Observer Research Foundation. (Retrieved from: https://www.orfonline.org/wp-content/uploads/2020/11/ORF_OccasionalPaper_283_SolidWasteManagement_FinalForUpload-2.pdf) [Accessed on 10 May 2023]

The Hindu. *Order issued for expansion of Chennai to 5,904 sq. km; over 1,200 villages set to be added*. (Retrieved from: https://www.thehindu.com/news/cities/chennai/government-issues-order-for-expanding-chennai-metropolitan-planning-area-to-5904-sq-km/article66041750.ece) [Accessed on 10 May 2023].

The New Indian Express. *Plastic, debris choking city's arterial drains*. (Retrieved from: https://www.newindianexpress.com/cities/bengaluru/2018/sep/26/plastic-debris-choking-citys-arterial-drains-1877238.html) [Accessed on 10 May 2023].

The Times of India. (2020, June 19). *Gujarat: Municipal limits of six cities expanded*. (Retrieved from: https://timesofindia.indiatimes.com/city/ahmedabad/gujarat-municipal-limits-of-six-cities-expanded/articleshow/76459795.cms) [Accessed on 10 May 2023].

The Times of India. *Chennai expands five-fold, officially (by Vivek Narayanan)*. (Retrieved from: https://timesofindia.indiatimes.com/city/chennai/move-will-ensure-balanced-urban-development-in-chennai-region-chennai-metropolitan-development-authority/articleshow/95027003.cms) [Accessed on 10 May 2023a].

The Times of India. Not just kite string, plastic kills birds too in Ahmedabad. (Retrieved from: https://timesofindia.indiatimes.com/city/ahmedabad/not-just-kite-string-plastic-kills-birds-too-in-ahmedabad/articleshow/97042642.cms) [Accessed on 10 May 2023b].

The Times of India. *Plastic waste choking storm water drains in Trichy city*. (Retrieved from: https://timesofindia.indiatimes.com/city/trichy/plastic-waste-choking-storm-water-drains-in-trichy-city/articleshow/87725356.cms) [Accessed on 10 May 2023c].

The Times of India. *Haryana: How startups, waste management companies can plug gaps in plastic recycling*. (Retrieved from: https://timesofindia.indiatimes.com/city/gurgaon/haryana-how-startups-waste-management-companies-can-plug-gaps-in-plastic-recycling/articleshow/92480524.cms) [Accessed on 10 May 2023d].

The Times of India. *West Bengal: High-temperature torch to convert waste into fuel*. (Retrieved from: https://timesofindia.indiatimes.com/city/kolkata/west-bengal-high-temperature-torch-to-convert-waste-into-fuel/articleshow/96122935.cms) [Accessed on 10 May 2023e].

The World Bank Data. *GDP per capita (current US$)—India*. (Retrieved from: https://data.worldbank.org/indicator/NY.GDP.PCAP.CD?locations=IN) [Accessed on 10 May 2023a].

The World Bank Data. *GDP per capita, PPP (current international $)—India*. (Retrieved from: https://data.worldbank.org/indicator/NY.GDP.PCAP.PP.CD?locations=IN) [Accessed on 10 May 2023b].

The World Bank Data. *GDP per capita, PPP (current international $)- Global average*. (Retrieved from: https://data.worldbank.org/indicator/NY.GDP.PCAP.PP.CD) [Accessed on 10 May 2023c].

The World Bank Data. *Urban population (% of total population)—India*. (Retrieved from: https://data.worldbank.org/indicator/SP.URB.TOTL.IN.ZS?locations=IN) [Accessed on 10 May 2023d].

Tsakona, M. et al. (2021). *Drowning in plastics: Marine litter and plastic waste vital graphics*, Supported by Norwegian Ministry of Climate and Environment & Norwegian Agency for Development Cooperation (Norad), United Nations Environment Programme (UNEP). (Retrieved from: https://wedocs.unep.org/xmlui/bitstream/handle/20.500.11822/36964/VITGRAPH.pdf) [Accessed on 10 May 2023].

Urban Agglomerations Census. (2011). (Retrieved from: https://www.census2011.co.in/urbanagglomeration.php) [Last Accessed on 10 May 2023].

US EPA. *National overview: Facts and figures on materials, wastes and recycling*; United States Environmental Protection Agency (US EPA). (Retrieved from: https://www.epa.gov/facts-and-figures-about-materials-waste-and-recycling/national-overview-facts-and-figures-materials) [Accessed on 10 May 2023].

Vaid, M., Mehra, K., & Gupta, A. (2021). Microplastics as contaminants in Indian environment: A review. *Environmental Science and Pollution Research, 28*, 68025–68052.

Valavanidis, A., Iliopoulos, N., Gotsis, G., & Fiotakis, K. (2008). Persistent free radicals, heavy metals and PAHs generated in particulate soot emissions and residue ash from controlled combustion of common types of plastic. *Journal of Hazardous Materials, 156*(1–3), 277–284.

van Doorn, L., Arnold, A., & Rapoport, E. (2019). In the age of cities: The impact of urbanisation on house prices and affordability. In R. Nijskens et al. (Eds.), *Hot property*. Springer. https://doi.org/10.1007/978-3-030-11674-3_1

Veolia. *Huafei plastic recycling plant*. (Retrieved from: https://www.veolia.cn/en/business-cases/huafei-plastic-recycling-plant) [Accessed on 10 May 2023].

Verma, R., Vinoda, K. S., Papireddy, M., & Gowda, A. N. S. (2016). Toxic pollutants from plastic waste - A review. *Procedia Environmental Sciences, 35*, 701–708.

Viaroli, S., Lancia, M., & Re, V. (2022). Microplastics contamination of groundwater: Current evidence and future perspectives. A review. *Science of the Total Environment, 824*, Article 153851. https://doi.org/10.1016/j.scitotenv.2022.153851

Vij, D. (2012). Urbanization and solid waste management in India: Present practices and future challenges. *Procedia - Social and Behavioral Sciences, 37*, 437–447.

Annual and Seasonal Variation of Particulate Matter: A Case Study Over Southeast Part of West Bengal, India

Jayashree Saha, Varun Narayan Mishra, Maya Kumari, Ram Pravesh Kumar, and Mousumi Majumdar

Abstract Urban air pollution has gained a lot of attention from policymakers because of the unprecedented rate of population and industrial activities around the world. Exposure to particulate matter (PM) air pollution adversely affects human health in many large cities around the world. Most of the large cities are extremely heterogeneous in population density and socioeconomic situation. So, mapping and monitoring urban air pollutants accurately is crucial for environmental risk and epidemiological studies. The present study investigates the spatial and temporal distribution of PM during 2021–2022 over the southeast part of West Bengal, India. Annual and seasonal spatial distribution maps of $PM_{2.5}$ and PM_{10} concentrations were produced by applying the inverse distance weighting (IDW) method. The results showed high variability of both pollutants. The pollution hotspots of PM were mainly concentrated in Kolkata, Howrah, the East Midnapore region, and their surroundings. This analysis points out significant variations in $PM_{2.5}$ and PM_{10} concentrations regarding the seasonal changes. Assessing air pollutants may help in policy formulation and decision-making.

Keywords Air pollution · Particulate matter · GIS · Interpolation

J. Saha · V. N. Mishra (✉) · M. Majumdar
Amity Institute of Geoinformatics and Remote Sensing (AIGIRS), Amity University, Noida, India
e-mail: vnmishra@amity.edu

M. Kumari
Amity School of Natural Resources and Sustainable Development (ASNRSD), Amity University, Noida, India

R. P. Kumar
School of Environmental Sciences, Jawaharlal Nehru University, New Delhi, India

P. K. Joshi et al. (eds.), *Geospatial Science for Urban Ecosystems*, The Urban Book Series, https://doi.org/10.1007/978-3-032-07155-2_11

1 Introduction

Rapid urban growth and industrial development have led to a rise in the existence of air pollution around the world. In recent years, India has witnessed rapid urbanization, large-scale industrialization, and a rise in energy consumption (Kaur & Pandey, 2021; Kamboj et al., 2022; Yadav et al., 2023). There is high concern in the most polluted urban areas regarding the decrease in the quality of life and chronic and acute diseases because of exposure to air pollutants (Lelieveld et al., 2015; Chen et al., 2020; Baca-López et al., 2021). It is considered one of the most serious environmental issues of urban areas in many developed and developing countries worldwide. According to the WHO, more than 90% of the global population is affected by high levels of air pollution, which causes around 7 million early deaths yearly (WHO, 2021; Wang et al., 2021). The hazard of urban air pollution to the living environment and physical and mental health of human beings has become increasingly prominent (Zhang, 2019; Tui et al., 2021).

At present, the regular occurrence of substantial pollution incidents initiated by particulate matter (PM) has developed an urgent concern in India. Among PM of varying sizes, respirable fraction (PM_{10}) and fine fraction ($PM_{2.5}$) are deemed the most hazardous air pollutants (Mukherjee & Agrawal, 2017) with recognized effects on human health, economic development, and sustainable environment of the society (Guan et al., 2016; Luo et al., 2021; Jin et al., 2022). Guidelines have been established to prescribe the acceptable limits of air pollution concentration levels for PM_{10} and $PM_{2.5,}$ that is, 15 and 5 $\mu g/m^{3,}$ individually (WHO, 2021). Additionally, the concentrations of NO_2, CO, and SO_2 pollutants are significantly contributing to the air pollution levels of an urban area. Pollution level tends to rise in city centers or urban localities due to their vicinity to industrial regions and the existence of highly concentrated air pollutants over the urban-rural gradient (Marsh & Grossa, 2002; Manisalidis et al., 2020; Wang et al., 2022). Several other emission sources are led by anthropogenic activities at the local level, such as fossil fuel, biofuel, industrial dust, etc., indicating the degradation of air quality in and around urban areas (Beig et al., 2020; Parde et al., 2020). Air pollutants play a critical role in controlling the chemical processes of the atmosphere and the mixing ratio of greenhouse gases (Beig et al., 2020). Out of the most polluted 100 cities in the world, 39 Indian cities were on the list as per the World Air Quality Report of 2019. Indian cities mostly suffer from the high concentration of pollutants in the form of NO_2, CO, SO_2, and PM, which leads to air pollution. There are many sources of these pollutants, such as transportation, industry, and domestic combustion of coal (Emami et al., 2018; Kamboj et al., 2022). Apart from this, an excessive concentration of these pollutants not only reduces visibility but also causes significant negative impacts on soil and water quality and crop yield (Ji & Ma, 2022).

The resultant variation of the pollutants' concentration differs with varying topographical and meteorological variables at a given location and instance (Bhunia &

Ding, 2020). So, analyzing the spatio-temporal distribution of air pollution has received considerable interest. It is now needed to use a coherent effort to locate the prospective sources of these pollutants in urbanized regions to avoid problems related to environmental air quality. In several studies, the relationship between air pollution and meteorological conditions, mostly focusing on cities and urban centers, is analyzed over different countries of the world (Hu et al., 2017; Shi et al., 2018; Adame et al., 2019; Luo et al., 2021; Chen & Lin, 2022; Yadav et al., 2023). There are evident differences in the distribution of air pollutants over time in different seasons (Bai et al., 2018; Tui et al., 2021). Many countries have been monitoring air quality, commonly known as the air quality index (AQI), using remotely sensed data or ground monitoring stations to improve the policy-related decision-making procedure (Alvarez-Mendoza et al., 2019). It is required to use continuous data for mapping and monitoring of air pollutants, along with establishing the correlation between variables. For this, Geographic Information System (GIS) offers tremendous capabilities in generating continuous surface data through spatial interpolation of scattered point data over large-scale regions (Yadav et al., 2023). The target air pollutants in this work are PM_{10} and $PM_{2.5}$ analyzed over the southeast part of West Bengal, India. The current study aims to analyze the spatial and temporal distribution of PM_{10} and $PM_{2.5}$ on an annual and seasonal level during the two years 2021 and 2022. The findings of this work may provide critical scientific and practical suggestions at the regional level for designing and implementing appropriate intervention policies.

2 Study Area

In this chapter, eight districts are chosen as study sites situated in the southeast region of West Bengal, India. The names of the eight districts are Nadia, Hooghly, Howrah, Kolkata, North 24 and South 24 Parganas, and East and West Midnapore. The district lies between 21°00′ N–24°00′ N latitude and 87°00′ E–89°00′ E longitude, covering an area of about 29449.84 km^2 (Fig. 1). It falls under the tropic of savanna climate, and the temperature ranges between 35 °C and 38 °C during summer, while in winter it varies between 15 °C and 9 °C. The study area has five separate seasons. Spring starts in March and ends in April, summer is from May to June, monsoon begins in mid-June and continues till September, autumn is from October to November, and finally, winter stretches between December and February. This southeastern district of West Bengal is covered with the coastal alluvial region (south), mangroves in the delta region, and densely forested areas.

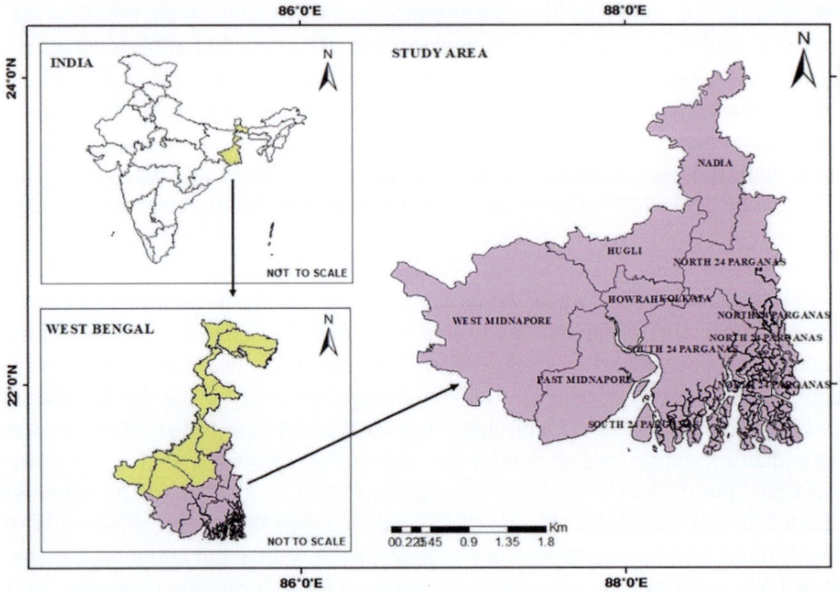

Fig. 1 Location map of the study area

3 Materials and Methodology

The information on $PM_{2.5}$ and PM_{10} over the study area was acquired from the Central Pollution Control Board (CPCB) and the West Bengal Pollution Control Board (WBPCB) during the period 2021–2022. In India, CPCB established a network of ground stations for monitoring the air quality. A total of 16 monitoring stations are located at Krishnanagar, Chinsurah, the University of Kalyani, the Bandel Thermal Power Station, Serampore College, Ghusuri, Rabindra Bharati University, Victoria, Panskura Banamali College, Haldia, Vidyasagar University, Basirhat College, West Bengal State University, Diamond Harbour Women's University, the Gangasagar Development Authority, and Sundarban Hazi Desarat College, covering eight districts of the southeast part of West Bengal. The secondary data of these pollutants were analyzed and processed using ArcGIS (v. 10.8.2) software. The methodology includes (1) generation of ground-based secondary data, (2) spatial interpolation using the inverse distance weighting (IDW) method, and (3) statistical analysis.

3.1 Database Creation

In this study, in situ data of $PM_{2.5}$ and PM_{10} were collected from the CPCB and WBPCB for the years 2021 and 2022. The data were distributed into different seasons, viz., pre-monsoon (March–May), monsoon (June–September), and post-monsoon (October–November) to investigate seasonal variation.

3.2 Spatial Interpolation

The interpolation approach is used to get the values of unknown points by using the known points. The spatial interpolation method basically worked between two estimated points, which were a sample randomly chosen from other points. All these sample points have different values. These values indicate the nearest random points for the interpolation. The spatial interpolation is the process of extracting the new information by using the known points. By using the Geographic Information System (GIS) tool, this process is much easier to estimate the unknown values. The IDW method was applied using the Spatial Analyst Extension of ArcGIS software. It is a relatively simple and broadly used method to create continuous surfaces from point data, particularly when working with spatially distributed data (Jones et al., 2003; Jumaah et al., 2019). For this study, IDW interpolation is used to generate spatially distributed maps of $PM_{2.5}$ and PM_{10} over terrain. In IDW, the values of measured points near unmeasured locations are used to estimate their values. IDW estimates the values of unknown points by assigning weights based on the distance between the unknown point and its neighboring data points (Wong et al., 2004). It applies a weighted linear combination set of sample points to generate surfaces and estimate the likelihoods of unmeasured points (Jumaah et al., 2019; Khouni et al., 2021).

4 Results and Discussion

The IDW interpolation method was used to prepare a spatial distribution map of PM concentration on yearly and season-wise scales through data acquired from monitoring stations.

4.1 Annual Variation of PM Concentration

The variation of $PM_{2.5}$ and PM_{10} concentrations annually for the years 2021 and 2022 is prepared using the IDW method. In the case of $PM_{2.5,}$ the highest and lowest values were found to be 160.25 (μg/m^3) and 7.86 (μg/m^3), respectively, in the year 2021 (Fig. 2a). Nadia and the South 24 Parganas district show the lowest concentration. Kolkata, Howrah, and West Midnapore show the highest concentration. For the year 2022, the highest and lowest values of $PM_{2.5}$ concentration were found to be 64.151 (μg/m^3) and 6.435 (μg/m^3), respectively (Fig. 2b). According to Fig. 2b, the high concentration zones are the districts of West Midnapore, Howrah, the eastern part of Hooghly, and Kolkata. The lowest concentration zones are the districts of East Midnapore and South 24 Pargana.

In the case of $PM_{10,}$ the highest and lowest values were found to be 160.26 (μg/m^3) and 15.07 (μg/m^3), respectively, in the year 2021 (Fig. 3a). The area of Kolkata, East Midnapore, and Howrah shows the highest concentration; West Midnapore shows a moderate concentration, and the Nadia district shows the lowest concentration of PM_{10}. For the year 2022, the high PM_{10} concentration was found to be 158.26 (μg/m^3) covering the area of Nadia and West Midnapore (Fig. 3b). Low PM_{10} concentration was found to be 27.29 (μg/m^3) covering the area of Hooghly, East Midnapore, and South 24 Parganas (Fig. 3b). The red color in the map shows the highest pollutant zone of $PM_{2.5}$ concentration, while the green color shows the lowest concentration zone.

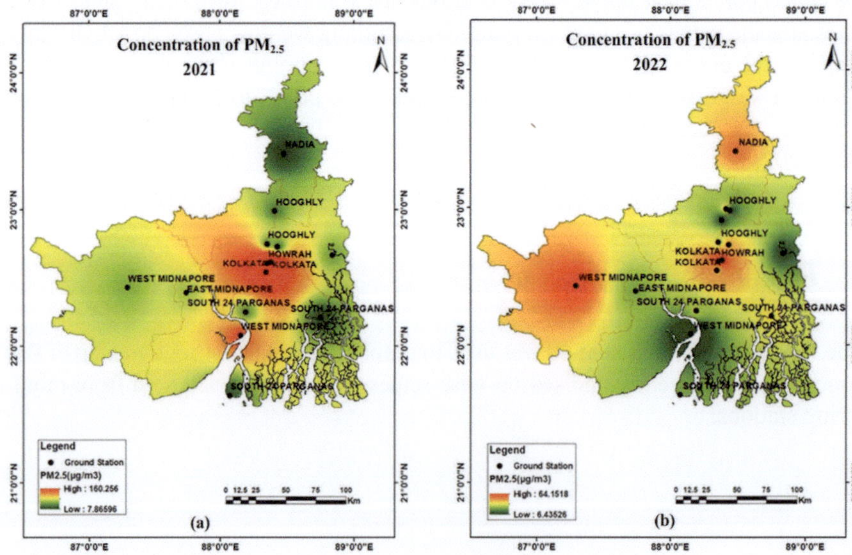

Fig. 2 Annual variation of $PM_{2.5}$ concentration in the year (**a**) 2021, (**b**) 2022

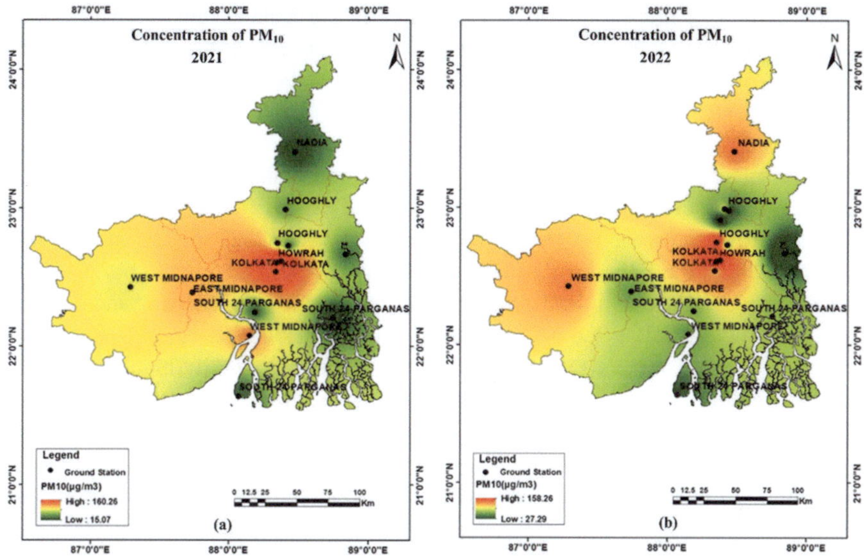

Fig. 3 Annual variation of PM_{10} concentration in the year (**a**) 2021, (**b**) 2022

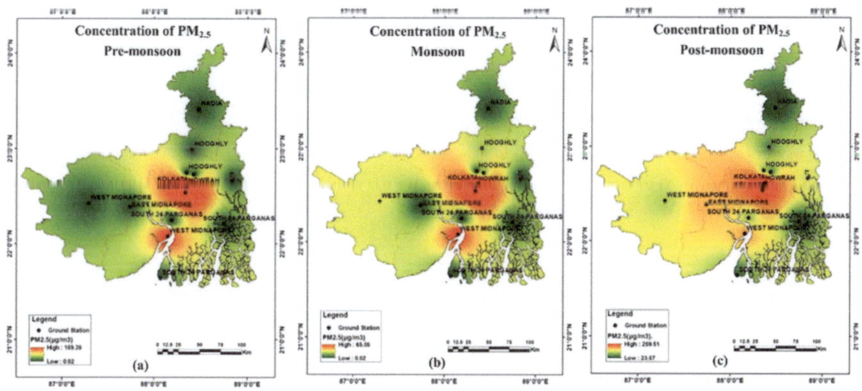

Fig. 4 Seasonal variation of $PM_{2.5}$ concentration in the year 2021

4.2 Seasonal Variation of PM Concentration

The variation of $PM_{2.5}$ and PM_{10} concentration seasonally (pre-monsoon, monsoon, and post-monsoon) for the years 2021 and 2022 is prepared using the IDW method. Seasonal variations of $PM_{2.5}$ in the year 2021 are shown in Fig. 4. In the pre-monsoon season, the highest and lowest values of $PM_{2.5}$ concentration were found to be 169.39 (µg/m^3) and 0.02(µg/m^3), respectively. The districts of West Midnapore, Hooghly, Nadia, and South 24 Parganas are under low $PM_{2.5}$ concentration. The highest $PM_{2.5}$ concentration zone covers the districts of Kolkata and East Midnapore. In the monsoon season map, the highest and lowest concentration range lies between 0.02 μg/m³ and 65.05 μg/m³. The western part of the study area is under a low concentration zone of $PM_{2.5}$ pollutant. Kolkata falls under the highest concentration zone of $PM_{2.5}$ in the monsoon season of the year 2021. In this winter season too, many people came to this city, and the city became rushed. In the post-monsoon season, the $PM_{2.5}$ ranges between 23.57 μg/m³ and 259.51 μg/m³ for the year 2021. Figure 5 shows the seasonal variation of $PM_{2.5}$ in the year 2022. In the pre-monsoon season of 2022, the $PM_{2.5}$ concentration lies between 0.05 μg/m³ and 68.79 μg/m³. The map shows that the highest concentration zone is the Nadia district and the lowest zone is the district of South 24 Parganas. In the monsoon season, the range of the highest and lowest concentrations of $PM_{2.5}$ ranges between 0.03 μg/m³ and 46.45 μg/m³. The spatio-temporal distribution map of $PM_{2.5}$ in the post-monsoon season of 2022 shows the range lies between 19.21 μg/m³ and 115.25 μg/m³, covering the Kolkata, Nadia, and South 24 Parganas districts. The rainfall is the main reason for the pollution range difference between the monsoon and the post-monsoon season because the rainfall easily forces down the air pollution—particulate matter ($PM_{2.5}$ and PM_{10}). The wet deposition process is basically improving the air quality.

The seasonal variation of PM_{10} concentration in the year 2021 is represented in Fig. 6. In the pre-monsoon season, the highest and lowest range of PM_{10} concentration was observed to be 0.02–169.39 μg/m³. The monsoon season shows the range between 0.02 μg/m³ and 65.05 μg/m³. In the post-monsoon season of 2021, PM_{10} concentration ranged from 0.06 μg/m³ to 259.54 μg/m³. The seasonal maps of PM_{10} concentration in the West Midnapore and East Midnapore districts show high variation because of the existence of iron and steel industries. These industrial sites cause more pollution in the city regions.

Figure 7 represents the seasonal variation of PM_{10} concentration in the year 2022. The PM_{10} concentration was found to be 0.75 μg/m³ (lowest) and 177.99 μg/

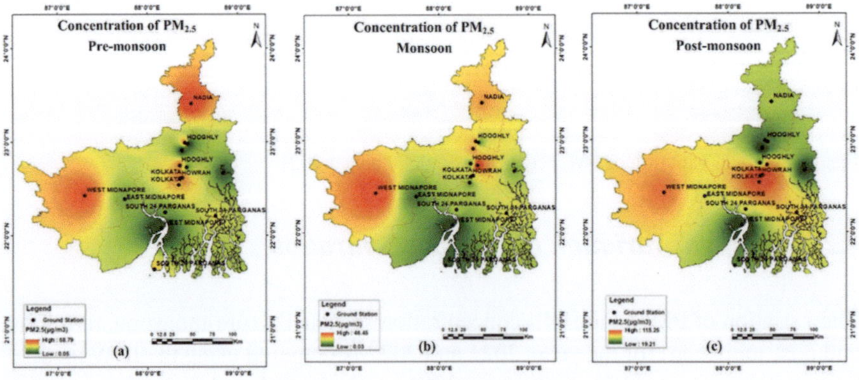

Fig. 5 Seasonal variation of $PM_{2.5}$ concentration in the year 2022

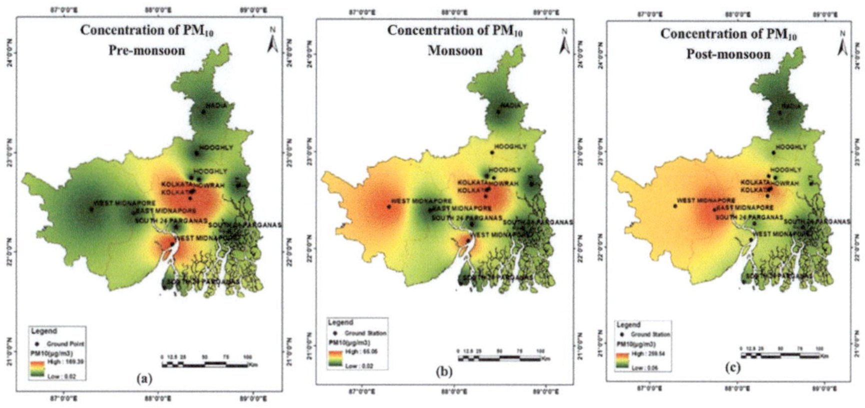

Fig. 6 Seasonal variation of PM_{10} concentration in the year 2021

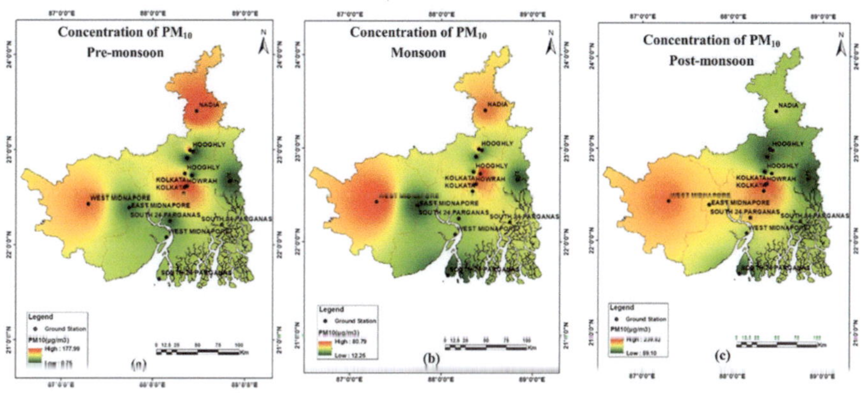

Fig. 7 Seasonal variation of PM_{10} concentration in the year 2022

m^3 (highest) in the pre-monsoon season of the year 2022. The monsoon season is showing a range between 12.25 μg/m^3 and 80.79 μg/m^3. In the year 2022, the post-monsoon season revealed PM_{10} concentration ranges of 69.10–239.92 μg/m^3. In the seasonal maps of the year 2022, Nadia district is showing the largest variation in PM_{10} concentration.

The monthly concentration of $PM_{2.5}$ and PM_{10} in the year 2022 is shown in Fig. 8. It is showing variation in the concentration in different seasons, like pre-monsoon (March–May), monsoon (June–September), and post-monsoon (October–November), over diverse stations located in the study area.

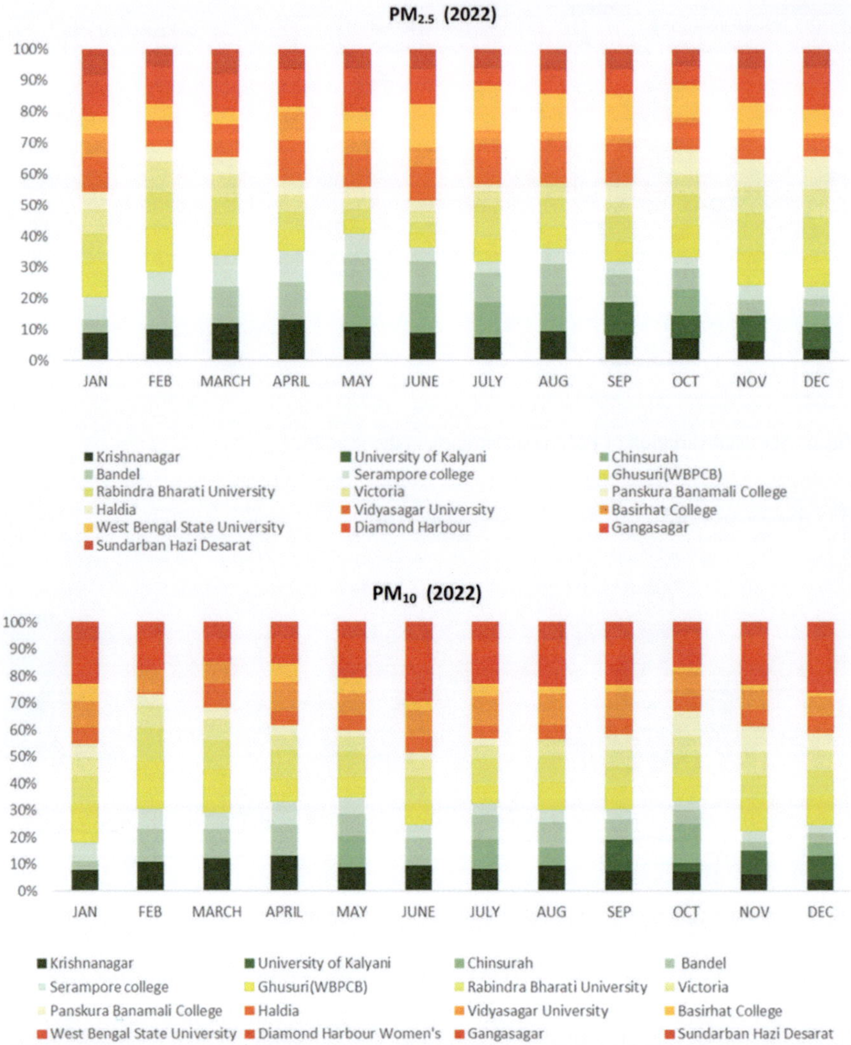

Fig. 8 Monthly station-wise concentration of $PM_{2.5}$ and PM_{10} in the year 2022

5 Conclusion

Air pollution is now considered a serious environmental issue. In this study, variation of $PM_{2.5}$ and PM10 concentration annually and seasonally was analyzed over the southeast part of West Bengal. An IDW-based interpolation method was used to map the spatial distribution of PM in the years 2021 and 2022. Spatial distribution of PM concentrations shows high variability on both annual and seasonal scales in a better way. The majority of areas, including Kolkata, Howrah, and the East Midnapore region, were observed as zones with high concentrations of PM, followed by poor air quality, and the district of South 24 Parganas is under a less concentrated zone. The outputs of this study would support identifying hotspot zones located across the study area, and accordingly, mitigation measures can be planned.

References

Adame, J. A., Notario, A., Cuevas, C. A., Lozano, A., Yela, M., & Saiz-Lopez, A. (2019). Recent increase in NO2 levels in the southeast of the Iberian Peninsula. *Science of the Total Environment, 693*, 133587.

Alvarez-Mendoza CI, Teodoro AC, Torres N, Vivanco V. (2019). Assessment of remote sensing data to model PM10 estimation in cities with a low number of air quality stations: A case of study in Quito, Ecuador. *Environment, 6*, 85.

Baca-López, K., Fresno, C., Espinal-Enríquez, J., Martínez-García, M., Camacho-López, M. A., Flores-Merino, M. V., & Hernández-Lemus, E. (2021). Spatio-temporal representativeness of air quality monitoring stations in Mexico City: Implications for public health. *Frontiers in Public Health, 8*, 536174.

Bai, J., de Leeuw, G., et al. (2018). Variations and photochemical transformations of atmospheric constituents in North China. *Atmospheric Environment, 189*, 213–226.

Beig, G., Sahu, S. K., Singh, V., Tikle, S., Sobhana, S. B., Gargeva, P., et al. (2020). Objective evaluation of stubble emission of North India and quantifying its impact on air quality of Delhi. *Science of the Total Environment, 709*, 136126.

Bhunia, G. S., & Ding, D. (2020). Temporal and spatial statistical analysis of ambient air quality of Assam (India). *Journal of the Air & Waste Management Association, 70*(8), 775–794.

Chen, P.-C., & Lin, Y.-T. (2022). Exposure assessment of PM2.5 using smart spatial interpolation on regulatory air quality stations with clustering of densely-deployed microsensors. *Environmental Pollution, 292*, 118401.

Chen, Y., Wild, O., Conibear, L., Ran, L., He, J., Wang, L., & Wang, Y. (2020). Local characteristics of and exposure to fine particulate matter (PM2.5) in four indian megacities. *Atmospheric Environment, X 5*, 100052.

Emami, F., Masiol, M., & Hopke, P. K. (2018). Air pollution at Rochester, NY: Long-term trends and multivariate analysis of upwind SO2 source impacts. *Science of the Total Environment, 612*, 1506–1515.

Guan, W. J., Zheng, X. Y., Chung, K. F., & Zhong, N. S. (2016). Impact of air pollution on the burden of chronic respiratory diseases in China: Time for urgent action. *Lancet, 388*, 1939–1951.

Hu, D., Wu, J., Tian, K., Liao, L., Xu, M., & Du, Y. (2017). Urban air quality, meteorology and traffic linkages: Evidence from a sixteen-day particulate matter pollution event in December 2015, Beijing. *Journal of Environmental Sciences, 59*, 30–38.

Ji, S., & Ma, S. (2022). The effects of industrial pollution on ecosystem service value: A case study in a heavy industrial area, China. *Environment, Development and Sustainability, 24*, 6804–6833.

Jin, Z., Velásquez Angel, M. A., Mura, I., & Franco, J. F. (2022). Enriched spatial analysis of air pollution: Application to the city of Bogotá, Colombia. *Frontiers in Environmental Science, 10*, 966560.

Jones, N. L., Davis, R. J., & Sabbah, W. (2003). A comparison of three-dimensional interpolation techniques for plume characterization. *Groundwater, 41*, 411–419.

Jumaah, H. J., Ameen, M. H., Kalantar, B., Rizeei, H. M., & Jumaah, S. J. (2019). Air quality index prediction using IDW geostatistical technique and OLS-based GIS technique in Kuala Lumpur, Malaysia. *Geomatics, Natural Hazards and Risk, 10*(1), 2185–2199.

Kamboj, K., Sisodiya, S., Mathur, A. K., et al. (2022). Assessment and spatial distribution mapping of criteria pollutants. *Water, Air, and Soil Pollution, 233*, 82.

Kaur, R., & Pandey, P. (2021). Air pollution, climate change, and human health in Indian cities: A brief review. *Frontiers in Sustainable Cities, 3*, 705131.

Khouni, I., Louhichi, G., & Ghrabi, A. (2021). Use of GIS based inverse distance weighted interpolation to assess surface water quality: Case of Wadi El Bey, Tunisia. *Environmental Technology & Innovation, 24*, 101892.

Lelieveld, J., Evans, J. S., Fnais, M., Giannadaki, D., & Pozzer, A. (2015). The contribution of outdoor air pollution sources to premature mortality on a global scale. *Nature, 525*, 367.

Luo, H., Han, Y., Cheng, X., Lu, C., & Wu, Y. (2021). Spatiotemporal variations in particulate matter and air quality over China: National, regional and urban scales. *Atmosphere, 12*, 43.

Manisalidis, I., Stavropoulou, E., Stavropoulos, A., & Bezirtzoglou, E. (2020). Environmental and health impacts of air pollution: A review. *Frontiers in Public Health, 8*, 14.

Marsh, W. M., & Grossa, J. M. J. (2002). *Environmental geography: Science, land use, and earth systems* (2nd ed.). Wiley.

Mukherjee, A., Agrawal, M. (2017). A Global Perspective of Fine Particulate Matter Pollution and Its Health Effects. In: de Voogt, P. (eds) Reviews of Environmental Contamination and Toxicology Volume 244. Reviews of Environmental Contamination and Toxicology, vol 244. Springer, Cham.

Parde, A. N., Ghude, S. D., Pithani, P., Dhangar, N. G., Nivdange, S., Krishna, G., Lal, D., Jenamani, R., Singh, P., Jena, C., Karumuri, R., Safai, P., & Chate, D. (2020). Estimation of surface particulate matter (PM2.5 and PM10) mass concentrations from ceilometer backscattered profiles. *Aerosol and Air Quality Research, 20*, 1640–1650.

Shi, C., Yuan, R., Wu, B., Meng, Y., Zhang, H., Zhang, H., & Gong, Z. (2018). Meteorological conditions conducive to PM2.5 pollution in winter 2016/2017 in the western Yangtze River Delta, China. *Science of the Total Environment, 642*, 1221–1232.

Tui, Y., Qiu, J., Wang, J., & Fang, C. (2021). Analysis of Spatio-temporal variation characteristics of main air pollutants in Shijiazhuang City. *Sustainability, 13*, 941.

Wang, Z., Shi, X., Pan, C., & Wang, S. (2021). Spatial and temporal characteristics of environmental air quality and its relationship with seasonal climatic conditions in Eastern China during 2015–2018. *International Journal of Environmental Research and Public Health, 18*, 4524.

Wang, L., Niu, D., Fan, H., & Long, X. (2022). Urban configuration and PM2.5 concentrations: Evidence from 330 Chinese cities. *Environment International, 161*, 107129.

Wong, D. W., Yuan, L., & Perlin, S. A. (2004). Comparison of spatial interpolation methods for the estimation of air quality data. *Journal of Exposure Science & Environmental Epidemiology, 14*, 404–415.

World Health Organization. (2021). *WHO global air quality guidelines: Particulate matter (PM2.5 and PM10), ozone, nitrogen dioxide, sulfur dioxide and carbon monoxide*. Available at: https://www.who.int/publications/i/item/9789240034228

Yadav, N. K., Mitra, S. S., Santra, A., et al. (2023). Understanding Responses of Atmospheric Pollution and its Variability to Contradicting Nexus of Urbanization–Industrial Emission Control in Haldia, an Industrial City of West Bengal. *Journal of the Indian Society of Remote Sensing, 51*, 625–646.

Zhang, Y. L. (2019). Dynamic effect analysis of meteorological conditions on air pollution: A case study from Beijing. *Science of the Total Environment, 684*, 178–185.

Shallow Groundwater Dynamics of an Urban Coastal Aquifer

Ande Bhuvaneshwari Devi, T. D. Aneesh, Reji Srinivas, and Archana M. Nair

Abstract We investigated the short-term changes in the shallow water table for the highly urbanized Cochin city of Kerala, India. The monthly water table observed at 97 Well locations for 16 months was used to generate a monthly water table fluctuation map and a seasonal recharge map. The majority of the coastal plains of the study area shows low recharge depths, while the eastern midland part shows moderate to high recharge. The quantification of the recharge affirms the proportionality of the recharge depths to precipitation. Seasonal recharge maps illustrated a significant recharge during the southwest monsoon period. Further, the monthly groundwater discharge flux rates using Darcy's tool suggest a high discharge from east to west. The mapped seepage velocity signifies a higher seepage rate from the midland portion of the study area. The spatial distribution patterns of total dissolved solids (TDS), electrical conductivity (EC), and pH concentrations display flush-out patterns that connect the freshwater availability to recharge from rainfall. Quality parameters are better in the midland region, indicating higher recharge in the midland than in the coastal lowland region and discharge from the midland to the lowland. Our study summarizes that the dynamics of the aquifer system in the midland region are more active due to laterite formation and steeper hydraulic gradients. The majority of enriched groundwater from the midland was draining out immediately after the monsoon period, which leads to a severe water crisis during the non-monsoon period. The study emphasizes the need to implement appropriate aquifer management based on the prevailing aquifer dynamics for sustainable groundwater resources.

Keywords Coastal aquifer · Cochin · Recharge · Water level fluctuation · Seepage velocity

---

A. B. Devi · A. M. Nair (✉)
Earth System Science and Engineering Group, Department of Civil Engineering, Indian Institute of Technology, Guwahati, India
e-mail: nair.archana@iitg.ac.in

T. D. Aneesh · R. Srinivas
National Centre for Earth Science Studies, Thiruvananthapuram, Kerala, India

P. K. Joshi et al. (eds.), *Geospatial Science for Urban Ecosystems*, The Urban Book Series, https://doi.org/10.1007/978-3-032-07155-2_12

1 Introduction

Coastal aquifer systems form an essential part of the global water resources. Approximately 70% of the world's population resides in the coastal regions. Millions of people across the globe are dependent on high-quality coastal aquifer systems for potable groundwater (Costall et al., 2020). Although coastal areas are the most developed areas in the world, social and economic development incorporates several hydrogeological issues with serious environmental hazards, which result in the risk of water level decline, contamination, and seawater intrusion (Lamichhane & Shakya, 2019; Liu et al., 2012; Mao et al., 2020; Telahigue et al., 2020; Zang & Li, 2021). The fragility and vulnerability to human activity render the necessity of an informed, professional management to ensure the survival of coastal aquifers. The complex and dynamic interface between the land and the sea enhances the challenges of a hydrogeologist in the evaluation of groundwater flow systems in human-impacted coastal aquifers (Baron & Williams, 2000). Therefore, a precise knowledge about the dynamics of the coastal aquifer system is crucial for sustainable groundwater resource management and planning (Keesari et al., 2014; Lathashri & Mahesha, 2016; Tabatabaei & Khozeymehnezhad, 2018; Werner et al., 2011).

A global assessment on groundwater consumption shows that countries such as India, China, Nepal, Bangladesh, and Pakistan are using nearly half the world's total groundwater resources (WWRD, 2015). Consequently, the conservation and management of groundwater resources are getting more attention in these regions. At present, a number of coastal regions in India are exposed to frequent adverse effects mainly due to groundwater exhaustion, such as the decrease in the long-term groundwater levels, reduction of aquifer areas, increased energy usage to raise water from increasingly deeper levels, and quality degradation due to saline water intrusion (Jha & Sinha, 2010). Various researchers have conducted studies on the hydrodynamics of coastal zones in different parts of India, like Southwest Bengal (Das & Pardeshi, 2018), Puri (Vijay & Mohapatra, 2016), Tamil Nadu (Singaraja et al., 2016), coastal Karnataka (Lathashri & Mahesha, 2016), and western Maharashtra (Keesari et al., 2014), addressing saline water intrusion and water level decline. Therefore, in this chapter, we illustrate the analysis carried out on the groundwater dynamics of Cochin, a coastal urban city in the southern part of the west coast of India.

Groundwater recharge is defined as the entry of water into the saturated zone made available at the water table surface, together with the associated flow away from the water table within the saturated zone (Healy & Cook, 2002). Transfer of water and energy between the lower atmosphere and the subsurface is a crucial process linked to the hydrological cycle. The quantification of feedback between the atmospheric boundary layer, the land surface, and the subsurface depends on characterizing this key process control mechanism (Kollet & Maxwell, 2008). Urbanization changes the topography and natural vegetation due to settlements resulting from the transformation of barren, fallow, agricultural, and forest cover. The urbanization process produces a different land surface process mechanism, creating disturbances in the ongoing hydrological process (Li et al., 2018; Owuor et al., 2016). This affects the surface and subsurface hydrological process of a region, displayed as increased stream flows and flooding characteristics (Fang et al., 2020; Oudin et al., 2018), a decrease in evapotranspiration, and deterioration in the water quality of surface streams and groundwater (Sharp, 2010; Hua, 2017; McGrane, 2016; Patra et al., 2018).

Groundwater recharge is a complex process affected by geomorphology, LULC, aquifer type, and climate. The recharge rate of an area is usually controlled by the climatic conditions, soil type, topography, and geological framework (Healy & Cook, 2002; Scanlon et al., 2002). The groundwater recharge process can be of two types: focused or diffused. In the process of diffuse recharge, a part of the water due to precipitation will be infiltrating through the unsaturated zone, resulting in percolation of an enormous amount of water from the land surface to the water table. Focused recharge happens in regions that are covered by water bodies. It is a known fact that climate has greater control on the rate of recharge in arid regions with high topographic relief, whereas, in humid climatic areas with low topographic relief, it is the geological framework that controls the rate of recharge (Sanford, 2002). However, the effect of land use and land cover changes on groundwater recharge, especially in diffuse recharge, is difficult to perceive. On the other hand, any change in the focused recharge is always noticeable, as it will directly impact the alterations in the existing water bodies. Consequently, in this study, we investigated the short-term diffuse recharge patterns of a highly urbanized coastal unconfined aquifer. Thus, this study focuses on recharge variability in an unconfined shallow aquifer with a humid tropical climate.

This chapter investigates the groundwater dynamics of the shallow unconfined aquifer of Cochin. Cochin belongs to the Southern State of Kerala in India on the west coast of the Arabian Sea. The state of Kerala, with about 590 km of coastal stretch, is blessed with ample water supplies, which comprise a network of backwaters, rivers, lakes, ponds, springs, and wells. Additionally, the state receives substantial rainfall throughout the year. However, the topography and soil characteristics of the area, merged with rapid urbanization, cause considerable stress on its groundwater levels (Jayasankar & Energy, 2017). The highest increase in rate of urban growth of 92.8 percent was recorded during 2001–2011 for the study area (Praveen Lal & Nair, 2017). The groundwater system plays a crucial role in the water supply to coastal communities and local wetland vegetation (Sreekesh et al., 2018). In the recent past, the study area has undergone extensive landscape modification due to urbanization and changed cropping patterns, leading to a decline in water level and saline water intrusion (Bhosale & Kumar, 2002; Pramada et al., 2018). However, very limited studies investigate the recharge conditions of the groundwater present in both shallow and deeper horizons, which is essential to analyze the degradation and strategize for the remedial measures to limit further damage (Keesari et al., 2014). With this motivation, we have conducted this study to investigate the short-term changes in the shallow water table in qualitative and quantitative aspects. Further, the in situ water quality parameters were analyzed for seasonal changes, whereas the spatial variability of the groundwater recharge is examined with field observation and followed by the estimation of groundwater discharge flux, seepage velocities, and direction with Darcy's law.

Numerous methods are available to quantify and monitor groundwater recharge that can be categorized as physical methods, numerical methods, water budget method, and the aquifer response method that includes both chemical tracer as well as analytical methods such as isotope techniques and chloride mass balance (Barron et al., 2013; Crosbie et al., 2013; Dams et al., 2008; Keese et al., 2005; Paul, 2006; Scanlon et al., 2002). However, the most widely used method for quantification of groundwater recharge is the water table fluctuation (WTF) method (Varni et al., 2013). The water table fluctuation method in groundwater quantification started in the early 1920s and is still reliable. Several research studies showed the use of the WTF method and proved the suitability of the WTF method for shallow unconfined aquifers (Ahmadi et al., 2012; Crosbie et al., 2013; da Silva et al., 2012). The water table fluctuation method can clearly pick the changes associated with spatial and temporal scale, unlike tracer techniques, representing the recharge variations associated with spatial scale but not in time scale (Scanlon et al., 2002). Hence, the WTF method was used to examine the spatio-temporal changes in groundwater recharge in this study. For this purpose, an analysis of the monthly observed water table fluctuations for 16 months was carried out to estimate and capture the spatial variability in the recharge rate. This will help analyze the controlling factors on recharge rate, which is an essential aspect of sustainable water management studies, especially under a climate change scenario.

2 Study Area

The study area has an expanse of 910 km^2 with a coastline of 46 km lying between N10° 15′ 21.51″ and N9° 47′ 14.52″ and E 76° 9′ 59.19″ and E 76° 28′ 39.28″ (Fig. 1). The Arabian Sea envelops the Cochin city all along the western boundary. The area belongs to the Periyar watershed toward the north and the Muvatupuzha watershed toward the south. Kottayam and Allapuzha districts surround the south, Idukki to the east, and Thrissur to the north. The region is highly urbanized and is the most populated part of Kerala. The city is rapidly growing northward with the announcement of major industrial projects in the last few years. Therefore, the study area is demarcated considering the urban sprawl that starts from the city center, expanding northward. A unique characteristic of the region is related to the transforming morphology within a narrow stretch. The topographic relief of the region ranges from very low to high, with a long coastal stretch toward its western part and a midland and highland linear stretch toward the central and eastern parts. The study area is confined to the lowland and midland regions. The elevation varies from <2 m amsl in the coastal region to 160 m amsl in the midland (Fig. 1a).

The study area is a part of a humid tropical zone experiencing a normal annual rainfall of 336 cm. The seasonal variability in rainfall can be accounted for by

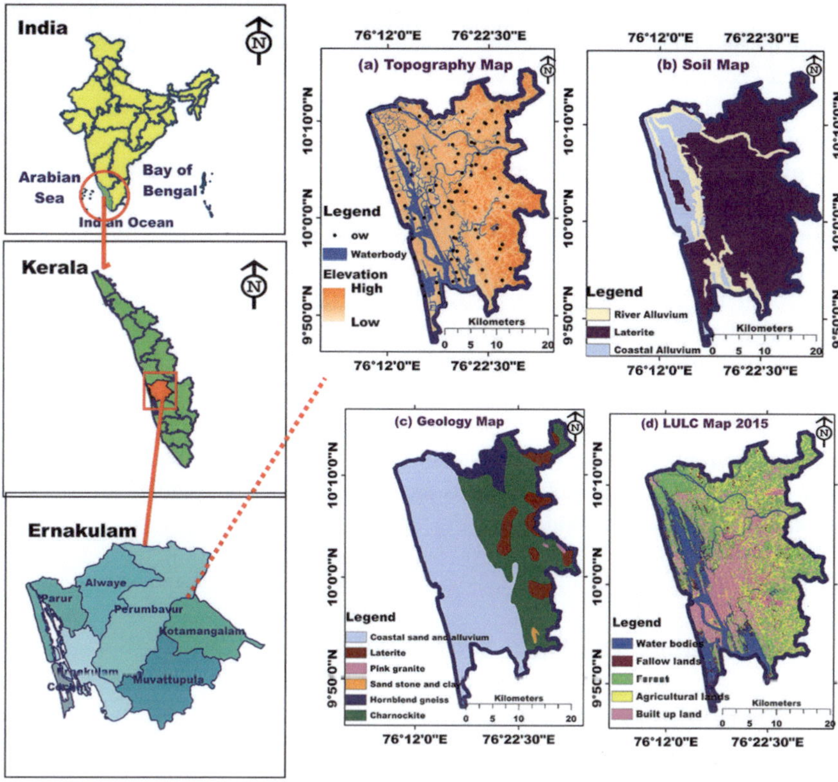

Fig. 1 Location map of the study area: (**a**) topography, (**b**) soil map, (**c**) geology map, and (**d**) LULC map of 2015

segregating the region's climate into four seasons: pre-monsoon (March to May), southwest monsoon period (June to August), northeast monsoon period (September to November), and post-monsoon (December to February). The monsoon periods contribute approximately 84% of the annual rainfall. The average minimum and maximum temperatures experienced in the region are approximately 23 and 31 °C, respectively. The monthly potential evapotranspiration (PET) varies from 0.095 to 1.56 m, and relative humidity ranges from 64% to 87% (Aneesh et al., 2019). The PET is less than the rainfall from May to November, representing the excess of water for recharge into the groundwater system (CGWB, 2013). Despite these favorable conditions, the region suffers extreme water shortage problems, which commence post-monsoon and are aggravated pre-monsoon with deficient summer showers.

Geologically, the study area consists of two major litho-units comprising Precambrian metamorphosed rocks in the eastern part and unconsolidated coastal alluvium in the western part (Fig. 1c). The region consists of sand, silt, and clay of the lagoon; backwater deposits; and shore deposits with alluvial formations along the coastal belt, laterites, charnockites, and charnockite gneiss rocks along the eastern midland (Fig. 1c). The porosity of the sand was observed to be 0.1–0.2 along this area (Pramada et al., 2018), while the laterite formations show a porosity of 0.27 to 0.3 and 30 m day^{-1} hydraulic conductivity (Bhosale & Kumar, 2002). Since groundwater exists under phreatic conditions in the coastal sediments, the weathered and fractured zone of the crystalline rocks from the aquifers is along the midland region (Sreekesh et al., 2018).

Based on morphological features and physicochemical properties, the major soil groups in the study area are lateritic, riverine alluvium, and coastal alluvium (Fig. 1b). The lateritic soil is the most predominant soil occupied by the region toward the central part, while coastal alluvium occupies the coastal stretch, and river alluvium is restricted to the banks of rivers and their tributaries.

Six major LULC features were evident in the region through analysis during the study period in 2015. About 29.29% is covered with built lands, forests cover 30.8%, agricultural lands cover 19%, fallow lands cover 6.2%, and water bodies cover 14.6%. Most of the built-up lands are from the lowland regions; the midland is more forest and agricultural lands (Fig. 1d).

3 Methodology and Data Acquisition

3.1 Groundwater Level Monitoring

The study region experiences monsoon rain for a considerable period from June to November. The monthly rainfall data was acquired from the regional meteorological department for 2014–2016. The data indicate variations in the precipitation pattern during the monsoon seasons. The monthly rainfall depth ranged from 585 to 1.65 mm during the concerned period of study, and a decrement was observed in 2015 compared to 2014 (Fig. 2). We have identified 97 spatially distributed existing dug wells (Fig. 1a) to characterize the water table fluctuation over the study area. The water level measurements were recorded manually in the first week of every month for 16 months, ranging from September 2014 to February 2016.

3.2 Natural Recharge Estimation Using Water Table Fluctuation Method (WTF)

The WTF method was utilized to estimate the groundwater recharge based on the groundwater budget principle (Healy & Cook, 2002). This method is quite common in coastal aquifer regions where shallow water tables display a steep rise in water levels and a decrease in dynamics within a short span (Tashie et al., 2016; Fang

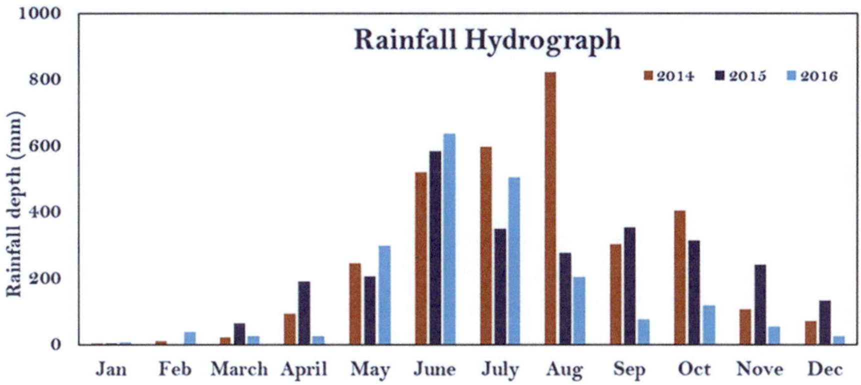

Fig. 2 Rainfall hydrograph of the study area during 2014 and 2015

et al., 2020). The method is based on the groundwater budget principle. The water balance equation of an aquifer can be given as:

$$\Delta S_{gw} = R - Q_{bf} - ET_{gw} - Q_{gw}^{off} + Q_{gw}^{on} \tag{1}$$

Here, ΔS_{gw}= change in groundwater storage = Aquifer recharge rate, Q_{bf}, is base flux,

ET_{gw} is evapotranspiration from the aquifer, and $Q_{gw}^{off}, Q_{gw}^{on}$ are aquifer inflow and outflow components of the aquifer included with pumping.

Since the model is based on the hypothesis that the rise in the groundwater table of an unconfined aquifer is due to the recharge, the rise in water level "Δh" is related to the recharge calculation as follows:

$$R = \frac{\Delta h}{\Delta t} S_y \tag{2}$$

Where R = Aquifer recharge rate, (LT^{-1}), Δh = change in water table height over time intervalΔt , S_y = specific yield, and Δt = time variation as considered in monthly time steps,

In this study, "R," the estimated recharge, measured in millimeters (mm) per month, is calculated using specific yield and water level fluctuation. The parameters were measured directly from the field. For an accurate assessment, the monthly time step is considered for the measurement of the differential water heads. The specific yield is purely dependent on the physical properties of aquifer material and is defined as the proportion of the volume of water in the aquifer to its own volume after saturation (Todd & Mays, 2004). Several methods, such as in situ field tests, laboratory tests (column drainage approach), and indirect methods based on textural characteristics of soil sediments (Varni et al., 2013), are utilized for the measurements. Based on the soil group in which the observation wells exist, the study area is divided into two groups: the Kalamasseri group of wells and the Kizhakkambalam group of wells. A pumping test was conducted at two locations, Kizhakkambalam and Kalamasseri, for 24 hours, representing these two clusters. From the pumping test data analysis, the resultant specific yield obtained at Kalamasseri was 0.29, and at Kizhakkambalam, it was 0.2. These values were assigned to the respective clusters. The monthly recharge depth was calculated using Eq. (2) at each location using a corresponding monthly water level fluctuation and specific yield. Based on the observed seasonal changes in rainfall pattern, the entire year was categorized into four seasons: pre-monsoon, monsoon (I), monsoon (II), and post-monsoon. The pre-monsoon period ranges from March to May; monsoon (I) is from June to August and is the primary rainfall season due to the southwest monsoon; monsoon (II) ranges from September to November, coinciding with the northeast monsoon; and post-monsoon is from December to February. In the entire 16 months of the study period, we had six seasons varying from post-monsoon 2014 to post-monsoon 2015, for which groundwater recharge was calculated.

3.3 *Geochemical Analysis*

For the deduction of the water quality parameters, a hydrochemical survey was carried out using a handheld instrument model no. Eutech Cyberscan PCD 650 at 43 well locations targeting the phreatic aquifer. Analysis of the field water quality using electrical conductivity (EC), total dissolved solids (TDS), and pH shows the effect of seawater on groundwater quality (Shin et al., 2020). Therefore, we measured these parameters for two seasons at the 43 locations during post-monsoon (December 2015) and pre-monsoon (May 2016). The spatial distribution of the various water quality parameters was prepared using the Inverse Distance Weighted (IDW) interpolation method in the ArcGIS platform.

3.4 *Modeling of Subsurface Flow Through Darcy's Velocity Tool in ArcGIS*

The conventional approach using Darcy's method was used to calculate the groundwater flow of the study area. Seepage velocity is the groundwater flowing velocity within the voids of the aquifer media and is less than Darcy's groundwater flux. Darcy's tool in the ArcGIS platform works on the principle of flow through the porous media, depending on the aquifers' hydraulic head gradient and flow properties. This tool provides the direction and magnitude of the seepage velocity. According to Darcy's law, the groundwater discharge flux or Darcy's velocity, is given by Eq. (3)

$$q = v_d = -K\frac{dh}{dl} = -Ki \tag{3}$$

$$i = \frac{dh}{dl} \tag{4}$$

$$K = \frac{T}{B} \tag{5}$$

The seepage velocity (v_s) represents the actual velocity of groundwater in aquifer media (Alabi et al., 2019), can be calculated by dividing Darcy's velocity by porosity (n) of the aquifer media, as represented in Eq. (6)

$$v_s = \frac{v_d}{n} \tag{6}$$

From Eqs. (2) and (4), seepage velocity can be rewritten as

$$v_s = -\frac{T}{Bn}\frac{dh}{dl} \tag{7}$$

where q is groundwater flux (m day^{-1}), v_d is Darcy's velocity, i ($i = \frac{dh}{dl}$) the hydraulic gradient, which signifies the change in hydraulic head per unit change in the length, K (m day^{-1}) is hydraulic conductivity, T (m^2 day^{-1}) is the transmissivity, and B(m) is the saturated thickness of the aquifer media.

Spatial variability of the seepage velocity of the aquifer system was mapped in ArcGIS using the modeling tool based on Eq. (7). Four major input parameters for seepage velocity analysis are groundwater elevation head, porosity, transmissivity, and saturated thickness of the study area. The data sets from the field observation of the wells consist of the location coordinates and static water heads (m bgl), representing the surface water level from the ground surface. As per the formation properties of the aquifer systems, the porosity of various formations ranges from 0.4 to 0.5 (sand), 0.27 to 0.3 (laterite), and 0.15 to 0.2 (Quilon beds) in the study area (Bhosale & Kumar, 2002). The saturation thickness (m) details of the study area were collected from reports of the Geological Survey of India, whereas the transmissivity (m^2 day^{-1}) values were collected at various locations from the reports of the Central Groundwater Board (CGWB, 2013). The Inverse Distance Weighting (IDW) interpolation technique was utilized to generate the porosity map, saturation thickness map, and transmissivity maps. The IDW method was used for the water level data of the well locations along with the static water level to generate water level fluctuation maps. The spatial variations of these parameters are illustrated in Fig. 3a–d. The groundwater elevation head map was produced using the digital

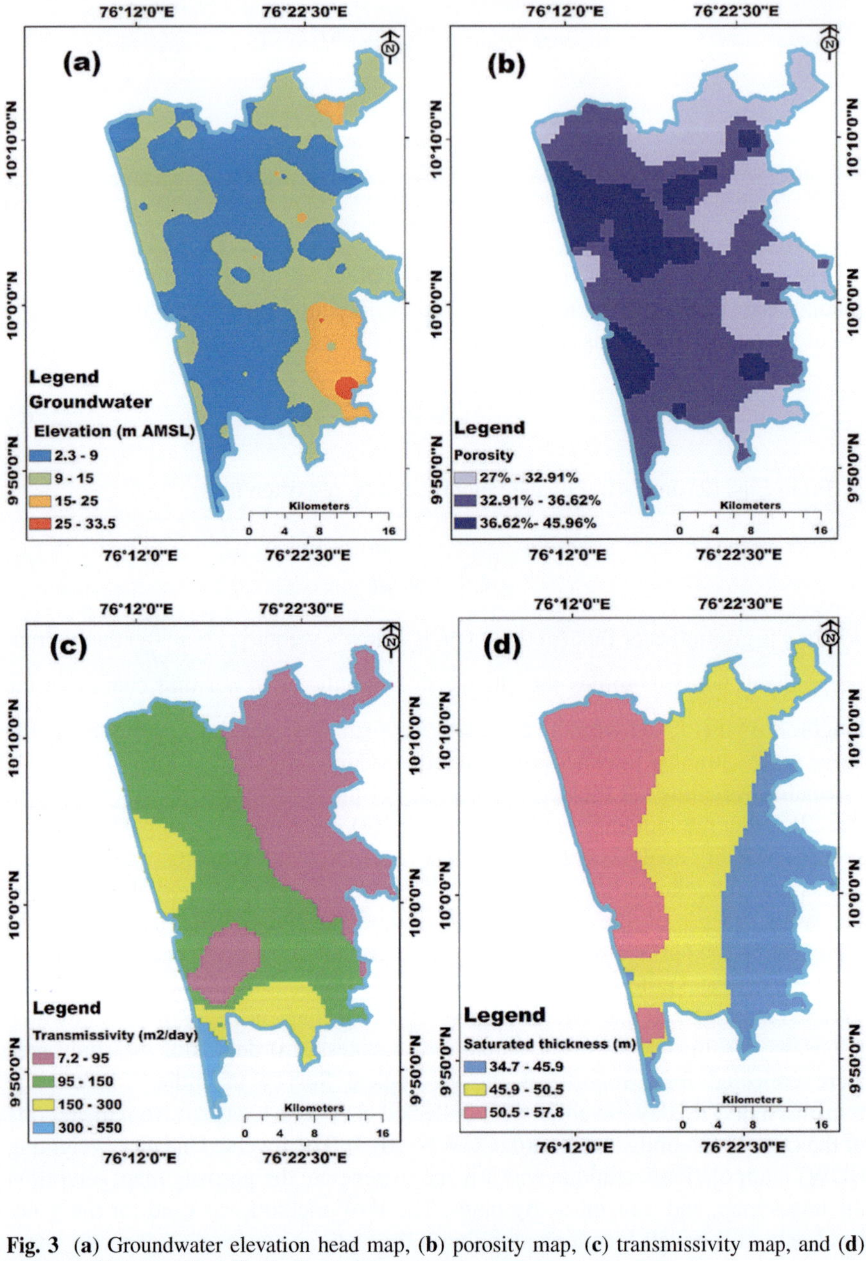

Fig. 3 (**a**) Groundwater elevation head map, (**b**) porosity map, (**c**) transmissivity map, and (**d**) saturated thickness of aquifer in the study area

elevation model and static water level map. The raster format of the input data sets was utilized in the ArcGIS platform to generate the magnitude and direction of seepage velocity maps. Figure 3a demonstrates the spatial distribution of the groundwater elevation map. Evaluation of the groundwater elevation heads relative to the mean sea level (MSL) shows the range between 33.5 m amsl to 2.3 m amsl in the study area, respectively. The porosity values vary between 45.96% and 27%, with the higher values along the western coastal alluvium and the lower values on the eastern midland region (Fig. 3b). The spatial distribution of transmissivity (Fig. 3c) depicts a minimum value of 7.2 m^2 day^{-1} and a maximum of 550 m^2 day^{-1}. The saturated thickness ranges from 34.7 to 57.8 m, with the maximum value along the lowland and the minimum along the midland area.

In addition to seepage velocity calculation, the monthly groundwater discharge flux, or Darcy's velocity, at three different sectional profiles from the north to south of the study area was derived. The sectional profile is named 1-1′ at the northern part, 2-2′ at the central region, and 3-3′ at the southern region of the study area. Groundwater discharge flux was estimated using Eq. (3) with hydraulic conductivity and gradient as input. For measuring the hydraulic gradient, adjacent well locations, water head elevation, and distance between corresponding well locations were considered. The approach adopted is explained in brief in Fig. 4.

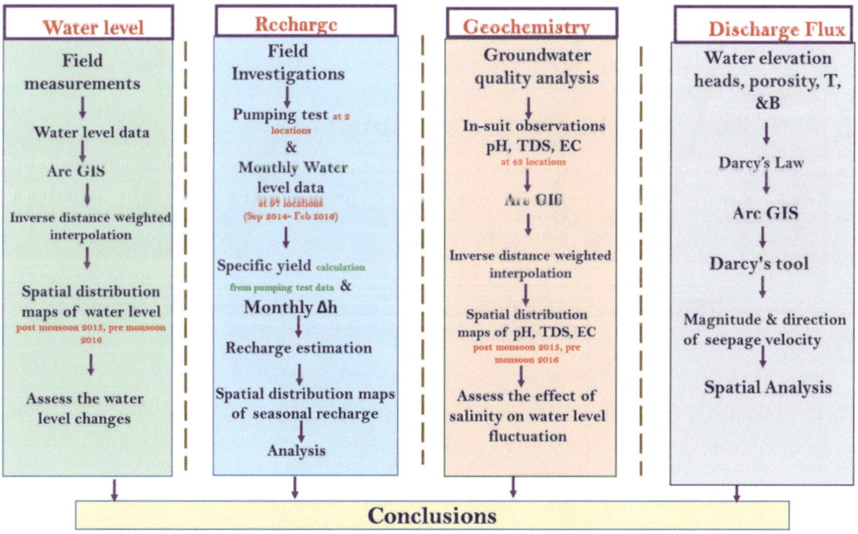

Fig. 4 Flow chart of the methodology

4 Results and Discussions

4.1 Water Level Changes

The water level for the pre-monsoon and post-monsoon of 2015 was monitored, and spatial distribution was mapped as illustrated in Figs. 5a, b. The groundwater levels in the pre-monsoon of the year 2015 range between 0.26 and 15.43 m bgl, and post-monsoon of 2015 between 1.13 and 14.13 m bgl. Therefore, the water level fluctuation ranges between −1.08 and + 5.77 m bgl, as represented in Fig. 5c. Since the annual groundwater replenishment is through monsoon rainfall, there is a more pronounced decline in water level during the pre-monsoon period. The lowland western coastal stretch shows the shallowest groundwater, whereas the midland eastern part shows a deeper water level. In this study, a rise in water level is a positive change, while a negative value indicates a fall in water level.

From the water table fluctuation analysis, it can be observed that the majority of the western coastal stretch shows maximum negative to low positive fluctuations.

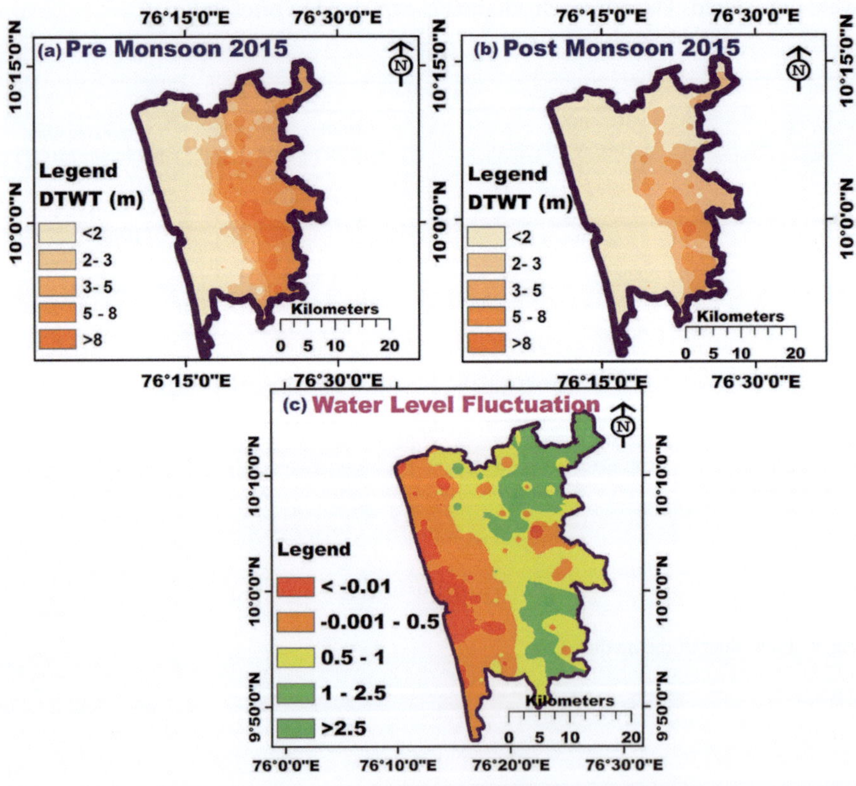

Fig. 5 Spatial distribution of water level (m bgl) (**a**) pre-monsoon 2015; (**b**) post-monsoon 2015; and (**c**) water table fluctuations between pre-monsoon and post-monsoon 2015

Significant positive changes can be observed along the midland in the southeast part and the northeast part (Fig. 5c) of the study area compared to the moderate positive changes along the east midland region. The highly urbanized region in the study area, covering Kochi, Parur, and Ernakulum talukas, shows negative to low positive changes in the water level fluctuations. From the LULC analysis, the variability can be attributed to the impervious land cover in addition to the higher rainfall in the eastern region (Aneesh et al., 2019). The groundwater table with respect to the mean sea level indicates that the deepest level is in the eastern part of the study area (Figs. 6a, b). The spatial pattern of the water level contours points to groundwater movement toward the western part of the study area. The 3 m contour interval of the water table shows a low hydraulic gradient in the lowland region and a high gradient in the midland region.

4.2 *Estimation of Groundwater Recharge Using the Water Table Fluctuation Method*

Evaluation of the spatio-temporal variations in recharge and the analysis of the rainfall hydrograph validate that precipitation is the major recharge factor. Analysis of the seasonal recharge depth maps (Figs. 7a–f) indicates the effect of rainfall patterns on the recharge process of an unconfined aquifer. The positive depth values in

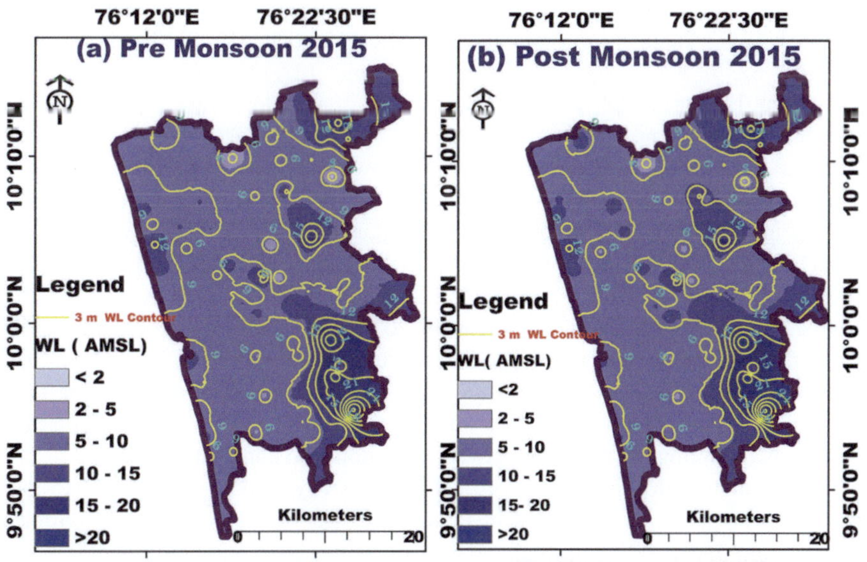

Fig. 6 Spatial distribution of groundwater level surface (m amsl): (**a**) pre-monsoon 2015 and (**b**) post-monsoon 2015

recharge maps indicate net recharge. In contrast, the negative values indicate a net loss in groundwater. Net recharge is shown mainly during monsoon I. When the pre-monsoon period received summer rainfall, as observed during the pre-monsoon of 2015, net recharge was observed in the pre-monsoon. However, in the post-monsoon of 2014 and 2015, the entire area endures a net loss in groundwater with a maximum net loss in the midland region, as illustrated in Fig. 7b, f. During the pre-monsoon of 2015, the whole region, except for a few parts toward the northeast, offers net positive recharge. Monsoon I 2015 shows a net positive value of recharge for the entire region, with the midland stretch showing the maximum positive value. Similar to the previous year, monsoon II and post-monsoon 2015 demonstrate a net loss in groundwater throughout the region, as demonstrated in Fig. 7e, f. The prominent variation in the recharge depths between the western coastal stretch and the eastern midland region can be attributed to the difference in the lithology as depicted in the geology map (Fig. 1c). Since the net recharge depth is proportional to the net actual water level depth, the analysis of the water level fluctuations (Fig. 5c) and recharge in monsoon I (Fig. 7d) corroborated with each other and reflected the changes in the groundwater table.

4.3 *In Situ Water Quality Assessment*

Total dissolved solids in the water reflect the contaminant availability in the water and decide the potability of water for consumption (Sajil Kumar, 2016). The contoured TDS for the post-monsoon 2015 and pre-monsoon 2016 has been presented in Figs. 8a, b with ranges between 36.63 and 727.42 mg/L and 96.61 and 916.58 mg/L. Lower TDS values were observed in most locations from Permbavur taluka toward the northeastern part of the study area for post-monsoon 2015 (Fig. 8a). Higher TDS values were identified toward the middle part of the western coastal stretch close to Kannamalli, Chellam, and Palluruthy during pre-monsoon 2016 (Fig. 8b). The permissible limit for TDS concentration is 1000 mg/L as per the WHO standards (2007). Since the samples in the study area are under the acceptable limit, they can be utilized for domestic and commercial purposes as a freshwater source (Mondal et al., 2008; Sajil Kumar, 2016). The western side of the region, along the lowland area adjacent to the seashore, exhibits a higher TDS content than the eastern part along the midland region. Comparing the spatial pattern of TDS maps in both seasons reveals a reduction in the area covered by freshwater in pre-monsoon, reflecting the recharge between pre-monsoon and post-monsoon.

Spatial distribution maps of electrical conductivity in post-monsoon 2015 and pre-monsoon 2016 are displayed in Figs. 9a, b. EC of the groundwater samples ranges from 39.91 to 788.9 μS/cm in post-monsoon 2015 and between 100.3 and 796.5 μS/cm in pre-monsoon 2016. The spatial distribution of EC has a similar pattern to TDS, with higher EC values in the western coastal region of the study for both seasons. There is a significant decrease in the areal extent of low EC values and

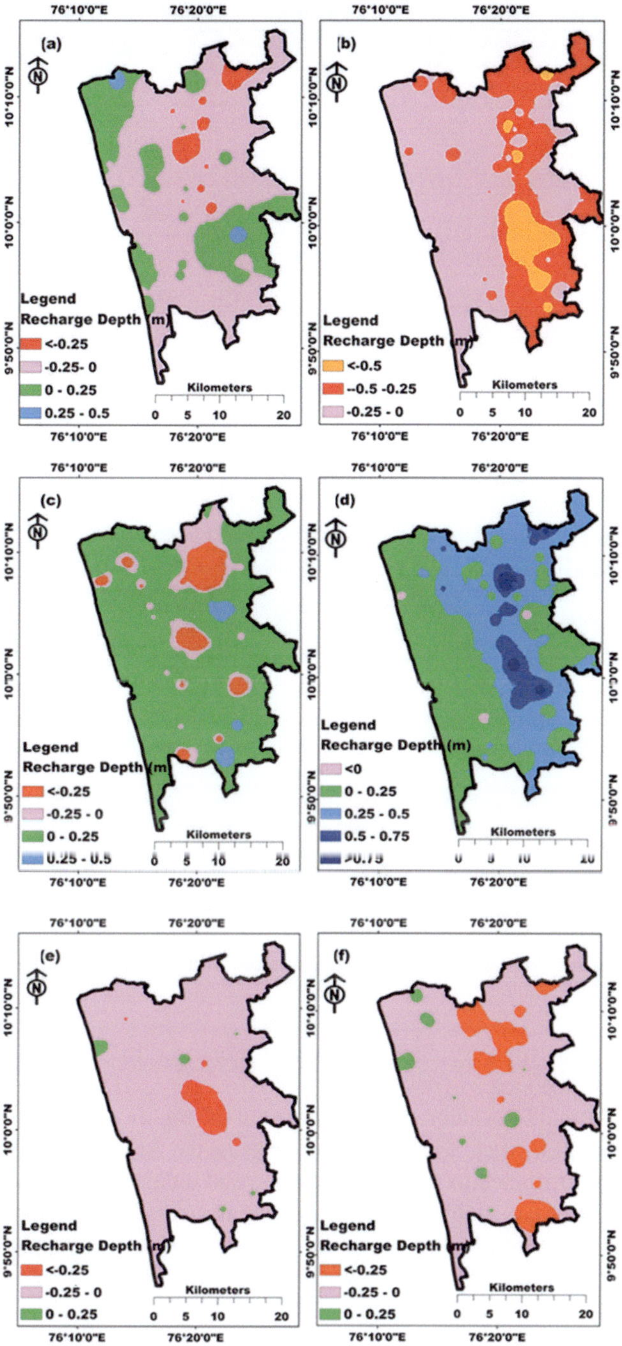

Fig. 7 Seasonal groundwater recharge depth maps for the study area: (**a**) monsoon II 2014, (**b**) post-monsoon 2014, (**c**) pre-monsoon 2015, (**d**) monsoon I 2015, (**e**) monsoon II 2015, and (**f**) post-monsoon 2015

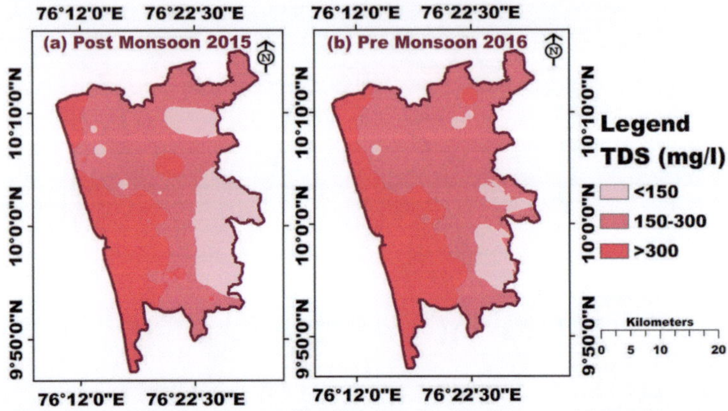

Fig. 8 Spatial distribution of total dissolved solids (TDS): (**a**) post-monsoon 2015 and (**b**) pre-monsoon 2016

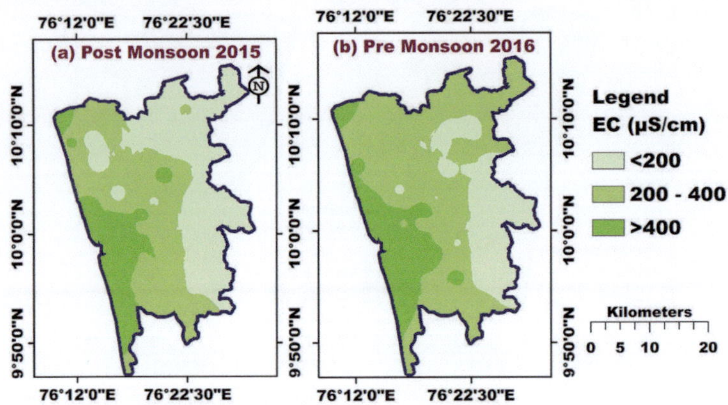

Fig. 9 Spatial distribution of electrical conductivity (EC): (**a**) post-monsoon 2015 and (**b**) pre-monsoon 2016

an increase in the high EC value from post-monsoon 15 to pre-monsoon 2016 (Figs. 9a, b) for the area.

Higher TDS and EC values were registered in the highly urbanized Kochi taluka of the Ernakulum district, within the proximity of the Arabian Sea. Water quality change may be attributed to saltwater intrusion and industrial and municipal waste disposal (Sajil Kumar, 2016) into aquifers at a seasonal scale. Relatively low values were identified in parts of the Aluva and Perumbavur talukas of the study area, belonging to the midland region toward the northeastern part. This suggests the rechargeability in the midland region, indicating good quality groundwater in the vicinity of agricultural and forest land cover as depicted in the LULC map of the study area (Fig. 1d).

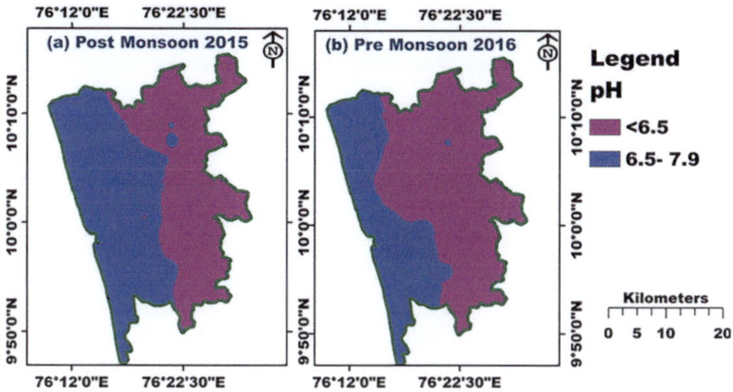

Fig. 10 Spatial distribution of pH: (**a**) post-monsoon 2015 and (**b**) pre-monsoon 2016

Another significant water quality parameter, pH, signifies the acidity of the water and the solubility of the toxic metals (Nienie et al., 2017). The pH in the region ranged between 5.36 and 7.21 for post-monsoon 2015 and 4.59 and 7.94 for pre-monsoon 2016. According to the WHO standards (2007), the permissible limit is 6.5–8.5, indicating that the water is within the acceptable limit for the study area. As shown in Figs. 10a, b, the spatial variation of pH demonstrates that the groundwater in the western part is alkaline. In contrast, the acidic nature is prominent in the eastern region due to the lateritic soils. The spatial map of pH during the post-monsoon 2015 shows a dilution generating a seasonal wedge in the pH value due to the rainfall-induced recharge for the monsoon period (Knutsson, 1994), as visible in Fig. 10a, b.

4.4 *Seepage Velocity Analysis*

As illustrated in Fig. 11, the seepage velocity map was prepared with the Darcy tool using inputs in raster format (Al-Madhlom et al., 2020; Kushwaha & Goyal, 2016). A significant area of the region (around 52.8%) exhibits the seepage velocity range from 5.45×10^{-5} to 8.6×10^{-4} m day$^{-1,}$ whereas around 31.2% displays a range between 6×10^{-4} and 5.15×10^{-3} m day^{-1} and the remaining area indicates a fluctuation between 9.28×10^{-7} and 8.6×10^{-4} m day^{-1}. From the flow direction map (Fig. 11), it was observed that a significant amount of seepage was toward the coastal region (western) from the midland (eastern) region. Despite the low porosity and saturated thickness values, the midland region exhibited high seepage magnitude due to the high groundwater elevation heads and laterite formation in the eastern part of the study area.

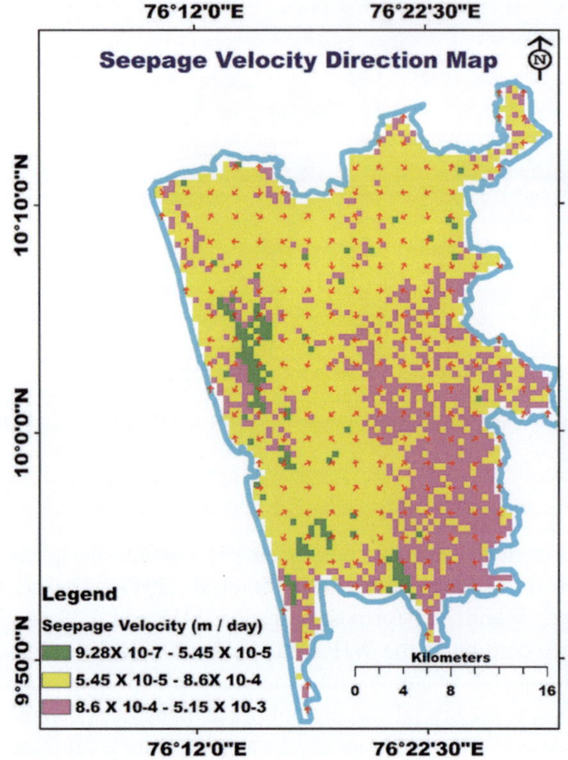

Fig. 11 Magnitude–direction map of seepage velocity

4.5 Estimation of Monthly Groundwater Discharge Flux

Three different sectional profiles across the study area, covering the coastal stretch and the midland region, were used to study the recharge-discharge dynamics of the region. The northern sectional profile (1-1′), the north central (2-2′), and the southern sectional profile (3-3′), as depicted in Fig. 12, were studied in detail. Five adjacent observation wells along each sectional profile were utilized in the calculation of discharge flux using Eq. (3). Figure 13a–f represents the groundwater table, topographic profile, and water table fluctuation profile along the three sectional profiles with mean sea level as datum zero.

The topographic profile of section 1–1′ along the northern part of the study region is represented in Fig. 13a. It shows a variation in elevation between 2.14 m amsl and 19.32 m amsl from west to east, with the water table ranging between 4.15 and 10.65 m. As visualized in Fig. 13b, the water table fluctuations represent a rise of 0.10 m and a fall of 1.71 m in this particular section. In section 2–2′ in the central part of the study area, elevation ranged between 0.03 and 15.4 m amsl and the water table elevation between 3.49 and 15.41 m as in Fig. 13c. A fall of 0.84 m and a rise

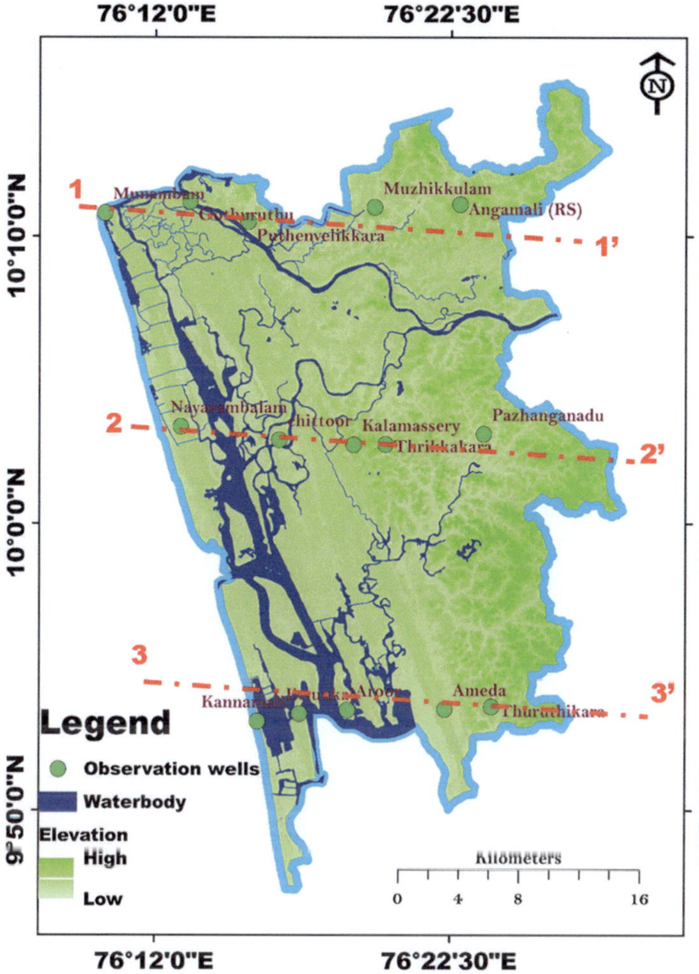

Fig. 12 Details of sections and observation well locations considered in groundwater discharge flux calculations

of 0.13 m were noticed in the average water table, as illustrated in Fig. 13d. Section 3-3′ along the southern part depicts an elevation range from 0.02 m to 58.04 m amsl with the water table elevation between 5.46 and 19.76 m as shown in Fig. 13e. Fluctuations of 1.27 m fall and 0.1 m rise were identified in the water table along the region, as shown in Fig. 13f.

As per the analysis, a low elevation was observed along the western section. A steeper elevation was observed along the eastern section, indicating a steep slope from east to west, justifying the maximum deviation in water levels on the eastern side, along with higher groundwater levels. The shallow water table was along the

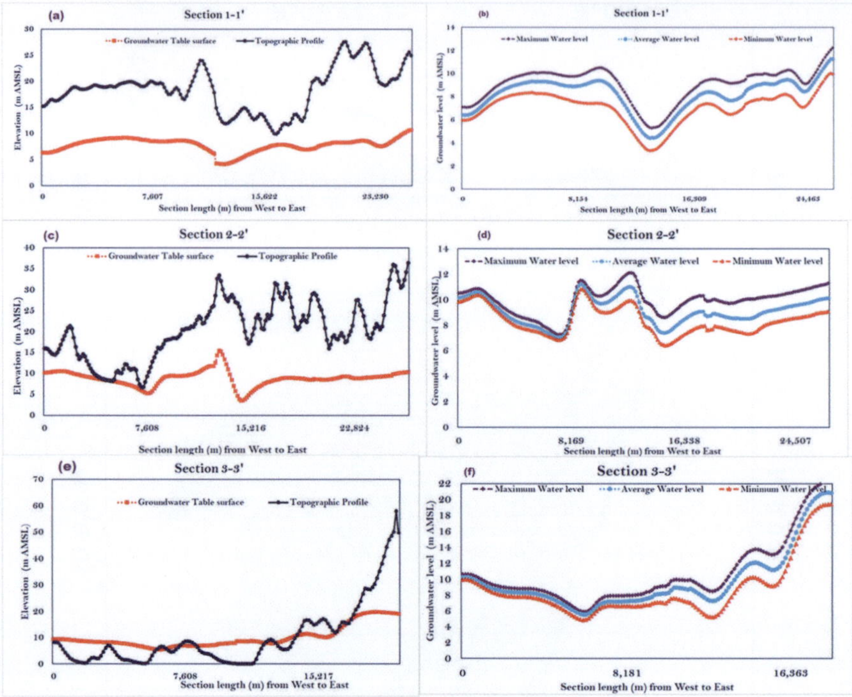

Fig. 13 (**a**) Topographic profile and groundwater level surface along section 1-1′, (**b**) groundwater fluctuations along section 1-1′, (**c**) topographic profile and groundwater level surface along section 2-2′, (**d**) groundwater fluctuations along section 2-2′, (**e**) topographic profile and groundwater level surface along section 3-3′, and (**f**) groundwater fluctuations along section 3-3′

western side, and hence the effect of topography on the water level and the water table fluctuations is clearly depicted.

4.6 Groundwater Discharge Flux

As previously mentioned, Darcy's law is used to determine the groundwater flux. Hydraulic gradient and conductivity are essential components to determine the groundwater flux. For the study on groundwater flux along each sectional profile, five scenarios were considered using five observation wells as displayed in Fig. 12. The details of each scenario along each sectional profile are detailed in Table 1. Based on the lithological properties of the aquifer system, varying hydraulic conductivity values ranging from 30 to 65 m day^{-1} were used. The hydraulic heads were obtained from the field data every month, and the distance between the well locations was determined with the ArcGIS measuring tool.

Table 1 Section and its detailing

	Section 1-1'	Section 2-2'	Section 3-3'
Case 1	Puthenvelikkara to Muamban	Kalamassery to Nayarambalam	Aroor to Kannamali
Case 2	Gothuruthu to Muamban	Chittoor to Nayarambalam	Ettunkal to Kannamali
Case 3	Angamalli to Muzhikkulam	Pazhanganadu to Thrikkakara	Thuruthikara to Ameda
Case 4	Angamali to Puthenvelikkara	Pazhanganadu to Kalamassery	Thuruthikara to Aroor
Case 5	Angamali to Muamban	Pazhanganadu to Nayarambalam	Thuruthikara to Kannamali

Table 2 Discharge flux (m day^{-1}) analyses along sectional profiles

Section and case	Section 1-1'	Section 2-2'	Section 3-3'
Case 1	-1.5×10^{-2} to -1.3×10^{-2}	-2.8×10^{-2} to -3.6×10^{-2}	1.6×10^{-3} to -1.4×10^{-2}
Case 2	-3.4×10^{-4} to -1.9×10^{-2}	-1.2×10^{-3} to -5.9×10^{-3}	-1.9×10^{-3} to -6.0×10^{-3}
Case 3	-6.9×10^{-4} to -1.0×10^{-2}	-7.1×10^{-3} to -1.6×10^{-2}	-6.9×10^{-4} to -1.1×10^{-2}
Case 4	-1.5×10^{-3} to 4.8×10^{-3}	7.8×10^{-3} to 1.5×10^{-3}	-2.1×10^{-2} to -4.1×10^{-2}
Case 5	-2.8×10^{-3} to -6.5×10^{-3}	-1.0×10^{-2} to -1.6×10^{-2}	-1.7×10^{-2} to -2.7×10^{-2}

As per the data available, for the case 1 scenario at section 1-1', where the observation wells were present at Puthenvelikkara and Muamban, the water level head at the former location was observed as h2 and at the latter as h1 in addition to the horizontal distance Δl. The vertical head difference for the time period is the water level differences at Muamban and Puthenvelikkara. The hydraulic gradient value is calculated with Eq. (4), and the discharge flux is calculated as per Eq. (3). The computed discharge flow for various sectional profiles is shown in Table 2.

A comprehensive analysis of the discharge flux indicated that the maximum discharge rate is observed from October to May, with the highest flux rates during April for all five cases (Fig. 14). The study concludes that a significant discharge is from the midland region toward the coastal area in all months. The high discharge rate was evident during the non-monsoon period, increasing the groundwater discharge from the hydraulically connected aquifers (Chidambaram et al., 2017). However, a contrasting result found in section 2-2′ for case 4 can be attributed to the undulating topography between the locations. The discharge rate with the aquifer characteristics can be utilized to analyze the quantity of groundwater recharge and its capacity.

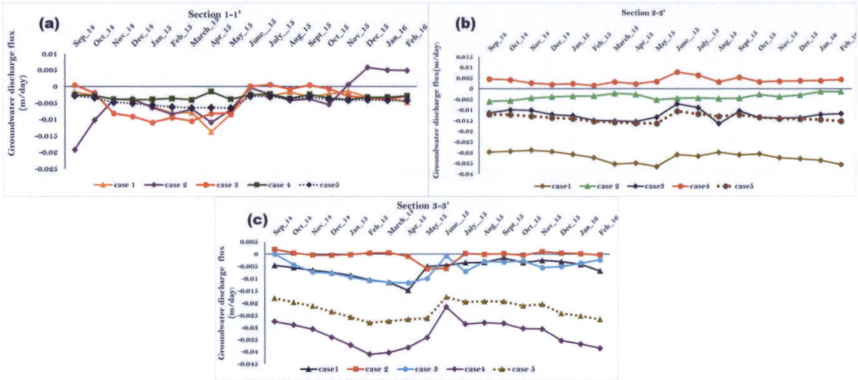

Fig. 14 (**a**) Groundwater discharge flux rates at section 1-1', (**b**) groundwater discharge flux rates at section 2-2', and (**c**) groundwater discharge flux rates at 3-3′

5 Conclusions

The study focuses on the short-term coastal aquifer dynamics that have undergone rapid urbanization. The monthly water table observed at 97 well locations for 16 months was used to generate a monthly water table fluctuation map and a seasonal recharge map. The seasonal fluctuations in water level range from −1.08 to +5.7729 m bgl. A positive sign indicates the rise, and a negative sign denotes the fall in water level between the two seasons. The groundwater recharge is directly proportional to water level fluctuations between pre- and post-monsoon. The seasonal recharge maps from the water table fluctuation method illustrated a significant recharge from June to August, the southwest monsoon period. The quantification of the recharge maps affirms the proportionality of the recharge depths to precipitation. The WTF method shows that a significant portion of the study area is moderately rechargeable. The majority of the coastal plains of the study area shows low recharge depths, while the eastern midland part of the study area shows moderate to high recharge. The water level change analysis shows discharge from the midland to the lowland and the lowland coastal stretch to the sea, especially during the post-monsoon period. The study depicts that the major recharge is happening in the midland region of the study area. At the same time, the discharge from midland to coastal areas is high in a short span, resulting in severe water scarcity during the pre-monsoon period. Geochemical analysis was carried out for 45 well locations to generate the spatial distribution pattern of TDS, EC, and pH. The fluctuations in TDS, EC, and pH concentrations indicate that the freshwater availability is directly connected to recharge from rainfall, as quality parameters are better in the midland region than in the coastal lowland region. The spatial distribution of geochemical parameters displays flush-out patterns indicating high recharge in the midland and discharge from the midland to the lowland. The mapped seepage velocity magnitude signifies a higher seepage rate from the midland portion of the study area. Further, the monthly groundwater discharge flux rates using Darcy's tool suggest a high discharge from east to west for the non-monsoon period. The analysis of topography, water level, and water table fluctuation profile shows the effect of the topography on the aquifer dynamics.

The study elaborates on a few of the ongoing processes leading to the water crisis problems during the non-monsoon seasons. Although the recharge is significant in the midland, groundwater movement due to the connectivity between the sea and the aquifer system involves heavy seepage from the midland toward the coastal region. Hence, an effort is required to hold the water in the midland region for future emergency usage. The study also points out the need to preserve natural land cover in the midland region, as significant recharge occurs along the midland. Active participation of the state government authorities and stakeholders is crucial in the land use and land cover of the region to achieve sustainable development.

Groundwater quality management is essential for urbanized locations due to the unreliability of the urban water supplies, leading to the overuse of the underground supplies. Appropriate use and reuse of water need to be popularized among the community. The critical condition of the resource is aggravated by the climatic and population expanse across the globe. Proper measures are required to avert acute crises over the years to ensure both qualitative and quantitative groundwater resources. Artificial and natural recharging mechanisms need to be cultivated in the minds of the population.

Acknowledgments The corresponding author acknowledges the Indian Institute of Technology, Guwahati, for providing facilities to conduct this study. The authors acknowledge the financial support received as a project grant from the National Centre for Earth Science Studies, Thiruvananthapuram. The authors extend their gratitude to Mr. Sreehari Sarangan, CGWB, for his valuable support for the study. The authors are thankful to organizations such as the Central Groundwater Board, Thiruvananthapuram, Indian Meteorological Department, Pune Geological Survey of India, Kerala Unit for maps and data, and USGS for the satellite images.

References

Ahmadi, T., Ziaei, A., Davary, K., Faridhosseini, A., Izadi, A., & Rasoulzadeh, A. (2012). Estimation of groundwater recharge using various methods in Neishaboor plain, Iran. In *Groundwater modeling and management under uncertainty*. https://doi.org/10.1201/b13167-4

Alabi, O. O., Ojurongbe, T. A., & Sedara, S. O. (2019). Gradients of seepage velocity model for contaminant transport prediction. *Scientific African, 4*, e00087. https://doi.org/10.1016/j.sciaf.2019.e00087

Al-Madhlom, Q., Al-Ansari, N., Abid, B., Laue, J., & Musa, H. (2020). Seepage velocity: Large scale mapping and the evaluation of two different aquifer conditions (silty clayey and sandy). *Hydrology, 7*(3), 1–30. https://doi.org/10.3390/HYDROLOGY7030060

Aneesh, T. D., Srinivas, R., Singh, A. T., Resmi, T. R., Nair, A. M., & Redkar, B. L. (2019). Stable water isotope signatures of dual monsoon precipitation: A case study of greater Cochin region, south-west coast of India. *Journal of Earth System Science, 128*, 8. https://doi.org/10.1007/s12040-019-1234-2

Baron, J. S., & Williams, M. W. (2000). Water resources research: Preface. *Water Resources Research, 36*(1), 11–12. https://doi.org/10.1029/1999WR900293

Barron, O. V., Donn, M. J., & Barr, A. D. (2013). Urbanisation and shallow groundwater: Predicting changes in catchment hydrological responses. *Water Resources Management*. https://doi.org/10.1007/s11269-012-0168-0

Bhosale, D. D., & Kumar, C. P. (2002). Simulation of seawater intrusion in Ernakulam coast. In *International conference on "hydrology and watershed management," January 2002* (pp. 390–399) http://www.angelfire.com/nh/cpkumar/publication/ernac.pdf

CGWB. (2013). *Ground water information booklet of Ernakulam District, Kerala state*. Ministry of Water Resources, Government of India.

Chidambaram, S., Nepolian, M., Ramanathan, A. L., Sarathidasan, J., Thilagavathi, R., Thivya, C., Prasanna, M. V., Srinivasamoorthy, K., Jacob, N., & Mohokar, H. (2017). An attempt to identify and estimate the subsurface groundwater discharge in the south east coast of India. *International Journal of Sustainable Built Environment, 6*(2), 421–433. https://doi.org/10.1016/j.ijsbe.2017.07.007

Costall, A. R., Harris, B. D., Teo, B., Schaa, R., Wagner, F. M., & Pigois, J. P. (2020). Groundwater throughflow and seawater intrusion in high quality coastal aquifers. *Scientific Reports, 10*(1), 1–33. https://doi.org/10.1038/s41598-020-66516-6

Crosbie, R. S., Scanlon, B. R., Mpelasoka, F. S., Reedy, R. C., Gates, J. B., & Zhang, L. (2013). Potential climate change effects on groundwater recharge in the High Plains aquifer, USA. *Water Resources Research, 49*(7), 3936–3951. https://doi.org/10.1002/wrcr.20292

da Silva, L. A., da Silva, A. M., Coelho, G., de Mello, C. R., & Pereira, D. D. R. (2012). Groundwater recharge estimate at alto Rio Grande—MG watershed. *Engenharia Agrícola*. https://doi.org/10.1590/S0100-69162012000600011

Dams, J., Woldeamlak, S. T., & Batelaan, O. (2008). Predicting land-use change and its impact on the groundwater system of the Kleine Nete catchment, Belgium. *Hydrology and Earth System Sciences*. https://doi.org/10.5194/hess-12-1369-2008

Das, S., & Pardeshi, S. D. (2018). Integration of different influencing factors in GIS to delineate groundwater potential areas using IF and FR techniques: A study of Pravara basin, Maharashtra, India. *Applied Water Science, 8*(7), 1–16. https://doi.org/10.1007/s13201-018-0848-x

Fang, D., Hao, L., Cao, Z., Huang, X., Qin, M., Hu, J., & Liu, Y. (2020). Combined effects of urbanisation and climate change on watershed evapotranspiration at multiple spatial scales. *Journal of Hydrology, 587*(March), 124869. https://doi.org/10.1016/j.jhydrol.2020.124869

Healy, R. W., & Cook, P. G. (2002). Using groundwater levels to estimate recharge. *Hydrogeology Journal, 10*(1), 91–109. https://doi.org/10.1007/s10040-001-0178-0

Hua, A. K. (2017). Land use land cover changes in detection of water quality: A study based on remote sensing and multivariate statistics. *Journal of Environmental and Public Health, 2017*, 5–7. https://doi.org/10.1155/2017/7515130

Jayasankar, P., & Sridar, B.M.N. (2017). An assessment of ground water potential of Kerala, India: A Case Study. *AEIJST, 5*(02),1–21.

Jha, B. M., & Sinha, S. K. (2010). Towards better management of ground water resources in India. *Water and Energy International, 67*(1), 2–18.

Keesari, T., Kulkarni, U. P., Jaryal, A., Mendhekar, G. N., Deshmukh, K. N., Hegde, A. G., & Kamble, S. N. (2014). Groundwater dynamics of a saline impacted coastal aquifer of western Maharashtra, India: Insights from a radiotracer study. *Journal of Radioanalytical and Nuclear Chemistry, 300*(1), 1–6. https://doi.org/10.1007/s10967-014-2940-5

Keese, K. E., Scanlon, B. R., & Reedy, R. C. (2005). Assessing controls on diffuse groundwater recharge using unsaturated flow modeling. *Water Resources Research, 41*(6), 1–12. https://doi.org/10.1029/2004WR003841

Knutsson, G. (1994). Acidification effects on groundwater—Prognosis of the risks for the future. *Future Groundwater Resources at Risk, 222*, 3–17.

Kollet, S. J., & Maxwell, R. M. (2008). Capturing the influence of groundwater dynamics on land surface processes using an integrated, distributed watershed model. *Water Resources Research, 44*(2), 1–18. https://doi.org/10.1029/2007WR006004

Kushwaha, K., & Goyal, R. (2016). Methodology for the estimation of groundwater flux across simplified boundary using GIS and groundwater levels. *Current Science, 110*(6), 1050–1058. https://doi.org/10.18520/cs/v110/i6/1050-1058

Lamichhane, S., & Shakya, N. M. (2019). Alteration of groundwater recharge areas due to land use/cover change in Kathmandu Valley, Nepal. *Journal of Hydrology: Regional Studies, 26*(October 2019), 100635. https://doi.org/10.1016/j.ejrh.2019.100635

Lathashri, U. A., & Mahesha, A. (2016). Groundwater sustainability assessment in coastal aquifers. *Journal of Earth System Science, 125*(6), 1103–1118. https://doi.org/10.1007/s12040-016-0719-5

Li, C., Liu, M., Hu, Y., Shi, T., Qu, X., & Walter, M. T. (2018). Effects of urbanisation on direct run-off characteristics in urban functional zones. *Science of the Total Environment, 643*, 301–311. https://doi.org/10.1016/j.scitotenv.2018.06.211

Liu, Y., Shang, S. H., & Mao, X. M. (2012). Tidal effects on groundwater dynamics in coastal aquifer under different beach slopes. *Journal of Hydrodynamics, 24*(1), 97–106. https://doi.org/10.1016/S1001-6058(11)60223-0

Mao, C., Tan, H., Song, Y., & Rao, W. (2020). Evolution of groundwater chemistry in coastal aquifers of the Jiangsu, East China: Insights from a multi-isotope (δ2H, δ18O, 87Sr/86Sr, and δ11B) approach. *Journal of Contaminant Hydrology, 235*(April), 103730. https://doi.org/10.1016/j.jconhyd.2020.103730

McGrane, S. J. (2016). Impacts of urbanisation on hydrological and water quality dynamics, and urban water management: A review. *Hydrological Sciences Journal, 61*(13), 2295–2311. https://doi.org/10.1080/02626667.2015.1128084

Mondal, N. C., Singh, V. S., Saxena, V. K., & Prasad, R. K. (2008). Improvement of groundwater quality due to fresh water ingress in Potharlanka Island, Krishna delta, India. *Environmental Geology, 55*(3), 595–603. https://doi.org/10.1007/s00254-007-1010-5

Nienie, A. B., Sivalingam, P., Laffite, A., Ngelinkoto, P., Otamonga, J. P., Matand, A., Mulaji, C. K., Mubedi, J. I., Mpiana, P. T., & Poté, J. (2017). Seasonal variability of water quality by physicochemical indexes and traceable metals in suburban area in Kikwit, Democratic Republic of the Congo. *International Soil and Water Conservation Research, 5*(2), 158–165. https://doi.org/10.1016/j.iswcr.2017.04.004

Oudin, L., Salavati, B., Furusho-Percot, C., Ribstein, P., & Saadi, M. (2018). Hydrological impacts of urbanisation at the catchment scale. *Journal of Hydrology, 559*, 774–786. https://doi.org/10.1016/j.jhydrol.2018.02.064

Owuor, S. O., Butterbach-Bahl, K., Guzha, A. C., Rufino, M. C., Pelster, D. E., Díaz-Pinés, E., & Breuer, L. (2016). Groundwater recharge rates and surface runoff response to land use and land cover changes in semi-arid environments. *Ecological Processes, 5*(1), 16. https://doi.org/10.1186/s13717-016-0060-6

Patra, S., Sahoo, S., Mishra, P., & Mahapatra, S. C. (2018). Impacts of urbanisation on land use / cover changes and its probable implications on local climate and groundwater level. *Journal of Urban Management, 7*(2), 70–84. https://doi.org/10.1016/j.jum.2018.04.006

Paul, M. J. (2006). Impact of land-use patterns on distributed groundwater recharge and discharge—A case study of western Jilin, China. *Chinese Geographical Science, 16*(3), 229–235. https://doi.org/10.1007/s11769-006-0229-5

Pramada, S. K., Minnu, K. P., & Roshni, T. (2018). Insight into sea water intrusion due to pumping: A case study of Ernakulam coast, India. *ISH Journal of Hydraulic Engineering*, 1–10. https://doi.org/10.1080/09715010.2018.1553642

Praveen Lal, C. S., & Nair, S. B. (2017). Urbanisation in Kerala—What does the census data reveal? *Indian Journal of Human Development, 11*(3), 356–386. https://doi.org/10.1177/0973703018763241

Sajil Kumar, P. J. (2016). Influence of water level fluctuation on groundwater solute content in a tropical south Indian region: A geochemical modelling approach. *Modeling Earth Systems and Environment, 2*(4), 1–9. https://doi.org/10.1007/s40808-016-0235-2

Sanford, W. (2002). Recharge and groundwater models: An overview. *Hydrogeology Journal, 10*(1), 110–120. https://doi.org/10.1007/s10040-001-0173-5

Scanlon, B. R., Healy, R. W., Scanlon, B. R., Healy, R. W., & Cook, P. G. (2002). Choosing appropriate techniques for quantifying groundwater recharge. *Hydrogeology Journal, 10*, 18–39. https://doi.org/10.1007/s10040-0010176-2

Sharp, J. M. (2010). The impacts of urbanisation on groundwater systems and recharge. *Aqua, 1*(may), 51–56. https://doi.org/10.4409/Am-004-10-0008

Shin, K., Koh, D. C., Jung, H., & Lee, J. (2020). The hydrogeochemical characteristics of groundwater subjected to seawater intrusion in the archipelago, Korea. *Water (Switzerland), 12*, 6. https://doi.org/10.3390/W12061542

Singaraja, C., Chidambaram, S., Jacob, N., Selvam, S., & Prasanna, M. V. (2016). Tidal effects on groundwater dynamics in shallow coastal aquifers—Southeast coast of Tamilnadu, India. *Arabian Journal of Geosciences, 9*, 6. https://doi.org/10.1007/s12517-016-2480-0

Sreekesh, S., Sreerama Naik, S. R., & Rani, S. (2018). Effect of sea level changes on the groundwater quality along the coast of Ernakulam District, Kerala. *Journal of Climate Change, 4*(2), 51–65. https://doi.org/10.3233/jcc-1800013

Tabatabaei, S. M., & Khozeymehnezhad, H. (2018). Studying the vulnerability factors of coastal aquifers due to sea saline water intrusion. *International Journal of Hydrology, 2*(6), 682–686. https://doi.org/10.15406/ijh.2018.02.00143

Tashie, A. M., Mirus, B. B., & Pavelsky, T. M. (2016). Identifying long-term empirical relationships between storm characteristics and episodic groundwater recharge. *Water Resources Research, 52*(1), 21–35.

Telahigue, F., Mejri, H., Mansouri, B., Souid, F., Agoubi, B., Chahlaoui, A., & Kharroubi, A. (2020). Assessing seawater intrusion in arid and semi-arid Mediterranean coastal aquifers using geochemical approaches. *Physics and Chemistry of the Earth, 115*(March 2019), 102811. https://doi.org/10.1016/j.pce.2019.102811

Todd, D. K., & Mays, L. W. (2004). *Groundwater hydrology*. John Wiley & Sons.

Varni, M., Comas, R., Weinzettel, P., & Dietrich, S. (2013). Application de la méthode de fluctuation du niveau piézométrique pour caractériser la recharge des eaux souterraines dans la plaine de la Pampa (Argentine). *Hydrological Sciences Journal, 58*(7), 1445–1455. https://doi.org/10.1080/02626667.2013.833663

Vijay, R., & Mohapatra, P. K. (2016). Hydrodynamic assessment of coastal aquifer against saltwater intrusion for city water supply of Puri, India. *Environmental Earth Sciences, 75*(7), 1–8. https://doi.org/10.1007/s12665-016-5357-3

Werner, A. D., Alcoe, D. W., Ordens, C. M., Hutson, J. L., Ward, J. D., & Simmons, C. T. (2011). Current practice and future challenges in coastal aquifer management: Flux-based and trigger-level approaches with application to an Australian case study. *Water Resources Management, 25*(7), 1831–1853. https://doi.org/10.1007/s11269-011-9777-2

WWRD. (2015). *Water for a sustainable world, the united nations world water development report 2015*. UNESCO.

Zang, Y., & Li, M. (2021). Numerical assessment of compressed air injection for mitigating seawater intrusion in a coastal unconfined aquifer. *Journal of Hydrology, 595*(December 2020), 125964. https://doi.org/10.1016/j.jhydrol.2021.125964

Land Use Land Cover and Hydrology: A Case of Teesta River Basin

Vicky Anand, Prithwi Pratim Guria, and Bakimchandra Oinam

Abstract Water resources and the ecosystem are significantly affected by changes in land use and land cover (LULC). It is essential to estimate the effects of change in LULC on the hydrology of the river basin and understand its mechanism. The main objective of this study is to analyze the effect of change in land use and land cover on the hydrological regime of the Teesta River basin using the comprehensive Soil and Water Assessment Tool (SWAT) model. Calibration and validation were carried out using the observed discharge data on the monthly time scale. The model showed good performance with $R^2 = 0.74$, Nash-Sutcliffe Efficiency (NSE) = 0.63, and Kling-Gupta Efficiency (KGE) = 0.58 during the calibration period and $R^2 = 0.75$, NSE = 0.69, and KGE = 0.66 during the validation period using the Sequential Uncertainty Fitting algorithm. The main contribution to the water yield arises from the northern part of the basin in the Himalayan ranges, which is covered with glaciers, forest, and scrubland. Due to changes in LULC in the past decade, a severe spatial and temporal shift was observed in the water balance components at the sub-basin scale. An increase in the surface runoff in the upper part of the catchment due to the melting of glaciers and the conversion of forest into agricultural fields may lead to an increase in events like flooding and landslides in the low-lying regions in the basin.

Keywords Sensitivity · Calibration · Validation · LULC · SWAT

V. Anand
Department of Civil Engineering, National Institute of Technology Manipur, Imphal, Manipur, India

River and Environmental Engineering Laboratory, Department of Civil Engineering, The University of Tokyo, Tokyo, Japan

P. P. Guria · B. Oinam (✉)
Department of Civil Engineering, National Institute of Technology Manipur, Imphal, Manipur, India
e-mail: bakim@nitmanipur.ac.in

P. K. Joshi et al. (eds.), *Geospatial Science for Urban Ecosystems*, The Urban Book Series, https://doi.org/10.1007/978-3-032-07155-2_13

1 Introduction

A growing population and an enhancement in living standards are resulting in an increase of water demand. As the amount of water enclosed in the hydrological cycle remains more or less the same, the increase in water demand has led to a decrease in per capita availability of water over the world (Boretti & Rosa, 2019; Kummu et al., 2016). The variability in the distribution of water resources on spatial and temporal scales makes the condition more complex. In India, the annual per capita availability of water resources has declined from 5176 m^3 in 1951 to about 1588 m^3 in 2010 and is likely to decline to about 1434 m^3 in 2025 (Central Water Commission, 2005). This alarming situation indicates the need for steps to be taken for proper planning and execution of various management programs to attain optimum utilization of available water resources.

In the past years, to assist the complexity associated with the hydrological processes in the hydrological system, numerous models have been developed (Arnold & Allen, 1996; Anand et al., 2020). A hydrological model simplifies the hydrological systems by forecasting hydrological phenomena and changes to compute the hydrological effects, which aids in formulating various policies on the management of water resources (Mirza., 2003). Based on the mutual relationship between climate and surface parameters, a mathematical relationship is established (Gosain et al., 2009). Among the various available hydrological models, the semi-distributed models are comprehensive and versatile from various aspects (Gosain et al., 2009). In the past, hydrological models like Water Erosion Prediction Project (WEPP) (Laflen et al., 1991), Agricultural Non-Point Source Pollution (AGNPS) (Young et al., 1989), and SWAT (Arnold et al., 1993; Anand & Oinam, 2019) have been extensively used. To assess the impact of change in climate on various hydrological processes and parameters, the SWAT model has been extensively used by the hydrologist (Bouraoui et al., 2005). This model is broadly used to estimate soil loss, assess water quality, predict surface runoff, and assess change in LULC (Gassman et al., 2007). The uncertainty associated with sediment and surface runoff in the Jingchuan River was quantified using the SWAT model (Shen et al., 2012). For water resource management and hydrological modeling, the SWAT model has been widely used across several river basins in India (e.g., Tapi River basin, Tungabhadra River basin in the southern part of India, Karnataka, and Tons River basin, a tributary of River Yamuna) (Shivhare et al., 2014; Sisay et al., 2017). SWAT could precisely simulate the runoff and other hydrological parameters in both wet and dry seasons (Govender & Everson, 2005). In some studies, it was found that streamflow downstream of the catchment is adversely affected by urbanization upstream, and underestimation in the extreme event simulated by the SWAT model was observed (Rosenthal et al., 1995). Quantification of freshwater availability was found to be successful using the modeling procedure of SWAT (Schuol et al., 2008).

The Teesta River (a right tributary of the Brahmaputra) basin is the largest river basin in Sikkim and plays a crucial role in the social and economic status of Sikkim and the northern parts of West Bengal. In the last few decades, there has been a significant change in the climate conditions due to various natural and anthropogenic factors (Gyamfi et al., 2016). The major part of this basin in India lies in dense forest cover and a snow-covered region. This river basin plays a vital role in hydroelectric power generation in India, as the basin lies in the high Himalayan region. The main objective of this study is (a) assessing LULC change in the Teesta River basin for three different time periods: 1994, 2004, and 2014; (b) simulating the effects of change in LULC by setting up a hydrological model; and (c) analyzing the effects of LULC change on water balance components like surface runoff, groundwater, evapotranspiration, and water yield in this catchment.

2 Study Area

The Teesta River basin lies in the middle Himalayan ranges in the state of Sikkim and the northern part of West Bengal. The geographical extent of the basin is between 26°36′N to 28°7′N and 87°59′E to 88°53′E and has a total catchment area of about 9450 km^2. The catchment is characterized by high Himalayan terrain with elevation ranges from 23 m to 8384 m above sea level and is surrounded by Tibet in the north and northeast, Bhutan and Nepal in the east and west, respectively, and the Jalpaiguri district of West Bengal in the south. Teesta originates in Pauhunri (7127 m above msl). The river is fed not only by precipitation but also by melting glaciers and snow as well as groundwater (Mukhopadhyay, 1982). Mean annual precipitation in the mountain area of the basin from 1963 to 2001 is 2218 mm, with 78% of the precipitation falling during the monsoon season (Mukhopadhyay, 1982), and the temperature lies between −4 and 41.3 °C. The maximum discharge observed in the basin is 18,150 m^3/s (Froehlich et al., 1990), whereas the minimum flow observed is 20 m^3/s in the dry season (Starkel & Basu, 2000). There are eight major tributaries of the Teesta River. Lachung, Dik, Chakung, Rangpo, and Rani Khola are the left bank tributaries, whereas the Zemu, Rangyong, and Rangit rivers are right bank tributaries (ENVIS, 2019). The geographical location of the Teesta River basin is shown in Fig. 1.

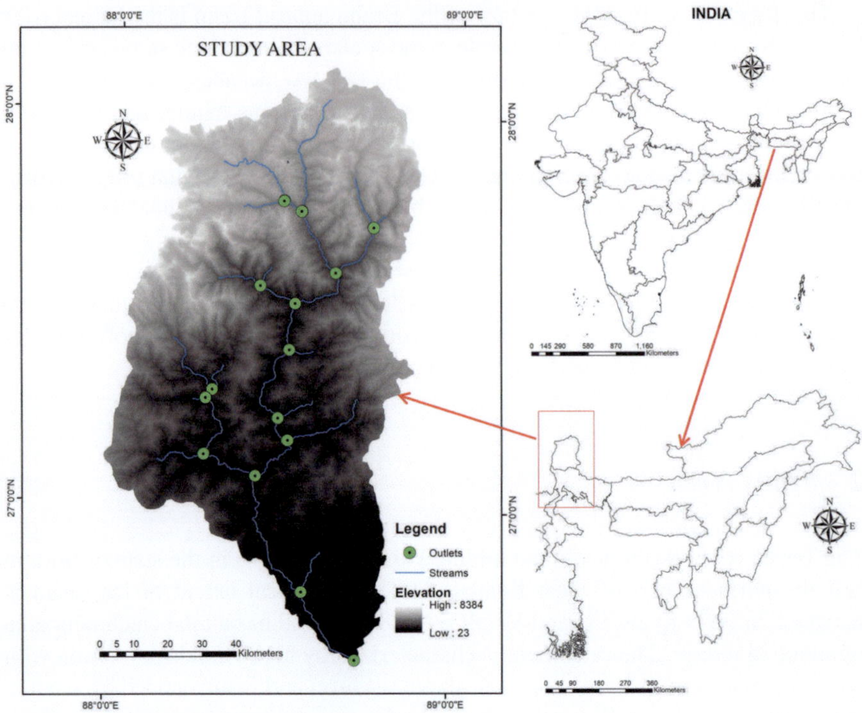

Fig. 1 Geographical location of the Teesta River basin

3 Materials and Methodology

3.1 Data Input and Evaluation

The SWAT model requires Digital Elevation Model (DEM), LULC, climate data, and soil data as model input to set up the model. Data was acquired from different organizations as mentioned in Table 1. Since the area is vast and cannot be covered by a single pass of sensors, multiple images were used and mosaicked to cover the study area. For 1994 and 2004, the Landsat 4–5 thematic mapper (TM) was used, and for the year 2014, Landsat 8 OLI C-1 level 1 images were used for land use land cover classification for different time periods. Due to the scarcity of good-quality data for a long duration, the model was calibrated and validated with the longest period of good-quality discharge data monitored from 1995 to 2005, obtained from the Irrigation and Waterways Department, Govt. of West Bengal.

3.2 SWAT Model

SWAT is a physically based semi-distributed model with explicit parameterization (Neitsch et al., 2011). In SWAT, the basin is divided into several smaller units called

Table 1 Input data and source

Data	Resolution	Sources
DEM	12.5 m	ALOS PALSAR, Alaska Satellite Facility
Landuse landcover[a]	30 m	Earth resource observation and science
Meteorological data	Daily	Indian Meteorological Department
Soil data[a]	1 km	NBSS & LUP
River discharge	Monthly	Irrigation and Waterways Department, Govt. of West Bengal

[a]Resampled to 12.5 m resolution

Hydrological Response Units (HRUs) based on soil type and LULC. Water balance in SWAT is based on the equation (Arnold & Fohrer, 2005).

$$\mathrm{SW}_t = \mathrm{SW}_0 + \sum_{i=0}^{n} R_{\mathrm{day}} + Q_{\mathrm{surf}} + E_a + W_{\mathrm{seep}} + Q_{gw}$$

where SW_t and SW_0 are the final and initial water contents on days *t* and *I*, respectively, in mm. R_{day}, Q_{surf}, E_{a}, W_{seep}, and Q_{gw} are the amounts of precipitation, surface runoff, evapotranspiration, percolation entering the vadose zone, and return flow, respectively, on day i (mm). In the present study, the Soil Conservation Service Curve Number (SCS-CN) method (USDA, 1972) was used for computing surface runoff.

3.3 Model Performance Evaluation

In order to accept the dependability of any model, a model performance evaluation is necessary. If the statistical indices of the model performance fall within the permissible range, then the model outputs are acceptable (Moriasi et al., 2007). There are several statistical models' performance indices to evaluate the model reliability, such as Nash-Sutcliffe Efficiency (NSE) (Nash & Sutcliffe, 1970), Coefficient of Determinacy (R^2), and Kling-Gupta Efficiency (KGE) (Gupta et al., 2009; Anand & Oinam, 2019).

(a) Nash Sutcliffe efficiency (NS), computed as Eq. (1).

$$NS = 1 - \frac{\sum_j \left(G_m - G_k\right)_j^2}{\sum_j \left(G_{m,j} - \overline{G_m}\right)^2} \tag{1}$$

where *G*: a variable, *m*: observed, *k*: simulated, and *j*: *j*th simulated value (Nash & Sutcliffe, 1970).

(b) Coefficient of Determination (R^2); computed as Eq. (2).

$$R^2 = \frac{\left[\sum_j \left(G_{n,j} - \overline{G_m}\right)\left(G_{k,j} - \overline{G_k}\right)\right]^2}{\sum_j \left(G_{n,j} - \overline{G_n}\right)^2 \sum_j \left(G_{k,j} - \overline{G_k}\right)^2} \quad (2)$$

where G: a variable, n: observed, k: simulated, and j: jth simulated value.

(c) Kling-Gupta Efficiency, computed as Eq. (3).

$$\text{KGE} = 1 - \sqrt{\left((r-1)^2 + (\alpha-1)^2 + (\beta-1)^2\right)} \quad (3)$$

where r is the coefficient of linear regression between simulated and observed variables, β is the ratio of the mean of simulated and observed data, and α is the ratio of the standard deviation of simulated and observed data (Gupta et al., 2009).

4 Results

4.1 LULC Classification and Change Analysis

LULC classification was carried out on Landsat 8 for 2014 and Landsat 4–5 for the years 2004 and 1994 using a maximum likelihood classifier in ERDAS Imagine. The land use land cover was classified in seven different categories, i.e., forest, glacier, scrub, water bodies, built-up, agriculture, and open area for the years 1994, 2004, and 2014 (Fig. 2). It was observed that some parts of agriculture, mainly tea gardens, were misclassified as forest and scrub. Due to a minor shift, a pixel lying in a particular class may get classified as another. This misclassification is mainly due to spectral class similarity in the older images. These problems were more detected in Landsat 4–5 TM images as compared to Landsat 8 images. To detect the accuracy of the classified image, accuracy assessment was carried out, and the overall accuracy was found to be 84%, 87%, and 92% for the years 1994, 2004, and 2014, respectively. It can be observed that the share of agricultural area has increased significantly from 2004 to 2014. This change was mainly concentrated in the southern part of the basin due to the conversion of open area into agriculture (mainly in the form of a tea garden). For all three different time periods, it was observed that there is an abrupt change in the water bodies and glaciers due to an increase in the temperature (Rahman et al., 2016) as an impact of the changing climate. Though the overall trends show there is a decrease in the area under glacier/snow from 1994 to 2014, which follows the trends observed in the Himalayan ranges (Maurer et al., 2019), but for the year 2004, an increase of 2.5% in area under glacier/snow cover

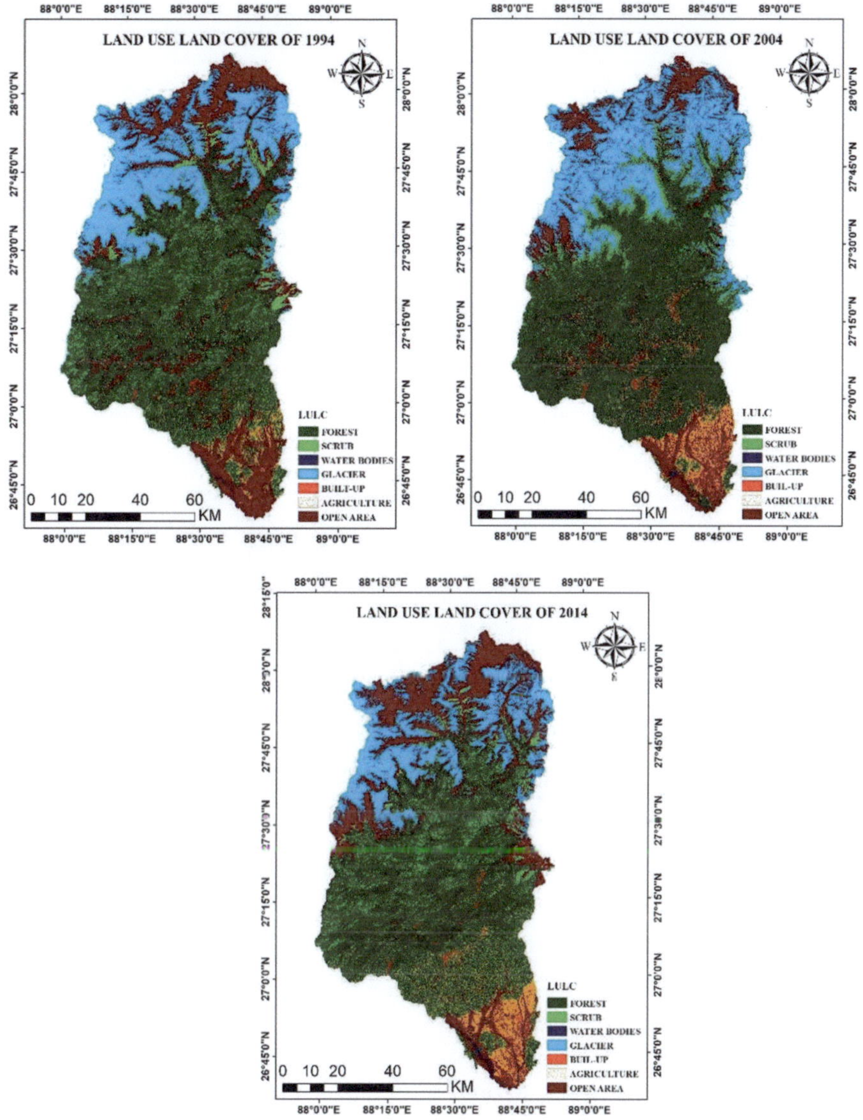

Fig. 2 LULC classifications for the years 1994, 2004, and 2014

was observed. The major cause behind this was that the satellite image used for the year 2004 was of the winter season, as it was the only OLI image available for the study area in 2004, which was cloud-free. Due to the winter season, the water bodies were frozen, and the area under the open category in the upper reaches of the Himalayan ranges was covered with snow and was classified as the snow/glacier

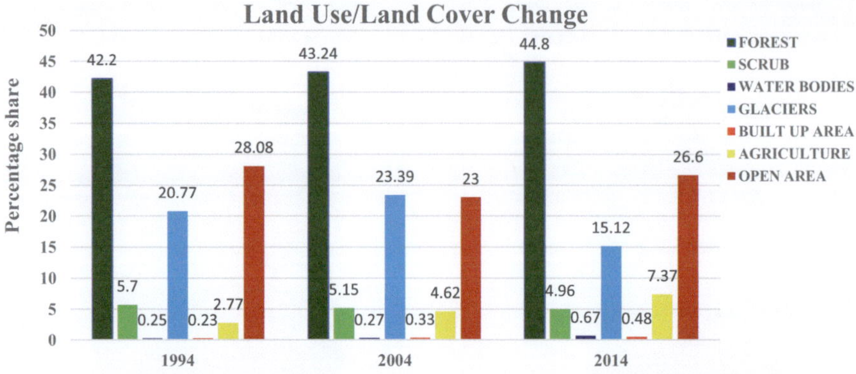

Fig. 3 Percentage share of the land use/land cover (LULC) classes in the Teesta River basin for three different time periods

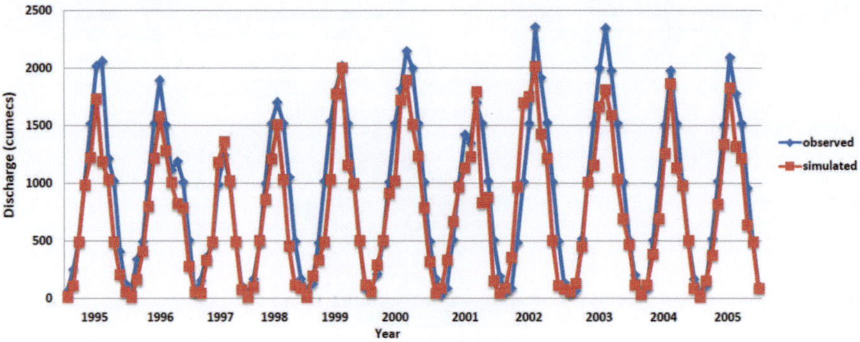

Fig. 4 Calibration and validation result of the SWAT model of the Teesta River basin

category. The percentage share of the area of various LULC for different time periods is shown in Fig. 3.

4.2 Sensitivity Analysis, Model Calibration and Validation

The hydrological model was set up for the time period 1994–2014 with the DEM, soil data, climate data, and LULC data. Sensitivity analysis was carried out on the various sets of parameters to identify the most influential parameters. The model was calibrated using the observed discharge data between the time periods 1995 and 2001, whereas validation was carried out between the time periods 2002 and 2005 (Fig. 4). The model performance showed good agreement between the observed and simulated values. Model performance indices have been shown in Table 2.

Table 2 Model performance indices

Process	R^2	NSE	KGE
Calibration (1995–2001)	0.74	0.63	0.58
Validation (2002–2005)	0.75	0.69	0.66

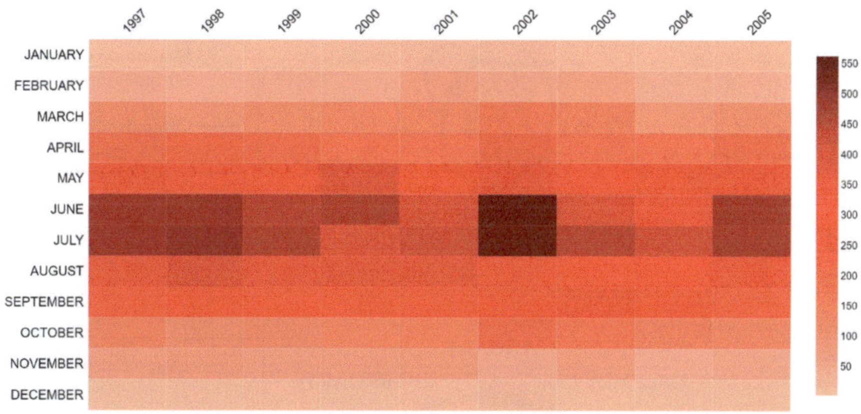

Fig. 5 Matrix illustrating the mean monthly precipitation (mm) patterns within the calibration and validation period for the Teesta River basin

Underprediction of the model results can be observed in certain years in peak values during wet seasons. As precipitation data is the key driving factor in the rainfall and runoff process, there is a higher uncertainty in the precipitation measurement in data-infrequent regions. As the northern portion of the basin lies in the higher altitude of the Himalayan ranges, dominated by snow and glaciers, there are very few or negligible gauging stations due to severe weather conditions and life risk factors, but in the southern portion of the basin, there are significant numbers of stations. Due to the inconsistency in measuring the precipitation in the upper reaches of the basin, it leads to underprediction of the runoff simulated values in the wet seasons.

A matrix illustrating the mean monthly precipitation patterns within the calibration and validation period for the Teesta River basin is shown in Fig. 5. Figure 6a represents a radar flow chart of various water balance components based on the baseline scenario LULC of 1994. It can be observed from the radar flow chart (Fig. 6a) that the various water balance components such as runoff, water yield, and groundwater have their peaks in the months of June, July, and August due to the prevalence of monsoon in those months, whereas actual evapotranspiration (AET) and potential evapotranspiration (PET) have their peaks in the months of March, April, and May.

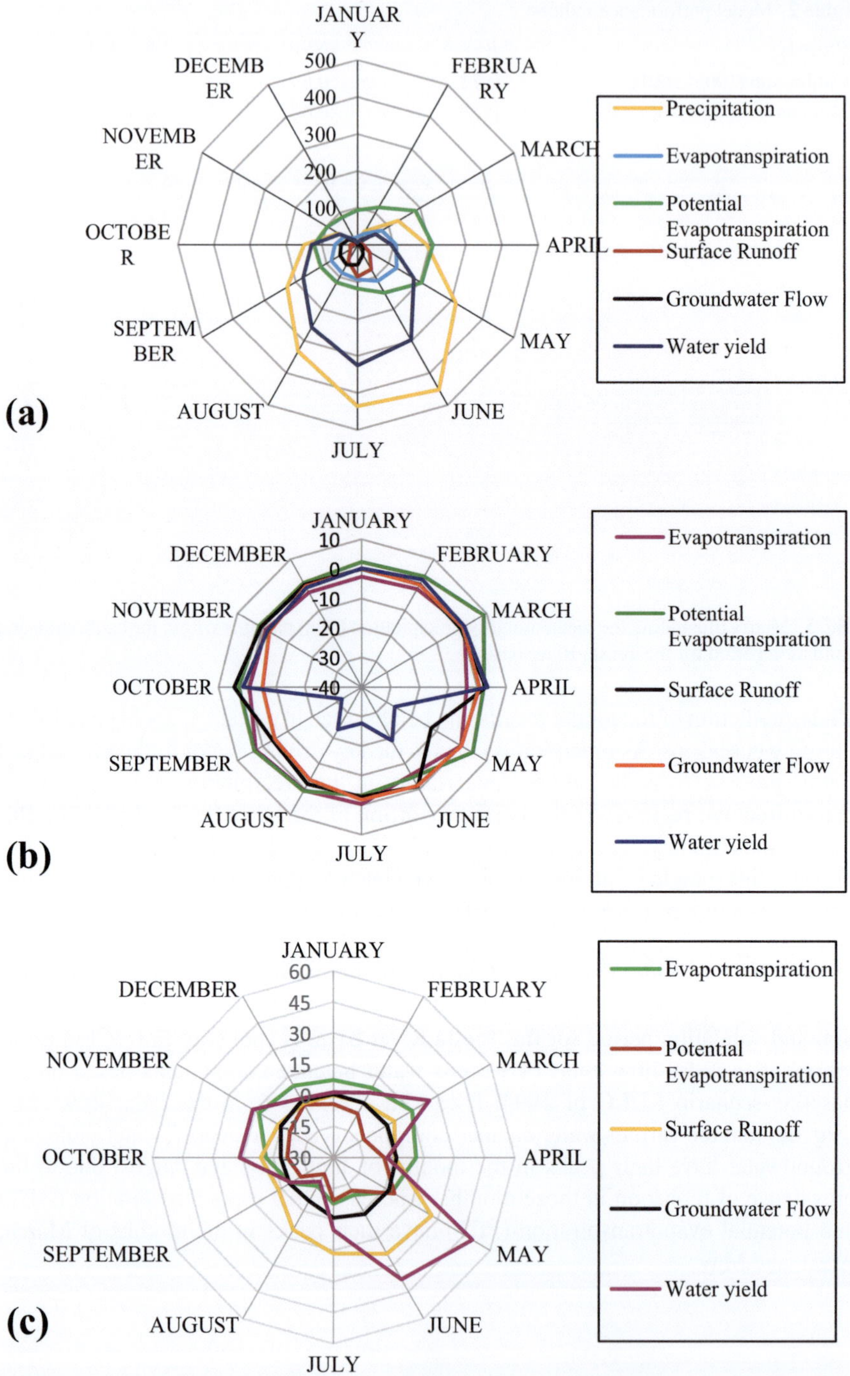

Fig. 6 (**a**) Water balance radar chart for 1994 LULC, (**b**) shift in water balance components from 1994 to 2004, and (**c**) shift in water balance components from 2004 to 2014

5 Discussions

5.1 *Spatial and Temporal Change in Water Balance Components Due to Change in LULC*

Fluctuations in various water balance components for two decades, from 1994 to 2004 and 2004 to 2014, are shown in Fig. 6b and Fig. 6c, respectively. It was observed that there was a reduction in water yield between April and October annually, whereas the other water balance components show similar and uniform changes in pattern between the time period of 1994 and 2004. There was no abrupt or drastic change in the water balance components observed during the time period 2004 to 2014, with surface runoff being the exception. Change in the surface runoff was found to be on the higher side between May and September during the monsoon season. It was observed that surface runoff in the higher altitude was found to be 6% higher as compared to the average catchment surface runoff. Melting of glaciers due to an increase in temperature in the past few decades and conversion of forest to agricultural area (mainly shifting agriculture commonly practiced in the hilly regions in India) were found to be the main reasons behind the increase in surface runoff.

Figure 7 shows the spatial variation of changes in surface runoff, evapotranspiration, and water yield on the sub-basin scale due to changes in LULC between 1994 and 2014. Some of the sub basins in the eastern part indicate a drastic decrease in the groundwater contribution because of the conversion of the area under scrub into either agriculture or built-up. Because a decrease in groundwater contribution in such sub-basins has led to an increase in evapotranspiration and surface runoff. The conversion of scrub into cropland (mainly due to shifting agriculture) and the melting of glaciers in the past decade due to an increase in temperature, especially in the upper steep hilly terrains of the basin, have led to a decrease in groundwater contribution and an increase in streamflow. A similar trend was observed in the center section of the basin, where melting glaciers and increasing surface runoff in the upper part of the catchment resulted in higher water yield.

There is a high resemblance between Landsat 4–5 and Landsat 8 sensors despite the different technologies. During the classification, it was found that band definitions differ just slightly for most bands, and the effect found was minimal. Because of the slight difference in the spectral reflectance of scrub, forest, and agriculture, mainly the tea gardens are prone to error during classification, which leads to a higher amount of uncertainty. However, supervised classification using maximum likelihood classifiers results in more accurate and logically consistent results. Any natural classes getting converted into agriculture show strong spectral changes, but even with techniques based on spectral bands or supervised classification techniques, modification of forests by single tree extraction cannot be adequately resolved. It was found that a hydrological model like SWAT can be used as a versatile tool even in the river basins, which have very diverse LULC with dense forests and glaciers. As shown in Fig. 7, spatial changes in the hydrological parameters and an increase in the surface runoff in the northern section of the Teesta River basin may lead to flooding in the central and especially in the southern section of the basin. A slight increase in the area under forest in the entire catchment may lead to a shift in

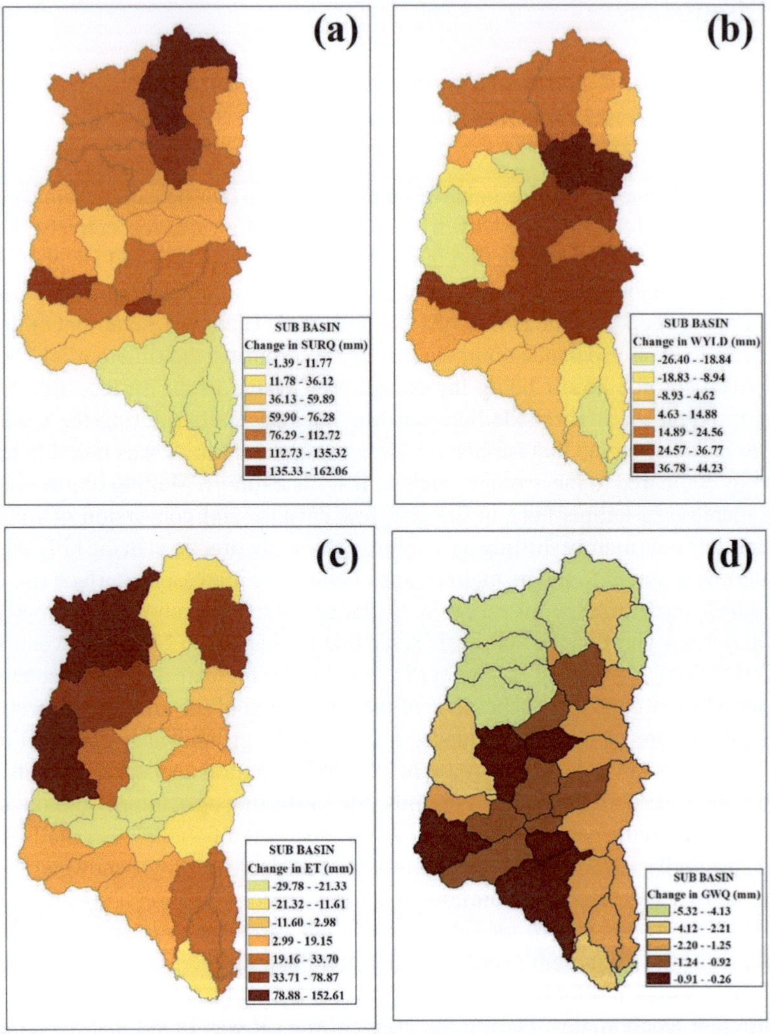

Fig. 7 Average shifts in water balance components in the Teesta River basin, comparing changes from the 1994 LULC map with the 2014 LULC map. Deviations in (**a**) surface runoff, (**b**) water yield, (**c**) evapotranspiration, and (**d**) Groundwater

groundwater contribution from the uplands. A significant increase in the water yield in the central part of the basin makes it more vulnerable to hazardous events like landslides. It can also be noted that agricultural activities and questions related to food security, vegetation, and wetlands can benefit from the high groundwater fluxes from the higher elevation and valley region of the catchment throughout the year.

The main limitation of this study is that it is based on LULC changes in the past. There is a future scope of a study that can be done based on analyzing the impacts of future projected LULC on the hydrological system or the processes in the basin. Future LULC of the basin using algorithms such as artificial neural network (ANN) and the Markov chain (MC) (Anand and Oinam 2020) can be projected, which will aid in planning the future and sustainable use of the water resources.

6 Conclusion

Changing LULC patterns in the past few decades due to various natural and anthropogenic factors is a serious concern in the Himalayan ranges. Melting of glaciers in the upper Himalayan ranges with high altitude and steep slope can have serious impacts on the Teesta River basin. Hydrological models like SWAT can be used as a versatile tool to model and analyze the impact of changes in LULC dynamics. As this study shows the good performance of the SWAT model with $R^2 = 0.74$, NSE = 0.63 in the calibration and $R^2 = 0.75$, NSE = 0.69 in the validation period, strongly suggests the applicability of SWAT even in the basins that have extremely diverse LULC. There is a future scope for a study that can be done using the future projected LULC with high-resolution datasets to quantify the associated risk of change in LULC on the hydrological regime of the Teesta River basin in the coming future.

Acknowledgments The authors gratefully acknowledge the valuable databases from Alaska Satellite Facility, NBSS, and LUP; Earth Resource Observation and Science, Irrigation and Waterways Department, Govt. of West Bengal; and the financial assistance provided by the Ministry of Education (MoE), Govt. of India.

References

Anand, V., & Oinam, B. (2019). Future climate change impact on hydrological regime of river basin using SWAT model. *Global Journal of Environmental Science & Management, 5*(4), 471–484.

Anand, V., & Oinam, B. (2020). Future land use land cover prediction with special emphasis on urbanization and wetlands. *Remote Sensing Letters, 11*(3), 225–234.

Anand, V., Oinam, B., & Parida, B. R. (2020). Uncertainty in hydrological analysis using multi-GCM predictions and multi-parameters under RCP 2.6 and 8.5 scenarios in Manipur River basin, India. *Journal of Earth System Science, 129*, 223. https://doi.org/10.1007/s12040-020-01492-z

Arnold, J. G., & Allen, P. M. (1996). Estimating hydrologic budgets for three Illinois watersheds. *Journal of Hydrology, 176*(1–4), 57–77.

Arnold, J. G., & Fohrer, N. (2005). SWAT-2000: Current capabilities and research opportunities in applied water-shed modeling. *Hydrological Processes, 19*(3), 563–572.

Arnold, J. G., Allen, P. M., & Bernhardt, G. (1993). A comprehensive surface groundwater flow model. *Journal of Hydrology, 142*, 47–69.

Boretti, A., & Rosa, L. (2019). Reassessing the projections of the world water development report. *npj Clean Water, 2*, 15. https://doi.org/10.1038/s41545-019-0039-9

Bouraoui, F., Benabdallah, S., Jrad, A., & Bidoglio, G. (2005). Application of the SWAT model on the Medjerda river basin (Tunisia). *Physics and Chemistry of Earth Parts A/B/C, 30*(8–10), 497–507.

Central Water Commission. (2005). *Water data complete book*. http://www.cwc.nic.in/main/downloads/Water_Data_Complete_Book_2005.

ENVIS Centre: Sikkim. (2019). *Status of environment and related issues*. https://www.sikenvis.nic.in/Database/Rivers_781.aspx

Froehlich, W., Gil, E., Kasza, I., & Starkel, L. (1990). Thresholds in the transformation of slopes and river channels in the Darjeerling Himalaya, India. *Mountain Research and Development, 10*, 301–312.

Gassman, P. W., Reyes, M. R., Green, C. H., & Arnold, J. G. (2007). The soil and water assessment tool: Historical development, applications, and future research directions. *Transactions of the American Society of Agricultural and Biological Engineers, 50*(4), 1211–1250.

Gosain, A. K., Mani, A., & Dwivedi, C. (2009). *Hydrological modeling literature review: Report no. 1*. Indo-Norwegian Institutional Cooperation Program.

Govender, M., & Everson, C. S. (2005). Modeling stream flow from two small South African experimental catchments using the SWAT model. *Hydrological Processes, 19*(3), 683–692.

Gupta, H. V., Kling, H., Yilmaz, K. K., & Martinez, G. F. (2009). Decomposition of the mean squared error and NSE performance criteria: Implications for improving hydrological modeling. *Journal of Hydrology, 377*, 80–91.

Gyamfi, C., Ndambuki, J., & Salim, W. (2016). Hydrologicalresponses to land use/cover changes in the Olifants Basin, South Africa. *Water, 8*(12), 588. https://doi.org/10.3390/w8120588

Kummu, M., Guillaume, J. H. A., de Moel, H., Eisner, S., Flörke, M., Porkka, M., Siebert, S., Veldkamp, T. I. E., & Ward, P. J. (2016). The world's road to water scarcity: Shortage and stress in the 20th century and pathways towards sustainability. *Scientific Reports, 6*, 38495. https://doi.org/10.1038/srep38495

Laflen, J. M., Elliot, W. J., Simanton, J. R., Holzhey, S., & Kohl, K. D. (1991). WEPP soil erodibility experiments for rangeland and cropland soils. *Journal of Soil Water Conservation, 46*(1), 39–44.

Maurer, J. M., Schaefer, J. M., Rupper, S., & Corley, A. (2019). Acceleration of ice loss across the Himalayas over the past 40 years. *Science Advances, 5*, eaav7266.

Mirza. (2003). Climate change and extreme weather events: Can developing countries adapt? *Climate Policy, 3*(3), 233–248.

Moriasi, D. N., Arnold, J. G., Van Liew, M. W., Bingner, R. L., Harmel, R. D., & Veith, T. L. (2007). Model evaluation guidelines for systematic quantification of accuracy in watershed simulations. *American Society of Agricultural and Biological Engineers, 50*(3), 885–900.

Mukhopadhyay, S. C. (1982). *The Tista Basin. A study in fluvial geomorphology*, New Delhi.

Nash, J. E., & Sutcliffe, J. V. (1970). River flow forecasting through conceptual models. A discussion of principles. *Journal of Hydrology, 10*(3), 282–290.

Neitsch, S. L., Arnold, J. G., Kiniry, J. R., & Williams, J. R. (2011). *Soil and water assessment tool theoretical documentation version 2009*. Texas Water Resources Institute.

Rahman, H., Senapati, S. V., & Kumar, A. (2016). *An analysis of past three decade weather phenomenon in the mid hills of Sikkim and strategies for mitigating possible impact of climate change on agriculture*. ENVIS Sikkim.

Rosenthal, W. D., Srinivasan, R., & Arnold, J. G. (1995). Alternative river management using a linked GIS hydrology model. *Transactions of the American Society of Agricultural Engineering, 38*(3), 783–790.

Schuol, J., Abbaspour, K. C., Srinivasan, R., & Yang, H. (2008). Estimation of freshwater availability in the West African sub-continent using the SWAT hydrologic model. *Journal of Hydrology, 352*(1–2), 30–49.

Shen, Z. Y., Chen, L., & Chen, T. (2012). Analysis of parameter uncertainty in hydrological and sediment modeling using GLUE method: A case study of SWAT model applied to Three Gorges Reservoir Region, China. *Hydrology & Earth System Sciences, 16*, 121–132.

Shivhare, V., Goel, M. K., & Singh, C. K. (2014). Simulation of surface runoff for upper Tapi sub catchment area (Burhanpur watershed) using SWAT. *The International Archives of the Photogrammetry, Remote Sensing and Spatial Information Sciences, 40*(8), 391.

Sisay, E., Halefom, A., Khare, D., Singh, L., & Worku, T. (2017). Hydrological modelling of ungauged urban watershed using SWAT model. *Modeling Earth Systems and Environment, 3*(2), 693–702.

Starkel, L., & Basu, S. (2000). *Rains, landslides and floods in the Darjeeling Himalaya*. Indian National Science Academy.

USDA Soil Conservation Service. (1972). *Section 4. Hydrology. In National Engineering Handbook, US*. Department of Agriculture-Soil Conservation Service: Washington, DC.

Young, R. A., Onstad, C. A., Bosch, D. D., & Anderson, W. P. (1989). AGNPS: A nonpoint source pollution model for evaluating agricultural watershed. *Journal of Soil and Water Conservation, 44*, 168–173.

Morphological Patterns of Competing Urban Green Spaces in Megacity: A Case of Delhi

Shadman Nahid, Varun Singh, Ashish Suman, and Susanta Mahato

Abstract This chapter presents an exhaustive analysis of morphological patterns of competing Urban Green Spaces (UGS) in a megacity—Delhi, India, from 1998 to 2018. Leveraging Landsat satellite data, Geographic Information System (GIS) tools, and Census data, the research explores the changing landscape of UGS and its implications for the well-being of urban dwellers. The spatial and temporal Land Use Land Cover (LULC) analysis reveals a consistent increase in human-dominated land use, particularly dense and sparse built-up areas, at the expense of vital green spaces like forests and water bodies. The study further delves into the morphological patterns of UGS, employing Morphological Spatial Pattern Analysis (MSPA). The results highlight the fragmentation and complexity that UGS has undergone over the years, particularly along the Aravali hills. The distribution of UGS is mapped, revealing a decreasing trend over time, with significant disparities in availability across different regions of Delhi. To understand the importance of these to the residents of the city, the study introduces the concept of UGS per capita distribution and Greenness Index (GI). The analysis indicates a concerning decline in UGS per capita in various districts, with southern and southwestern regions exhibiting a more favorable scenario. The GI, correlating UGS availability with population density, highlights the need for strategic interventions to mitigate the adverse effects of rapid urbanization. The study concludes with an in-depth discussion that contextualizes the findings within the broader urban development narrative. The insights from this research offer valuable guidance for urban planners, policymakers, and residents in developing sustainable strategies to preserve and enhance UGS, ensuring a healthier and more ecologically balanced urban environment.

S. Nahid · V. Singh · A. Suman
School of Environmental Sciences (SES), Jawaharlal Nehru University, New Delhi, India

S. Mahato (✉)
Department of General and Applied Geography, School of Applied Sciences, Dr. Harisingh Gour Vishwavidyalaya, Sagar, Madhya Pradesh, India
e-mail: ap_susantam@dhsgsu.edu.in

P. K. Joshi et al. (eds.), *Geospatial Science for Urban Ecosystems*, The Urban Book Series, https://doi.org/10.1007/978-3-032-07155-2_14

Keywords Urban Green Spaces · Landscape Analysis · Land Use Land Cover (LULC) · Greenness Index (GI) · Morphological Spatial Pattern Analysis (MSPA) · Sustainable Urban Development

1 Introduction

Green spaces refer to any vegetated land adjoining an urban area and include bushland, nature reserves, national parks, outdoor sports fields, school playgrounds, and rural or semi-rural areas immediately adjoining an urban area. (Mahato et al., 2023). It is characterized by a significant presence of greenery (Holland et al., 2021; Taylor & Hochuli, 2017) and is becoming increasingly scarce in urban environments, a sentiment echoed by numerous observers (Beatley, 2011). Urban green spaces (UGS), encompassing artificial, semi-natural, and natural ecosystems dominated by vegetation within megacities, have drawn renewed attention in the wake of climate change. City dwellers are rediscovering the appeal of low-density areas boasting scenic beauty and recreational options (Labib et al., 2021; Mahato and Pal, 2022). While green spaces are not uncommon in cities (Beatley, 2011), their potential and importance to confer physical health benefits have a higher appreciation than rural green spaces (Cox et al., 2018). Given the ongoing urbanization trend, where an estimated 68 percent of the global population (6.7 billion) is projected to reside in urban areas by 2050 (UN-HABITAT, 2024), integrating the competing green spaces into urban planning becomes imperative for fostering healthier cities (Dye, 2008).

The UGS offers a multitude of ecosystem goods and services, including provisioning, regulatory, and cultural services. *Firstly*, they contribute raw materials, such as timber and leaf litter, and play a crucial role in replenishing water supplies to ensure the availability of potable water (Wolch et al., 2014). *Secondly*, UGS act as natural filters, mitigating air and water pollution, regulating microclimates, and preventing flooding (Vieira et al., 2018). *Thirdly*, these spaces serve as recreational hubs, encouraging outdoor physical activity and social interactions, thus contributing to the physical and psychological well-being of urban dwellers (Wood et al., 2018; Matos et al., 2019). And, *finally*, UGS play a pivotal role in promoting social justice by offering equitable and accessible spaces for the public (Braubach et al., 2017). Existing research suggests that UGS can help reduce health disparities by alleviating psychosocial stress in economically disadvantaged households (Kondo et al., 2018), mitigating environmental hazards in lower socioeconomic communities (Bhatnagar, 2017; Nahid et al., 2023), and improving the overall health of vulnerable groups (Hunter et al., 2019; Mahato et al., 2021). Thus, ensuring equal access to UGS for all citizens becomes crucial, with a growing focus on understanding the spatio-temporal patterns and inequalities in UGS services.

In the context of mega- and metro cities, most of these across the globe exhibit the lowest environmental quality concerning air quality and the highest population density (Arunadeepa, 2007; Senanayake et al., 2013). Similarly, in Delhi, the challenge of air pollution has intensified alongside urbanization, primarily driven by regional and local sources (Kumar et al., 2015), like anthropogenic emissions from vehicles, electricity generation, construction, road dust, waste burning, and household activities, which contribute significantly to air pollution (Mahato and Ghosh, 2020). In between, the phenomenon of the heat island effect further challenges the energy budget, air temperature, and various environmental processes. Studies have highlighted the negative correlation between the vegetation cover and ambient temperature (Bastiaanssen et al., 1998; Weng et al., 2004; Kumar et al., 2012). Conversely, the built-up demonstrates a positive correlation with temperature, indicating the influence of built-up areas on local temperatures (Kumari et al., 2018).

Beyond the realm of temperature regulation, the UGS and vegetation cover play a crucial role in reducing air pollution through their purification capacity (Bolund & Hunhammar, 1999). Modeling studies on criteria pollutants have highlighted the significance of green cover in mitigating the adverse impacts of air pollution (Venkataraman et al., 2018). The benefits of UGSs extend beyond environmental functions to encompass social interaction, recreation, aesthetics, cultural heritage preservation, and ecological contributions (Haaland & van Den Bosch, 2015; Priemus, 2004; Mell, 2009). Moreover, UGSs have been recognized for positively impacting the human psyche by alleviating mental stress (Faryadi & Taheri, 2009), with environmental quality exerting a substantial influence on the overall quality of life for urban inhabitants (Wolch et al., 2014). A growing body of literature has provided ample evidence of the association between health impacts and prevalent air pollutants, such as ozone, nitrogen oxides, and particulate matter (O'Lenick et al., 2019). In large cities where air pollution sources are abundant and temperatures are amplified as heat island effects (Zhou & Shepherd, 2010; Conlon et al., 2016; Mahato et al., 2023; Mahato et al., 2025), these environmental hazards coalesce, posing considerable health challenges.

This study is positioned to capture morphological patterns of competing UGSs, including their quality and distribution, which is critical for sustainable urban development. This aims to unveil spatial differentiation laws at various scales, providing a foundation for the formulation of effective green planning guides. As the national capital of India grapples with the complexities of urbanization, this research seeks to offer insights that can inform strategic interventions for sustainable urban development, acknowledging the interconnectedness of environmental quality, human well-being, and the broader fabric of the urban landscape.

2 Materials and Methods

2.1 Study Site

The study focuses on the National Capital of India, Delhi, which stands as the world's second-largest megacity and India's most extensive urban agglomeration. As of the 2021 Census of India, Delhi is home to a staggering 1.68 crore residents, exhibiting a notably high population density of 11,297 persons per square kilometer

(Rathi et al., 2023). Positioned between latitudes 28.41°N and 28.88°N and longitudes 76.84°E and 77.35°E, Delhi lies within the Indo-Gangetic region, encompassing an expansive area of 1484 km^2. Situated along the banks of the Yamuna River, a tributary of the Ganga, the city's elevation ranges between 213 and 290 m above sea level. Delhi experiences a semi-arid to sub-humid climate, characterized by a substantial saturation deficit and limited to moderate rainfall. The summers are marked by extreme heat, with temperatures soaring between 43 and 48 °C, while winters bring chilly conditions, with minimum temperatures dropping to 4–6 °C (Mohan et al., 2020). The city encounters an annual average rainfall ranging from 400 to 714 mm, portraying a pattern of minimal precipitation.

Delhi serves as the economic epicenter of the nation, attracting residents from diverse regions and contributing significantly to urban sprawl. Despite its status as the capital, Delhi grapples with insufficient urban planning and natural resource management. This deficit poses challenges to maintaining a high standard of living in the city. The rapid urbanization and population growth in Delhi pose multifaceted challenges. Urban sprawl has led to alterations in the city's landscape, resulting in a competing distribution of UGS. The increasing pressure on natural resources calls for a comprehensive understanding of the city's urbanization dynamics and green space characteristics to formulate effective strategies for sustainable urban development. The primary objectives of this study are to comprehensively gather information on Delhi's urbanization, analyze the distribution of UGS, and delineate their characteristics. This novelty lies in its integrated approach to estimating the Greenness Index (GI) and per capita UGS. The findings will hold implications for urban planning, environmental conservation, and sustainable development, guiding policymakers in formulating strategies to enhance the quality of life in megacities across the world.

2.2 Data Acquisition and Preprocessing

Landsat TM/ETM+/OLI satellite images spanning the two-decade period (1998–2018) were acquired from the United States Geological Survey (USGS) website (https://www.usgs.gov/) in GeoTIFF format, utilizing the UTM projection and WGS 84 datum. LULC maps for the years 1998 to 2011 were sourced from the comprehensive study conducted by Ghosh et al. (2014). The integration of phenological behavior evaluation in the multi-seasonal data collection process facilitated the consideration of intra-annual variability in land use practices, thereby minimizing confusion regarding distinct vegetation classes. The LULC classification encompasses eleven distinct classes, namely Rabi (winter), Kharif (monsoon), Zaid (summer), double crops, waterbody, dense built-up, sparse built-up, plantation, scrub, forest, and exposed area. The overall map accuracy for the years 1998, 2003, and 2011 stood at 83% (kappa accuracy 74%), 79% (69%), and 82% (73%), respectively. Employing a similar classification technique for the satellite data from 2018, the resultant map exhibited an overall accuracy of 84% (kappa accuracy 76%). To facilitate the quantification of UGS, the maps were aggregated into selected LULC categories, streamlining the analysis and interpretation process.

2.3 Estimation of NDVI

The red and infrared bands from Landsat data for the years 1998, 2003, 2011, and 2018 were utilized to calculate the Normalized Difference Vegetation Index (NDVI). NDVI [= (NIR-Red)/(NIR + Red)] serves as a key indicator of vegetation health, relying on the fact that healthy vegetation reflects well in the near-infrared spectrum. The ratio of near-infrared (NIR) and red bands is normalized from −1 to +1, partially accommodating variations in illumination and surface slope (Bindi et al., 2009).

2.4 Urban Green Space Extraction

For UGS extraction, LULC classes were regrouped into two categories: vegetated (1) and non-vegetated (0). Vegetated categories included forest, plantation, and scrub, while all other LULC classes were classified as non-vegetated. Simultaneously, NDVI was categorized into vegetated (1) and non-vegetated (0) using density slicing. The intersection of density-sliced NDVI maps with the vegetation and non-vegetated layers provided the distribution of UGSs.

2.5 Urban Green Space Per Capita Extraction

Population density, an indicator of urban living quality, was calculated using the Census of India 2011 data. This metric was crucial for evaluating green space per capita and determining the green space requirement for enhanced living standards. UGS maps, combined with Delhi's population data at the ward level, facilitated the computation of UGS per capita (sq m/capita). Adhering to World Health Organization (WHO) guidelines, the minimum green space required per capita for healthy living is 9.5 sqm/person (Kuchelmeister, 1998). The methodology is depicted in Fig. 1.

2.6 Estimation of Greenness Index

A Greenness Index (GI) was derived by relating UGS per capita to population density at the ward level. This index plays a pivotal role in gauging the quality of life for urban residents. Areas meeting WHO criteria are deemed "good" and require no

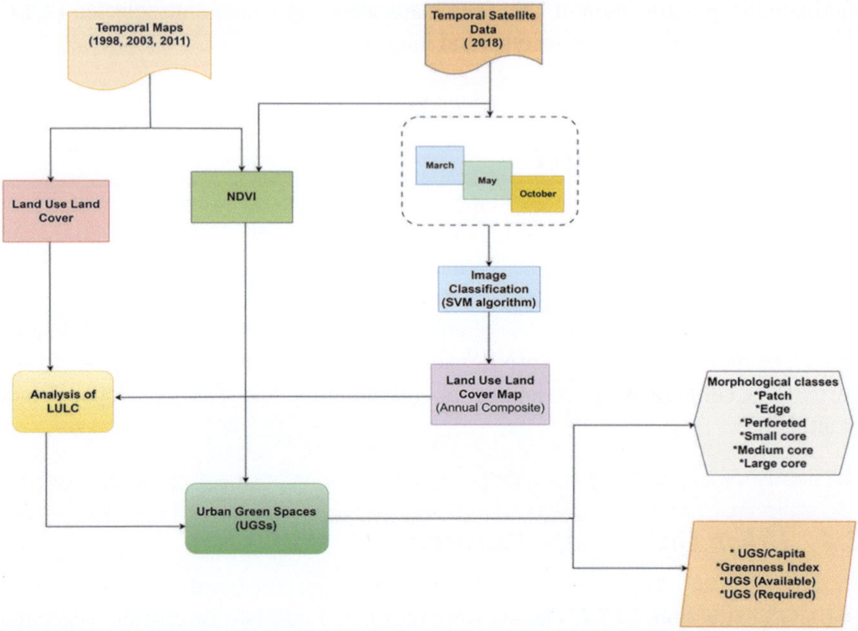

Fig. 1 Flowchart illustrating the comprehensive methodology of the study

additional plantation, while those falling short necessitate planting for sustained quality of life.

2.7 Urban Green Space Morphology Patterns

For a deeper understanding of UGS morphology, maps were subjected to Spatial Pattern Analysis (MSPA) using the MSPA toolbox in ArcGIS 10.8. This analysis classified UGSs into different patterns, such as core, edge, perforation, patch, and branch, providing valuable insights into the morphological dynamics of UGS within the city. The description of the different MSPA classes is given in Table 1.

3 Results

3.1 Spatio-Temporal LULC Analysis

The spatio-temporal mapping of Land Use Land Cover (LULC) in Delhi for the years 1998, 2003, 2011, and 2018 provides crucial insights into the city's evolving landscape (Fig. 2). The LULC classification reveals a consistent shift toward

Table 1 Definition of morphological spatial patterns (Wang et al., 2019; Vogt et al., 2007)

MSPA classes	Definitions
Core	UGS surrounded on all sides (8-connectivity) by UGS
Edge	Transition zone between green and gray infrastructure
Perforation	Transition zone between GI and gray infrastructure areas; interior regions of GI shaped like a doughnut with perforations (inner edges)
Patch	Unconnected class without a core

human-dominated land use, specifically dense and sparse built-up areas, accompanied by a substantial decline in forest, scrub, and water bodies. The trend in dense built-up areas is particularly noteworthy, showcasing a significant growth of 11% over the two decades. The expansion is evident with a progression from 4% (67.47 sq. km) in 1998 to 15% (235.70 sq. km) in 2018. Similarly, sparsely built-up areas display a temporal increase, reaching 34.23% of the total area in 2018. However, there is a decline observed in 2011, attributed to the transformation of sparse built-up areas into dense built-up regions (Table 2).

In contrast, scrub and forest areas consistently decrease, contributing 7.81% and 3.81% of the total area in 2018, respectively. Plantation, reflecting planned efforts to maintain green cover, increases by 2% over time, reaching 5.10% of the total area in 2018. However, water bodies witness a decline in both quantity and quality, emphasizing the environmental impact of urbanization on aquatic ecosystems. These results underscore the dynamic nature of urbanization in Delhi and highlight the urgent need for sustainable land use practices and conservation measures to preserve the city's ecological balance amid rapid development.

3.2 *Spatio-Temporal Distribution of Urban Green Spaces (UGSs)*

The spatio-temporal distribution of UGS in Delhi, as depicted in Fig. 3, reveals a concerning trend of decreasing UGS over the studied time period, primarily attributed to rapid urbanization. The analysis indicates that a substantial portion of UGS is concentrated in the southern part of Delhi, while other regions face a notable deficiency in green spaces. Notably, Central Delhi, administered by the New Delhi Municipal Corporation (NDMC), showcases a considerable amount of UGS, although it has become increasingly fragmented over time. The southern UGS, influenced by population growth, has also experienced heightened fragmentation. The distribution of UGSs exhibits a growing fragmentation pattern, signaling potential challenges for the city's environmental sustainability. The spatio-temporal analysis of UGS offers valuable insights into the evolving landscape of Delhi. The decreasing trend raises concerns about the potential environmental implications of urban expansion. The concentration of UGS in specific regions, particularly the

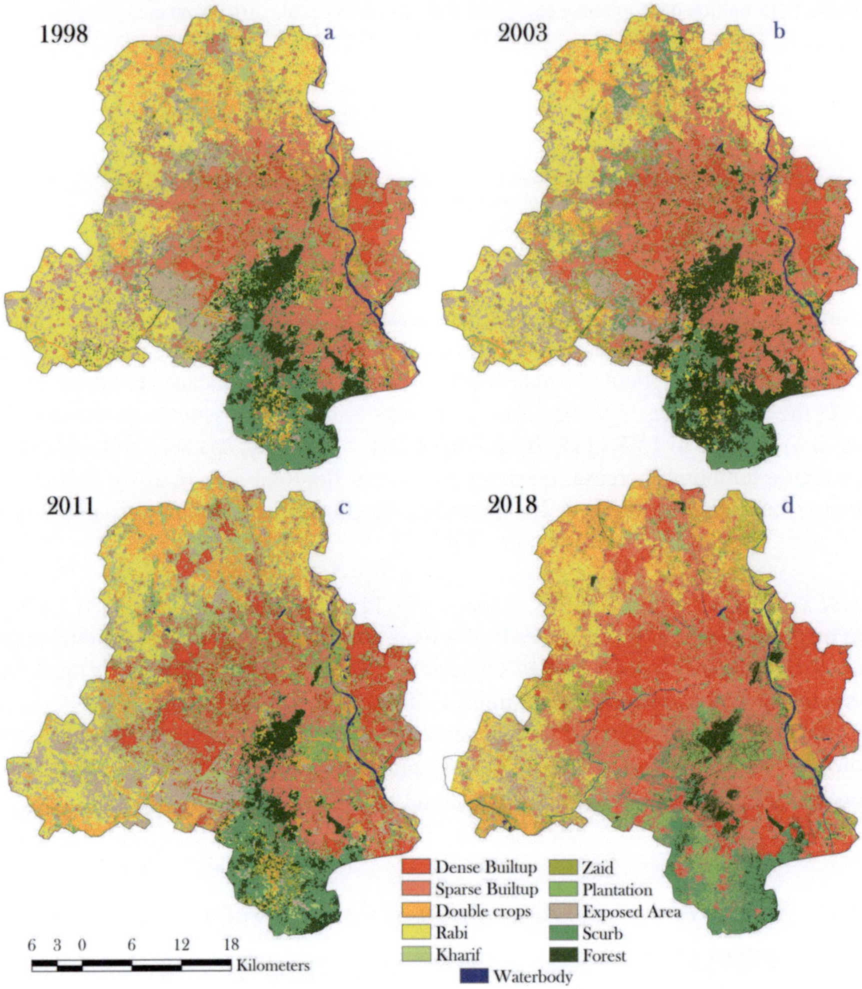

Fig. 2 Temporal LULC maps of Delhi derived from Landsat satellite data. Maps were prepared for each season (March, May, and October) and then integrated to form an annual map. LULC maps for the respective years (**a**) 1998, (**b**) 2003, (**c**) 2011, and (**d**) 2018

southern part, highlights disparities in green space distribution, which may contribute to environmental inequities across the city. The fragmentation observed in Central Delhi and the southern UGS underscores the need for strategic urban planning to maintain cohesive green spaces.

Table 2 Distribution of Land Use Land Cover (LULC) over the four time period

LULC classes	1998	2003	2011	2018
Dense built-up	67.47	130.34	181.27	235.70
Sparse built-up	373.45	415.38	260.12	518.08
Double crops	154	136.98	212.40	174.67
Rabi	276.99	276.26	183.96	176.94
Kharif	100.36	18.57	164.92	37.59
Zaid	23.74	34.46	31.42	34.43
Plantation	58.79	68.21	119.52	77.17
Exposed area	154.59	155.59	179.58	71.81
Scrub	127.78	98.70	103.62	118.19
Forest	122.06	176.29	74.34	57.70
Waterbody	11.41	8.38	8.07	10.92
Total	1471.19	1519.20	1519.28	1513.25

3.3 Competing Urban Green Space

Figure 4a presents the UGS per capita distribution at the ward level in Delhi. The World Health Organization (WHO) recommends a minimum of 9.5 sqm/person for healthy living. The analysis reveals a stark disparity in UGS availability across districts, with Eastern, Central, Northern, and Western districts experiencing considerably lower UGS per capita. In contrast, the southern and southwestern districts demonstrate sufficient UGS, attributed to well-planned urbanization. Out of 282 wards, 175 fall below the WHO standard, emphasizing the need for increased green spaces in these areas. This underscores the significance of strategic urban planning to enhance UGS per capita in underserved regions. The Greenness Index, reflecting UGS per capita to population density, is illustrated in Fig. 4b. Higher greenness indices are observed in the South, Southwest, and parts of Central and Northern Delhi, indicating favorable green space availability. Conversely, the Eastern, Northeastern, and Western parts exhibit lower greenness indices, emphasizing the need for interventions to enhance green cover. This analysis reinforces the importance of considering population density in assessing the quality of green spaces, especially in areas with lower indices.

Figure 4c showcases the spatio-temporal distribution of UGS in Delhi, revealing a decreasing trend over time. The southern part stands out as a major contributor to UGS, aligning with the findings of the spatio-temporal analysis. However, disparities persist, with other regions experiencing a scarcity of UGS. This underscores the challenge of maintaining a balance between urban development and preserving green spaces, particularly in rapidly urbanizing areas. The UGS requirement analysis in Fig. 4d delineates areas with varying needs for green spaces. Well-planned regions in the South and Southwest exhibit adequacy, while the North and East regions demand increased green cover. Notably, 57 wards have no immediate requirement, while 28 wards have sufficient but fragile green cover. However, 109 wards face a deficiency, 84 wards have significantly fewer green spaces, and 4 wards

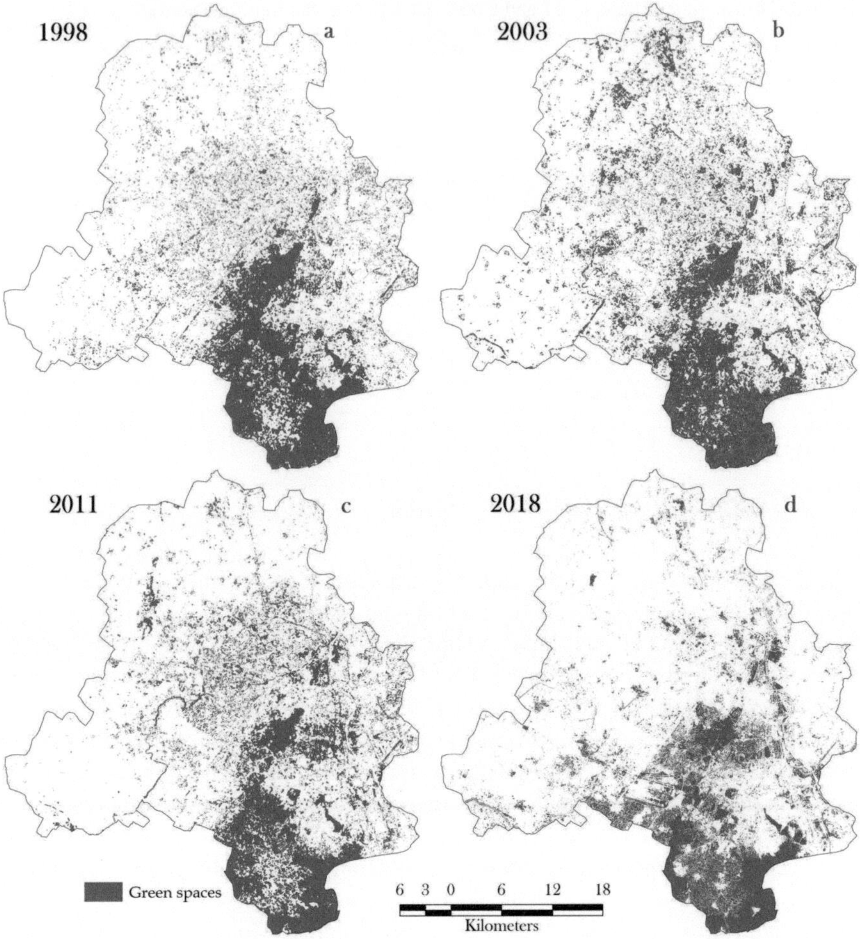

Fig. 3 Map showing changes in green spaces from 1998 to 2018 across four panels labeled (**a**), (**b**), (**c**), and (**d**). Each panel represents a different year: 1998, 2003, 2011, and 2018

require substantial green space augmentation. This analysis provides a basis for targeted interventions, emphasizing the urgency of addressing specific ward-level green space requirements for sustained urban quality and environmental health. The comprehensive examination of UGS distribution, per capita availability, greenness indices, and requirement analysis offers valuable insights for urban planners and policymakers. The observed disparities highlight the need for tailored strategies in different regions of Delhi. While certain areas showcase adequate green spaces, addressing the deficiencies in others requires a nuanced approach. The findings underscore the importance of sustainable urban planning to ensure equitable access to green spaces, promoting environmental well-being and a higher quality of life for Delhi's residents. The integration of population density considerations further

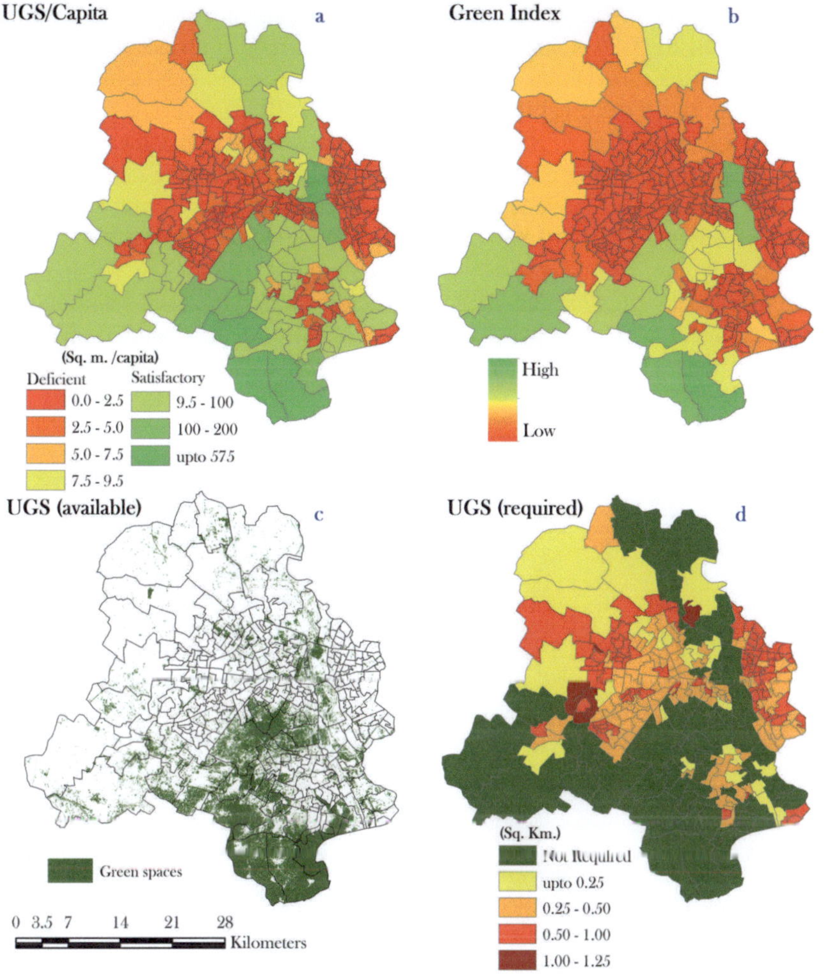

Fig. 4 (**a**) Urban green spaces (UGS) per capita, (**b**) Green Index, (**c**) distribution of UGS, and (**d**) UGS required

refines the assessment, allowing for targeted interventions to address specific needs in the urban landscape.

3.4 Morphological Patterns of Urban Green Spaces (UGS)

Figure 5 presents a spatio-temporal analysis of UGS morphology in Delhi, aimed at enhancing our understanding of UGS characterization. This analysis distinguishes six morphological classes, namely patch, edge, perforated, small core, medium core, and large core, each exhibiting different distribution patterns over four time periods (1998, 2003, 2011, and 2018) (Table 3). The analysis reveals several intriguing trends in the distribution of UGS morphological classes. Notably, the patch category exhibits a marginal increase of 1.16% over the two-decade period, rising from 16.72% in 1998 to 17.88% in 2018. The edge category displays a more pronounced increase of 2%, growing from 22.46% to 25.41% over the same timeframe. The most remarkable change is observed in the perforated class, with a substantial 17% increase, soaring from 3.86% in 1998 to 20.93% in 2018. In contrast, the Small Core and Medium Core categories experience modest increases of 5% and 0.8%, respectively, between 1998 and 2018. However, the large core class stands out with a stark decline, plummeting by 27% over the same period, decreasing from 51.91%

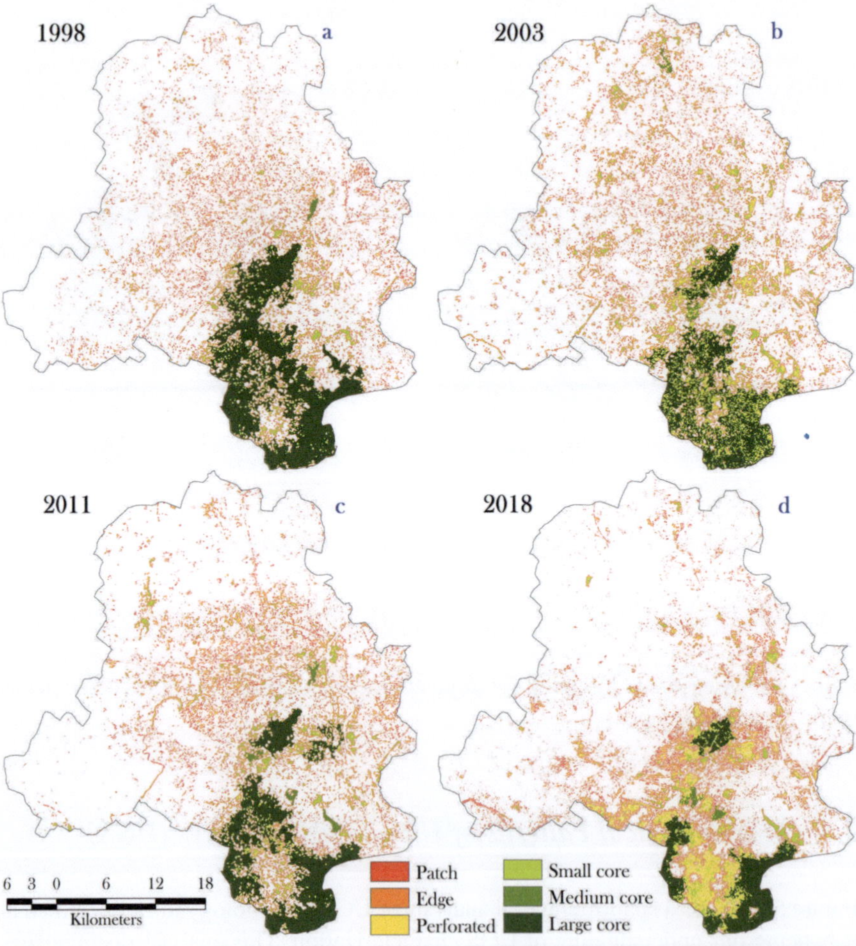

Fig. 5 Morphological patterns of UGS of the respective years (**a**) 1998, (**b**) 2003, (**c**) 2011, and (**d**) 2018

Table 3 Distribution of UGS morphological classes over the four time period

Morphological classes	1998	2003	2011	2018
Patch	16.72	14.48	16.52	17.88
Edge	22.46	29.89	31.65	25.41
Perforated	3.86	17.55	4.82	20.93
Small Core	4.63	9.85	8.40	10.46
Medium Core	0.42	1.07	1.80	1.26
Large Core	51.91	27.16	36.81	24.06

in 1998 to 24.06% in 2018. These findings underline the dynamic nature of UGS morphology in Delhi. Importantly, this dynamic shift is not uniform across all morphological classes, which is a key observation from this analysis. These results suggest that the composition and distribution of UGS morphological patterns in Delhi are changing over time. Such changes may be indicative of various urban development and planning processes, environmental pressures, and the city's evolving ecological and sociological dynamics. A deeper investigation into the factors driving these changes is warranted to understand their implications for sustainable urban development and green space preservation in Delhi.

4 Discussion

4.1 *Urban Green Space Morphology Dynamics of Urban Green Spaces (UGS)*

The study's comprehensive analysis of UGS morphology in Delhi over the years reveals intricate dynamics influenced by rapid urbanization. The morphological classes exhibit shifts, with substantial changes in edge, perforated, and core types. The observed decrease in large core-type UGS emphasizes potential fragmentation, raising concerns about the overall connectivity and accessibility of green spaces. The rise in small core-type UGS indicates efforts to enhance compact green areas, possibly reflecting the implementation of localized green initiatives. These findings underscore the need for adaptive strategies in urban planning to sustain a balance between expansive and compact green spaces, promoting ecological resilience and improved quality of life (Artmann et al., 2019; Hansen et al., 2019).

4.2 *Implications for Urban Planning and Policy*

UGS dynamics hold significant implications for urban planning and policy formulation (Kim & Kwon, 2021). The consistent decrease in UGS over time, particularly in central and southern regions, calls for strategic interventions. Urban planners and policymakers need to address the challenges posed by rapid urbanization, such as increased population density and shrinking green spaces. The study emphasizes the importance of preserving UGS in densely populated areas to mitigate adverse effects on physical, mental, and emotional well-being. Policies promoting sustainable urban development, green infrastructure, and community engagement are crucial to ensuring the continued availability and accessibility of green spaces in Delhi.

4.3 Population Density and Urban Green Space Availability

The study extends to the broader context of population density and its impact on UGS availability in Delhi. The study aligns with previous research highlighting the inverse relationship between population density and green space availability (Sikorska et al., 2020; Song et al., 2021). Delhi's high population density, as indicated by census data, contributes to the reduced availability of UGS. This is a critical concern given the profound impact of green spaces on urban dwellers' well-being. Despite being the capital of India, the study emphasizes the urgent need for attention to this escalating issue. Strategic planning, regulatory measures, and community involvement are imperative to address the challenges posed by overpopulation and illegal settlements (Grangxabe et al., 2023).

4.4 Comparative Analysis and Global Perspectives

The study draws parallels with international research, emphasizing the global significance of UGS in urban areas. Comparative analyses with studies in Poland highlight common challenges related to insufficient UGS in densely populated cities (Sikorska et al., 2020). Lessons from international practices, such as the Urban Neighborhood Green Index (UNGI) in Poland, offer insights into identifying critical areas and formulating improvement policies (Badach & Raszeja, 2019). The focus on individual green assets, such as terrace gardens and private lawns, presents an alternative approach to enhancing greenery in densely built-up areas. The study contributes to the discourse on global urbanization challenges, emphasizing the need for context-specific strategies while learning from successful international initiatives (Frantzeskaki et al., 2018; Croese & Duminy, 2023).

4.5 Urban Heat Islands (UHI) and Energy Consumption

The discussion delves into the environmental implications of reduced greenery, particularly the formation of Urban Heat Islands (UHI) and increased energy consumption. The densely populated built-up areas in Delhi, lacking sufficient green cover, contribute to the annual rise in electricity consumption. The study's findings align with research linking UHI formation to decreased greenery. Strategies that focus on individual green assets are proposed to counteract UHI effects (Ulpiani, 2021). This underscores the interconnectedness of urban planning, environmental sustainability, and energy consumption, emphasizing the need for holistic policies that address multiple facets of urban development.

4.6 Recommendations for Future Urban Green Space Management

The study calls for continued monitoring and conservation efforts to ensure the sustainable development of USG in Delhi. It recommends a multifaceted approach to future UGS management, involving adaptive urban planning, community engagement, and policy interventions. The findings serve as a foundation for informed decision-making, guiding stakeholders toward a resilient and green urban future in the face of rapid urbanization challenges.

5 Conclusion

UGS play a pivotal role in enhancing the well-being of urban dwellers, influencing various aspects of health, social interactions, physical activity, and environmental quality. This study provides valuable insights into the dynamic nature of UGS in Delhi, shedding light on the changing landscape over the years. The findings reveal a consistent decline in UGS and GI in certain areas, emphasizing the urgent need for strategic interventions to address the challenges posed by rapid urbanization. Despite the numerous benefits associated with UGS, the study underscores the environmental and health implications of inadequate green spaces. The negative impacts include the Urban Heat Island (UHI) effect, increased electricity consumption, and poor air quality, all of which contribute to the degradation of residents' health and well-being. The characterization of green spaces exposes the fragmentation and complexity that have intensified over the years, particularly along the Aravali hills, indicating the need for targeted conservation efforts. The study also highlights the concerning environmental quality in the National Capital Region (NCR), with poor air quality, high population density, and unauthorized construction contributing to the degradation of green spaces. Residential areas located away from the city center exhibit better environmental quality, underlining the influence of unregulated construction and illegal settlements.

In light of these findings, the study suggests implementing mitigation strategies at the ward level, focusing on afforestation programs and spatial analysis. The promotion of smart options and the development of ward-level plans for urban sustainability are deemed crucial in a developing country where resources are scarce. The study's analysis serves as a foundation for informed decision-making by urban planners, policymakers, and residents, guiding future efforts to enhance and sustain UGS in Delhi. While this study provides valuable insights, it is essential to acknowledge its limitations. The analysis relies on available data, and any gaps or inaccuracies in the data sources may influence the study's outcomes. Additionally, the study focuses on a specific time frame, and the dynamic nature of urban environments necessitates continuous monitoring to capture ongoing changes accurately. The complexity of factors influencing UGS, such as socioeconomic aspects and policy implementation, requires a more in-depth investigation for a comprehensive understanding. Despite these limitations, this study serves as a valuable contribution to the discourse on UGS and lays the groundwork for future research and interventions in sustainable urban development.

References

Artmann, M., Inostroza, L., & Fan, P. (2019). Urban sprawl, compact urban development and green cities. How much do we know, how much do we agree? *Ecological Indicators, 96*, 3–9.

Arunadeepa, K. P. W. (2007). *Factors influencing the success of urban green spaces in Sri Lanka* (Doctoral dissertation).

Badach, J., & Raszeja, E. (2019). Developing a framework for the implementation of landscape and greenspace indicators in sustainable urban planning. Waterfront landscape management: Case studies in Gdańsk, Poznań and Bristol. *Sustainability, 11*(8), 2291.

Bastiaanssen, W. G., Menenti, M., Feddes, R. A., & Holtslag, A. A. M. (1998). A remote sensing surface energy balance algorithm for land (SEBAL). 1. Formulation. *Journal of Hydrology, 212*, 198–212.

Beatley, T. (2011). Biophilic cities: what are they?. *Biophilic Cities: Integrating Nature into Urban Design and Planning*, 45–81.

Bhatnagar, A. (2017). Environmental determinants of cardiovascular disease. *Circulation Research, 121*(2), 162–180.

Bindi, M., Brandani, G., Dessì, A., Dibari, C., Ferrise, R., Moriondo, M., & Trombi, G. (2009). Impact of climate change on agricultural and natural ecosystems. *American Journal of Environmental Sciences, 5*(5), 633–638.

Bolund, P., & Hunhammar, S. (1999). Ecosystem services in urban areas. *Ecological Economics, 29*(2), 293–301.

Braubach, M., Egorov, A., Mudu, P., Wolf, T., Ward Thompson, C., & Martuzzi, M. (2017). Effects of urban green space on environmental health, equity and resilience. *Nature-Based Solutions to Climate Change Adaptation in Urban Areas: Linkages between Science, Policy and Practice*, 187–205.

Conlon, K., Monaghan, A., Hayden, M., & Wilhelmi, O. (2016). Potential impacts of future warming and land use changes on intra-urban heat exposure in Houston, Texas. *PloS one, 11*(2), e0148890.

Cox, D. T., Shanahan, D. F., Hudson, H. L., Fuller, R. A., & Gaston, K. J. (2018). The impact of urbanisation on nature dose and the implications for human health. *Landscape and Urban Planning, 179*, 72–80.

Croese, S., & Duminy, J. (2023). Co-producing urban expertise for SDG localization: The history and practices of urban knowledge production in South Africa. *Urban Geography, 44*(3), 538–557.

Dye, C. (2008). Health and urban living. *Science, 319*(5864), 766–769.
Faryadi, S. and Taheri, S., (2009). Interconnections of Urban Green Spaces and Environmental Quality of Tehran. *Int. J. Environ. Res, 3*(2), pp.199–208.
Frantzeskaki, N., Bach, M., & Mguni, P. (2018). Understanding the urban context and its challenges. *Co-creating sustainable urban futures: A primer on applying transition management in cities*, 43–61.
Ghosh, A., Sharma, R. and Joshi, P.K., 2014. Random forest classification of urban landscape using Landsat archive and ancillary data: Combining seasonal maps with decision level fusion. *Applied Geography, 48*, pp.31-41.
Grangxabe, X. S., Maphanga, T., Madonsela, B. S., Gqomfa, B., Phungela, T. T., Malakane, K. C., et al. (2023). The escalation of informal settlement and the high levels of illegal dumping post-apartheid: Systematic review. *Challenges, 14*(3), 38.
Haaland, C., & van Den Bosch, C. K. (2015). Challenges and strategies for urban green-space planning in cities undergoing densification: A review. *Urban Forestry & Urban Greening, 14*(4), 760–771.
Hansen, R., Olafsson, A. S., Van Der Jagt, A. P., Rall, E., & Pauleit, S. (2019). Planning multi-functional green infrastructure for compact cities: What is the state of practice? *Ecological Indicators, 96*, 99–110.
Holland, I., DeVille, N. V., Browning, M. H., Buehler, R. M., Hart, J. E., Hipp, J. A., et al. (2021). Measuring nature contact: A narrative review. *International Journal of Environmental Research and Public Health, 18*(8), 4092.
Hunter, R. F., de la Haye, K., Murray, J. M., Badham, J., Valente, T. W., Clarke, M., & Kee, F. (2019). Social network interventions for health behaviours and outcomes: A systematic review and meta-analysis. *PLoS Medicine, 16*(9), e1002890.
Kim, I., & Kwon, H. (2021). Assessing the impacts of urban land use changes on regional ecosystem services according to urban green space policies via the patch-based cellular automata model. *Environmental Management, 67*(1), 192–204.
Kondo, M. C., Fluehr, J. M., McKeon, T., & Branas, C. C. (2018). Urban green space and its impact on human health. *International Journal of Environmental Research and Public Health, 15*(3), 445.
Kuchelmeister, G. (1998). *Urban forestry: present situation and prospects in the Asia and Pacific region, FAO Asia-Pacific Forestry Sector Outlook Stud, Rome*. Food and Agriculture Organization of the United Nations. Forestry Policy and Planning Division.
Kumar, K. S., Bhaskar, P. U., & Padmakumari, K. (2012). Estimation of land surface temperature to study urban heat Island effect using Landsat ETM+ image. *International journal of Engineering Science and technology, 4*(2), 771–778.
Kumar, R., Barth, M. C., Pfister, G. G., Nair, V. S., Ghude, S. D., & Ojha, N. (2015). What controls the seasonal cycle of black carbon aerosols in India? *Journal of Geophysical Research: Atmospheres, 120*(15), 7788–7812.
Kumari, B., Tayyab, M., Shahfahad, S., Mallick, J., Khan, M. F., & Rahman, A. (2018). Satellite-driven land surface temperature (LST) using Landsat 5, 7 (TM/ETM+ SLC) and Landsat 8 (OLI/TIRS) data and its association with built-up and green cover over urban Delhi, India. *Remote Sensing in Earth Systems Sciences, 1*, 63–78.
Labib, S. M., Lindley, S., & Huck, J. J. (2021). Estimating multiple greenspace exposure types and their associations with neighbourhood premature mortality: A socioecological study. *Science of the Total Environment, 789*, 147919.
Matos, P., Vieira, J., Rocha, B., Branquinho, C., & Pinho, P. (2019). Modeling the provision of air-quality regulation ecosystem service provided by urban green spaces using lichens as ecological indicators. *Science of the Total Environment, 665*, 521–530.
Mahato, S. and Ghosh, K.G. (2020). Short-term exposure to ambient air quality of the most polluted Indian cities due to lockdown amid SARS-CoV-2. *Environmental Research, 188*, p.109835.
Mahato, S. and Pal, S. (2022). Revisiting air quality during lockdown persuaded by second surge of COVID-19 of megacity Delhi, India. *Urban climate, 41*, p.101082.

Mahato, S., Kundu, B., Makwana, N. and Joshi, P.K. (2023). Early summer temperature anomalies and potential impacts on achieving Sustainable Development Goals (SDGs) in National Capital Region (NCR) of India. *Urban Climate, 52*, p.101705.

Mahato, S., Pal, S., Joshi, P.K., Matzarakis, A., Tarolli, P. and Anand, V. (2025). Early summer warming amplification threats towards sustainable development goals (SDGs) in India. *Ecological Informatics*, p.103156.

Mahato, S., Talukdar, S., Pal, S. and Debanshi, S. (2021). How far climatic parameters associated with air quality induced risk state (AQiRS) during COVID-19 persuaded lockdown in India. *Environmental Pollution, 280*, p.116975.

Mell, I. C. (2009). Can green infrastructure promote urban sustainability? In *Proceedings of the institution of civil engineers-engineering sustainability (Vol. 162, no. 1)* (pp. 23–34). Thomas Telford Ltd.

Mohan, M., Sati, A. P., & Bhati, S. (2020). Urban sprawl during five decadal period over National Capital Region of India: Impact on urban heat Island and thermal comfort. *Urban Climate, 33*, 100647.

Nahid, S., Mahato, S., Anees, M.M., Pandey, D.N. and Joshi, P.K., 2023. Air pollutants-induced environmental critical zones in capital city of India. In Advancements in urban environmental studies: Application of geospatial technology and artificial intelligence in urban studies (pp. 283–299). Cham: Springer International Publishing.

O'Lenick, C. R., Wilhelmi, O. V., Michael, R., Hayden, M. H., Baniassadi, A., Wiedinmyer, C., et al. (2019). Urban heat and air pollution: A framework for integrating population vulnerability and indoor exposure in health risk analyses. *Science of the Total Environment, 660*, 715–723.

Priemus, H. (2004). The future of social housing: From social Housing Estates. *Research Bulletin*.

Rathi, N., Mahato, S., Pandey, D.N. and Joshi, P.K., 2023. Identifying the Flood Hazard Zones in Urban Area Using Flood Hazard Index (FHI)—A Case of Capital City of India. In Advancements in Urban Environmental Studies: Application of Geospatial Technology and Artificial Intelligence in Urban Studies (pp. 341–357). Cham: Springer International Publishing.

Senanayake, I. P., Welivitiya, W. D. D. P., & Nadeeka, P. M. (2013). Remote sensing based analysis of urban heat islands with vegetation cover in Colombo city, Sri Lanka using Landsat-7 ETM+ data. *Urban Climate, 5*, 19–35.

Sikorska, D., Łaszkiewicz, E., Krauze, K., & Sikorski, P. (2020). The role of informal green spaces in reducing inequalities in urban green space availability to children and seniors. *Environmental Science & Policy, 108*, 144–154.

Song, Y., Chen, B., Ho, H. C., Kwan, M. P., Liu, D., Wang, F., et al. (2021). Observed inequality in urban greenspace exposure in China. *Environment International, 156*, 106778.

Taylor, L., & Hochuli, D. F. (2017). Defining greenspace: Multiple uses across multiple disciplines. *Landscape and Urban Planning, 158*, 25–38.

Ulpiani, G. (2021). On the linkage between urban heat Island and urban pollution Island: Three-decade literature review towards a conceptual framework. *Science of the Total Environment, 751*, 141727.

UN-HABITAT, 2024. World cities report 2024: Cities and climate action. Stylus Publishing, LLC.

Venkataraman, C., Brauer, M., Tibrewal, K., Sadavarte, P., Ma, Q., Cohen, A., et al. (2018). Source influence on emission pathways and ambient PM 2.5 pollution over India (2015–2050). *Atmospheric Chemistry and Physics, 18*(11), 8017–8039.

Vieira, J., Matos, P., Mexia, T., Silva, P., Lopes, N., Freitas, C., et al. (2018). Green spaces are not all the same for the provision of air purification and climate regulation services: The case of urban parks. *Environmental Research, 160*, 306–313.

Vogt, P., Riitters, K. H., Estreguil, C., Kozak, J., Wade, T. G., & Wickham, J. D. (2007). Mapping spatial patterns with morphological image processing. *Landscape Ecology, 22*, 171–177.

Wang, J., Xu, C., Pauleit, S., Kindler, A., & Banzhaf, E. (2019). Spatial patterns of urban green infrastructure for equity: A novel exploration. *Journal of Cleaner Production, 238*, 117858.

Weng, Q., Lu, D., & Schubring, J. (2004). Estimation of land surface temperature–vegetation abundance relationship for urban heat Island studies. *Remote Sensing of Environment, 89*(4), 467–483.

Wolch, J. R., Byrne, J., & Newell, J. P. (2014). Urban green space, public health, and environmental justice: The challenge of making cities 'just green enough'. *Landscape and Urban Planning, 125*, 234–244.

Wood, E., Harsant, A., Dallimer, M., Cronin de Chavez, A., McEachan, R. R., & Hassall, C. (2018). Not all green space is created equal: Biodiversity predicts psychological restorative benefits from urban green space. *Frontiers in Psychology, 9*, 2320.

Zhou, Y., & Shepherd, J. M. (2010). Atlanta's urban heat Island under extreme heat conditions and potential mitigation strategies. *Natural Hazards, 52*, 639–668.

Urban Ecosystem Dynamics and Microclimate: A Case of Delhi, India

Shubham Kumar, Rakesh Kadaverugu, and Laxmi Kant Sharma

Abstract Rapid urbanization and changing climatic conditions play a significant role in increasing temperature in urban environments due to the phenomenon of the Urban Heat Island (UHI) effect. In the future, this phenomenon will take place in megacities like Delhi on a large scale with a high population growth rate. In the current study, we analyzed the seasonal changes in green spaces and water bodies and their associated land surface temperature (LST) in the years 1992 and 2023 using Google Earth Engine (GEE). The % area change of the green spaces (GS) significantly increased in every season, whereas the % change in water bodies decreased in every season except the post-monsoon season. One important finding concluded that the LST of GS and water bodies (WB) and the average LST of the Delhi region continuously increased in the last 31 years. This indicates that anthropogenic activities significantly contributed to UHI and surpassed the urban cooling island (UCI) impacts of GS and WB. The results concluded that the post-monsoon season was found with a higher share of GS (431.81 km^2) and WB (13.29 km^2) than that of other seasons in the year 2022–2023. The landscape matrices of urban GS like Mean Patch Size (MPS), Percent Cover (PC), and Largest Patch Index (LPI) were found to be highest (0.13 ha, 29.06%, and 8.8%, respectively) in the post-monsoon season and lowest (0.06 ha, 6.31%, and 0.33%, respectively) in the pre-monsoon season. The results of the study showed significant relevance with "India's Smart Cities Mission" (ISCM), in which the area of GS and WB can be specified over a certain population. The study provides insight into the decreasing efficacy of GS and WB due to the influence of anthropogenic and changing climatic conditions. This leads to more areas of GS and WB being required in the coming years due to changing climatic conditions and anthropogenic burdens over the city.

S. Kumar · L. K. Sharma (✉)
Department of Environmental Science, School of Earth Sciences, Central University of Rajasthan, Ajmer, Rajasthan, India
e-mail: laxmikant_evs@curaj.ac.in

R. Kadaverugu
Cleaner Technology and Modelling Division, Council of Scientific and Industrial Research-National Environmental Engineering Research Institute, Nagpur, Maharashtra, India

P. K. Joshi et al. (eds.), *Geospatial Science for Urban Ecosystems*, The Urban Book Series, https://doi.org/10.1007/978-3-032-07155-2_15

Keywords Google Earth Engine (GEE) · Urban Heat Island (UHI) · Urban green spaces · Urban ecosystems · Urban wetland

1 Introduction

According to the United Nations World Urbanization Prospects Report (2018), almost 55% of the urban population is going to rise to 60% by 2050. Africa and Asia, both continents, would be witnesses to this increase in urbanization. This development will destroy and/or fragment seminatural and natural habitats, resulting in loss of ecosystem services and biodiversity and negatively impacting human well-being (Grimm et al., 2008). This will lead to considerable land use and land cover change, decreasing the area of water bodies (WB) and urban green spaces (GS) (Yu et al., 2018). Urbanization would ultimately cause considerable changes in the urban meteorological conditions in such a way that it creates a typical phenomenon of microclimate warming known as Urban Heat Island (UHI). In simple words, UHI may be precisely defined as the development of warmer and drier conditions (considerably higher temperature) in urban settings due to the replacement of natural land cover with dense buildings and other constructions that considerably absorb and retain heat (Oke, 1973). Further, the severity and frequency of extreme weather events, like heat waves, are predicted to increase due to changing climatic conditions. Also, it is more prominent at night than during the day (Oke, 1982). The climate change phenomena are anticipated to be amplified by such events when combined with the phenomenon of UHI. The effect of UHI on socio-economic status becomes more prominent in summer than in wintertime (Han et al., 2022). The UHI effect may increase the electricity cost (from air conditioners), air pollution, heat-related illness, and mortality. The role of urban GS and WB is still uncertain (Grilo et al., 2020). There is a high need to evaluate the actual efficacy of these urban ecosystems. Therefore, proper assessment of urban GS and WB will provide valuable insight into overcoming UHI phenomena, especially in highly developed Asian and African cities. The interlinking between the urban climate and land use caught the eyes of city planners and policymakers to improve the well-being of the urban environment (Gunawardena et al., 2017). The current study will help to use and develop the green infrastructure and WB as a way to promote the urban cooling island (UCI) effect.

Urban GS and WB provide multiple social, aesthetic, environmental, economic, and recreational values to urban areas. In recent years, GS and WB have emerged as prominent nature-based solutions (NbS) for making cities resilient to climate-induced hazards like flash floods (Keeler et al., 2019). NbS, if it is supported by vegetation and water availability, avoids heat buildup by the increase in relative humidity through evapotranspiration and the creation of shadows (Voogt & Oke, 2003). GS and WB ensure the availability of biodiversity habitats (Pinho et al., 2016) and the reduction of air pollution as other important ecosystem services.

However, different types of GS provide a different proportion of ecosystem services. Not all GS and WB are likely to provide the same level of UCI effect. This may vary with different geographical, hydrological, and climatic (such as temperature, wind turbulence, humidity, wind direction, and wind speed) vegetation types, shape and size of patches, and distances from the construction sites, etc. UHI phenomena can be minimized by effective planning and management of the urban ecology and landscape (Du et al., 2017). Temperature, relative humidity, and wind speed can have an impact on UHI by altering bioclimatic comfort. These changes can also have an impact on the cooling effect by altering how water vaporizes (Wu et al., 2021).

Remote sensing and Geographic Information Systems (GIS) offer a unique advantage in the past, present, and future assessment of UCI. More importantly, it depicts a clear trend and status of variation in land surface temperature (LST). In comparison, the remotely sensed LST is considered one of the prominent factors affecting the energy transfer in the near-surface layers of the urban atmosphere. Landsat data provides a good agreement with the assessment of UCI effect and LST with a spatial resolution of 30 meters. Therefore, we can easily and effectively evaluate the efficacy of any urban natural ecosystem to minimize the effect of UHI in highly populated cities. For example, in a study of the capital of India, the past and present changes in the natural urban ecosystems were evaluated along with the efficacy of the GS and WB. Further, the actual potential of these ecosystems to minimize the UHI was also evaluated.

2 Methodology

2.1 *Study Area*

Delhi is situated on the bank of the holy Yamuna River at the periphery of the Gangetic plains with a geographical area of 1483 km^2 (length and width of 51.90 and 48.48 km, respectively) (Fig. 1). The population of Delhi was 16.78 million in 2011, out of which 97.5% live in urban areas.

There is a difference in elevation all over the study area. Some of the portions of Delhi are a part of the Aravalli, which is called the Lungs of Delhi, with high vegetation cover. The study area consists of thorny scrub vegetation or ridge forest, which is found in arid and semi-arid zones. According to Champion and Seth (1968), it is categorized as a "Semi-Arid Open Scrub" or "Tropical Thorn Forest." As per the Koppen-Geiger climate classification, the study area falls under the "dry-winter humid subtropical climate" (C*wa*) bordering on a "hot semi-arid climate" (*BSh*) with a high variation of seasonal temperature and rainfall. The Delhi region receives ample rainfall, generally about 790 mm annually (65.8 mm monthly). The temperature ranges from 46 °C in summer to 0 °C in winter.

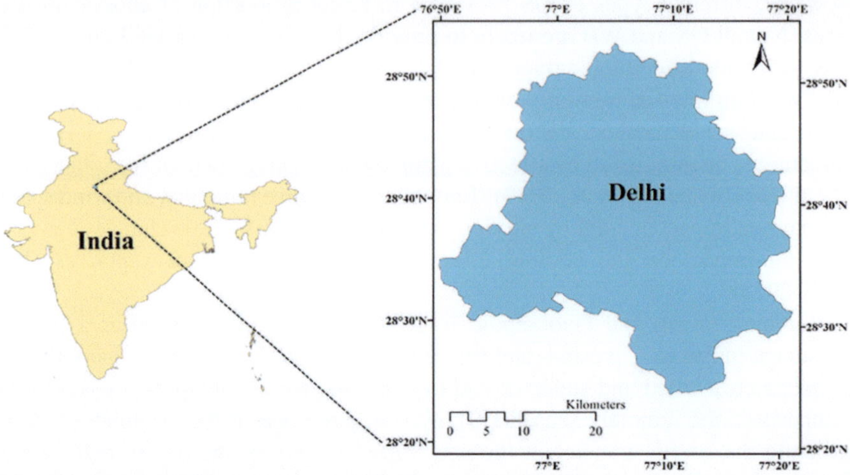

Fig. 1 Study area map of the National Capital Territory of India, Delhi

2.2 Satellite Data

The "Google Earth Engine" (GEE) platform was used to acquire spatio-temporal Landsat-5 and Landsat-9 visible and thermal bands. The data were obtained from the Landsat Thermal Infrared Sensor (TIRS) and the "Operational Land Imager" (OLI). The data were collected seasonally in two time periods covering the years 1992 and 2023. There are two major reasons for the analysis of the microclimate conditions for UHI. The first reason is greater spatial variation in the UHI effect in different urban environments (Nichol et al., 2009). Whereas the second reason is a stronger surface UHI due to the larger impacts of solar radiation on surface warming (Klok et al., 2012). Normalized Difference Vegetation Index (NDVI) and LST were calculated at a 30-meter visible and 100-meter thermal band, respectively. The thermal bands were also resampled to 30 m spatial resolution.

2.3 Data Analysis

Landsat data consists of several errors in itself. Therefore, data preprocessing was performed before processing the data in a useful form. Then, the Modified Normalized Difference Water Index (MNDWI), NDVI, and LST were calculated seasonally for the years 1992–1993 and 2022–2023. The calculation of the GS and WB areas was estimated. Similarly, LST data were classified into three major classes. UCI effects were further analyzed concerning GS and WB patches. The efficacy of urban GS and WB was analyzed from the above results.

3 Results and Discussion

3.1 *Status of Green Spaces and Water Bodies*

The GS during 1992–93 was found in a very limited area (Fig. 2). The WB was also in limited spaces. The high seasonal variation in the GS area was found. The GS and WB areas were observed to be highest during the winter season in and around the study area.

Further, the following results were observed from the study area that indicated a considerable change in the urban GS and WB (Fig. 3) from 1993 to 2023. It was observed that GS and WB areas were higher in September–October-November (SON) months, followed by December–January-February (DJF), June–July-August (JJA), and MAM (March–April-May), respectively (Table 1).

3.2 *Temperature Profiling*

During 1992–1993, it was observed that the higher temperature classes were found in the southern and southwestern parts of Delhi. The area of GS increased in the post-monsoon season. This indicates the role of GS. Therefore, the intensity of the UCI effect was in a similar pattern to the GS and WB, which were present in almost every season. The area of GS increased in a very considerable manner in the year 2022–2023. This neutralizes the effect of UHI with the UCI effect of GS to an extent, specifically in the summer season. But the UCI effect could not overcome the ever-increasing average LST in the study area. The GS LST category had lower temperatures in 2022–2023, but it was higher than that of 1992–1993.

Artificial construction and human activity are one of the main reasons for global warming (Masson-Delmotte et al., 2022). This is also the major cause of UHI. The average LST shows a continuous increase whereas the increasing area of GS was making the city more resilient in terms of the UHI effect (Fig. 4). The JJA season was found the highest LST followed by MAM, SON and DJF, respectively (Table 1).

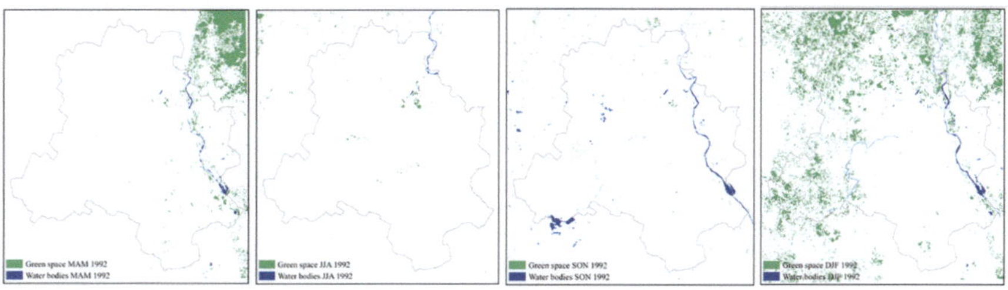

Fig. 2 Urban green spaces and water bodies in Delhi, India, from 1992 to 1993

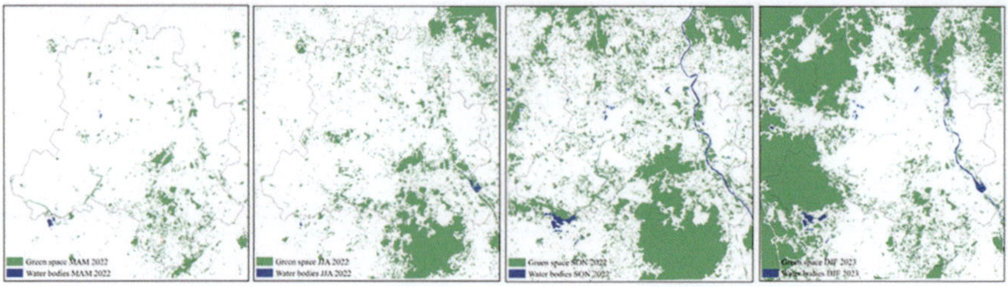

Fig. 3 Urban green spaces and water bodies in Delhi, India, during 2022–2023

Table 1 LST concerning the area in different seasons

Year	Season	WB	GS	Average LST	GS LST	WB LST
1992–1993	MAM	4.48	12.28	31.16 ± 4.13	24.03 ± 1.97	20.99 ± 2.13
	JJA	1.23	4.75	44.04 ± 3.05	37.37 ± 2.47	35.41 ± 2.49
	SON	12.28	12.28	23.24 ± 1.84	21.39 ± 1.29	19.88 ± 1.13
	DJF	10.52	10.52	20.05 ± 1.86	18.30 ± 1.06	17.76 ± 1.55
2022–2023	MAM	0.5	93.72	38.59 ± 2.42	35.75 ± 1.62	30.6 ± 1.06
	JJA	0.77	252.4	41.1 ± 2.81	37.7 ± 1.72	31.7 ± 1.28
	SON	13.29	431.81	31.9 ± 1.95	30.53 ± 1.35	26.85 ± 1.6
	DJF	10.21	425.36	20.55 ± 1.44	19.76 ± 0.86	18.07 ± 0.8

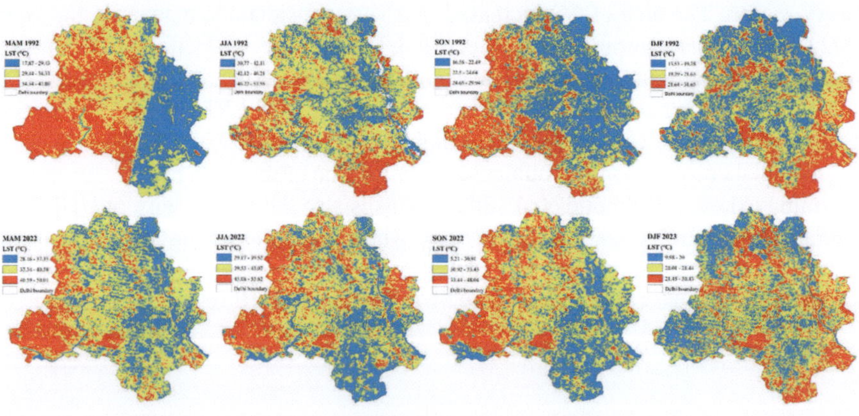

Fig. 4 LST profile of Delhi, India, during 1992–1993 and 2022–2023

3.3 Landscape Metrics of Green Spaces during 2022–2023

The calculation of landscape metrics gave an insight into the role and impact of urban GS in the neighboring environment. Therefore, we calculated three metrics, including Mean Patch Size (MPS), Percent Cover (PER), and Largest Patch Size (LPI). MPI is defined as "the average number of pixels of the patches in a

Table 2 Landscape metrics of urban GS during 2022–2023

Season	Mean patch size	Percent cover	Largest patch size
MAM	0.06	6.31	0.33
JJA	0.11	16.99	3.71
SON	0.13	29.06	8.8
DJF	0.09	28.63	7.59

landscape." This was one of the simplistic spatial attributes for the examination of the landscape. Further, PER is the "area of green spaces over the total study area," and LPI is the "area of the largest patch of the corresponding patch type divided by the total landscape area." The MPS, PER, and LPI were calculated for each season of the year 2022–2023. The highest MPS, PER, and LPI values were observed in the SON (0.13 ha, 29.06%, and 8.8%, respectively) (Table 2). Similarly, the minimum MPS, PER, and LPI values were found in the MAM (0.06 ha, 6.31%, and 0.33%, respectively). Thus, the landscape metrics showed a good agreement with the seasonal variation in the area dynamics of urban green spaces.

3.4 *Seasonal Efficacy of Green Spaces and Water Bodies*

The seasonal efficacy of GS and WB was changed considerably for the % area change between the years 1992–1993 and 2022–2023. The positive change in the GS area was observed in every season of the year 2022–2023 compared to that of the year 1992–1993, and the post-monsoon season (SON) witnessed the highest increase in GS (Table 3). Whereas the negative change in the WB area was observed except during the post-monsoon season. The drastic change in the WB area was seen in the pre-monsoon season, in which around 89% of the area in the year 2022–2023 decreased concerning 2022–2023.

Similarly, a considerable change in the efficacy of the GS and WB was observed between the years 1992–1993 and 2022–2023 as the area of GS and WB changed. The results indicated that the efficacy of the GS and WB decreased since 1992–1993. Only in the monsoon did it show the positive change in efficacy (i.e., negative change in LST) of WB, resulting in an improved urban cooling island effect. The findings of the current study demonstrated that a positive change in average LST was observed in every season except the monsoon season.

The study highlighted the effectiveness of green spaces and WB in MAM, JJA, SON, and DJF seasons, and the temperature differences between GS, LST, and average LST were 8.13, 6.67, 1.85, and 1.75 °C in the year 1992–1993, respectively. The important fact was that temperature differences between GS and LST, along with average LST, reduced in the year 2022–2023 by 2.84, 3.4, 1.37, and 0.79 °C, respectively. Similarly, the difference in WB LST and average LST showed comparatively high-temperature differences with values of 11.17, 8.63, 3.36, and 2.29 °C in the year 1992–1993, respectively. But it also observed a considerable

Table 3 Change in percentage area and LST difference between years 1992–1993 and 2022–2023

Season	GS (% area change)	WB (% area change)	Average LST (change)	LST GS (change)	LST WB (change)
MAM	66.32	−88.84	6.43	11.72	9.61
JJA	5231	−37.4	−2.11	0.33	−3.71
SON	11,602	8.22	8.66	9.14	6.97
DJF	266.97	−2.95	3	1.46	0.28

Table 4 Efficacy of green spaces and water bodies for urban cooling island effect (unit in °C)

Year	Season	LST different between average LST and GS	LST different between average LST and WB	LST different between average GS and WB
1992–1993	MAM	8.13	11.17	3.04
	JJA	6.67	8.63	1.96
	SON	1.85	3.36	1.51
	DJF	1.75	2.29	0.54
2022–2023	MAM	2.84	7.99	5.15
	JJA	3.4	9.4	6
	SON	1.37	5.05	4.05
	DJF	0.76	2.48	1.69

decrease in temperature difference between WB LST and average LST except in the post-monsoon season with values of 7.99, 9.4, 5.05, and 2.47 °C, respectively. It was observed that the difference between average LST and GS LST decreased since the year 1992–1993 (Table 4). The difference in average LST and WB LST has also decreased but improved in the post-monsoon season since the year 1992–1993. It was important to notice that the LST difference between WB and GS between the years 1992–1993 and 2022–2023 increased in every season. Thus, the findings of the study concluded that ever-increasing LST decreased the efficacy of the GS and WB during 1992–1993 to 2022–2023. Also, a huge increase in the GS and WB areas considerably minimized the effect of UHI in the Delhi region, but was insufficient to overcome it fully.

4 Conclusion

The urban WB showed a considerably higher variation in LST in every season than that of urban GS. It was highlighted that the difference in average GS as well as WB LST between the years 1992 and 2023 had a strong increase, except for the post-monsoon season. It is a sign of ever-increasing average LST and decreasing efficacy of the green spaces and WB of the city. Further, the planning and management practices must be improved to ensure Sustainable Development Goal-11 (SDG-11). More urban GS and WB should be developed in Delhi to ensure it is a sustainable city. This will also help to create habitats for biodiversity outside the protected areas, along with the mitigation of urban flash floods and water level recharge. The findings concluded that the efficacy of GS and WB was reduced. Therefore, more GS and WB areas would be required to minimize the effect of UHI in highly populated cities like Delhi.

Acknowledgments We are grateful to the DST-FIST-supported Remote Sensing and GIS Lab, Department of Environmental Science, Central University of Rajasthan, India, for providing the necessary lab facilities for carrying out the study. We also want to acknowledge the support for the study by IIT Tirupati Navavishkar I-Hub Foundation under the NAV-i-GEE Programme in collaboration with CSIR-NEERI, Nagpur, and IIT Tirupati.

References

Champion, H. G., & Seth, S. K. (1968). *A revised survey of the forest types of India*. Manager of Publications.

Du, H., Cai, W., Yanqing, X., Wang, Z., Wang, Y., & Cai, Y. (2017). Quantifying the cool Island effects of urban green spaces using remote sensing data. *Urban Forestry & Urban Greening, 27*, 24–31.

Grilo, F., Pinho, P., Aleixo, C., Catita, C., Silva, P., Lopes, N., Freitas, C., Santos-Reis, M., McPhearson, T., & Branquinho, C. (2020). Using green to cool the grey: Modelling the cooling effect of green spaces with a high spatial resolution. *Science of the Total Environment, 724*, 138182.

Grimm, N. B., Faeth, S. H., Golubiewski, N. E., Redman, C. L., Jianguo, W., Bai, X., & Briggs, J. M. (2008). Global change and the ecology of cities. *Science, 319*(5864), 756–760.

Gunawardena, K. R., Wells, M. J., & Kershaw, T. (2017). Utilising green and bluespace to mitigate urban heat Island intensity. *Science of the Total Environment, 584*, 1040–1055.

Han, W., Tao, Z., Li, Z., Cheng, M., Fan, H., Cribb, M., & Wang, Q. (2022). Effect of urban built-up area expansion on the urban heat islands in different seasons in 34 metropolitan regions across China. *Remote Sensing, 15*(1), 248.

Keeler, B. L., Hamel, P., McPhearson, T., Hamann, M. H., Donahue, M. L., Meza Prado, K. A., Arkema, K. K., et al. (2019). Social-ecological and technological factors moderate the value of urban nature. *Nature Sustainability, 2*(1), 29–38.

Klok, L., Zwart, S., Verhagen, H., & Mauri, E. (2012). The surface heat Island of Rotterdam and its relationship with urban surface characteristics. *Resources, Conservation and Recycling, 64*, 23–29.

Masson-Delmotte, V., Zhai, P., Pörtner, H.-O., Roberts, D., Skea, J., & Shukla, P. R. (2022). *Global warming of 1.5 C: IPCC special report on impacts of global warming of 1.5 C above pre-industrial levels in context of strengthening response to climate change, sustainable development, and efforts to eradicate poverty*. Cambridge University Press.

Nichol, J. E., Fung, W. Y., Lam, K.-s., & Wong, M. S. (2009). Urban heat Island diagnosis using ASTER satellite images and 'in situ'air temperature. *Atmospheric Research, 94*(2), 276–284.

Oke, T. R. (1973). City size and the urban heat Island. *Atmospheric Environment (1967), 7*(8), 769–779.

Oke, T. R. (1982). The energetic basis of the urban heat Island. *Quarterly Journal of the Royal Meteorological Society, 108*(455), 1–24.

Pinho, P., Correia, O., Lecoq, M., Munzi, S., Vasconcelos, S., Gonçalves, P., Rebelo, R., et al. (2016). Evaluating green infrastructure in urban environments using a multi-taxa and functional diversity approach. *Environmental Research, 147*, 601–610.

United Nations, Department of Economic and Social Affairs, Population Division (2018). World Urbanization Prospects: The 2018 Revision, *Methodology*. Working Paper No. ESA/P/WP.252. New York: United Nations.

Voogt, J. A., & Oke, T. R. (2003). Thermal remote sensing of urban climates. *Remote Sensing of Environment, 86*(3), 370–384.

Wu, S., Yang, H., Luo, P., Luo, C., Li, H., Liu, M., Ruan, Y., et al. (2021). The effects of the cooling efficiency of urban wetlands in an inland megacity: A case study of Chengdu, Southwest China. *Building and Environment, 204*, 108128.

Yu, Z., Shaobin, X., Zhang, Y., Jørgensen, G., & Vejre, H. (2018). Strong contributions of local background climate to the cooling effect of urban green vegetation. *Scientific Reports, 8*(1), 6798.

Artificial Neural Network and Landscape Approach for Pattern of Green Spaces in Municipal Area: A Case of Siliguri

Jayanta Mondal and Arijit Das

Abstract Green spaces are becoming widely degraded as a result of the massive urbanization that is occurring worldwide. Urban Green Spaces (UGS) are vital for improving urban quality of life because they provide vital ecosystem services. These include a variety of vegetated areas found inside cities, such as parks, green roofs, and street vegetation. This chapter describes a pilot project that used geographic information systems (GIS) and remote sensing to map and delineate urban green cover in the Siliguri municipal region of India. The study's objective was to create mapping and evaluation methods that are suitable and scalable for measuring the amount of green plant cover in the area. Tests were conducted using Landsat OLI and TM satellite images to identify the vegetation cover and a consistent technique for measuring the vegetation's extent. In an effort to distinguish vegetation cover from remotely sensed images, image processing techniques, including various vegetation indices, were used. For such procedures, specialized GIS scripts were created to make them quick, automated, and repeatable. Lastly, the urban green space has been implemented utilizing an artificial neural network (ANN), a machine learning technique, with various vegetation indices utilized as an input parameter. At a five-year temporal interval, the vegetation cover was evaluated and mapped from 1990 to 2020. The accuracy of the categorization was evaluated in comparison to field observations and aerial picture interpretation. Depending on the year, the photo-based classification's overall kappa accuracy ranged from 68% to 82%.

Keywords Urban green space · Geographical information system · Landsat OLI and TM · Artificial neural network · Kappa

J. Mondal (✉) · A. Das
Department of Geography, University of Gour Banga, Malda, India

P. K. Joshi et al. (eds.), *Geospatial Science for Urban Ecosystems*, The Urban Book Series, https://doi.org/10.1007/978-3-032-07155-2_16

1 Introduction

Green space degradation with rapid urbanization is a common phenomenon (Chen et al., 2006) for every urban area in the world. Urban Green Space (UGS) refers to the green patches within an urban area, including vegetation in the street, park greenery in the terrestrial space between buildings, and green roofs, (Tan et al., 2013). Urban green has provided many services to maintain the quality of life in the urban areas (Wang et al., 2017). Most of the cities have faced a serious issue related to environmental stress due to the lack of green cover and its declining trend (De Ridder, 2004). The investigation and monitoring of urban vegetation is therefore essential for mitigating environmental problems.

Historically, the area of UGS has been traced using field observation and aerial photograph interpretation (API), which is quite common (Yang, 2007). These conventional techniques are subjective, time-consuming, and difficult to replicate. Such conventional techniques have a limitation and are considered a disadvantage for use in a larger area like the Siliguri municipal area (40.1 sq. km), where continuous monitoring of green space is essential. Remote sensing and geographical information system (GIS) techniques are very useful tools for mapping the green space and its management nowadays (Muller et al., 2011; Narumalani et al., 1997; Congalton et al., 2002). Advanced technologies of remote sensing and image analysis, jointly with conventional ground assessment, provide a powerful tool for temporal analysis of the changing scenario of vegetation zones in Australia (Apan et al., 2002). There are different types of satellite images available for mapping the urban green space, having different spatial resolution (ground area per pixel) and spectral resolution (number of bands). Among them, Landsat, LISS, and SPOT images are commonly used to determine the vegetation cover in the studied region. Those images are used in vegetation studies, including green clustering, classification, image masking, etc. (Narumalani et al., 1997; Nagler et al., 2005; Weber & Dunno, 2001).

In this chapter, the various vegetation cover indices within the Siliguri municipal region were identified and quantified using Landsat data. There are several techniques to delineate the vegetation. The Normalized Difference Vegetation Index (NDVI) is widely used for detecting vegetation cover. This index is basically a ratio difference between the vegetation reflectance in red and near infrared bands, respectively (Nageswara Rao et al., 2005). This is a very powerful and globally accepted index to indicate the vegetation (Gandhi et al., 2015; Ali-Bik et al., 2018). Although NDVI is used by different researchers, it has saturated in high leaf area and is limited to identifying the variety of the vegetation type (Huete et al., 2002). To accurately define the vegetation cover, however, several indices were also computed in this study, including the NDVI: Soil Adjusted Vegetation Index (SAVI), Enhanced Vegetation Index (EVI), Transformed Vegetation Index (TVI), and Optimized Soil Adjusted Vegetation Index (OSAVI). EVI is frequently used as an alternative to NDVI as it overcomes the limitation of NDVI and requires the information of blue reflectance (Rocha & Shaver, 2009). Reflectance of sand and soil is much higher than vegetation in some places; thus, it is sometimes difficult to identify the vegetation. Therefore, soil-adjusted indices were introduced as SAVI, OSAVI, modified

SAVI, etc. (Karnieli et al., 2001; Gilabert et al., 2002). TVI, introduced by Deering (1978), is a transformed index of NDVI. It transforms the negative value of the NDVI histogram into a normal distribution (Abd-El Monsef et al., 2017).

Finally, machine learning (ML) was then employed on the indices to prepare the UGS extent for the region of interest. Different MLs like artificial neural network (ANN), support vector machine (SVM), and random forest (RF) have been used to make image-based classifications with the input parameters (Jean & Cho, 2019). In this case, ANN has been introduced to prepare UGS mapping, and the different vegetation indices have been used as input parameters, as ANN has a powerful technique to make decisions rather than the other MLs. ANN is getting popular day by day as it is an alternative to the statistical method for classifying the remotely sensed data (Mahmon & Ya'acob, 2014). ANN has used the back propagation of the k-means algorithm and used several approaches to classify images. This study also evaluated the accuracy of the products that were generated through satellite images before further analysis. Landscape matrices were used as the indicators of UGS, as it has a capacity to analyze the configuration of the concerned feature (Li et al., 2001). Different dimensions of UGS were assessed in this chapter through the different indicators. The main objective of this chapter is to delineate the UGS of the Siliguri municipal area.

1.1 About the Study Area

Siliguri is known as Northeast India's gateway, situated at the junction of the Jalpaiguri and Darjeeling districts beside the river Mahananda at the foothills of the Himalayas at a location of 88.43 degrees E longitude and 26.71 degrees N latitude (Fig. 1). It is the third biggest urban agglomeration, following Kolkata and Asansol

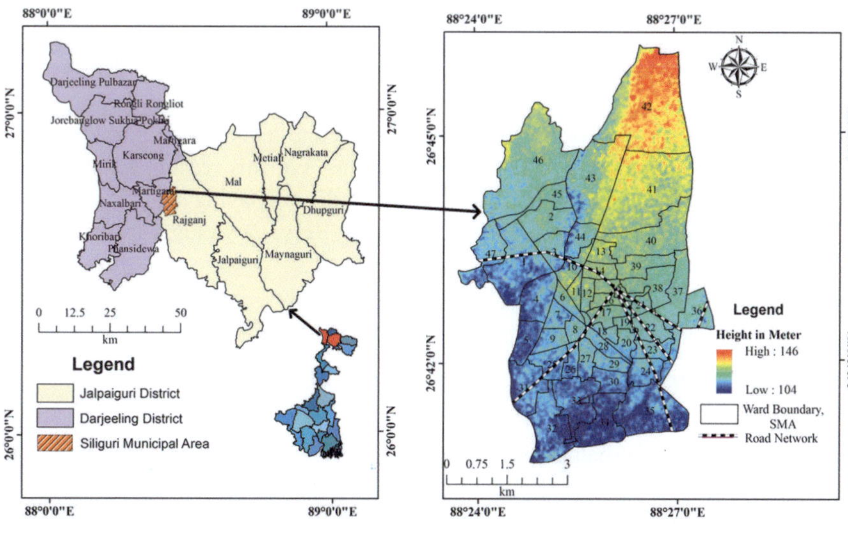

Fig. 1 Location of the study area

Table 1 Spectral characteristics of multi-temporal satellite data

Satellite	Sensor	Path/Row	Year	Resolution	Source
Landsat 5	TM	139/41	1990, 1995 2000, 2005 2010	30 m	U.S Geological Survey
Landsat 8	OLI	139/41	2015, 2020	30 m	U.S Geological Survey

Note: *OLI* Operational Land Imager, *TM* Thematic Mapper

in West Bengal. It is surrounded by Bangladesh, Uttar Dinajpur, and Bihar in the south and the Himalayan range on the north side. On the west lies Nepal, and the Jalpaiguri and Kalimpong districts form its eastern boundary. Siliguri subdivision falls under the Darjeeling district, consisting of the Naxalbari, Matigara, Siliguri Municipal Corporation, Phansidewa, and Kharibari C.D. Blocks. Siliguri Municipal Corporation covers an area of 41.60km^2.

2 Materials

The analysis of UGS‘s spatio-temporal components across time and its modeling needs moderate to high-resolution satellite imagery. In this sense, Landsat imagery is beneficial, as it offers effective assistance for extensive landscape research, and this type of data is publicly available and updated on a regular basis (Nkeki, 2016). Cloud-free multi-temporal and multispectral Landsat images of Siliguri municipality and its environs were used as the study's data (Table 1), which were divided into seven temporal periods (1990, 1995, 2000, 2005, 2010, 2015, and 2020).

3 Methods

This study's primary goal is to give a geographical and temporal knowledge of the Siliguri Municipal Region's UGS scenario pattern. For this reason, Landsat data were derived from remotely sensed photos, essentially displaying the UGS‘s temporal deterioration. This study's whole methodology (Fig. 2) yields a fresh way for defining the UGS.

3.1 Execution of Satellite Image-Based Indices

For the delineation of urban green spaces, several vegetation indices have been calculated using satellite images. Previous studies also demonstrated the use of several indices for this purpose. In this study, indices like NDVI, SAVI, EVI, OSAVI, and

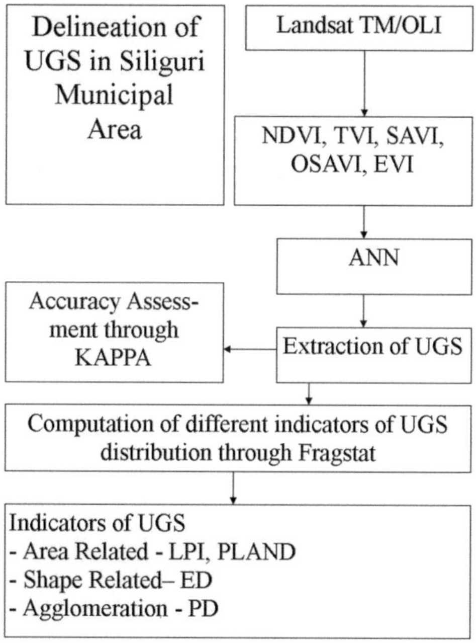

Fig. 2 Flow chart of the methodological framework

TVI have been calculated using Grass GIS 7.0.4. Rouse et al. (1974) introduced NDVI for the first time. This index is basically used for the determination of the state of a plant's health. It is chlorophyll sensitive. NDVI value ranges between −1 and +1. If the values are negative, then they correspond to water. Values between −0.1 and 0.1 indicate rock, barren areas, snow, and sand. A 0.2–0.4 value signifies low vegetation cover, more precisely grasslands or shrubs, and higher positive values mean dense and healthy vegetation cover as a whole (Deering, 1978). It has various advantages, such as eliminating topographic effect, sun illumination angles, and haze as well (Mróz & Sobieraj, 2004). It needs higher remote sensing calibration, as this index is quite sensitive to cloud cover, soil color, brightness, and leaf canopy density. NDVI is expressed as Eq. 1.

$$\text{NDVI} = \frac{bNIR - bR}{bNIR + bR} \tag{1}$$

where *bNIR* represents the infrared band, and *bR* shows the red band.

Huete (1988) propounded the concept of SAVI. It is basically used in the areas where vegetation cover is low and adjusted to soil brightness by introducing the soil adjustment factor "L." Values of L are −1 for low vegetation cover, 0.5 indicates intermediate vegetation cover (most widely used values), and 0.25 to 0 indicates high vegetation cover. SAVI is expressed as Eq. 2.

$$\text{SAVI} = \left[\frac{(bNIR - bR)}{bNIR + bR + L}\right](1 + L) \tag{2}$$

where *bNIR* stands for near infrared bands and *bR* for red bands, while *L* signifies the soil adjustment factor, which has an empirical value of 1.

OSAVI, i.e., Optimized Soil Adjusted Vegetation Index, is a newly modified form of SAVI. It is basically featured to accommodate a huge variation due to dense soil backgrounds. OSAVI does not depend on soil line and greatly eliminates soil background effects (Xue & Su, 2017). Like other vegetation indices, the OSAVI value ranges between −1 and +1. +1 is indicative of healthier vegetation, whereas −1 signifies vegetation with less vigor. OSAVI is encoded by the following equation:

$$\mathrm{OSAVI} = \left[\frac{(bNIR - bR)}{bNIR + bR + 0.16} \right] \tag{3}$$

Where, *bNIR* signifies the near infrared band and *bR* the red band, 0.16 is considered as the soil-adjusted coefficient.

Deering (1978) proposed the concept of the Transformed Vegetation Index. It basically eliminates negative values and gives a normal distribution to NDVI histograms. It is not calculated when the value of NDVI is <−0.5.

$$\mathrm{TVI} = \sqrt{\mathrm{NDVI}} + 0.5 \tag{4}$$

In early 2006, Huete et al. successfully applied the Enhanced Vegetation Index. This index is optimized for vegetation indices, which are subjected to an enhanced vegetation signal (Eq. 4). It is basically designed for high biomass regions (Amazon forest) and helpful for reducing atmospheric influences. EVI is sensitive to canopy density, structural variations, plant physiognomy, leaf area index, etc. EVI increases in the presence of snow (Huete et al., 2002).

$$\mathrm{EVI} = \frac{xbNIR - xbR}{xbNIR + k1xR - k2xbB + L}(1 + L) \tag{5}$$

where x is the reflectance of "top-of-atmosphere," *bNIR* is the near infrared band, *bR* is the red band, *bB*is the blue band, $k1$ and $k2$ is used as a coefficient, which means the utility of the blue band as an atmospheric correction of the red band, and the values of $k1$ and $k2$are 6.0 and 7.5.

3.2 Delineation of UGS

Scholars have widely used machine learning techniques such as support vector machines, random forests, and artificial neural networks in various fields of science. In this study, an ANN is used to classify the UGS. The satellite-based indices have been used to identify the green cover properly. A band composition technique has been adopted to merge all the indices (NDVI, EVI, TVI, SAVI, and OSAVI), and

ANN has been applied to it to delineate the urban green space. ANN is extensively accepted as a technology that gives the alternative way to tackle complex and ill-defined problems. The techniques are worked in parallel to the human brain. Numerous neurons are connected in a complex network in the human body. Consequently, the brain sends a signal via neurons to other organs. This is the brain function of the human body. Similar to the human body, artificial neural networks (ANNs) are constructed using an association of interconnected junctions and an algorithm for assigning weights and inauguration functions to individual nodes. The neurons get information, analyze it, and send the output to a different neural network. As a result, many neural network architectures are employed while making decisions or addressing issues. The training samples are utilized to determine which data are most useful and how many algorithms are being employed in the network. Because ANNs are fault-tolerant, they can process noisy and inconsistent data. Using nonlinear combinations of the input variables, "Multi-Layer Perceptron" (MLP) has been accustomed to building representations that represent the system. With the help of training data, it can quickly and effectively forecast and generalize the data. According to Mabel and Fernandez (2008), the artificial neural network (ANN) is a more potent tool for pattern detection, forecasting, and more accurate data set optimization. Thus, by combining the aforementioned indices, an ANN classifier has been employed to designate UGS.

This study uses an artificial neural network (ANN) as a feed-forward network, where the hidden layers have sigmoid activation functions and the output node has linear activation functions. The model was performed in R Studio software. More than one hidden layer is often not necessary. Our architectures have only one hidden layer. A backpropagation technique with gradient descent and momentum terms is accustomed to train the artificial neural network. The maximum training epochs, learning rate, and number of nodes in a single hidden layer must all be defined for the ANN. We have applied the optimum number error technique in this investigation. The learning rate has fluctuated from 0.01 to 1.0 in steps of 0.05, while the hidden layer's node count has ranged from 5 to 23. Equation (Eq. 6) expresses the neural network with an MLP. The hidden layer and output have revealed the following equation: this stands for unit *i*'s net input.

$$Z = \sum_{j=1}^{k} W_{ji} y_j + \theta_i \tag{6}$$

Y_j represents the output generated from *j; the error of unit i* is presented through Y_i, *and* W_{ji} represents the vector. *k* shows the neuron number. The algorithm then transfers to Z (weighted sum). $\hat{y}_i$ represents the incoming signals of unit *i*. The following formula (Eq. 7) shows the f_i (sigmoid function) of *i*.

$$\hat{y} = \frac{1}{1 + e^{-z}} \tag{7}$$

3.3 Accuracy Assessment

There are various statistical tools to validate the classification. In this case, the Kappa coefficient (Eq. 8) has been used to assess the accuracy of UGS. It is a statistical technique that is used for measuring accuracy between user-assigned ratings and producer ratings. Finally, overall accuracy has been assessed according to the Kappa coefficient. Kappa statistics is expressed as follows:

$$k = \frac{N\sum_{i=1}^{r} xii - \sum_{i=1}^{r}\left(x_i + \mathrm{y}_i\right)}{N^2 - \sum_{i=1}^{r}\left(x_i + \mathrm{y}_i\right)} \tag{8}$$

Where N refers to the total training sample, xi and yi represent row and column, respectively.

Monserud and Leemans (1992) proposed five degrees of agreement: <0.4 denotes bad type, values between 0.4 and 0.55 correspond to fair type, 0.55 to 0.7 denotes good, 0.7 to 0.85 denotes very good, and >0.85 denotes outstanding agreement. In order to determine the accuracy for UGS, 150 ground control points and 90 training samples were used in the calculation.

3.4 Calculation of Different Indicators of Urban Green Space (UGS)

UGS has been used to execute the different indicators of green space with class metrics (Table 2) using Fragstat 4.2 software. Class metrics compute every class of greenery in the landscape or every patch type, which generates an output file. The file contains columns that are representative of individual metrics, and a row represents every class.

4 Result

A manual thresholding was performed on the vegetation-based indices after calculating. Google Earth images and ground points were taken as the validation point. After preparing the vegetation maps based on NDVI, SAVI, OSAVI, TVI, and EVI, they have been used for further processing to delineate the UGS mapping.

More for all indices green covered area gradually decreases with time. OSAVI covered the maximum vegetated area among all the indices in 1990, 2015, and 2020, where EVI was measured in 1995 and 2010. SAVI recorded its highest in 2000. NDVI depicts a maximum in 2005 (Table 3).

Table 2 Different Landscape metrics and their description

Metrics	Abbreviations	Description	Unit	Equation
Largest Patch Index	LPI	Simple measure of dominance. It generally quantifies the total landscape area in percentage that is surrounded by the biggest patch. LPI = 0 indicates largest patch becomes increasingly small of corresponding patch and LPI = 100 means entire landscape is dominated by a single patch in corresponding patches	Percent	$\mathrm{LPI} = \frac{\max\limits_{m}^{n=1}\left(a_{mn}\right)}{A}(100)$ a_{mn} = Area of patch $_{mn}$(m^2), A = Landscape area(m^2)
Edge Density	ED	ED = 0 means absence of edge in landscape	Meters per hectare	$\mathrm{ED} = \frac{X}{A}(10{,}000)$ X = Length of edge (m) A = Landscape area (m^2)
Patch Density	PD	It is a fundamental aspect of landscape pattern. Cell size is the ultimate determinant of highest number of patches per unit of area	Number per 100 hectares	$\mathrm{PD} = \frac{X_i}{A}(10{,}000)(100)$ X_i = Number of total patches in each class (patch type i) A = Landscape area(m^2)
Percentage of Landscape	PLAND	Proportional plenty of each patch in the landscape. If PLAND = 100 it means the entire landscape is dominated by a single patch, whereas if PLAND = 0 then each class becomes rare in the landscape	Percent	$Pi = \frac{\sum_{n=1}^{m} a_{mn}}{A}(100)$ Pi = percentage of landscape occupied by patch, a_{mn} = area (m^2) of $_{mn}$ patch, A = Landscape area (m^2)

Source: McGarial and Marks, 1995

4.1 *Spatio-Temporal Pattern of EVI*

The Siliguri municipal area is surrounded by scattered vegetation and tea plantations mainly. Thirty years of monitoring (1990–2020) have been traced in the region from EVI (Fig. 3).

The value of EVI ranges from 0.02 to 0.96 for the study region in the concerned period of time. Vegetation cover is gradually reducing day by day in the study

Table 3 Green covered area from different indices

Green covered area in hector					
Year	NDVI	SAVI	OSAVI	EVI	TVI
1990	789.3	794.79	820.89	807.3	779.4
1995	765.27	751.77	770.76	785.43	760.77
2000	670.32	702.81	691.65	687.24	697.14
2005	570.15	525.6	521.82	526.41	496.89
2010	402.39	388.53	413.64	434.79	401.4
2015	248.04	229.23	260.73	211.05	229.41
2020	150.12	156.69	166.23	141.03	130.14

Source: Calculated by Author

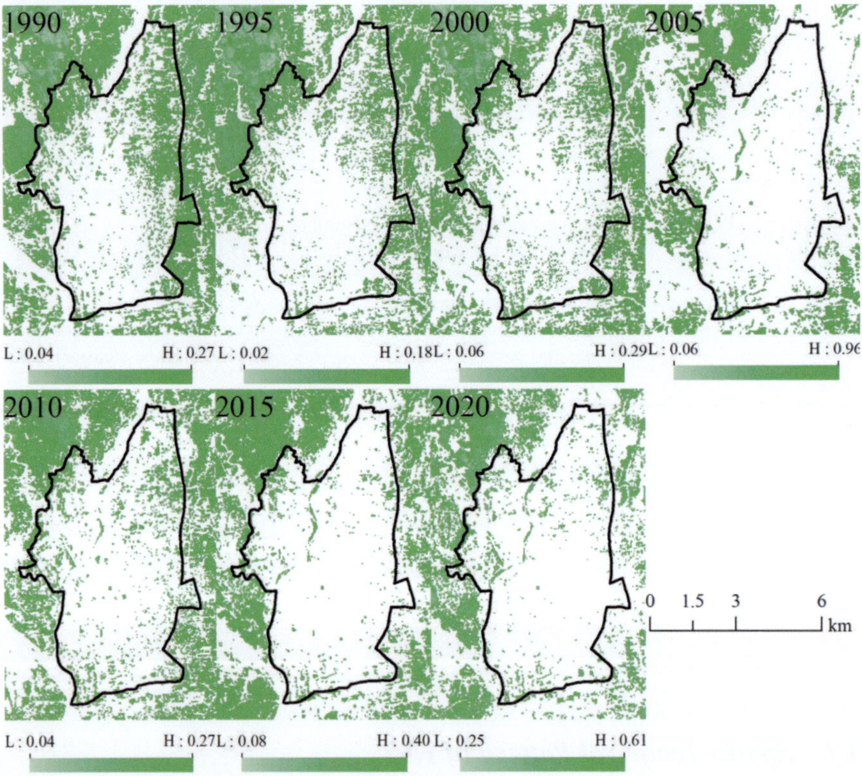

Fig. 3 Enhanced Vegetation Index (EVI) of different years

region. 35% of the total area was occupied by the green cover in 1990, as per the calculation of EVI. It was reduced to 29% and 27% in 1995 and 2000, respectively. After 2000, a huge reduction in vegetation was found over the region. In 1990, the green area occupied 807.3 hectares; by 2020, it will still be 141.03 hectares.

4.2 *Spatio-Temporal Pattern of NDVI*

NDVI is a very popular and simple method to identify the vegetation cover. The values of NDVI have ranged from 0.17 to 1.00 for the study region. The northeastern region of the Siliguri municipality was covered in dense vegetation in 1990, according to data obtained by thresholding the NDVI. After 2010, it was reduced in terms of areal extent (Fig. 4).

The green cover was found to be 789.3 hectares in 1990 as per the NDVI indices, but in 1995 it was 670.32 hectares. NDVI of 2010 and 2020 have 402.39 and 150.12 hectares of green covered area, respectively, which indicates a "linear declining" trend in vegetation cover. After 2010, the green coverage sharply declined as recorded by NDVI.

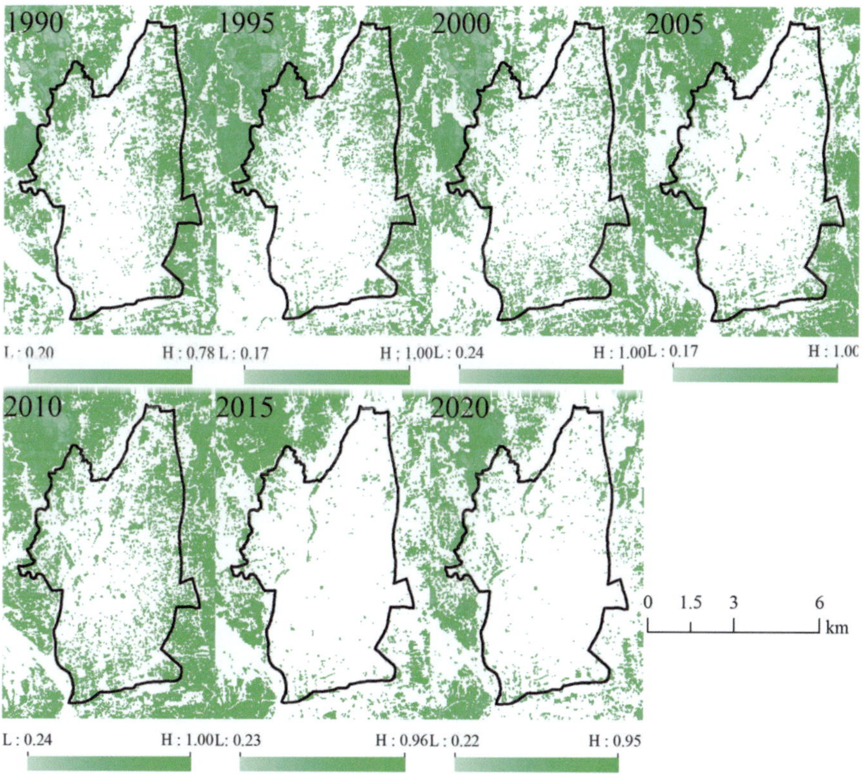

Fig. 4 Normalized Differentiate Vegetation Index (NDVI) of different years

4.3 Spatio-Temporal Pattern of SAVI

Soil values are adjusted in the SAVI index to delineate the vegetation. The value of SAVI ranges from 0.14 to 1.95 for the Siliguri municipal area. It has been found from SAVI that the agglomerated forests have been transformed into scattered trees after 2005 (Fig. 5).

SAVI indicates that the vegetation cover was suddenly decreased after 2010. 794.79 hectares of green cover were found in 1990. In 1995 green cover area became 751.77 hectares. From 2000 to 2020, at five-year intervals, the vegetation areas were 702.81, 525.6, 388.53, 229.23, and 156.69, respectively. In the case of SAVI, the vegetation area was declining with time, like other vegetation indices. A dense patch of vegetation cover was wiped out in 2020 in the northeastern portion of Siliguri municipality.

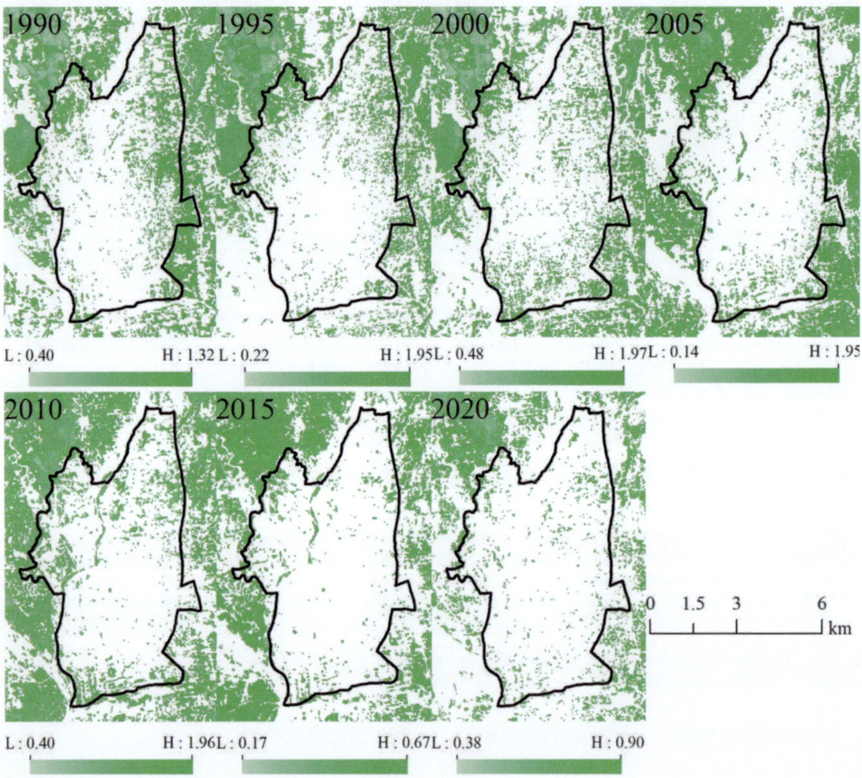

Fig. 5 Soil Adjusted Vegetation Index (SAVI) of different years

4.4 Spatio-Temporal Pattern of OSAVI

OSAVI is the optimized form of SAVI; it gives a more prominent areal coverage of vegetation cover. OSAVI value of the particular region has ranged from 0.08 to 0.99. The more in value indicates the dense cover in vegetation. A major green cover suddenly vanished after 2000 (Fig. 6).

OSAVI shows 820.89 hectares of area occupied by green cover in 1990. In the case of 1995, it was 770.76 hectares. For 2000, 2005, and 2010, the green cover area was 691.65, 521.82, and 413.64 hectares, respectively. After 2005, the vegetation cover was suddenly decreased at a severe rate. In 2005, the vegetation cover was 260.73 hectares (Table 3). 2015 and 2020 had 260.73 and 166.23 hectares of green area in the study area.

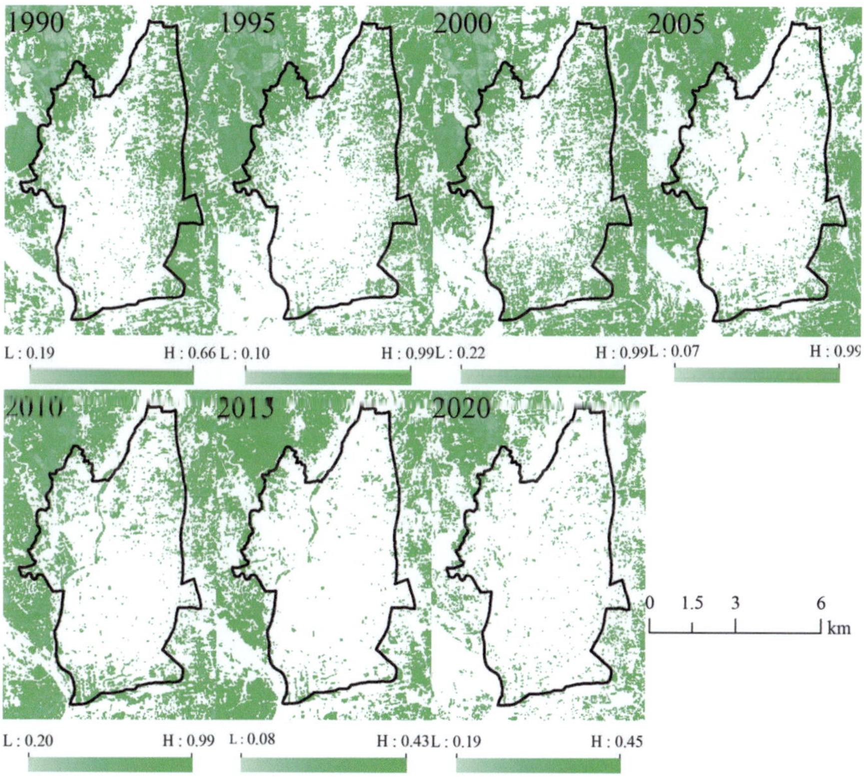

Fig. 6 Optimized Soil Adjusted Vegetation Index (OSAVI) of different years

4.5 Spatio-Temporal Pattern of TVI

As per TVI, the vegetated area had remarkably declined in 2020. Not only in area extent but also in terms of density, the vegetation area was fragmented. TVI value ranges from 0.77 to 1.22 for the concerned study area. Vegetation cover was reducing at a high rate after 2000 (Fig. 7).

779.4 hectares of vegetation area were covered in 1990 in Siliguri municipality as per TVI. In 1995, it became 760.77 hectares. In 2000, 2005, and 2010, the areal coverage of vegetated land was 697.14, 496.89, and 401.4 hectares, respectively. After 2010, a rapid reduction in vegetation cover was found as per the TVI. The green area in 2015 was 229.41 hectares, which was approximately half of the vegetated land of 2010.

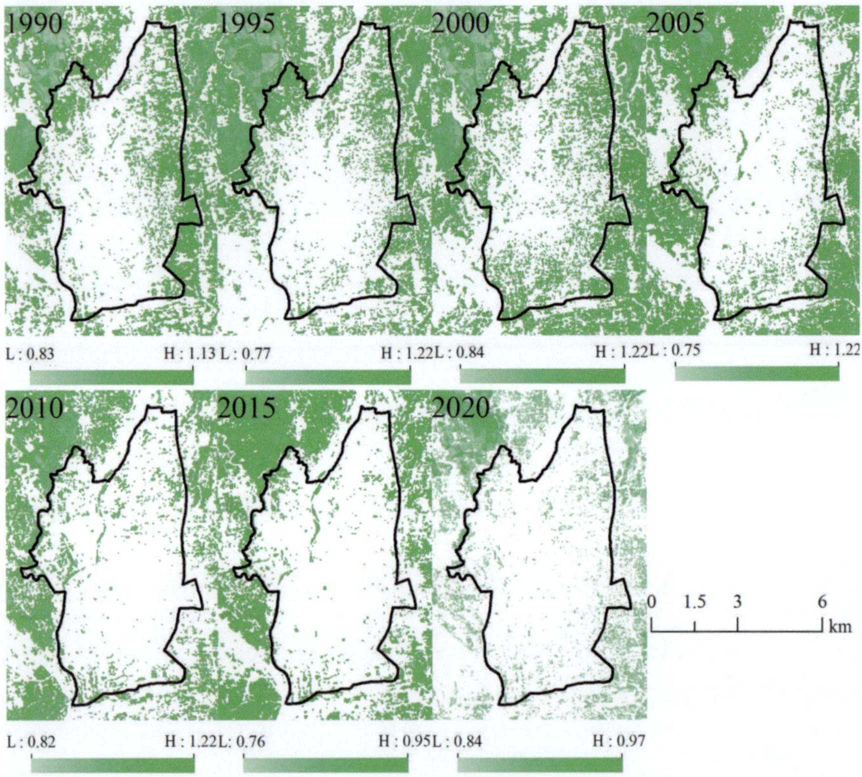

Fig. 7 Transformed Vegetation Index (TVI) of different years

4.6 Spatio-Temporal Pattern of UGS

Combining all the indices mentioned before, a composite file had been prepared. 100 training signatures were collected from both the UGS and non-UGS features, and an ANN was performed on the combined indices to prepare the UGS map. An overall accuracy based on the Kappa coefficient was calculated for each year to validate the UGS extent with the ground truth. 72% and 78% overall accuracy were found for 1990 and 1995, respectively. For 2000, 2005, 2010, 2015, and 2020, the overall accuracy was 68%, 74%, 82%, 79%, and 71%, respectively. Sixty samples for both the UGS and non-UGS sections from ground truth to validate the maps.

A dense UGS surrounds the entire Siliguri municipal border area, where a bunch of patches of green space were found in the central area in 1990. In 2000, dense UGS remained only in the northeast and western portions of Siliguri. After 2005, the UGS extent decreased till 2020 (Fig. 8). In 2020, there was a massive reduction in the extent of UGS.

In 1990, the total UGS was about 34% of the total area of the Siliguri municipal area. The UGS was reduced to about 24% in 1995, and in 2000 it became 22%. In the years 2005, 2010, 2015, and 2020, the UGS area reduced by about 18%, 17%, 10%, and 7%, respectively (Fig. 9). A linear declining trend was found in the temporal change in the UGS area extent. A high degree of magnitude (R^2: 0.960) was found between time and areal UGS extinction.

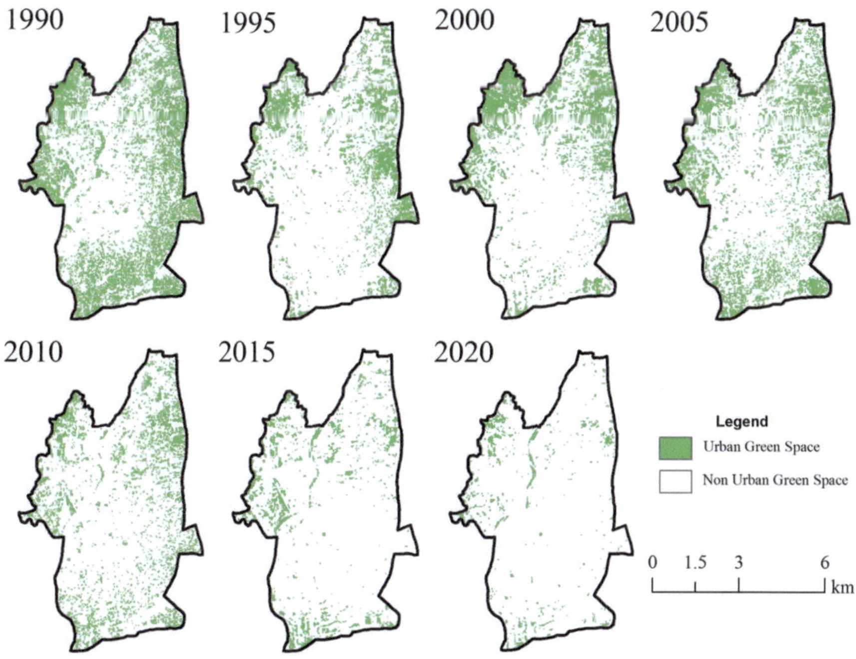

Fig. 8 Urban green space (UGS) of different years

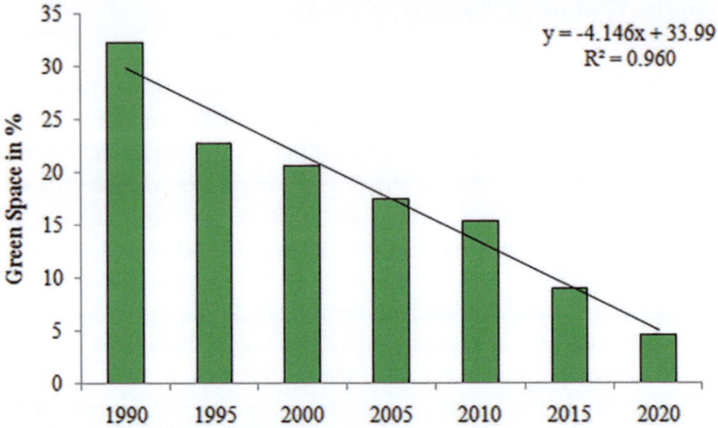

Fig. 9 Temporal distribution of green space in Siliguri municipality

Table 4 Landscape matrices of UGS

Year	NP	PD	ED	LPI	PLAND
1990	338	8.827	168.029	36.25	32.33
1995	25	0.652	49.969	24.42	22.72
2000	118	3.081	89.283	22.47	20.63
2005	741	1.351	87.113	19.84	17.43
2010	37	0.966	104.483	15.49	15.261
2015	25	0.652	49.969	9.16	9.002
2020	15	0.391	28.216	4.31	4.551

Source: Author Produced

4.7 *Indicators of UGS*

Different landscape matrices were calculated from the UGS to describe the green space in different dimensions. A variety of characteristics of green space were assessed through the different matrices (Table 4). Each matrix was used as the indicator of UGS. Patch density, edge density, largest patch density, and percentage of landscape were taken as the landscape pattern indicators of UGS. In this study. PLAND and LPI are considered area-related matrices. ED is the shape complexity-related matrices, and PD is related to aggregation and fragmentation matrices. Such a three-dimensional landscape pattern was assessed in this chapter.

The number of patches of UGS is decreasing day by day in the study region. That means the agglomeration of the UGS reduced with time. In 1990, it was 338, but in 2020, it became only 15. From the areal dimension, it was found that PLAND represents the percentage of green covered area. From 32.33 in 1990, the PLAND became 4.551 in 2020. The LPI is also decreasing with time; the largest patch was

36.25% in 1990, but in 2020, the largest patch remained 1.31%. The shape of the UGS is also declining with time, as it is shown by ED. The value of ED in 1990 was 168.029, but in 2020 it became 28.216 only. The agglomeration of UGS was indicated by PD. A maximum of 8.827% of UGS seems to be agglomerated or clustered in the study region. The agglomeration of UGS is gradually decreasing day by day in the study region. From all points of view, the UGS status is declining with time in Siliguri municipality.

5 Discussion

NDVI is commonly used to detect the vegetation cover. It has some limitations because it focuses on dense vegetation (broadleaf) areas; it fails to identify the versatility of vegetation type (Wu et al., 2014). To cover up this limitation, five vegetation indices were used in this work with versatile criteria to detect vegetation cover properly. TVI omitted the negative value of NDVI and standardized the index value (Bohrer et al., 2014). EVI has the power to detect the different types of vegetation as per the spectral reflectance (Matsushita et al., 2007). Sometimes NDVI fails to differentiate the soil reflectance and vegetation reflectance (Gilabert et al., 2002). Thus, SAVI and OSAVI have been used as soil-adjusted indices. These indices have been used as input parameters in the ANN classifier to predict the UGS. The UGS pattern in the Siliguri municipal area suddenly decreases after 2005 at a high rate. Acceleration is built up as urbanization causes massive vegetation damage (Chakraborty et al., 2018). PLAND of the UGS is decreasing with time in the Siliguri municipal area due to rapid urbanization. It indicates that the total vegetated coverage over the total municipal area was reduced. LPI has a similar condition to PLAND. The other class matrix, like the number of patches, declined at a remarkable rate. This indicates a shrinking of the vegetated area due to urbanization (Seto & Fragkias, 2005). ED and PD values of UGS, when they decrease, indicate a reduction in green space over an area. ED and PD of UGS declined at a homogeneous rate. This is an indication of a significant reduction in green space due to the urban expansion in Siliguri (Darehshouri et al., 2020). The study has unequivocally shown that the Siliguri municipal area's green areas are gradually being encroached upon as a result of the area's fast built-up growth.

6 Conclusion

The urban region around the Siliguri municipal area holds a complex land use pattern and socio-economic character. It has been found from the study that this region has gone through several land use transformations over the time period from 1990 to 2020. This rapid change in land use has caused a deep impact on the environment as well as the socio-economic conditions in Siliguri. Greening activities should concentrate on providing the conditions required for UGS configuration to affect UHI (e.g., more equal distribution) in cities where configuration is not a determinant of UHI, as well as increasing UGS. Increased greenery alone does not yield optimum results where both composition and design already affect UHI; attention should also be paid to the layout and optimization of the UGS pattern.

Despite this, the research has important consequences for green space planning. The concept of the work can be repeated in a cost-effective manner at desired time scales to produce an inventory of neighborhood-level green space distribution indicators, since the distribution indicators are obtained from remote sensing data. This will be particularly useful in the case of cities in developed countries with high spatial and temporal urban dynamism. Furthermore, the work design will aid planners in identifying neighborhoods that need special consideration, especially in developed countries. The study attempted to identify the urban green spaces within the Siliguri municipal area and their impact on the thermal environment, ecosystem, and socio-economic status for implementing proper planning and management policy regarding green space distribution within the Siliguri municipality. The study is also useful for research at the micro level or peri-urban areas around small and medium towns like Siliguri. Though some limitations are there in this study, like the unavailability of high-resolution satellite images and diversified social data, the images of medium resolution and the census data have been used to fulfill the purpose of the study; however, the proper research and improvement in methodology, the whole work can be very useful.

References

Abd-El Monsef, H., Hassan, M. A., & Shata, S. (2017). Using spatial data analysis for delineating existing mangroves stands and siting suitable locations for mangroves plantation. *Computers and Electronics in Agriculture, 141*, 310–326.

Ali-Bik, M. W., Hassan, S. M., Abou El Maaty, M. A., Abd El Rahim, S. H., Abayazeed, S. D., & Wahab, W. A. (2018). The late Neoproterozoic pan-African low-grade metamorphic ophiolitic and Island-arc assemblages at Gebel Zabara area, central Eastern Desert, Egypt: Petrogenesis and remote sensing-based geologic mapping. *Journal of African Earth Sciences, 144*, 17–40.

Apan, A. A., Raine, S. R., & Paterson, M. S. (2002). Mapping and analysis of changes in the riparian landscape structure of the Lockyer Valley catchment, Queensland, Australia. *Landscape and Urban Planning, 59*(1), 43–57.

Bohrer, G., Beck, P. S., Ngene, S. M., Skidmore, A. K., & Douglas-Hamilton, I. (2014). Elephant movement closely tracks precipitation-driven vegetation dynamics in a Kenyan forest-savanna landscape. *Movement Ecology, 2*(1), 1–12.

Chakraborty, A., Seshasai, M. V. R., Reddy, C. S., & Dadhwal, V. K. (2018). Persistent negative changes in seasonal greenness over different forest types of India using MODIS time series NDVI data (2001–2014). *Ecological Indicators, 85*, 887–903.

Chen, X. L., Zhao, H. M., Li, P. X., & Yin, Z. Y. (2006). Remote sensing image-based analysis of the relationship between urban heat Island and land use/cover changes. *Remote Sensing of Environment, 104*(2), 133–146.

Congalton, R. G., Birch, K., Jones, R., & Schriever, J. (2002). Evaluating remotely sensed techniques for mapping riparian vegetation. *Computers and Electronics in Agriculture, 37*(1–3), 113–126.

Darehshouri, S., Michelsen, N., Schüth, C., & Schulz, S. (2020). A low-cost environmental chamber to simulate warm climatic conditions. *Vadose Zone Journal, 19*(1), e20023.

De Ridder, K. (2004). *Benefits of urban green space (BUGS).* Final Report. Section, 6, 53

Deering, D. W. (1978). *Rangeland reflectance characteristics measured by aircraft and spacecraft sensors* (Doctoral dissertation, Texas A&M University. Libraries).

Gandhi, G. M., Parthiban, S., Thummalu, N., & Christy, A. (2015). Ndvi: Vegetation change detection using remote sensing and gis–A case study of Vellore District. *Procedia Computer Science, 57*, 1199–1210.

Gilabert, M. A., González-Piqueras, J., Garcıa-Haro, F. J., & Meliá, J. (2002). A generalized soil-adjusted vegetation index. *Remote Sensing of Environment, 82*(2–3), 303–310.

Huete, A. R. (1988). A soil-adjusted vegetation index (SAVI). *Remote Sensing of Environment, 25*(3), 295–309.

Huete, A., Didan, K., Miura, T., Rodriguez, E. P., Gao, X., & Ferreira, L. G. (2002). Overview of the radiometric and biophysical performance of the MODIS vegetation indices. *Remote Sensing of Environment, 83*(1–2), 195–213.

Huete, A. R., Didan, K., Shimabukuro, Y. E., Ratana, P., Saleska, S. R., Hutyra, L. R., et al. (2006). Amazon rainforests green-up with sunlight in dry season. *Geophysical Research Letters, 33*, 6.

Jean, S., & Cho, K. (2019). Context-aware learning for neural machine translation. *arXiv. preprint arXiv:1903.04715.*

Karnieli, A., Kaufman, Y. J., Remer, L., & Wald, A. (2001). AFRI—Aerosol free vegetation index. *Remote Sensing of Environment, 77*(1), 10–21.

Li, L., Weinberg, C. R., Darden, T. A., & Pedersen, L. G. (2001). Gene selection for sample classification based on gene expression data: Study of sensitivity to choice of parameters of the GA/KNN method. *Bioinformatics, 17*(12), 1131–1142.

Mabel, M. C., & Fernandez, E. (2008). Analysis of wind power generation and prediction using ANN: A case study. *Renewable Energy, 33*(5), 986–992.

Mahmon, N. A., & Ya'acob, N. (2014). A review on classification of satellite image using artificial neural network (ANN). In *2014 IEEE 5th control and system graduate research colloquium* (pp. 153–157). IEEE.

Matsushita, B., Yang, W., Chen, J., Onda, Y., & Qiu, G. (2007). Sensitivity of the enhanced vegetation index (EVI) and normalized difference vegetation index (NDVI) to topographic effects: A case study in high-density cypress forest. *Sensors, 7*(11), 2636–2651.

McGarial, K., & Marks, B. J. (1995). FRAGSTAT: Spatial pattern analysis program for quantifying landscape structure. Portland, OR: Department of Agriculture, Forest Service, Pacific Northwest Research Station. https://doi.org/10.2737/PNW-GTR-351

Monserud, R. A., & Leemans, R. (1992). Comparing global vegetation maps with the kappa statistic. *Ecological Modelling, 62*(4), 275–293.

Mróz, M., & Sobieraj, A. (2004). Comparison of several vegetation indices calculated on the basis of a seasonal SPOT XS time series, and their suitability for land cover and agricultural crop identification. *Technical Sciences, 7*(7), 39–66.

Muller, A., Reiter, J., & Weiland, U. (2011). Assessment of urban vulnerability towards floods using an indicator-based approach—A case study for Santiago de Chile. *Natural Hazards and Earth System Sciences, 11*, 2107–2123. https://doi.org/10.5194/nhess-11-2107-2011

Nageswara Rao, P. P., Shobha, S. V., Ramesh, K. S., & Somashekhar, R. K. (2005). Satellite-based assessment of agricultural drought in Karnataka state. *Journal of the Indian Society of Remote Sensing, 33*(3), 429–434.

Nagler, P. L., Cleverly, J., Glenn, E., Lampkin, D., Huete, A., & Wan, Z. (2005). Predicting riparian evapotranspiration from MODIS vegetation indices and meteorological data. *Remote Sensing of Environment, 94*(1), 17–30.

Narumalani, S., Zhou, Y., & Jensen, J. R. (1997). Application of remote sensing and geographic information systems to the delineation and analysis of riparian buffer zones. *Aquatic Botany, 58*(3–4), 393–409.

Nkeki, F. N. (2016). Spatio-temporal analysis of land use transition and urban growth characterization in Benin metropolitan region, Nigeria. *Remote Sensing Applications: Society and Environment, 4*, 119–137.

Rocha, A. V., & Shaver, G. R. (2009). Advantages of a two band EVI calculated from solar and photosynthetically active radiation fluxes. *Agricultural and Forest Meteorology, 149*(9), 1560–1563.

Rouse, J. W., Haas, R. H., Schell, J. A., Dering, D. W., & Harlan, J. C. (1974). Monitoring the Vernal Advancement of Retrogradation of Natural Vegetation, Type III, Final Report. MD, USA: NASA/GSFC.

Seto, K. C., & Fragkias, M. (2005). Quantifying spatiotemporal patterns of urban land-use change in four cities of China with time series landscape metrics. *Landscape Ecology, 20*(7), 871–888.

Tan, P. Y., Wang, J., & Sia, A. (2013). Perspectives on five decades of the urban greening of Singapore. *Cities, 32*, 24–32.

Wang, Q., Jiang, N., Yin, S., Li, X., Yu, F., Guo, Y., & Zhang, R. (2017). Carbonaceous species in PM2. 5 and PM10 in urban area of Zhengzhou in China: Seasonal variations and source apportionment. *Atmospheric Research, 191*, 1–11.

Weber, R. M., & Dunno, G. A. (2001). Riparian vegetation mapping and image processing techniques, Hopi Indian Reservation, Arizona. *Photogrammetric Engineering and Remote Sensing, 67*(2), 179–186.

Wu, D., Wu, H., Zhao, X., Zhou, T., Tang, B., Zhao, W., & Jia, K. (2014). Evaluation of spatiotemporal variations of global fractional vegetation cover based on GIMMS NDVI data from 1982 to 2011. *Remote Sensing, 6*(5), 4217–4239.

Xue, J., & Su, B. (2017). Significant remote sensing vegetation indices: A review of developments and applications. *Journal of Sensors, 2017*.

Yang, X. (2007). Integrated use of remote sensing and geographic information systems in riparian vegetation delineation and mapping. *International Journal of Remote Sensing, 28*(2), 353–370.

Dynamics of Ecosystem Services Values (ESVs) in Response to Land Use/Land Cover (LU/LC) Change in Central NCR

Mohd Rihan, Shahfahad, Mohd Waseem Naikoo, Swapan Talukdar, and Atiqur Rahman

Abstract Natural ecosystems provide vital resources for wildlife as well as human wellness through many ecosystem services (ES). ES are greatly impacted by changes in land use/land cover (LU/LC), which are influenced by both human and natural processes. This study focused on examining the micro-level variations in ES due to LU/LC change in the selected sub-districts of the Central National Capital Region (CNCR) between 1991 and 2021. The LU/LC classification was made using a hybrid machine learning (ML) model integrating support vector machine (SVM), artificial neural network (ANN), and random forest (RF) algorithms. By using multi-temporal LU/LC maps, ES were quantified for different LU/LC classes using the global coefficient of 2014. The evaluation of the LU/LC change's impact on ES valuation was conducted through elasticity approaches. The result shows a substantial increase of 181.49% in the total built-up area from 1991 to 2021, while the total

M. Rihan
Department of Geography, Faculty of Sciences, Jamia Millia Islamia, New Delhi, India

Urban Innovation and Sustainability, Department of Development and Sustainability, Asian Institute of Technology, Pathum Thani, Thailand

Shahfahad
Department of Geography, Faculty of Sciences, Jamia Millia Islamia, New Delhi, India

Department of Geography, Institute of Science, Banaras Hindu University, Varanasi, Uttar Pradesh, India
e-mail: arahman2@jmi.ac.in

M. W. Naikoo
Department of Geography and Disaster Management, University of Kashmir, Srinagar, Jammu & Kashmir, India

S. Talukdar
Department of Geography, Asutosh College, University of Calcutta, Kolkata, West Bengal, India

A. Rahman (✉)
Department of Geography, Faculty of Sciences, Jamia Millia Islamia, New Delhi, India
e-mail: arahman2@jmi.ac.in

P. K. Joshi et al. (eds.), *Geospatial Science for Urban Ecosystems*, The Urban Book Series, https://doi.org/10.1007/978-3-032-07155-2_17

cropland experienced a significant decline of 33.09% in the study area, including a faster reduction in cropland. During 1991–2001, the ecosystem services values (ESVs) have witnessed an increase of 11.85% in CNCR. The maximum increase in ESVs has been noted from the Gurgaon, Faridabad, and Ghaziabad sub-districts. Moreover, among all LU/LC classes, cropland exhibited the maximum decrease in ESVs during the same period in Gurgaon, Gautama Buddha Nagar, and Najafgarh. The change in ES in urban areas affects air quality, the urban heat island effect, water infiltration, and recreational opportunities. The results can help planners and researchers to draw up long-term management plans for local ecosystems, considering the consequences of changes in LU/LC on the ES.

Keywords Ecosystem services values · Hybrid machine learning · Land use/land cover · Artificial neural network · Central National Capital Region

1 Introduction

Ecosystem services (ES) are the essential services that ecosystems provide to people. These include resources like food and water, regulating environmental processes, providing cultural and recreational opportunities, and fulfilling fundamental supporting functions (Chen et al., 2023; Petsch et al., 2022; MEA, 2005). ES are those advantages that people get from ecological processes that improve their quality of life. These services not only maintain the equilibrium of ecological functions but also provide necessary resources for organisms (Larson et al., 2023; Jamean & Abas, 2023). Each ecosystem provides distinct and heterogeneous services, exemplified by the contrasting services offered by forest ecosystems compared to desert or grassland ecosystems (Costanza et al., 1997). These natural ecosystems are currently being altered to become human-dominated ecosystems (Jacinto et al., 2023). Recently, land use/land cover (LU/LC) changes caused by urbanization have significantly impacted the health of global ecosystems (Fang et al., 2022; Ayieko et al., 2023). Urban growth often results in areas transitioning from a natural state with abundant natural landscape (Mourya et al., 2021; Zhou et al., 2023) to a state with reduced and fragmented natural land cover (Jiang et al., 2023). The increasing demand for fuel, food, materials, and water has led to rapid and extensive changes in ecosystems (MEA, 2005; Entrekin et al., 2011).

Technological advancements, population growth, and personal lifestyles indirectly affect biodiversity, altering ecosystems and modifying the ecosystems' benefits, ultimately influencing human wellness (Sandifer et al., 2015; Tang et al., 2023). Although urban regions cover only a small part of the surface of Earth, they exert significant ecological impacts and indirect effects on ecosystems (Reid et al., 2019). Ecosystem degradation and biodiversity loss will likely affect ecosystem functionality and resilience to disturbance (Cappelli et al., 2022; Ibanez et al., 2022). This deterioration will disrupt the services that ecosystems provide to people now and in the future (MEA, 2005). In addition, changes in LU/LC, urbanization, overexploitation of resources, forest fires (Shahfahad et al., 2022a; Rihan et al., 2023), and catastrophic events are all human-caused factors altering ES (Latinopoulos et al., 2021). This destructive behavior, as well as the exploitation of the environment, puts the health and happiness of future generations at risk (Szocik & Reiss, 2023). As reported by the MEA (2005), about 60 percent of ES have been lost, severely affecting ecosystems and their components. Shahfahad et al. (2022b) emphasize that changes in LU/LC, primarily caused by the increasing demand for land due to population growth, are significantly affecting ecosystem services.

The LU/LC change patterns in India exhibit greater complexity and uniqueness compared to other nations (Rahman et al., 2012; Naikoo et al., 2023). The accelerated population growth in the country drives the establishment of new settlements and urban expansion, resulting in alterations in LU/LC (Talukdar et al., 2020). This transformation involves the conversion of productive land like forests, croplands, grasslands, and water bodies into built-up areas that are incapable of delivering ES (Rahman et al., 2011). Several scientific research teams, including Costanza et al. (2014), Talukdar et al. (2020), and Morya and Punia (2022), have investigated the influence of LU/LC alterations on ecosystem service values (ESVs). These studies have consistently revealed similar outcomes and findings. In addition, a number of methods, such as the InVEST-based model, the cost-based method, and the value transfer method (VTM), have been utilized by scientists to calculate and evaluate the ES and their values in terms of LU/LC change (Zhan et al., 2019; Talukdar et al., 2020). By combining VTM along with the cost-based method, Costanza et al. (1997) created an innovative approach to implementing ESVs on a worldwide scale. Wangai et al. (2019) used the ES matrix technique in their study to assess and quantify how ES transformation responds to alterations in LU/LC. While other studies have used the VTM to calculate the ESVs in response to changes in LU/LC. For example, Song and Deng (2017), Bryan et al. (2018), and Talukdar et al. (2020) have used this method in their respective studies.

The study of how LU/LC change affects ESVs has recently gained considerable attention and importance (Wu et al., 2013; Pham & Lin, 2023). In India, there is limited research on assessing changes in the ESVs (Acharya et al., 2019; Imdad et al., 2022). Mourya et al. (2021) pointed out that the rapid change in LU/LC due to urbanization is the main factor for the decline in ES and their values in the Delhi NCR of India. Similarly, Sharma et al. (2020) found that extensive urbanization and the resulting rapid changes in LU/LC are significantly affecting ESVs in NCR. They pointed out that the conversion of arable land to cultivated and open land around Delhi is the main cause of loss of ES. Previous studies have focused on how urbanization influences changes in LU/LC in Delhi, which is a big metropolitan area. However, no studies have focused on how these changes impact ES. Therefore, our objective is to examine the impact of LU/LC change on ES and their values in 43 sub-districts. By subdividing the study area in this way, we aim to capture the nuanced variations in LU/LC changes and their consequent effects on ES values. The detailed insights gained from this analysis will help planners and scientists develop management strategies for ecosystem restoration and sustainable use.

2 Study Area

The selected area for this study is the peri-urban region extending from Delhi to the Central National Capital Region (CNCR) (Fig. 1), which is recognized as one of the four policy zones of the NCR regional plan 2021. The CNCR includes six interconnected urban areas: Gurgaon-Manesar in Gurgaon, Faridabad-Ballabhgarh in Faridabad, Sonipat-Kundli in Sonipat, and Bahadurgarh in Jhajjar, all located in Haryana, as well as Ghaziabad-Loni in Ghaziabad and Noida in Gautama Buddha Nagar in Uttar Pradesh (Sharma & Sen, 2011). These areas are known for their rapidly changing LU/LC and are expected to expand into larger urban centers (Dutta et al., 2020; Naikoo et al., 2020; Shahfahad et al., 2022a). The population census shows that from 2001 to 2011, the population density in the NCT of Delhi increased from 9340 to 11,297 people per square kilometer, a growth rate much higher than the national average or any other state or union territory in India (Census of India, 2001, 2011).

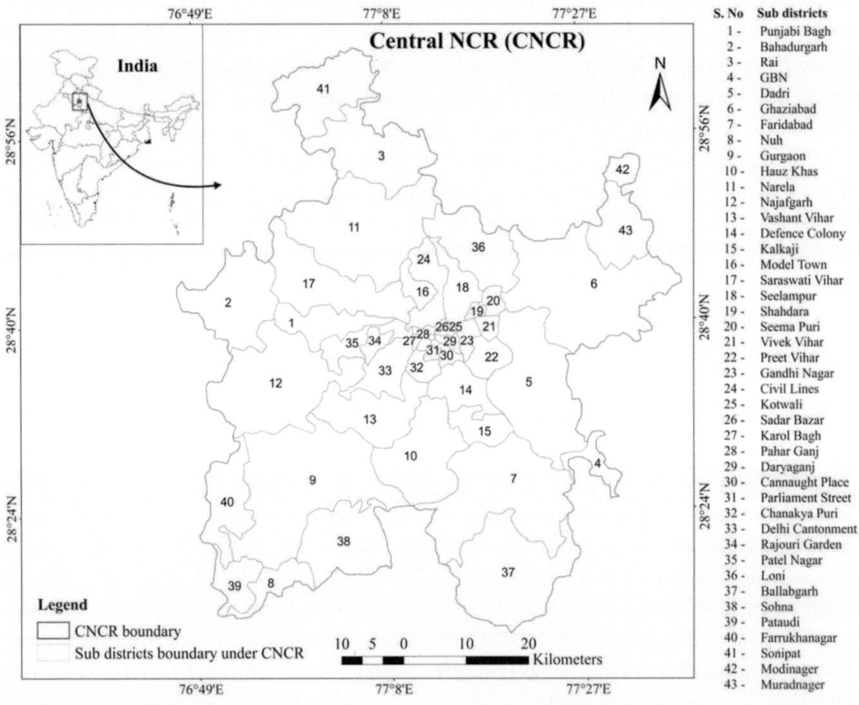

Fig. 1 Study area location map

3 Data Sources and Methods

3.1 *Data Sources*

The Earth Explorer Portal of the United States Geological Survey (USGS) provided the Landsat 4, 5 (TM) data (1991, 2001, and 2011) and the Landsat 8 (OLI/TIRS) data (2021) used in this study. The data for each year were downloaded for the same season, specifically the months of September to December. Table 1 contains the details of the satellite data used in this research. Furthermore, the map of the study area was taken from the NCR Planning Board report.

3.2 *Methods*

3.2.1 Method for LU/LC Classification

To classify the accurate LU/LC map, we used a hybrid machine learning (ML) technique integrating artificial neural network (ANN), support vector machine (SVM), and random forest (RF) in ArcGIS 10.8 software. Based on the Level 1 classification scheme of the National Remote Sensing Centre (NRSC) in India, LU/LC are classified into eight classes: water bodies, vegetation, scrubland, cropland, built-up areas, open land, barren land, and river beds. To ensure accuracy, the LU/LC categories were validated using Google Earth imagery and compared with the results of previous literature.

3.2.2 Accuracy Assessment

In the domain of LU/LC classification, the evaluation of post-classification accuracy assumes great significance (Naikoo et al., 2020). It plays a vital role in analyzing the accuracy of classified maps and determining their suitability for specific applications. Generally, a classified map should exhibit an accuracy level of at least 80% to ensure precise interpretation and identification (Ventura et al., 2018). In a previous study, methods like the indices-based scores, error matrix, and kappa

Table 1 Detailed overview of the satellite data

Month/Year of acquisition	Satellite	Sensor	Spatial resolution	Path/ Row	Average cloud cover
Nov–Dec 1991	Landsat 4 & 5	TM	30 m	146/40, 147/40	7%
Feb–March 2001					2%
Sept–Oct 2011					4%
Oct–Nov 2021	Landsat 8	OLI/ TIRS	30 m	146/40, 147/40	0.13%

coefficient were used to determine the accuracy of LU/LC (Talukdar et al., 2021). In this ongoing study, the accuracy of LU/LC maps with respect to CNCR was calculated using the kappa coefficient approach. A total of 240 random points were systematically selected to conduct this assessment, ensuring equitable representation of each LU/LC class across the entire study area. As field data were unavailable for 1991 and 2001, ground observations were acquired through high-resolution satellite imagery. For the period between 2011 and 2021, a combination of Google Earth Pro and field visits was used in regions with limited accessibility or the absence of field information. The results of the accuracy assessment show overall accuracies of 89.34% for 1991, 88.21% for 2001, 91.12% for 2011, and 90.04% for 2021. These findings substantiate the quality of the LU/LC maps generated in this study and offer valuable insights for their practical utilization.

3.2.3 Evaluation of Ecosystem Services Values (ESVs)

The concept of ESVs has evolved significantly since the late twentieth century, beginning with the foundational work of Costanza et al. (1997). This concept was further developed by the MEA (2003, 2005), which helped to spread ESVs globally. Building on this framework, Xie et al. (2008) adapted the ESV classifications for China and categorized ES into four main types and nine subtypes, using methods from both Costanza et al. (1997) and the MEA (2003, 2005). In a later refinement, Costanza et al. (2014) enhanced their ESV calculation method and adjusted the values for certain land use classes. This updated method was applied in our study to estimate ESVs for different LU/LC classes in forty-three sub-districts (tehsils) of the study area. Using ArcGIS 10.8, we created spatio-temporal maps to visualize the ESVs for each sub-district to illustrate the distribution and value of ES across the study area.

3.2.4 Elasticity of ESVs Change

The elasticity of ESVs reflects their sensitivity to changes in LU/LC. It is used to quantify the extent of variation in ESVs due to shifts in LU/LC patterns. In their study at the district and national levels, Song and Deng (2017) measured the absolute change in ESVs resulting from a proportional change in LU/LC, as described in Eqs. 1 and 2.

$$\mathrm{EEL} = \left| \frac{\left(\mathrm{ESV}_{\mathrm{end}} - \mathrm{ESV}_{\mathrm{start}}\right) / \mathrm{ESV}_{\mathrm{start}} \times 100\%}{\mathrm{LCP}} \right| \tag{1}$$

$$\mathrm{LCP} = \frac{\sum_{i=1}^{6} \Delta \mathrm{LUT}_t}{\sum_{i=1}^{6} \Delta \mathrm{LUT}_t} \times \frac{1}{T} \times 100\% \tag{2}$$

where EEL denotes the elasticity of ESV alteration in relation to the shift LU/LC. $\mathrm{ESV}_{\mathrm{end}}$ and $\mathrm{ESV}_{\mathrm{start,}}$ respectively, show the ESV at the end and start of the study period. LCP reveals the change percentage of land area, reflecting the speed and sensitivity of LU/LC change. $\Delta\mathrm{LUT}_i$ represents the changed area of the *I* type of LU/LC, while LUT_i signifies the area of the *I* type of LU/LC. T represents the time gap of the study period.

4 Results and Discussion

4.1 Estimation of Spatio-Temporal LU/LC Classification

The LU/LC for the periods 1991, 2001, 2011, and 2021 were generated using a combined hybrid ML method that integrates ANN, RF, and SVM. These maps classified the landscape into eight distinct classes shown in Figs. 2a–d. Covering 394,740.16 hectares and encompassing 43 sub-districts, the study area shows an uneven distribution of LU/LC types, which contributes to a diverse range of ES. To formulate a management strategy, it is crucial to comprehend this diversity at the sub-district level.

The overall analysis of LU/LC shows that the cropland areas have consistently decreased, from 235,744 hectares in 1991 to 157,729 hectares in 2021, indicating potential challenges to food security and emphasizing the necessity for sustainable land and agricultural policies. On the other hand, built-up land has markedly expanded, from 60,025 hectares in 1991 to 168,967 hectares in 2021, reflecting rapid urbanization that necessitates strategic urban planning to tackle infrastructure, environmental, and social issues. Water body extents have varied, beginning at 2805 hectares in 1991, peaking at 4686 hectares in 2011, and then decreasing to 4020 hectares by 2021. Vegetation coverage also changed significantly, dropping from 29,658 hectares in 1991 to 10,412 hectares in 2011, then increasing to 21,819 hectares in 2021 (Table 2). This increase in vegetation is largely due to extensive greening efforts by the state government of Delhi and local authorities during the period, as documented in recent studies (Shahfahad et al., 2022b; Paul & Nagendra, 2015).

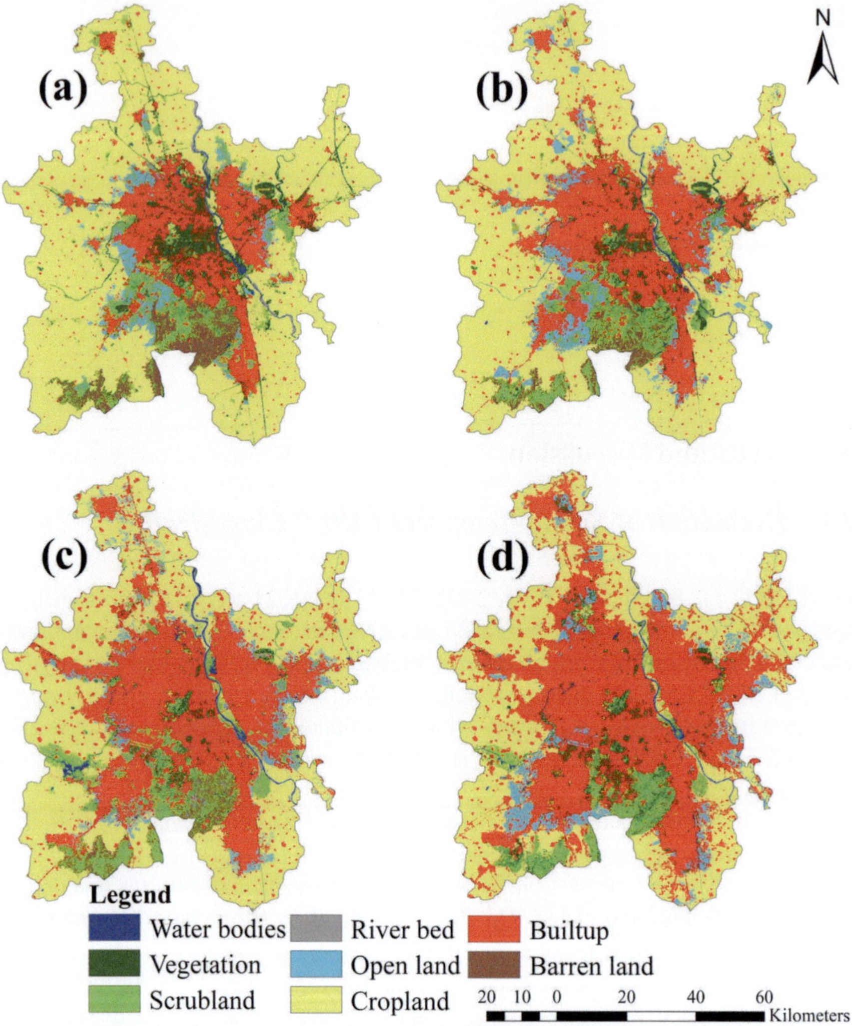

Fig. 2 Land use/land cover maps of Central NCR (**a**) 1991, (**b**) 2001, (**c**) 2011, and (**d**) 2021

Table 2 LU/LC classes area during 1991–2021 (area in hectares)

LU/LC classes	1991	2001	2011	2021
Water bodies	2805.060	2890.470	4686.000	4020.200
Vegetation	29658.037	20676.047	10412.400	21819.300
Scrubland	31434.400	26899.100	35505.157	20338.307
Cropland	235744.000	216959.000	182269.000	157729.000
Built-up	60024.990	91025.420	129268.620	168966.630
Open land	15638.400	22994.500	21947.300	19769.000
Barren land	17590.000	12088.000	9964.350	1481.400
Riverbed	1845.270	1207.620	687.330	616.320
Total	**394740.157**	**394740.157**	**394740.157**	**394740.157**

4.2 Evaluation of ESVs

The total ESVs derived from LU/LC classes for each sub-district were calculated through the global coefficient values from Costanza et al. (2014) for the periods 1991, 2001, 2011, and 2021, as illustrated in Figs. 3 and 4. The sub-district level

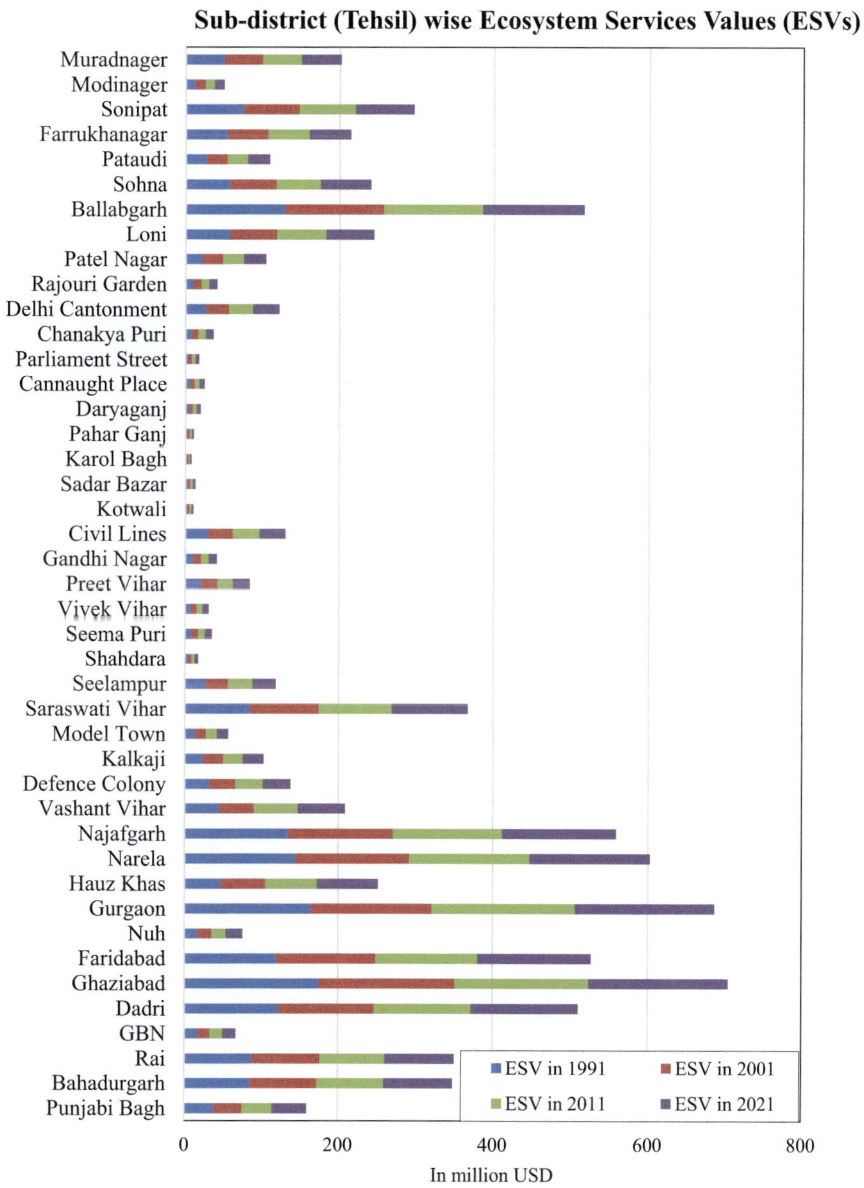

Fig. 3 Sub-district wise variation of ESVs during 1991–2021

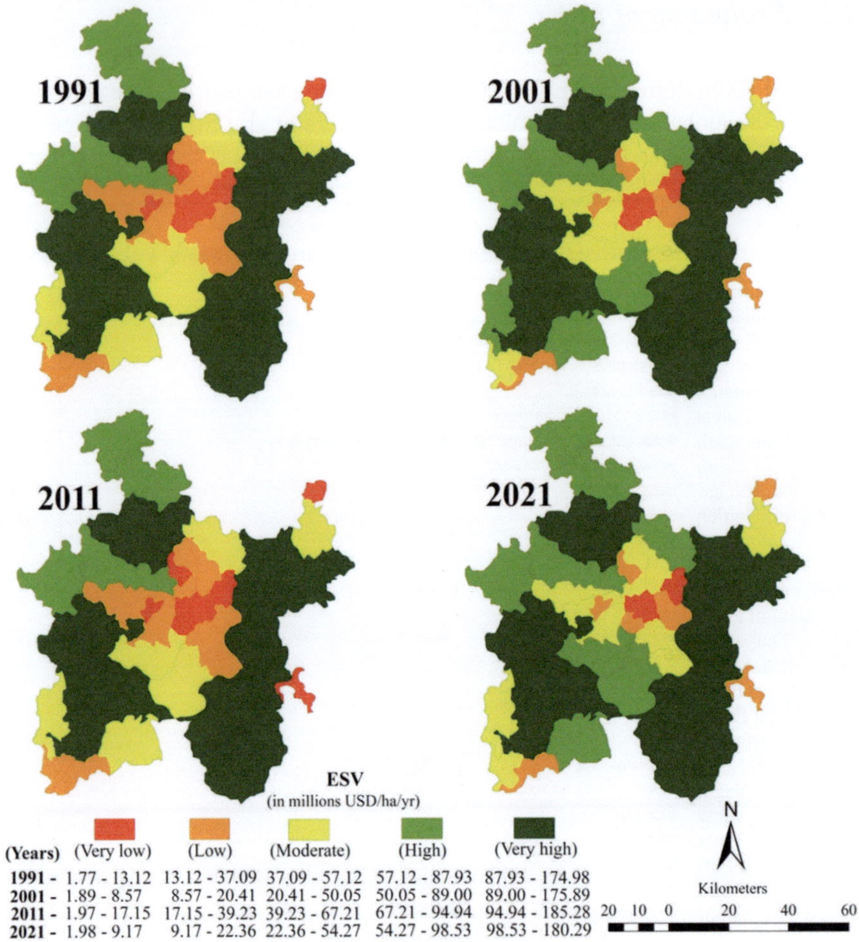

Fig. 4 Sub-district wise distribution of Ecosystem Service Values (ESVs) from 1991 to 2021

analysis shows that Ghaziabad ($174,975,684), Gurgaon ($164,282,761), Narela ($144,594,524), Najafgarh ($134,288,011), Ballabgarh ($129,092,363), and Faridabad ($119,234,636) recorded the highest ESVs in 1991. Over the subsequent decade, the ESVs in most sub-districts either remained stable or exhibited minor fluctuations. By 2011, significant ESV increases were observed in sub-districts like Ghaziabad ($175,891,215), Narela ($147,040,773), and Najafgarh ($136,435,570), attributed to increased street vending activities. Conversely, sub-districts such as Gurgaon ($156,441,340), Faridabad ($128,642,000), and Ballabgarh ($127,895,747) saw declines in their ESVs. By 2021, significant ESV increases were noted in Ghaziabad ($180,294,097), Najafgarh ($146,469,487), Faridabad ($145,947,441), and Dadri ($138,324,799), largely due to the alteration of open lands into built-up land (Naikoo et al., 2020). Meanwhile, ESVs decreased in Gurgaon ($180,264,662), Narela ($155,057,283), and Ballabgarh ($130,384,317).

4.3 *Change in Ecosystem Services Values (ESVs)*

The sub-district-wise analysis from 1991 to 2001 shows that Gautam Buddha Nagar, Gurgaon, Sonipat, and Ballabgarh experienced a decline in the total value of ESVs (Fig. 5). Among these sub-districts, the study shows the highest rate of decline in

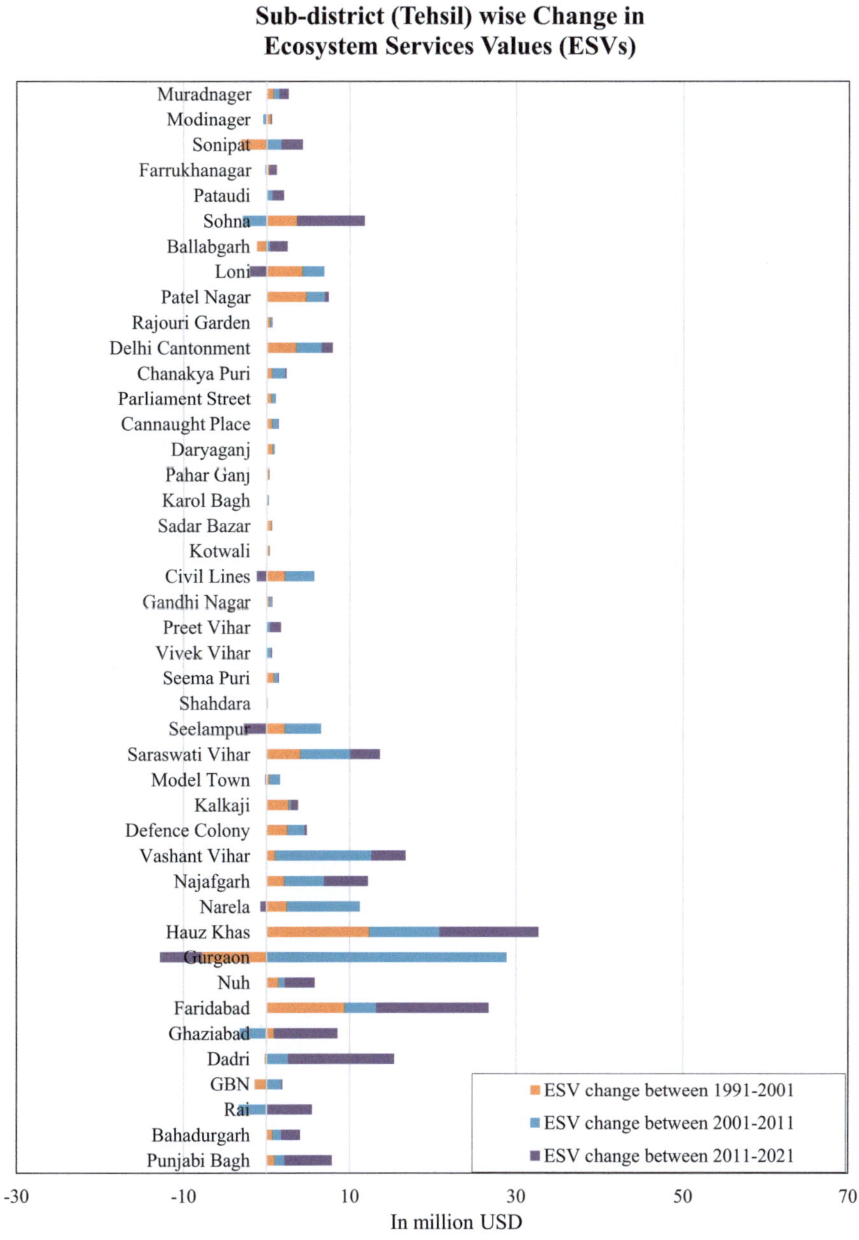

Fig. 5 Sub-district wise Ecosystem Service Values (ESVs) change during 1991–2021

Gautam Buddha Nagar, Gurgaon, Sonipat, and Ballabgarh, with declines of 8.53% ($1,437,589), 4.77% ($784,142), 4.16% ($311,765) and 0.93% ($119,661), respectively. On the other hand, Hauz Khas and Patel Nagar recorded the highest growth rates of 26.57% ($12,331,242) and 22.27% ($4,739,781), respectively. In the period from 2001 to 2011, only Sohna, Rai, and Modinagar recorded a decline in total ESVs. The study indicates a moderate decline in these sub-districts: Sohna recorded a decline of 4.87% ($2,943,920), Rai a decline of 3.76% ($3,304,675), and Modinagar a decline of 3.57% ($451,870). Conversely, Vasant Vihar and Gurgaon recorded the highest growth rates of 25.37% ($11,573,938) and 18.43% ($28,837,190), respectively. From 2011 to 2021, only Seelampur and Gurgaon recorded a decline in total ESVs. These sub-districts recorded a high rate of decline, with Seelampur recording a decline of 8.33% ($2,751,955) and Gurgaon a decline of 2.71% ($5,013,868). Over the entire study period of 30 years, Hauz Khas shows the highest rate of increase, with an increase of 70% ($32,653,806 USD) in the ESVs.

5 Conclusion

This study comprehensively examines the impact of LU/LC changes on the valuation of ES and their values within the CNCR of India. The findings reveal that the predominant driver for the reduction in ESVs is the extensive transformation of cropland into built-up land, attributed to rapid urbanization. From 1991 to 2021, built-up areas expanded significantly from 60,025 hectares to 168,967 hectares, whereas other LU/LC categories experienced notable declines. Sub-district-level analyses, encompassing regions such as Gurgaon, Faridabad, Ghaziabad, GBM, Narela, Najafgarh, and Ballabhgarh, consistently indicated a decline in cropland concomitant with urban expansion over the same period. Quantitatively, the ESVs for major sub-districts exhibited dynamic changes: in Gurgaon, ESVs increased from $164,282,761 to $180,264,662; in Faridabad, from $119,234,636 to $145,947,441; and in Ghaziabad, from $174,975,684 to $180,294,097. These fluctuations underscore the intricate relationship between urban growth and ecosystem service provision, with decade-specific variations observed between 1991 and 2021.

This research underscores the critical need for improved policy frameworks and urban planning to mitigate adverse environmental and ecological effects. Additionally, the study emphasizes the significance of scaling this methodology to a national level, addressing the current gap in localized studies within India. Implementing such an approach on a broader scale could yield vital insights into the regional and national implications of LU/LC transformations on ecosystem services, thereby informing sustainable development strategies.

References

Acharya, R. P., Maraseni, T., & Cockfield, G. (2019). Global trend of forest ecosystem services valuation–an analysis of publications. *Ecosystem Services, 39*, 100979.

Ayieko, A., Moses, G., Godfrey, M., Kimwatu, D., & Mwangi, A. (2023). Spatial modeling of groundwater across land use land cover and climate change gradient using SWAT and Logan's method: A case study of Mbagathi sub-catchment. *Modeling Earth Systems and Environment*, 1–17.

Bryan, B. A., Ye, Y., Zhang, J., & Connor, J. D. (2018). Land-use change impacts on ecosystem services value: Incorporating the scarcity effects of supply and demand dynamics. *Ecosystem Services, 32*, 144–157.

Cappelli, S. L., Domeignoz-Horta, L. A., Loaiza, V., & Laine, A. L. (2022). Plant biodiversity promotes sustainable agriculture directly and via belowground effects. *Trends in Plant Science, 27*(7), 674–687

Census of India (2001). Town directory, NCT of Delhi - 2001. Census Digital Library. https://censusindia.gov.in/nada/index.php/catalog/23308 [Accessed July 2023].

Census of India (2011). Districts of Delhi , Population Census. https://www.census2011.co.in/census/state/districtlist/delhi.html [Accessed June 2023].

Chen, C., Deng, W., Ignatieva, M., Bi, L., Du, A., & Yang, L. (2023). Synergy of urban green space planning and ecosystem services provision: A longitudinal exploration of China's development. *Urban Forestry & Urban Greening, 127997.*

Costanza, R., d'Arge, R., De Groot, R., Farber, S., Grasso, M., Hannon, B., et al. (1997). The value of the world's ecosystem services and natural capital. *Nature, 387*(6630), 253–260.

Costanza, R., De Groot, R., Sutton, P., Van der Ploeg, S., Anderson, S. J., Kubiszewski, I., et al. (2014). Changes in the global value of ecosystem services. *Global Environmental Change, 26*, 152–158.

Dutta, D., Rahman, A., Paul, S. K., & Kundu, A. (2020). Estimating urban growth in peri-urban areas and its interrelationships with built-up density using earth observation datasets. *The Annals of Regional Science, 65*, 67–82.

Entrekin, S., Evans-White, M., Johnson, B., & Hagenbuch, E. (2011). Rapid expansion of natural gas development poses a threat to surface waters. *Frontiers in Ecology and the Environment, 9*(9), 503–511.

Fang, Z., Ding, T., Chen, J., Xue, S., Zhou, Q., Wang, Y., et al. (2022). Impacts of land use/land cover changes on ecosystem services in ecologically fragile regions. *Science of the Total Environment, 831*, 154967.

Ibanez, T., Platt, W. J., Bellingham, P. J., Vieilledent, G., Franklin, J., Martin, P. H., et al. (2022). Altered cyclone–fire interactions are changing ecosystems. *Trends in Plant Science, 27*(12), 1218–1230.

Imdad, K., Rihan, M., Sahana, M., Parween, S., Ahmed, R., Costache, R., et al. (2022). *Wetland health, water quality, and resident perceptions of declining ecosystem services: a case study of Mount Abu, Rajasthan, India* (pp. 1–27). Environmental Science and Pollution Research.

Jacinto, E., Fangue, N. A., Cocherell, D. E., Kiernan, J. D., Moyle, P. B., & Rypel, A. L. (2023). Increasing stability of a native freshwater fish assemblage following flow rehabilitation. *Ecological Applications, e2868.*

Jamean, E. S., & Abas, A. (2023). Valuation of visitor perception of urban forest ecosystem services in Kuala Lumpur. *Land, 12*(3), 572.

Jiang, L., Wang, Z., Zuo, Q., & Du, H. (2023). Simulating the impact of land use change on ecosystem services in agricultural production areas with multiple scenarios considering ecosystem service richness. *Journal of Cleaner Production, 397*, 136485.

Larson, S., Jarvis, D., Stoeckl, N., Barrowei, R., Coleman, B., Groves, D., et al. (2023). Piecemeal stewardship activities miss numerous social and environmental benefits associated with culturally appropriate ways of caring for country. *Journal of Environmental Management, 326*, 116750.

Latinopoulos, D., Koulouri, M., & Kagalou, I. (2021). How historical land use/land cover changes affected ecosystem services in lake Pamvotis, Greece. *Human and Ecological Risk Assessment: An International Journal, 27*(6), 1472–1491.
Millennium Ecosystem Assessment (MEA). (2003). *Ecosystems and human Well-being: Current state and trends*. Island Press.
Millennium Ecosystem Assessment (MEA). (2005). *Ecosystems and human well-being: Synthesis*. Island Press.
Morya, C. P., & Punia, M. (2022). Impact of urbanization processes on availability of ecosystem services in National Capital Region of Delhi (1992–2010). *Environment, Development and Sustainability, 24*(5), 7324–7348.
Mourya, M., Kumari, B., Tayyab, M., Paarcha, A. A., & Rahman, A. (2021). Indices based assessment of built-up density and urban expansion of fast growing Surat city using multi-temporal Landsat data sets. *GeoJournal, 86*, 1607–1623.
Naikoo, M. W., Rihan, M., & Ishtiaque, M. (2020). Analyses of land use land cover (LULC) change and built-up expansion in the suburb of a metropolitan city: Spatio-temporal analysis of Delhi NCR using landsat datasets. *Journal of Urban Management, 9*(3), 347–359.
Naikoo, M. W., Talukdar, S., Ishtiaq, M., & Rahman, A. (2023). Modelling built-up land expansion probability using the integrated fuzzy logic and coupling coordination degree model. *Journal of Environmental Management, 325*, 116441.
Paul, S., & Nagendra, H. (2015). Vegetation change and fragmentation in the mega city of Delhi: Mapping 25 years of change. *Applied Geography, 58*, 153–166.
Petsch, D. K., Cionek, V. D. M., Thomaz, S. M., & dos Santos, N. C. L. (2022). Ecosystem services provided by river-floodplain ecosystems. *Hydrobiologia*, 1–22.
Pham, K. T., & Lin, T. H. (2023). Effects of urbanisation on ecosystem service values: A case study of Nha Trang, Vietnam. *Land Use Policy, 128*, 106599.
Rahman, A., Aggarwal, S. P., Netzband, M., & Fazal, S. (2011). Monitoring urban sprawl using remote sensing and GIS techniques of a fast growing Urban Centre, India. *IEEE Journal of Selected Topics in Applied Earth Observations and Remote Sensing, 4*(1), 56–64.
Rahman, A., Kumar, S., Fazal, S., & Siddiqui, M. A. (2012). Assessment of land use/land cover change in the North-West District of Delhi using remote sensing and GIS techniques. *Journal of the Indian Society of Remote Sensing, 40*(4), 689–697.
Reid, A. J., Carlson, A. K., Creed, I. F., Eliason, E. J., Gell, P. A., Johnson, P. T., et al. (2019). Emerging threats and persistent conservation challenges for freshwater biodiversity. *Biological Reviews, 94*(3), 849–873.
Rihan, M., Ali Bindajam, A., Talukdar, S., Waseem Naikoo, M., Mallick, J., & Rahman, A. (2023). Forest fire susceptibility mapping with sensitivity and uncertainty analysis using machine learning and deep learning algorithms. *Advances in Space Research, 72*(2), 426–443.
Sandifer, P. A., Sutton-Grier, A. E., & Ward, B. P. (2015). Exploring connections among nature, biodiversity, ecosystem services, and human health and well-being: Opportunities to enhance health and biodiversity conservation. *Ecosystem Services, 12*, 1–15.
Shahfahad, Naikoo, M. W., Islam, A. R. M. T., Mallick, J., & Rahman, A. (2022a). Land use/land cover change and its impact on surface urban heat Island and urban thermal comfort in a metropolitan city. *Urban Climate, 41*, 101052.
Shahfahad, Naikoo, M. W., Das, T., Talukdar, S., Asgher, M. S., Asif, & Rahman, A. (2022b). Prediction of land use changes at a metropolitan city using integrated cellular automata: Past and future. *Geology, Ecology, and Landscapes*, 1–19.
Sharma, S., & Sen, A. (2011). Levels of development in central National Capital Region 1981—2001. *Institute of Town Planners, India Journal, 8-2*, 86–104.
Sharma, S., Nahid, S., Sharma, M., Sannigrahi, S., Anees, M. M., Sharma, R., ... & Joshi, P. K. (2020). A long-term and comprehensive assessment of urbanization-induced impacts on ecosystem services in the capital city of India. *City and Environment Interactions, 7*, 100047.
Song, W., & Deng, X. (2017). Land-use/land-cover change and ecosystem service provision in China. *Science of the Total Environment, 576*, 705–719.

Szocik, K., & Reiss, M. J. (2023). Why space exploitation may provide sustainable development: Climate ethics and the human future as a multi-planetary species. *Futures, 147*, 103110.

Talukdar, S., Singha, P., Mahato, S., Praveen, B., & Rahman, A. (2020). Dynamics of ecosystem services (ESs) in response to land use land cover (LU/LC) changes in the lower Gangetic plain of India. *Ecological Indicators, 112*, 106121.

Talukdar, S., Eibek, K. U., Akhter, S., Ziaul, S. K., Islam, A. R. M. T., & Mallick, J. (2021). Modeling fragmentation probability of land-use and land-cover using the bagging, random forest and random subspace in the Teesta River Basin, Bangladesh. *Ecological Indicators, 126*, 107612.

Tang, Z., Xie, M., Chen, B., Xu, M., & Ji, Y. (2023). Do social and ecological indicators have the same effect on the subjective Well-being of residents? *Applied Geography, 157*, 102994.

Ventura, D., Bonifazi, A., Gravina, M. F., Belluscio, A., & Ardizzone, G. (2018). Mapping and classification of ecologically sensitive marine habitats using unmanned aerial vehicle (UAV) imagery and object-based image analysis (OBIA). *Remote Sensing, 10*(9), 1331.

Wangai, P. W., Burkhard, B., & Müller, F. (2019). Quantifying and mapping land use changes and regulating ecosystem service potentials in a data-scarce peri-urban region in Kenya. *Ecosystems and people, 15*(1), 11–32.

Wu, K. Y., Ye, X. Y., Qi, Z. F., & Zhang, H. (2013). Impacts of land use/land cover change and socioeconomic development on regional ecosystem services: The case of fast-growing Hangzhou metropolitan area, China. *Cities, 31*, 276–284.

Xie, G. D., Zhen, L., Lu, C. X., Xiao, Y., & Chen, C. (2008). Expert knowledge based valuation method of ecosystem services in China. *Journal of Natural Resources, 23*(5), 911–919.

Zhan, J., Zhang, F., Chu, X., Liu, W., & Zhang, Y. (2019). Ecosystem services assessment based on emergy accounting in Chongming Island, eastern China. *Ecological Indicators, 105*, 464–473.

Zhou, T., Liu, H., Gou, P., & Xu, N. (2023). Conflict or coordination? Measuring the relationships between urbanization and vegetation cover in China. *Ecological Indicators, 147*, 109993.

Ecology–Urbanization Nexus: Assessing Urban Ecological Security in the Kolkata Metropolitan Area, India

Subrata Ghosh, Santanu Dinda, and Nilanjana Das Chatterjee

Abstract The chapter provides an in-depth evaluation of Urban Ecological Security (UES), emphasizing the impacts of urbanization pressure and ecological sensitivity within the context of the interconnections between urban ecology and development in Kolkata Metropolitan Area (KMA). Assessing UES is crucial for promoting sustainable urban planning and ecological health monitoring. The primary aims is to assess the ecological security status of the study area which is one of the rapidly expanding metropolitan areas of eastern India using advanced geospatial modeling techniques. The outcomes of this research contribute to the advancement of methodologies in the realm of UES assessment. Moreover, the results offer valuable and scientifically sound insights that can guide the optimization of land-use planning strategies, potentially leading to enhanced ecological security in urban areas. In addition, this study underscores the importance of considering ecological security as a fundamental aspect of urban development, aligning with the broader goals of sustainability and environmental preservation.

Keywords Urbanization pressure · Ecological vulnerability · DEMATEL-ANP · Sustainable urban development · Ecological resilience

1 Introduction

1.1 Conceptualizing Urban Ecology

Ecology is the scientific discipline dedicated to the study of how living organisms interact with each other and their surrounding environment, encompassing the examination of patterns, processes, and relationships (Odum, 1975). Ecology, as defined by the Cary Institute of Ecosystem Studies, embodies a comprehensive and holistic approach (Verma et al., 2020). It can be succinctly described as the

S. Ghosh (✉) · S. Dinda · N. Das Chatterjee
Department of Geography, Vidyasagar University, Midnapore, West Bengal, India

P. K. Joshi et al. (eds.), *Geospatial Science for Urban Ecosystems*, The Urban Book Series, https://doi.org/10.1007/978-3-032-07155-2_18

scientific examination of the intricate dynamics shaping the presence and numbers of living beings, the interplay among these life forms, and their intricate connections with the conversion and flow of energy and substances in the natural world. Cities now serve as catalysts for development while simultaneously acting as forces propelling environmental transformation (Alberti, 2008). Expanding upon the foundational understanding of ecology, urban ecology can be defined as the study of the intricate relationships among living organisms and their environment within urban areas. This discipline encompasses a broad range of factors, including spatial distribution, population densities, interactions between different species, and the complex dynamics of energy and matter. Urban ecology delves into how these elements interact and evolve in response to human activities, infrastructure development, and environmental changes occurring within urban landscapes (Xiao & Chen, 2002; Sahani & Raghavaswamy, 2018). In this context, an urban ecosystem emerges as a dynamic result of human population growth and the simultaneous evolution of urban centers, which include cities, towns, clustered agglomerations, and expansive megacities. The concept of 'urban ecology' originated with the Chicago School of Sociology in the 1920s, aiming to apply principles from ecosystem ecology to better understand the complexities of urban environments and their impacts on social dynamics, biodiversity, and environmental sustainability. This interdisciplinary approach continues to evolve, integrating ecological principles with urban planning and social sciences to address the challenges and opportunities posed by urbanization worldwide (Burgess, 2012).

Urban areas represent complicated ecological systems primarily governed by human influence (Batten, 2001). Human elements distinguish urban ecosystems from natural ecosystems in multiple distinct ways. Successfully managing these systems requires a deep understanding of the complex connections that intertwine human activities with ecological processes, governing their dynamics and evolutionary trajectories (Hultman, 1993; Forman, 2014; Koroso et al., 2021). Given that ecological systems inherently undergo constant transformations, the resilience and adaptability of urban ecosystems to these alterations emerge as pivotal elements in the pursuit of long-term sustainability for cities.

1.2 Urbanization Challenges and Implications

Currently, over 50% of the global population resides within cities of diverse dimensions (United Nations, 2018). It is projected that by the year 2050, approximately two-thirds of the global population will live in urban areas (World Bank, 2021). Rapid urban expansion stands as a defining trait for numerous cities worldwide, particularly those in the proximity of major urban hubs (Turner et al., 2003; Gong et al., 2009). Urbanization is a multifaceted and dynamic process influenced by numerous socio-economic and demographic factors (Cook, 2002). It is characterized by the transformation of natural landscapes into altered human-made environments, accompanied by population growth. Consequently, urbanization leads to substantial modifications in the land use and land cover (LULC) of urban regions (Morimoto & Negishi, 2019). This rapid urban expansion and consequential LULC changes pose significant challenges for urban planners, as they have far-reaching impacts on sustainable city development (Grimm et al., 2008; Tripathy & Kumar, 2019). The alterations in LULC resulting from rapid urban growth contribute to various environmental deteriorations, such as the diminishment of green spaces, soil erosion, degradation of urban ecosystems, loss of biodiversity, modifications in urban microclimates, and heightened environmental pollution, all occurring across various spatial and temporal scales (Dinda et al., 2020).

1.3 Urbanization-Induced Environmental Stress

Development and environmental transformations have become intricately interwoven within urban settings. Within this context, the assessment of urban LULC changes and their repercussions on the urban environment stands as a frequently examined phenomenon. This scrutiny yields valuable insights into historical patterns of LULC alterations, contributing to a more profound comprehension of effective management strategies (Das & Angadi, 2021). Urban landscapes, in their typical manifestation, embody a complex mosaic of diverse and interconnected ecosystem patches (Verma et al., 2020). Such landscapes exhibit intricate structures that bestow a range of ecosystem services, supporting the socio-economic framework while simultaneously preserving the equilibrium between environmental preservation and developmental pursuits (Parker et al., 2003). Critical constituents of a sustainable city encompass the presence of robust local biodiversity and the accompanying ecosystem services it provides. Regrettably, the relentless march of urbanization continues to erode natural urban ecosystems at unprecedented rates. Startling estimates suggest that the rapid urbanization witnessed in developing nations may obliterate as much as 88% of urban biodiversity, thereby severely curtailing their capacity to adapt to the impending pressures of urban expansion (Wu, 2008). The contemporary pace of urbanization, along with the associated ecological vulnerabilities in expansive urban areas, is undeniably reshaping the functioning of ecosystems and, consequently, the delivery of ecosystem services (Costanza et al., 2014).

Therefore, comprehending the intricate patterns of urban ecological security (UES) and assessing them on both global and regional scales holds significant importance for urban planners and the academic community alike (Feng et al., 2017). The analysis of UES constitutes a multifaceted process that delves into the correlation between the degree of urbanization and the diverse pressures arising from various environmental activities that are deemed crucial for delivering an array of ecosystem services and upholding the ecological sustainability of a city.

2 Concept of Urban Ecological Security (UES)

2.1 Definitions and Rationale

Ecological security is widely recognized as a fundamental aspect of regional stability. It also holds a vital role in fostering sustainable urban development and has garnered increased global attention in recent decades (Nuissl et al., 2009; Wang et al., 2018). Ecological security refers to the landscape configuration that provides a sufficient supply of ecosystem services to sustain the socio-economic system while maintaining a delicate balance between environmental preservation and economic development (Feng et al., 2017). The late 1970s marked the beginning of heightened concern regarding ecological security, as the environment faced a multitude of threats, ranging from natural disasters like earthquakes, floods, and massive landslides to human-induced crises such as pollution, radioactive leaks, and other anthropogenic disasters. As a result, the concept of ecological security has gained prominence, offering a framework for evaluating the fundamental structure and functionality of ecosystems, gauging their safety levels, and assessing environmental susceptibility (Han et al., 2015).

Gong et al. (2009) have delineated ecological security as the configuration of a landscape capable of supplying adequate ecosystem services to sustain the socio-economic system while preserving equilibrium between environmental preservation and development. On a similar note, Feng et al. (2018) have posited that ecological security encapsulates the concept of 'environmental health'and emphasizes an ecosystem's ability to effectively provide ecological services. This idea was first introduced in the global ecological security monitoring framework by the Insurance Accounting and System Association (IASA) in 1989 (Xiao & Chen, 2002). The unchecked exploitation of natural resources, rapid industrial growth, and accelerated urbanization collectively pose significant challenges to global ecological security (Wu & Hu, 2020). Therefore, the term "urban ecological security" (UES) refers to the general well-being and stability of urban environments (Li et al., 2020).

The evaluation undertaken by the UES assessment involves a systematic analysis of the interplay between evolving LULC patterns and the vital ecological functions that play a pivotal role in furnishing diverse ecosystem services and upholding the environmental sustainability of a given city (Fig. 1). In essence, the measurement of UES comprehensively elucidates the myriad pressures exerted by urbanization on its surrounding environment through various mechanisms, encompassing the scrutiny of alterations in land use, encroachment upon natural areas, depletion of urban green spaces, fragmentation of landscapes, pollution, and the escalating burden of population.

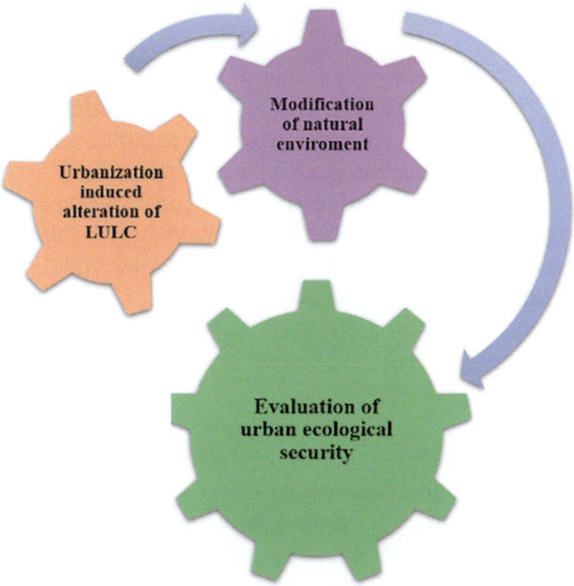

Fig. 1 Conceptual framework of urban development and urban ecological security (UES)

2.2 *Dimensions and Applicability of UES*

The UES is a multifaceted concept composed of several interrelated components that collectively aim to ensure the sustainability and resilience capacity of urban environments.

2.2.1 Environmental Quality

UES relies on upholding rigorous environmental standards within cities. This necessitates continuous oversight and enhancement of air quality to mitigate pollutants detrimental to both human well-being and ecosystems. Likewise, the preservation of safe and uncontaminated water sources within urban locales is of utmost importance, encompassing drinking water, rivers, lakes, and groundwater. Furthermore, the preservation of robust urban soils holds significance, serving not only to support vegetation but also as a bulwark against erosion and a foundation for stable urban development (Cohen, 2006).

2.2.2 Biodiversity Conservation

Safeguarding and elevating ecosystems and biodiversity constitutes a fundamental element of UES (Bhattacharya, 2017). This aspect encompasses the safeguarding of urban natural and semi-natural environments, such as parks, wetlands, forests, and green pathways. Furthermore, efforts in habitat restoration strive to counteract habitat deterioration, thus fostering an environment conducive to the sustenance of indigenous plant and animal life. The advocacy for urban green areas, which encompasses parks and urban woodlands, not only serves as vital wildlife habitats but also augments the urban ecological framework as a whole.

2.2.3 Resource Efficiency

UES hinges on resource efficiency, which involves reducing resource use and waste production. Cities should prioritize energy efficiency in building design, leverage technological advancements, and incorporate renewable energy sources. Effective water management, including conservation and wastewater recycling, is crucial for sustainability (Sun et al., 2018). Additionally, waste reduction initiatives, emphasizing recycling and composting, contribute to reduced solid waste generation and less reliance on landfills.

2.2.4 Climate Resilience

Urban centers must proactively prepare themselves for the consequences of climate change to safeguard ecological security. This entails crafting plans and bolstering infrastructure to cope with escalating temperatures, severe weather occurrences, and the encroachment of rising sea levels. Moreover, strengthening both preparedness and response measures for natural calamities such as urban floods, water scarcity, and heat waves is imperative for upholding urban robustness when confronted with climate-induced adversities (Hoffmann et al., 2012).

2.2.5 Sustainable Urban planning and Design

The imperative for UES hinges upon the adoption of sustainable practices in urban planning and design. Integrating green spaces and infrastructure into city development is pivotal in ameliorating ecological impacts. Promoting compact, multifunctional urban layouts curbs extensive commuting and diminishes energy demands. Furthermore, bolstering and enhancing public transportation networks is crucial to diminishing the dependency on private vehicles, thereby cutting emissions and mitigating traffic congestion (Das & Jain, 2022).

2.2.6 Policy and Governance

The implementation and reinforcement of environmental regulations are central to ensuring urban ecological stability. It is imperative to have policies in place that safeguard the environment and facilitate sustainable urban growth. Aligning urban initiatives with global sustainability objectives, like the Sustainable Development Goals (SDGs), ensures that cities actively contribute to worldwide ecological stability (Punzo et al., 2022). Furthermore, bolstering local government institutions overseeing urban planning, environmental stewardship, and disaster management strengthens urban resilience (Ghosh et al., 2023b).

2.2.7 Environmental Monitoring

The collection and evaluation of data pertaining to critical environmental factors like air and water quality, as well as biodiversity, enable well-informed choices and prompt actions when necessary (Yang et al., 2020). The development and application of performance metrics serve as a means to gauge advancements in attaining UES objectives, offering a foundation for adjustments and enhancements as needed.

2.2.8 Equity and Social Inclusion

Achieving UES necessitates a commitment to equity and social inclusion. This component ensures that all residents, including marginalized and vulnerable communities, benefit from ecological security efforts. Addressing disparities in environmental risks and benefits among different social groups promotes environmental justice and fosters a sense of shared responsibility for urban ecology.

2.3 *Methodological Overview of UES Assessment*

UES has emerged as a pressing issue for urban planners and the academic realm. The assessment of the UES assessment represents a vast and intricate evaluation framework necessitating the seamless integration of diverse scientific disciplines, expert insights, and collaboration among pertinent governmental bodies and departments. Various approaches are employed to assess UES, including mathematical and statistical modeling (Liu et al., 2005), index-based analysis (Li et al., 2019), GIS-based techniques (Feng et al., 2017), and innovative methods such as entropy assessment (Zhang et al., 2009), landscape matrix evaluation (Gao et al., 2013), GIS-based overlay analysis (Feng et al., 2018), and ecological footprint modeling (Wu & Hu, 2020). Ecological security hinges on a multitude of factors and their intricate interplay, rendering a singular model inadequate for assessing UES within complex urban extents (Li et al., 2019). Consequently, the prevailing method for UES evaluation is the index-based approach. Recent research underscores the enhanced precision and efficiency achieved by integrating multiple models with various factors into the ecological security assessment framework. These approaches encompass landscape pattern analysis, fuzzy inference techniques, analytical hierarchy processes, ecological footprint analysis, and more (Sun et al., 2018; Yang & Cai, 2020).

Furthermore, the utilization of multi-criteria-based decision-making techniques (MCDM) in evaluating UES has long held the interest of researchers and decision-makers alike (Wang et al., 2019). MCDM is widely recognized as a decisive tool for making informed decisions and selecting appropriate criteria across various disciplines, including socio-economics, geography, and the environment. The fusion of geographic information systems (GIS) with MCDM remains a prevalent approach in the assessment of UES. Examples include the application of techniques such as the Pontious matrix overlay within GIS (Feng et al., 2018), spatial principal component analysis (SPCA), and various algorithmic methodologies (Wang et al., 2019), among others. Urban ecosystems differ significantly from their natural counterparts due to their predominant influence by human activities. Urban environments are continually evolving with changes in land use, rapid expansion of built-up areas, infrastructure development, and a regrettable reduction in natural elements. Consequently, assessing UES necessitates the consideration of diverse factors, including socio-economic, demographic, urbanization, and environmental aspects (Fig. 2 and Table 1).

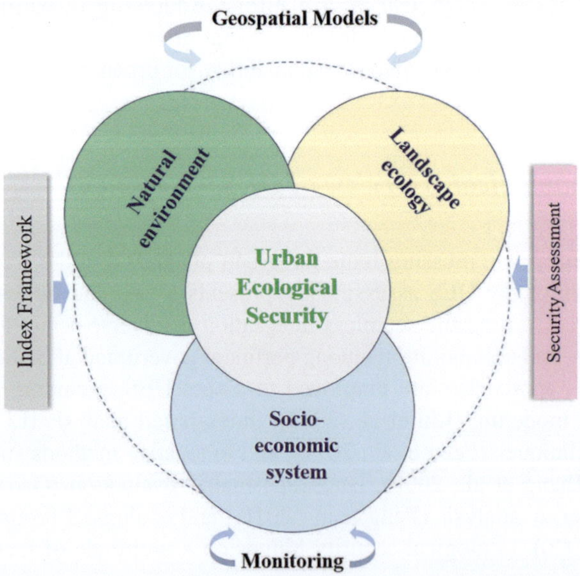

Fig. 2 Relationship between ecological security and natural environment, landscape ecology, and socio-economic systems of an urban area

Table 1 Variables and methods used in different research on UES assessment in the past decade (Source: Ghosh, 2023)

Author	Research objectives	Case study area	Method	Significant variables
Hua et al. (2011)	Land ecological security assessment	Bai Autonomous Prefecture of Dali, China	Entropy and composite index method	Population density and growth rate, per capita cultivated land and production, urbanization level, gross domestic product, and proportion of ecological land
Han et al. (2015)	UES assessment	Beijing–Tianjin–Hebei metropolitan region, China	Fuzzy-entropy methods	Per capita consumption, green space, road pavement, built-up area, industry, sewage, health care, and education facilities
Feng et al. (2018)	Land ecological security and its driving factors	Ningbo coastal city, China	Generalized additive models	Proximity to city center, district center, transport, pollution source, NDVI, and built-up area
Sun et al. (2018)	Urban ecological security assessment	Pearl River Delta urban agglomeration, China	Mamdani fuzzy method	Population density and growth rate, per capita arable land and water resource, air quality, percentage of green area and green area change rate, and proportion of industries
Li et al. (2019)	Urban ecological security assessment	Pearl River Delta urban agglomeration, China	DEMATEL method	Industry per unit area, population density, water area, forest area, proportion of construction land, NDVI, ecosystem service, patch and edge density, Shannon's diversity
Das et al. (2020)	Urban ecological health assessment	Kolkata Metropolitan Area, India	Ecosystem vigor, organization and resilience (V-O-R)-based assessment	NDVI, landscape connectivity and heterogeneity, and landscape shape index
Wu and Hu (2020)	Ecological security	Zhoushan Island,	Partial least	Population, ecologically productive land area, ecological carrying capacity and ecological footprint

2.4 UES Evaluation Index System

The urban ecosystem components diverge from those found in natural ecosystems due to the emergence of various models in recent times for ecological security assessment. Prominent among these are the Pressure-State-Response (PSR) model, the Driving Force-State-Response (DFSR) model, and the Driving

Force-Pressure-State-Influence-Response (DPSIR) model, which enjoy widespread application. The PSR model effectively captures the intricate connection between human activity and the ecological environment (Dinda et al., 2022). The DFSR model introduces a noteworthy alteration to the PSR model by substituting the concept of "pressure" with the term "driving force." This fundamental modification aims to enhance the comprehensiveness of the PSR model's metrics, encompassing additional indicators like societal activities, economic structures, and other human-centric variables (Wen & Hou, 2021). The DPSIR model builds upon the PSR framework by introducing two additional components: driving force and influence indicators (Sobhani et al., 2023). This model primarily elucidates the intricate interplay between economic circumstances and ecological dynamics. These models are ill-equipped to precisely measure the intricacies and unpredictability inherent in the connections among the factors.

The ecological "pressure-state-response" (PSR) framework, in its adapted form, is extensively employed to assess potential susceptibility within an urban setting (Yang et al., 2021; Ghosh et al., 2021). By incorporating various spatial factors, it demonstrates robust applicability at a regional level. Ecological "pressure" denotes the external stressors and disturbances, while ecological "state" pertains to the outcomes resulting from these external stressors, particularly in the context of local ecological concerns. On the other hand, ecological "response" encompasses an ecosystem's inherent capability to autonomously regulate and rebound from damage inflicted by external forces, serving as a measure of its adaptability during the urbanization process and reflecting the degree of coordination between urban development and ecological health.

2.5 UES Applications in Urban Planning and Management

The assessment of UES carries profound implications and applications, transcending conventional urban planning paradigms (Hodson & Marvin, 2009). In an era marked by rapid urbanization, understanding the ecological dynamics of cities assumes paramount importance. Firstly, it guides cities toward sustainable development by revealing the intricate balance between urban growth and environmental integrity. Such insights empower urban planners and policymakers to craft strategies that harness economic prosperity while safeguarding ecosystems (Dhyani et al., 2018). Furthermore, urban ecological assessments provide the foundation for climate resilience, equipping cities with the foresight to adapt to climate change impacts. By identifying vulnerable areas and enhancing disaster preparedness, cities can ensure the continuity of essential services during extreme weather events. Beyond environmental considerations, these assessments promote public health. By pinpointing environmental health risks associated with air and water quality, they facilitate targeted interventions to protect the well-being of urban populations. The assessments also serve as a catalyst for biodiversity conservation in urban areas, identifying critical habitats and species, thus facilitating conservation initiatives (Sridhar, 2020). Overall, UES assessments transcend urban boundaries, guiding cities toward holistic sustainability, resilience, and enhanced quality of life.

2.6 Linking UES and Sustainable Development Goals (SDGs)

UES assessments serve as a pivotal instrument in the pursuit of Sustainable Development Goals (SDGs). These assessments are not merely a technical exercise; rather, they represent a sophisticated approach to aligning urban development with the global sustainability agenda. By meticulously scrutinizing urban ecosystems, these assessments address several critical SDGs (Mudau et al., 2020). They contribute to Goal 11, which calls for sustainable cities and communities, by guiding urban planners in creating environmentally responsible and resilient cities. Furthermore, they support Goal 3 by improving public health through identifying and mitigating environmental health risks. By pinpointing resource inefficiencies and pollution sources, these assessments also advance Goal 12, promoting responsible consumption and production. Moreover, they foster biodiversity conservation and ecosystem restoration, thus aiding in the realization of Goal 15—Life on Land (Park & Han, 2021). In essence, UES assessments are a linchpin in translating the ideals of the SDGs into tangible, sustainable urban development outcomes.

3 Assessing UES in the Indian Context: A Case Study

3.1 Evaluating UES of the Kolkata Metropolitan Area

The primary aim of this case study is to create a comprehensive model for evaluating UES using advanced integrated methodologies, specifically for a rapidly expanding urban area in eastern India. The Kolkata Metropolitan Area (KMA), also known as the Kolkata Urban Agglomeration (KUA), ranks as the third-largest urban agglomeration in India (Census, 2011). On a global scale, it ranks as the 13th largest metropolis. Notably, it holds the distinction of being the largest urban agglomeration in eastern India. Situated along the Hooghly River's north-south axis, the KMA held a population of 14.77 million in 2011, comprising approximately 16% of West Bengal's total population. Projections suggest that by 2025, the population of the KMA will surge to 20.1 million, propelled by an annual growth rate of 1.8% since the 2011 Census. The KMA exhibits a remarkable average population density of 7950 individuals per square km, highlighting its significant urban density. This vibrant region accommodates diverse commercial, educational, healthcare, and other institutions, serving not only its own needs but also those of the broader state and nation. With a relatively high per capita income and a robust gross domestic product, the KMA has become a breeding ground for the expansion of service sectors and industrialization, fostering ample opportunities for infrastructure development. Accordingly, it has witnessed rapid migration and the continual expansion of built-up areas (Nath & Acharjee, 2013). The unchecked and swift expansion of urban areas is placing rural surroundings at heightened risk, leading to habitat degradation, soil and water contamination, and, in turn, the degradation of the city's ecological well-being (Bhatta, 2009; Mandal et al., 2019; Das & Jain, 2022; Dinda et al., 2022). Therefore, it becomes imperative to evaluate the ecological security of the KMA and project its potential to optimize sustainable administration.

3.2 Research Design and Methodology

Ecological processes within urban areas exhibit intricate, interconnected complexities. Therefore, a proficient analytical framework is essential for understanding UES. In the present case study, the Decision-Making Trial and Evaluation Laboratory (DEMATEL) model was used, combined with the analytic network process (ANP) and GIS methodologies to evaluate the UES. The DEMATEL model, widely utilized in diverse domains, proves invaluable for deciphering intricate causal relationships among variables (Sumrit & Anuntavoranich, 2013). DEMATEL allows for the assessment of interrelationships between chosen criteria, while ANP aids in determining the significance of these criteria in UESA. The combined DEMATEL-ANP model proves to be a robust tool for decision-making, effectively addressing real-world scenarios (Shakeri & Khalilzadeh, 2020).

The evaluation of ecological susceptibility and ecological security involves a complex analysis, incorporating various elements. This study utilized a combination of vector and remote sensing data. The methodology employed classified LULC maps, Google Earth imagery, Open Street Map data, official government reports, and information collected through field surveys. Selected indicators have been classified into three distinct categories: ecological pressure, ecological sensitivity, and ecological resilience. A total of 12 factors have been chosen and grouped within these three index systems, including land change rate, built-up area density, green area proportion, green area change intensity, proportion of water bodies, available open space, landscape connectivity, micro-habitat quality, land surface temperature, transport proximity, industry proportion, and population density.

To create an ecological security map for a chosen region, a sophisticated hybrid multi-criteria decision support model was devised. Priority weights, obtained through DEMATEL-ANP analysis, were incorporated into the ecological security map through the utilization of ArcGIS 10.7 software. Spatial integration, achieved via the following equation, was employed to produce UES maps (Feng et al., 2018; Ghosh, 2023):

$$\mathrm{UESA} = \sum_{i=1}^{n} f_i \times W_i \qquad (1)$$

Where f_i and W_i denote specific normalized criteria and the respective weights of each criterion, respectively. The model's validation and the evaluation of the ecological security map were conducted using the Receiver Operator Characteristics (ROC) curve. The overall methodological workflow of the case study is presented in Fig. 3.

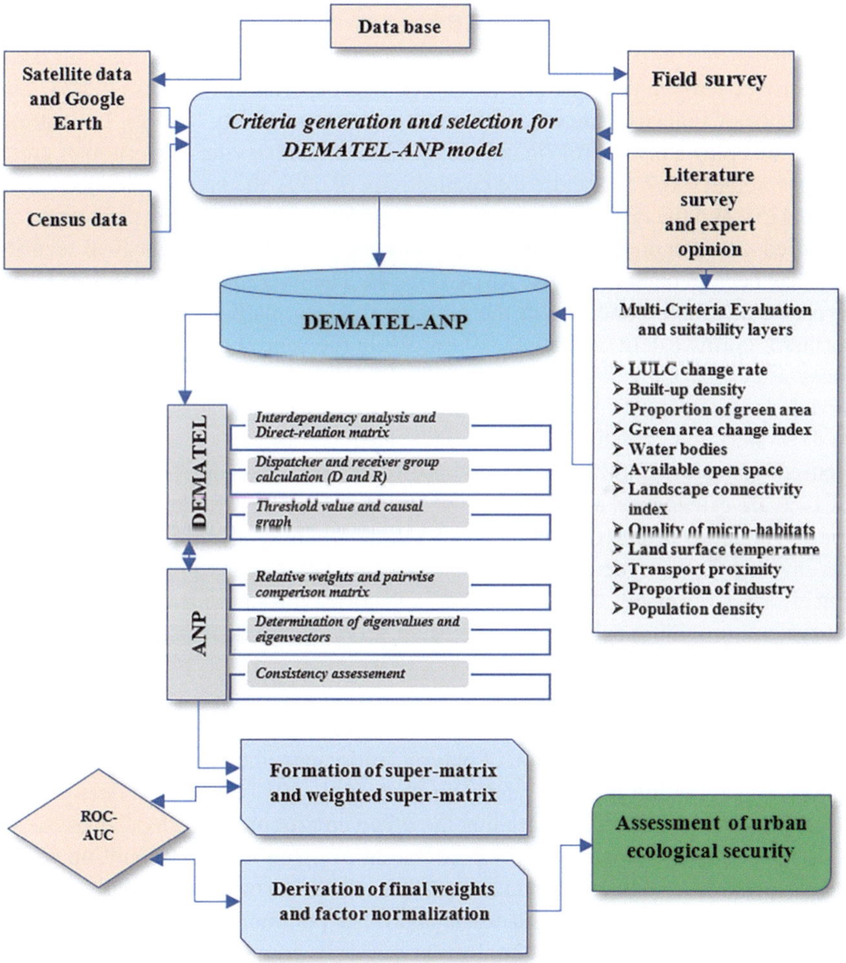

Fig. 3 Methodological workflow of the case study (adapted from Ghosh et al., 2023a, b)

3.3 Results and Discussion

The DEMATEL technique was employed to evaluate the direct connections between factors, with a focus on gauging their interdependencies by estimating dispatcher (D) and receiver (R) values derived from the relational matrix. The most significant indicators among those chosen include the rate of change in LULC, built-up density, intensity of change in green areas, landscape connectivity index, proximity to transportation networks, and the proportion of industrial activity. In accordance with their significance and causal relationships, ecological security factors can be categorized into two distinct groups: ecological risk factors and ecological security factors. The primary drivers impacting UES include the rate of LULC change (0.721), built-up density (0.688), intensity of green area change rate (0.551), magnitude of land surface temperature (0.401), proximity to transportation networks (0.631), the prevalence of industrial areas (0.425), and population density (0.472). In contrast, factors that play a pivotal role in upholding UES encompass the proportion of green spaces (weight = 0.709), landscape connectivity (0.699), the presence of water bodies (0.475), and the quality of micro-habitats (0.592) (Table 2).

Based on a comprehensive assessment of multiple factors, an ecological security map has been generated for the KMA (Fig. 4). The overall state of UES in KMA reveals a favorable condition in the peripheral zones, particularly in the areas of northern, northwestern, and southeastern, while the central part of the city experiences a less favorable situation. In terms of summary statistics derived from the model, approximately 19.18% of the area falls into the "very unsecured" category, 16.41% into the "relatively unsecured" category, 14.21% into the "generally secured" category, 22.08% into the "relatively secured" category, and the remaining 28.12% are categorized as "very secured" as of 2019. The "very unsecured" classification signifies a state of profound ecosystem structure degradation, highlighting the environment's vulnerability and necessitating urgent conservation and restoration efforts. Conversely, the "very secure" categorization signifies a robustly healthy ecosystem, providing substantial ecosystem services and demonstrating resilience in the face of disturbances.

The swift alteration of LULC represents a significant cause for ecological concern within the KMA. As previously mentioned, the rapid expansion of urban development has led to the contraction of natural landscapes, including vegetation, open spaces, and agricultural fallow lands (Bardhan et al., 2015). This relentless process of urbanization is gradually degrading the urban environment and local ecosystems (Das et al., 2020). Consequently, the alteration of green areas into built-up zones, the disappearance of micro-habitats, an uptick in land surface temperatures, and the ongoing generation of municipal waste and other pollutants (Ghosh et al., 2023a) (Fig. 5). These outcomes of urbanization not only compromise ecological well-being but also erode ecological security.

Table 2 Normalized weights of the selected parameters of UES assessment (Source: Ghosh et al., 2021)

Parameters	Normalized weights	Function
LULC change rate	0.721	−*ve*
Built-up area density	0.688	−*ve*
Proportion of green spaces	0.709	+*ve*
Green space change intensity index	0.551	−*ve*
Proportion of water bodies	0.475	+*ve*
Available open space	0.466	+*ve*
Landscape connectivity index	0.699	+*ve*
Quality of micro-habitats	0.592	−*ve*
Land surface temperature	0.401	−*ve*
Transport proximity	0.631	−*ve*
Proportion of industry	0.425	−*ve*
Population density	0.472	−*ve*

'+*ve*' denotes ecological risk factor and '−*ve*' denotes ecological safety factor

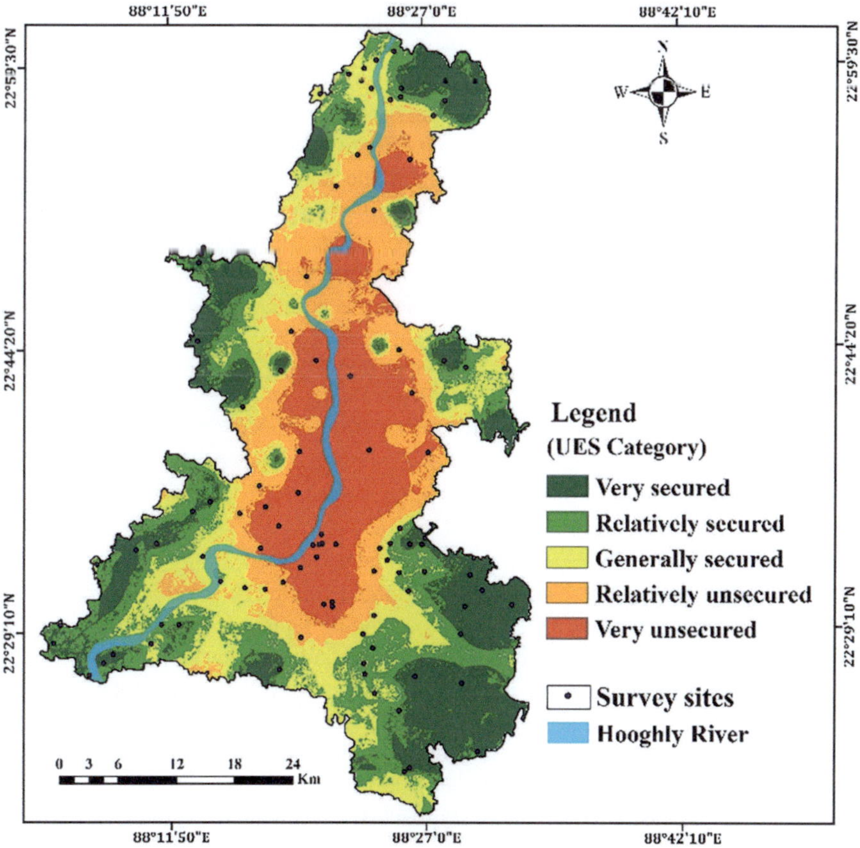

Fig. 4 Urban ecological security map of the KMA (Source: Ghosh, 2023)

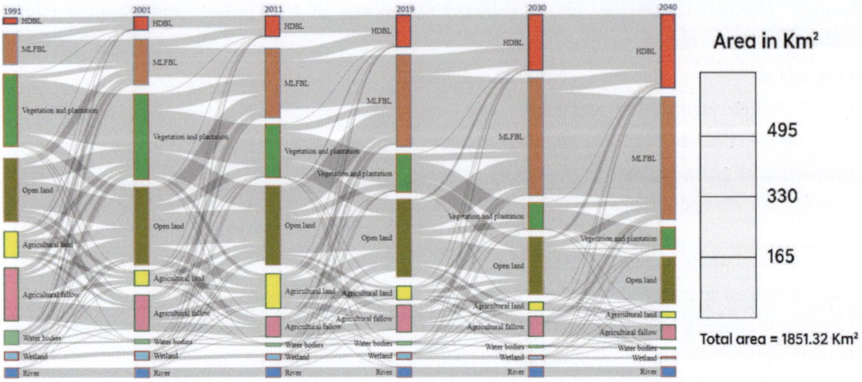

Fig. 5 Pattern and magnitude of land-use and land-cover conversion of KMA during 1991 to 2019, and predicted for 2030 and 2040 using ANN-Markov model (*HDBL-high-density built-up land; MLFBL-moderate to low and fragmented built-up land;* adapted from Ghosh et al., 2023a, b)

3.4 Model Efficiency, Applications and Future Frontier

Identifying ecological security patterns is vital for maintaining ecological health and promoting urban sustainability. Predicting future LULC patterns and their impact on ecological security is a key step in evidence-based environmental management. The present study employed an integrated approach to evaluate ecological security in the case study area, offering an effective method for assessing complex relationships among various criteria and aiding decision-making. The integration of DEMATEL-ANP significantly improved the model's performance. This framework provides valuable insights and practical implications for researchers and managers across various aspects. DEMATEL helps uncover cause-and-effect relationships among selected criteria, while ANP enables the adjustment of criteria weights, enhancing the model's accuracy. Furthermore, incorporating expert judgment and field surveys into the DEMATEL-ANP model yielded superior results.

4 Conclusion

The burgeoning field of urban expansion and its associated challenges is emerging as a paramount focus of research in the imminent future. This research imperative holds particular significance in regions experiencing rapid urban growth, notably in developing countries such as India. An analysis conducted by Verma et al. (2020) revealed a noteworthy surge in interest in subjects like "urban ecology," "urban biodiversity management," "urban ecosystem services," and "urban ecological security." Their study reported a substantial 1089 research articles published on these topics in esteemed journals between 2009 and 2019. The case study has introduced a clear and easily reproducible framework to evaluate the ecological security of an urban environment, offering valuable insights for city administrators involved in land management. The incorporation of the DEMATEL approach within the ANP framework effectively captures the intricate relationships among these variables. The rising vulnerability of ecologically insecure regions indicates that urbanization's adverse effects on urban ecological well-being are becoming more pronounced. This research will aid urban planners and policymakers in comprehending the city's ecological resilience in the face of ongoing urban expansion, enabling them to devise tailored policies in response.

References

Alberti, M. (2008). Advances in urban ecology: integrating humans and ecological processes in urban ecosystems. Boston, MA: Springer US.

Bardhan, R., Kurisu, K., & Hanaki, K. (2015). Does compact urban forms relate to good quality of life in high density cities of India? Case of Kolkata. *Cities, 48*, 55–65.

Batten, D. F. (2001). Complex landscapes of spatial interaction. *The Annals of Regional Science, 35*, 81–111.

Bhatta, B. (2009). Analysis of urban growth pattern using remote sensing and GIS: A case study of Kolkata, India. *International Journal of Remote Sensing, 30*(18), 4733–4746.

Bhattacharya, T. R. (2017). Comparative assessment of ecosystem and biodiversity conservation measures in Indian smart cities: A City biodiversity index approach. *International Journal of Sustainable Future for Human Security, 5*(2), 18.

Burgess, E. W. (2012). The growth of the city: An introduction to a research project. In R. Park et al. (Eds.), The City (1925) *The urban sociology reader* (pp. 91–99). Routledge.

Census. (2011). *Census of India 2011*. Government of India. http://censusindia.gov.in/DigitalLibrary/Archive_home.aspx

Cohen, B. (2006). Urbanization in developing countries: Current trends, future projections, and key challenges for sustainability. *Technology in Society, 28*(1–2), 63–80.

Cook, E. A. (2002). Landscape structure indices for assessing urban ecological networks. *Landscape and Urban Planning, 58*(2–4), 269–280.

Costanza, R., De Groot, R., Sutton, P., Van der Ploeg, S., Anderson, S. J., Kubiszewski, I., et al. (2014). Changes in the global value of ecosystem services. *Global Environmental Change, 26*, 152–158.

Das, S., & Angadi, D. P. (2021). Assessment of urban sprawl using landscape metrics and Shannon's entropy model approach in town level of Barrackpore sub-divisional region, India. *Modeling Earth Systems and Environment, 7*(2), 1071–1095.

Das, S., & Jain, G. V. (2022). Assessment and prediction of urban expansion using CA-based SLEUTH urban growth model: A case study of Kolkata metropolitan area (KMA), West Bengal, India. *Journal of the Indian Society of Remote Sensing*, 1–26.

Das, M., Das, A., & Mandal, A. (2020). Research note: Ecosystem Health (EH) assessment of a rapidly urbanizing metropolitan city region of eastern India–a study on Kolkata Metropolitan Area. *Landscape and Urban Planning, 204*, 103938.

Dhyani, S., Lahoti, S., Khare, S., Pujari, P., & Verma, P. (2018). Ecosystem based disaster risk reduction approaches (EbDRR) as a prerequisite for inclusive urban transformation of Nagpur City, India. *International Journal of Disaster Risk Reduction, 32*, 95–105.

Dinda, S., Chatterjee, N. D., & Ghosh, S. (2020). An integrated simulation approach to the assessment of urban growth pattern and loss in urban green space in Kolkata, India: A GIS-based analysis. *Ecological Indicators, 121*, 107178.

Dinda, S., Chatterjee, N. D., & Ghosh, S. (2022). Modelling the future vulnerability of urban green space for priority-based management and green prosperity strategy planning in Kolkataku, India: A PSR-based analysis using AHP-FCE and ANN-Markov model. *Geocarto International, 37*(22), 6551–6578.

Feng, Y., Liu, Y., & Liu, Y. (2017). Spatially explicit assessment of land ecological security with spatial variables and logistic regression modeling in Shanghai, China. *Stochastic Environmental Research and Risk Assessment, 31*(9), 2235–2249.

Feng, Y., Yang, Q., Tong, X., & Chen, L. (2018). Evaluating land ecological security and examining its relationships with driving factors using GIS and generalized additive model. *Science of the Total Environment, 633*, 1469–1479.

Forman, R. T. (2014). *Urban ecology: Science of cities*. Cambridge University Press.

Gao, J. J., Lv, S. H., Zheng, Z. R., Feng, C. Y., Ye, S. X., & Zhang, C. (2013). Ecological safety assessment in Hulunbeier steppe based on the landscape pattern. *Advanced Materials Research, 726*, 1121–1129.

Ghosh, S. (2023). *Ecological sensitivity analysis in Kolkata Metropolitan Area: A step towards sustainable growth. Doctoral thesis*. Vidyasagar University. http://hdl.handle.net/10603/512827

Ghosh, S., Chatterjee, N. D., & Dinda, S. (2021). Urban ecological security assessment and forecasting using integrated DEMATEL-ANP and CA-Markov models: A case study on Kolkata Metropolitan Area, India. *Sustainable Cities and Society, 68*, 102773.

Ghosh, S., Dinda, S., Das Chatterjee, N., & Dutta, S. (2023a). Assessing the impact of urban landuse dynamics on the ecological environment of East Kolkata: A study for sustainable urban development. In *Environmental management and sustainability in India: Case studies from West Bengal* (pp. 523–540). Springer International Publishing.

Ghosh, S., Dinda, S., Chatterjee, N. D., & Bera, D. (2023b). Linking ecological vulnerability and ecosystem service value in a fast-growing metropolitan area of eastern India: a scenario-based sustainability approach. *Environment, Development and Sustainability*, 1–31.

Gong, J. Z., Liu, Y. S., Xia, B. C., & Zhao, G. W. (2009). Urban ecological security assessment and forecasting, based on a cellular automata model: A case study of Guangzhou, China. *Ecological Modelling, 220*(24), 3612–3620.

Grimm, N. B., Faeth, S. H., Golubiewski, N. E., Redman, C. L., Wu, J., Bai, X., & Briggs, J. M. (2008). Global change and the ecology of cities. *Science, 319*(5864), 756–760. tripathy.

Han, B., Liu, H., & Wang, R. (2015). Urban ecological security assessment for cities in the Beijing–Tianjin–Hebei metropolitan region based on fuzzy and entropy methods. *Ecological Modelling, 318*, 217–225.

Hodson, M., & Marvin, S. (2009). 'Urban ecological security': A new urban paradigm? *International Journal of Urban and Regional Research, 33*(1), 193–215.

Hoffmann, P., Krueger, O., & Schlünzen, K. H. (2012). A statistical model for the urban heat Island and its application to a climate change scenario. *International Journal of Climatology, 32*(8), 1238–1248. https://doi.org/10.1002/joc.2348

Hua, Y. E., Yan, M. A., & Limin, D. O. N. G. (2011). Land ecological security assessment for Bai autonomous prefecture of Dali based using PSR model—With data in 2009 as case. *Energy Procedia, 5*, 2172–2177.

Hultman, J. (1993). Approaches and methods in urban ecology. *Geografiska Annaler: Series B, Human Geography, 75*(1), 41–49.

Koroso, N. H., Lengoiboni, M., & Zevenbergen, J. A. (2021). Urbanization and urban land use efficiency: Evidence from regional and Addis Ababa satellite cities, Ethiopia. *Habitat International, 117*, 102437.

Li, Z. T., Yuan, M. J., Hu, M. M., Wang, Y. F., & Xia, B. C. (2019). Evaluation of ecological security and influencing factors analysis based on robustness analysis and the BP-DEMALTE model: A case study of the Pearl River Delta urban agglomeration. *Ecological Indicators, 101*, 595–602.

Li, Z. T., Li, M., & Xia, B. C. (2020). Spatio-temporal dynamics of ecological security pattern of the Pearl River Delta urban agglomeration based on LUCC simulation. *Ecological Indicators, 114*, 106319.

Liu, H., Wang, H., & Liu, K. (2005). A review of ecological security assessment and relevant methods in China. *Nature Ecology and Conservation, 8*, 34–37.

Mandal, J., Ghosh, N., & Mukhopadhyay, A. (2019). Urban growth dynamics and changing land-use land-cover of megacity Kolkata and its environs. *Journal of the Indian Society of Remote Sensing, 47*(10), 1707–1725.

Morimoto, J., & Negishi, J. (2019). Ecological resilience of ecosystems to human impacts: Resilience of plants and animals. *Landscape and Ecological Engineering, 15*(2), 131–132.

Mudau, N., Mwaniki, D., Tsoeleng, L., Mashalane, M., Beguy, D., & Ndugwa, R. (2020). Assessment of SDG indicator 11.3. 1 and urban growth trends of major and small cities in South Africa. *Sustainability, 12*(17), 7063.

Nath, B., & Acharjee, S. (2013). Urban municipal growth and landuse change monitoring using high resolution satellite imageries and secondary data: A geospatial study on Kolkata Municipal Corporation, Kolkata, India. *Studies in Surveying and Mapping Science, 1*(3), 43–54.

Nuissl, H., Haase, D., Lanzendorf, M., & Wittmer, H. (2009). Environmental impact assessment of urban land use transitions—A context-sensitive approach. *Land Use Policy, 26*(2), 414–424.

Odum, E. P. (1975). *Ecology, the link between the natural and the social sciences*. Oxford and IBH Publishing.

Park, S. C., & Han, B. H. (2021). Using the city biodiversity index as a method to protect biodiversity in Korean Cities. *Sustainability, 13*(20), 11284.

Parker, D. C., Manson, S. M., Janssen, M. A., Hoffmann, M. J., & Deadman, P. (2003). Multi-agent systems for the simulation of land-use and land-cover change: A review. *Annals of the Association of American Geographers, 93*(2), 314–337.

Punzo, G., Castellano, R., & Bruno, E. (2022). Using geographically weighted regressions to explore spatial heterogeneity of land use influencing factors in Campania (Southern Italy). *Land Use Policy, 112*, 105853.

Sahani, S., & Raghavaswamy, V. (2018). Analyzing urban landscape with City Biodiversity Index for sustainable urban growth. *Environmental Monitoring and Assessment, 190*(8), 1–18.

Shakeri, H., & Khalilzadeh, M. (2020). Analysis of factors affecting project communications with a hybrid DEMATEL-ISM approach (a case study in Iran). *Heliyon, 6*(8), e04430.

Sobhani, P., Esmaeilzadeh, H., Wolf, I. D., Deljouei, A., Marcu, M. V., & Sadeghi, S. M. M. (2023). Evaluating the ecological security of ecotourism in protected area based on the DPSIR model. *Ecological Indicators, 155*, 110957.

Sridhar, K. S. (2020). Is India's urbanization really too low? *Area Development and Policy, 5*(1), 32–49.

Sumrit, D., & Anuntavoranich, P. (2013). Using DEMATEL method to analyze the causal relations on technological innovation capability evaluation factors in Thai technology-based firms. *International Transaction Journal of Engineering, Management, & Applied Sciences & Technologies, 4*(2), 81–103.

Sun, J., Li, Y. P., Gao, P. P., & Xia, B. C. (2018). A Mamdani fuzzy inference approach for assessing ecological security in the Pearl River Delta urban agglomeration, China. *Ecological Indicators, 94*, 386–396.

Tripathy, P., & Kumar, A. (2019). Monitoring and modelling spatio-temporal urban growth of Delhi using Cellular Automata and geoinformatics. *Cities, 90*, 52–63.

Turner, B. L., Kasperson, R. E., Matson, P. A., McCarthy, J. J., Corell, R. W., Christensen, L., et al. (2003). A framework for vulnerability analysis in sustainability science. *Proceedings of the National Academy of Sciences, 100*(14), 8074–8079.

United Nations. (2018). *World urbanization prospects United Nations*, Retrieved March 16, 2022, from https://www.un.org/development/desa/en/news/population/2018-revision-of-world-urbanization-prospects.html

Verma, P., Singh, R., Singh, P., & Raghubanshi, A. S. (2020). Critical assessment and future dimensions for the urban ecological systems. In *Urban ecology* (pp. 479–497). Elsevier.

Wang, Z., Deng, X., Wong, C., Li, Z., & Chen, J. (2018). Learning urban resilience from a social-economic-ecological system perspective: A case study of Beijing from 1978 to 2015. *Journal of Cleaner Production, 183*, 343–357.

Wang, S., Zhang, X., Wu, T., & Yang, Y. (2019). The evolution of landscape ecological security in Beijing under the influence of different policies in recent decades. *Science of the Total Environment, 646*, 49–57.

Wen, J., & Hou, K. (2021). Research on the progress of regional ecological security evaluation and optimization of its common limitations. *Ecological Indicators, 127*, 107797.

World Bank. (2021). *Urban development prospects*. Retrieved March 17, 2022, from https://www.worldbank.org/en/topic/urbandevelopment/overview#:~:text=Today%2C%20some%2056%25%20of%20the,people%20will%20live%20in%20cities

Wu, W. W. (2008). Choosing knowledge management strategies by using a combined ANP and DEMATEL approach. *Expert Systems with Applications, 35*(3), 828–835.

Wu, X., & Hu, F. (2020). Analysis of ecological carrying capacity using a fuzzy comprehensive evaluation method. *Ecological Indicators, 113*, 106243.

Xiao, D., & Chen, W. (2002). On the basic concepts and contents of ecological security. *The Journal of Applied Ecology, 13*(3), 354–358.

Yang, Y., & Cai, Z. (2020). Ecological security assessment of the Guanzhong Plain urban agglomeration based on an adapted ecological footprint model. *Journal of Cleaner Production, 120973*.

Yang, T., Zhang, Q., Wan, X., Li, X., Wang, Y., & Wang, W. (2020). Comprehensive ecological risk assessment for semi-arid basin based on conceptual model of risk response and improved TOPSIS model-a case study of Wei River Basin, China. *Science of the Total Environment, 719*, 137502.

Yang, H., Zhai, G., & Zhang, Y. (2021). Ecological vulnerability assessment and spatial pattern optimization of resource-based cities: A case study of Huaibei City, China. *Human and Ecological Risk Assessment: An International Journal, 27*(3), 606–625.

Zhang, Y., Odeh, I. O. A., & Han, C. (2009). Bi-temporal characterization of land surface temperature in relation to impervious surface area, NDVI and NDBI, using a sub-pixel image analysis. *International Journal of Applied Earth Observation and Geoinformation, 11*(4), 256–264. https://doi.org/10.1016/j.jag.2009.03.001

Land Acquisition for Urban Housing Developments: A Multi-Stakeholder Analysis in Aligarh City

Md Kaikubad Ali and Shahab Fazal

Abstract Providing adequate and affordable housing to the growing population in urban areas is a major challenge for urban planning and sustainability in South Asia, especially in India. Urban development authorities are striving to control the growth of informal settlements over the peri-urban spaces and accommodate the increasing population by managing urban housing projects. These authorities have resorted to traditional land acquisition policy to acquire land for urban development works, especially housing projects for the lower-and middle-income population. The present chapter, being qualitative research, investigates the issues and challenges of the land acquisition process for urban housing projects in a rapidly expanding city in western Uttar Pradesh in North India. Taking the case of the Aligarh City Master Plan, the study examines the ground realities of the practicing land acquisition policy, which more or less replicates the Land Acquisition Act of 1894. Finally, it has adopted a multi-stakeholder analysis approach to explore the issues and challenges of the land acquisition process by gathering information through in-depth interviews and focused group discussions with different stakeholders. The findings expose the gaps between housing demand and the government supply mechanisms. Aligarh Development Authority (ADA) planned and developed only 5216 housing plots over the last four decades, extending over six major and seven minor housing projects, but failed to acquire the required land and build adequate housing plots due to the lengthy and complex land acquisition process. The multi-stakeholder analysis reveals that issues like the lengthy acquisition process, the delayed land transfer, ownership disputes, the unwillingness of landowners, and the competing real estate market pose serious questions about the viability and applicability of the current land acquisition policy.

M. K. Ali
Department of Geography, Jamia Millia Islamia, New Delhi, India

S. Fazal (✉)
Department of Geography, Aligarh Muslim University, Aligarh, India

P. K. Joshi et al. (eds.), *Geospatial Science for Urban Ecosystems*, The Urban Book Series, https://doi.org/10.1007/978-3-032-07155-2_19

Keywords Land acquisition · Urban housing · Urban development · Master plan · Urban land use planning

1 Introduction

Urbanization in developing countries like India follows a lopsided pattern (Datta, 2006; Fazal & Ali, 2018) where the population is rapidly growing in a few pockets, especially in the existing big cities. The substantial increase in urban population can be attributed to significant migration from rural to urban areas, as highlighted by Bhagat (2004, 2017) and Kundu (2011). As cities are the center of the economy and employment (Cohen, 2006), they offer multifaceted income and livelihood opportunities that attract thousands of people from poverty-ridden rural areas. The transition of population from rural to urban areas collectively leads to a surge in urban population, exerting pressure on both land and infrastructure. This ultimately drives the expansion of urban built-up areas, encompassing both vertical and horizontal dimensions, as noted by He et al. (2017) and Longyu et al. (2009). However, unlike Western cities, cities in India expand horizontally in the surrounding space, sprawling outward and encroaching on the adjoining rural hinterland (Fazal, 2013). The outward expansion of urban built-up areas, known as the horizontal spillover of urban population, constitutes a significant aspect of urban growth during the post-liberalization phase, as indicated by Shaw (2005), Chadchan and Shankar (2012), and Shaw and Das (2018).

Therefore, the real growth of the urban population in Indian cities is taking place in the peripheral areas, where the in-migrated population from the rural hinterland is contesting for space. The demands of space for the housing of a growing population, as well as economic activities, have intensified, especially after the liberalization of the economy in 1991, resulting in urban sprawl, growth of urban slums, and deterioration of urban land and environment. Of all the urban problems, housing for in-migrants, which amplifies the city's population, is the critical one (Yun & Kim, 2022). For the disposal of affordable housing for the growing population, the prime task of the developmental authorities is to plan areas for devoted residential colonies, especially for lower-income and medium-income group communities. The demand for housing for the growing population is the major reason development authorities emerged alongside the urban local bodies.

To facilitate the planned development of cities and the execution of urban land policies, development authorities in Indian cities have adopted the strategy of extensive land acquisition and development, as observed by Diwakar (1991) and Bentinck (2000). Land acquisition has been used to carry out a number of planning and development activities by the state and private sectors. It serves as the principal approach to fulfill the escalating demand for land driven by swift economic and urban expansion, typically employed by governments (Hui & Bao, 2013). Government has to acquire or take over private lands either by force or by the consent of the landowner—by offering measurable compensation which is more than the market value because of the compulsory character of the acquisition—for the extension of infrastructures like transport networks, industrial corridors, urban development, etc., which are meant for public welfare in particular and inclusive growth and development of the country in general.

The first effort at large-scale land acquisition was undertaken in Delhi for the Delhi Development Authority to secure the planned development of the city. Currently, the initiation of urban development projects in most cities relies on the application of "eminent domain" by the state, with land acquisition occurring under the singular Land Acquisition Act of 1894. There has been extensive research examining urban expansion in Indian cities in recent years, as evidenced by studies conducted by Fazal (2000), Kantakumar et al. (2016), Kumar et al. (2011), Narain et al. (2014), Goswami and Khire (2016), Lata et al. (2001), Rahman et al. (2011), Ramachandra et al. (2014), and Verma (2017). However, less focus has been on examining the land acquisition process for urban land use planning and urban development, especially in the smaller cities (Tier II and Tier III cities), where rural-to-urban migration has been phenomenal in the post-liberalization period. Moreover, most studies on land acquisition in India focus on examining the issues and challenges related to land acquisition for large industries (Nielsen, 2010; Sarkar, 2007) and infrastructural development (Trivedi, 2020). There is a lack of research on small-scale land acquisition for urban housing projects by development authorities in Indian cities. Hence, it is crucial to examine the urban expansion process and the ensuing land acquisition for housing development in these small cities in order to evaluate the extent and spatial variations among them.

1.1 Land Acquisition Policy in India

India serves as a significant case for comprehending the processes of urban land expansion and the policies related to urban development, as highlighted by Johnson and Chakravarty (2013). In the colonial period, especially after the establishment of rail transportation, the Britishers started acquiring private lands for rail line expansion. This paved the way for the acquisition of land for additional development projects, such as the establishment of industries and other developmental activities. In many cases, Britishers started facing obstructions when local people resisted giving up land. This necessitated the formation of a land acquisition law through which they could forcibly acquire private lands in the name of public works. The Land Acquisition Act of 1894 was enacted by the government in the British Parliament, granting it the authority to acquire private land under the principle of "eminent domain." This legislation primarily addressed the rights of landowners, the compensation provided for land acquisition, and the objectives intended to be achieved through such actions, as outlined by Mukerji (2017). The Land Acquisition Act of 1894 has gone through more than a dozen amendments since then (viz., 1914, 1919, 1920, 1921, 1923, 1933, 1938, 1951, 1984, 2007, 2011, and 2013), but even after several amendments, its original nature remained intact (Verma, 2015).

Land acquisition falls within the concurrent list under the Indian Constitution. Article 254 (2) of the Constitution permits the state governments to amend acts under the concurrent list after the approval of the central government. However, most states follow the same land acquisition law as enjoyed by the central government. And the land acquisition laws in India, even after 125 years of the first law, more or less follow the Land Acquisition Act, 1984. In 2011, the United Progressive Alliance (UPA) government introduced the Rehabilitation and Resettlement component for individuals affected by land acquisition, replacing the Land Acquisition Act of 1984. Subsequently, in 2013, the legal framework underwent a transformation, leading to the enactment of the Right to Fair Compensation and Transparency in Land Acquisition, Rehabilitation, and Resettlement Act, 2013, commonly referred to as the Land Acquisition Act, 2013. Inadequacies and complexities in the century-old Land Acquisition Act delayed many development projects. In 2015, the newly formed NDA government proposed some major amendments to the LAA 2013 to ease the hurdles to completing various development projects. But the bill was not passed in the upper house (*Rajya Sabha*) of the parliament and consequently lapsed.

2 Study Area

Aligarh City is one of the 63 million-plus cities in India, with a population of nearly 1.25 million in 2022 (Population Census, 2023). In recent times, the city has experienced a rapid increase in population that led to the asymmetric expansion of urban built-up on the peripheries, prominently along the major roads (Banu & Fazal, 2013; Ali, 2021). The city expansion is also fueled by industrial, infrastructural, administrative, and educational development. The presence of vibrant industries, especially agro-based ones and lock industries, has acted as an urban pull factor, attracting many laborers and job seekers from the surrounding region and other states. Moreover, the presence of Aligarh Muslim University in the city has accentuated urban growth. According to the 2011 census, the city added 203,488 people since the 2001 census, out of which 93 percent were migrants (Table 1). A field visit to a meat processing industry in Talaspur found the presence of laborers from Bihar and West Bengal, as well as from Assam.

The increasing population, the need for migrant housing, infrastructure demands, and the expansion of industrial and commercial centers have compelled the city to extend beyond its municipal boundaries, leading to unforeseen changes in land use.

Table 1 Trends of in-migration in Aligarh City (2001–2011)

Census	2001	2011	Change (%)
Total population	669,087	872,575	203,488 (30.41)
In-migrants	154,965	343,818	188,853 (121.87)

Source: Census of India, 2001 and 2011

Effectively integrating the expanding population into the urban landscape has consistently been a focal point for urban planners and policymakers. The scarcity of housing reveals another dimension of pressure on land resources. The Master Plan of Aligarh estimates the need for 179,443 houses by the year 2021 for an estimated 1.3 million inhabitants. Aligarh Development Authority, therefore, carried out some major housing projects to ease increasing housing demands in the city and to provide affordable houses to the growing city population, which was mainly due to the migration of rural people from the surrounding region. This required a large-scale acquisition of lands in the fringe areas. This paper, therefore, examines an important aspect of urban expansion and urban development, i.e., the land acquisition process.

3 Objectives

The prime objective of the present research is to examine the land acquisition process for urban housing development in Aligarh City. It also aims to investigate the issues and challenges of the land acquisition process for urban housing development using a multi-stakeholder analysis approach.

4 Methodology

This study, adopting a qualitative research method, is carried out through empirical field visits and interviews of different stakeholders, directly and indirectly, involved in the land acquisition process in the study area. The field observation found three key stakeholders. The first is the ADA, which acquires land for urban housing projects. The second stakeholder is the land-givers, mostly farmers, who gave up their land. The third group includes private builders or property dealers, the brokers between housing projects and land-givers. Therefore, a multi-stakeholder analysis approach (Saha et al., 2021), a widely used approach in qualitative research, has been adopted to investigate the issues and challenges related to the land acquisition process faced by these three stakeholders. Much of the information has been retrieved from ADA. Two ADA officers, including the land acquisition officer and cartographer, were interviewed to discuss the land acquisition process in the city, different housing projects under the master plan, their implementations, problems, and prospects. Interviews were conducted with fifteen local residents, particularly those aged between 40 and 60, to gather insights from the perspective of land givers. The focus was on understanding their knowledge and awareness regarding the land acquisition process and the impacts it has on their lives and livelihoods. The residents were identified and selected from the villages where ADA acquired land for housing projects. Local real estate businessmen and property dealers were also interviewed regarding their role in the land acquisition process, land market, and land prices around the city.

5 Results and Discussion

5.1 Land Acquisition in Aligarh City

The rapid urbanization during the early decades of independence envisaged uncontrolled land transformation and the development of illegal colonies and slums in the city. This necessitated the need for planned and regulated development in the municipal area. Under the purview of the Uttar Pradesh Urban Planning and Development Act, 1973, Aligarh City started working on urban planning measures to tackle the growing population and demand for land for city expansion. The city had already become a municipality in 1885 with the establishment of the Aligarh Municipal Board (*Nagar Palika*). The Aligarh Municipal Board initiated the first master plan for the city's development in 1980. The main objective of the plan was to overcome the uncontrolled and unplanned land-use changes, illegal constructions, and unauthorized developmental changes in the city and make a zonal plan for prospective urban development in the region. This institutionalized the creation of the Aligarh Development Authority (ADA) in 1981 for the execution and monitoring of the proposed Master Plan.

The Aligarh Development Authority (ADA) holds exclusive jurisdiction over the planning and development of the city region within the study area. According to information retrieved from its official website, the ADA is guided by several key objectives. First, it is tasked with the preparation of a Master Plan to facilitate planned urban development. Subsequently, the ADA is responsible for overseeing the development and control processes in alignment with the Master Plan. Land acquisition and its management for housing and urban development constitute another pivotal role, emphasizing the authority's commitment to addressing housing needs. Additionally, the ADA is actively involved in the construction of housing and overall urban development. Lastly, the authority is dedicated to the provision of both physical and social infrastructure, underlining its comprehensive approach to fostering sustainable and well-rounded urban growth in the region.

The plan included the then municipal area and all the land within the 8 km buffer of the municipal boundary. In 2001, when the first master plan period was over, a new master plan was again proposed from 2001 to 2021. The ADA aimed at construction and housing development in the region to provide affordable housing to different income groups. For this, the master plan prepared by ADA allocated some lands for housing projects, which required land acquisition. Over the past years, the ADA attempted to acquire lands from surrounding villages to extend urban housing development.

5.2 Process and Functions of Land Acquisition

The land acquisition process in India is generally composed of three macro processes, viz., (a) identification of affected parties, (b) notification of acquisition to the stakeholders and consideration of their grievances, and (c) distribution of an acceptable compensation package (Mahalingam & Vyas, 2011).

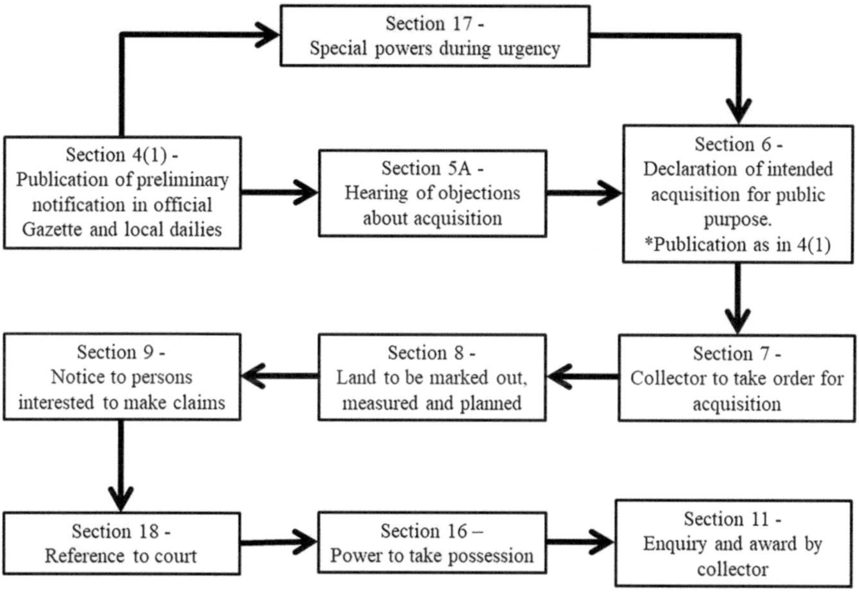

Fig. 1 Process of land acquisition (Source: Land Acquisition Act, 1984)

The land acquisition process in Aligarh City is executed under the guidelines of the LAA of 1984 (Fig. 1). However, the ADA does not acquire land directly from the landowners. It is the state government that, after the request of the development authority, acquires the proposed land for the development authority. The process of land acquisition in the study area is: *First*, the housing project is proposed, and the area is identified where land is to be acquired for development. *Second*, the development authority formulates the land acquisition proposal for planned development. *Third*, the ADA sends land acquisition proposals to the state government through the District Magistrate or Collector for issuing notifications under sections 4 & 6 of the LAA. *Fourth*, Section 4 and Section 6 are published in newspapers, and target landowners are notified. *Fifth*, the grievances of the landowners are then addressed by the land acquisition officer. *Sixth*, the possession of the notified land is taken over, and the award of the notified land is declared by the competent authority. *Seventh*, ADA pays the compensation amount to the District Collector for the disbursement of compensation to the landowners. *Eighth*, the Collector transfers ownership (mutation) of the acquired land to the development authority. Finally, on the *ninth*, ADA starts working on the project after the completion of the land acquisition.

5.3 *Formal Land Acquisition*

As stated above, ADA is the prime institution in the city to look after the city's development and future urban planning. The ADA, shortly after its inception, started urban development with some major housing projects to provide affordable housing

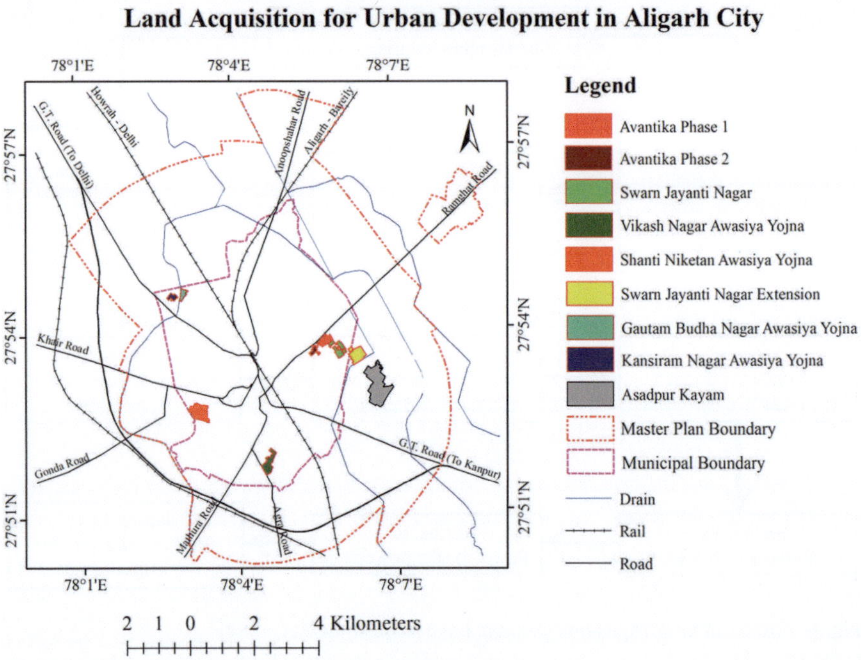

Fig. 2 Sites of land acquisition for ADA housing projects

to the growing city population, most of whom were migrants from the other part of the Aligarh district, as well as from other districts of the state. The housing projects taken up by ADA (Fig. 2) and for which land acquisitions were made are discussed below.

5.3.1 Avantika Colony Phase 1 and Phase 2

After the formulation of the first master plan in 1980, the first housing project undertaken by ADA was Avantika Colony Awasiya Yojana. For this, the ADA identified land in the northeast of Kisanpur village, along the Ramghat Road on the eastern side of the built-up city. During the 1980s, Kisanpur village was one of the big villages in the city periphery. As it was subject to urban encroachment shortly, it had been experiencing large-scale land transformation and abandonment of agricultural land to vacant lands. Besides, there was a demand for houses for government officials, doctors, and bank officials along the Ramghat Road. The development authority, therefore, found the area surrounding the village suitable and justifiable for planned urban development. ADA proposed 15.24 ha of land for the housing project named Avantika Colony Awasiya Yojana (Table 2). The proposed land was acquired from the villages in two phases, viz., Phase 1: 31/7/1983 and Phase 2: 11/9/1985. Avantika Colony Phase 1 and 2 schemes mainly focused on providing

Table 2 Share of land use in urban housing projects

Housing project	Area proposed	Area acquired	Residential	Commercial	Community services	Health	Education	Park/ Open space	Others
Avantika Colony (1 & 2)	15.2	15.2	53.4	5.3				9.8	21.5
Vikash Nagar Awasiya Yojana	14.5	11.8	56.6	1.3	2.1			9	21
Santiniketan Colony	17.7	16.1	59.4	1.9			3.8	7.5	27.4
Swarn Jayanti Nagar	10.0	8.7	63.2	1.6	3.2		1.2	6.3	24.5
Swarn Jayanti Nagar extension	14.3	11.7	59.2	1.3	1.3	0.7	1.4	7.2	28.9
Asadpur Qayam	53.2	31.3	Abandoned						

Source: Aligarh Development Authority

Table 3 Plots distribution among different income groups

Housing project	HIG	MIG	LIG	Economically Weaker Section (EWS)	Others	Total
Avantika Colony (1 & 2)	139	138	83			360
Vikash Nagar Awasiya Yojana	7	112	109	450		678
Santiniketan Colony	24	120	412	817		1373
Swarn Jayanti Nagar	28	50	108	280	448	914
Swarn Jayanti Nagar extension	39	68	132	372		611
Total	237	488	844	1919	448	

Source: Aligarh Development Authority

housing plots for higher-and medium-income group residences. A total of 360 plots and houses were constructed and developed under this project, of which 139 units were for HIG, and 138 and 83 were for MIG and LIG, respectively (Table 3).

5.3.2 Vikash Nagar Awasiya Yojana

The second housing project was assigned along Agra Road, named Vikash Nagar. For this new project, a total of 14.5 ha of land area was proposed to be acquired around the village of Kasbakol. However, only 11.8 ha were acquired in two phases: (a) Phase 1: 20/10/1983 and Phase 2: 23/11/1985 (Table 2). The final award of the land came on 20/09/1986. The remaining land could not be acquired, as landowners resisted giving up the land. There had been some small-scale industries along the Agra road. The demand for housing and the presence of vast open land in the Kasbakol village offered a suitable location for housing projects for ADA. This scheme was mainly focused on providing houses to the medium and low-income groups, as well as the Economically Weaker Section (EWS) people. Therefore, ADA proposed Vikash Nagar Awasiya Yojana with 678 housing units, out of which 450 were for EWS (Table 3).

5.3.3 Santiniketan Colony Awasiya Yojana

Santiniketan Colony Awasiya Yojana is the third housing project by ADA. This project aimed to ease the growing population pressure in the city built where population density had always been high. ADA identified land in the western fringe of the then-city built-up along the Khair Bypass road and proposed 17.7 ha of land for acquisition. However, it could acquire 16.1 ha in two phases. ADA acquired 15.3 ha of land on 29 January 1985, and the remaining land 12 years later on 04 October 1997. Santiniketan Colony Awasiya Yojana also emphasized providing houses mainly to the LIG and EWS. Around 1373 plots and houses were constructed under this project (Table 3).

5.3.4 Swarn Jayanti Nagar Awasiya Yojana

The first three housing projects by ADA were initiated just after the master plan came into effect and ADA was established. It took more than 15 years for the completion of these three projects. The fourth housing project was proposed in the Kisanpur village, beside the Avantika Colony. In this housing project, ADA proposed 10 ha for acquisition. But only 8.7 ha of land were acquired in several phases, as there was unwillingness among the villagers and farmers against the acquisition. The first phase of land acquisition was done on 12/8/1998, and the remaining were acquired on several occasions, including 26/07/1999, 05/10/2004, 2/12/2004, 29/08/2006, and 7/04/2007.

Conflicts of interest and demand for high compensation were some reasons landowners resisted giving up their land. Besides, ADA acquired some lands owned by *gram samaj,* which were reserved against any acquisition. The land area under registered society or colony, or temples, mosques, graveyards, etc., which is the land of *gram samaj,* is exempted from acquisition under LAA. Therefore, landowners filed a case against the land acquisition, some of which are still pending. This resulted in a delay in land transfer, as it took 9 years to acquire the land. It led to the delay of the project implementation. By that time, the second master plan of the city, Aligarh Mahayojana 2001–2021, came into effect, and Swarn Jayanti Nagar Awasiya Yojana continued in the next plan. The construction of the houses was completed in 2009. This project made provision for housing for all income groups, with a total of 914 plots and houses (Table 3).

5.3.5 Swarn Jayanti Nagar Extension

This is the extension of the Swarn Jayanti Nagar Awasiya Yojana. After the implementation of the second master plan, ADA proposed to extend the area of the existing Swarn Jayanti Nagar and identified the vast vacant land along the G.T. Bypass road between Quarsi and Kisanpur village for this project. It proposed a total of 14.3 ha of land for acquisition, but only 11.7 ha of land was acquired (Table 2). The final award of the land acquisition came on 10/12/2010. ADA proposed to develop a total of 611 housing units, mostly for LIG and EWS (Table 3).

5.3.6 Asadpur Qayam

The city has been expanding at its own pace, and the lengthy acquisition process could not cope with it. Besides, the three major housing projects on the eastern side, i.e., Avantika Colony, Swarn Jayanti Nagar, and Swarn Jayanti Nagar Extension, have resulted in high demand for land in the surrounding area, and private apartments and housing projects have started to grow. This again amplified the problem of unplanned urban expansion. Thus, ADA planned to acquire a large area with a future perspective of built-up expansion, a village in the fringes that is subject to being consumed by urban built-up in the near future, for planned urban development on a large scale. Asadpur Qayam village, situated on the eastern fringe along the Etah—Aligarh Bypass road, was identified as the best for it.

For the development of the housing project, ADA proposed a total of 53.23 ha of land for acquisition to develop the Aligarh Residential Colony. After notifying the villagers, ADA started acquisition, and 31.26 ha was acquired in two phases on 11/04/2011 and 18/10/2012. A major setback came in the process when landowners protested against the land acquisition. Due to the continuous protests of the villagers, the acquisition process was called off. The reason, when investigated, the ADA said, was that the villagers were unwilling to sell their agricultural lands, which are their primary source of income. They were also not happy with the amount of compensation.

5.3.7 Other Housing Projects

The incident of Asadpur Qayam was a jolt to the ADA. Besides, the process of land acquisition has been lengthy in all cases. Completion of projects was delayed due to the delayed land acquisition, hindering the supply of affordable houses to the growing population. Thereafter, ADA decided to either directly buy land from the villagers or take land from the land bank operated by Aligarh Nagar Nigam (ANN). ANN has given a total of 4.42 ha (44187.77 sq. m.) of land to the ADA for urban housing projects in different locations through the Urban Land Ceiling Act. These parcels of land have been used for constructing residential apartments for various income groups. The details of these small housing projects are given in Table 4. In most of the aforementioned projects, ADA has acquired land from Nagar Nigam. In addition to this, ADA has bought some land through direct negotiation with the landowners. Most of these projects were developed after 2012.

Table 4 Other small housing projects by ADA

Project	Location	Area	Description
Dodhpur 1	Near Zohra Bagh Colony	0.15 ha	8 flats for HIG
Dodhpur 2	Sir Syed Nagar	0.41 ha	20 flats for MIG
Dodhpur 3	Dodhpur Mafi	0.24 ha	12 plots for MIG
Rasulpur Samad	Near Zafrabad village, along G. T. Road	0.27 ha	103 single-story buildings for LIG
Brij Vihar	Near Zafrabad village, along G. T. Road	0.45 ha	60 single-story buildings for LIG
Kansiram Nagar Awasiya Yojana	Elampur Village, G.T. Road	3.9 ha	1283 units of houses for EWS
Gautam Buddha Nagar Awasiya Yojana	Baraula Zafrabad village, G. T. Road	4.64 ha	10 plots for HIG and 144 plots for EWS.

Source: Aligarh Development Authority

5.4 *Informal Land Acquisition*

Aligarh Development Authority could not meet the demand for housing in the city. When the public authority fails, the private builders grab the opportunity. Several property dealers and real estate companies started constructing individual apartments, group housing, and township projects, supplying affordable houses. The required land for these private projects was informally acquired, where they directly bought land from the landowners. Two such private township projects are discussed later.

5.4.1 Ozone City

Ozone City is one of the big private townships in Aligarh City. It is located on the eastern side of the city, in the Mahua Khera village, connected to G.T. Road in the south and Ramghat Road in the north with canal-side roads. The township is owned by Ozone Group, a Faridabad-based real estate company. The project covers around 81 ha of land and is approved by ADA. The construction started around 2010.

5.4.2 Sangwan City

Another big private township is Sangwan City, located near the Ozone City Township. It is also connected to G.T. Road in the south and Ramghat Road in the north with canal-side roads. The township is spanned over 29 ha of land. It is owned by Sangwan Landco Pvt. Ltd. The initial phase covering 25 acres, which includes plots, duplex and simplex houses, residential apartments, and a commercial market, has already been completed. Currently, Phase-II, encompassing another 25 acres, is in progress.

5.4.3 Industries at Talaspur Khurd

Land acquisition in the city has not only occurred for housing projects. Vast agricultural land has been acquired for industries in Talaspur Khurd Village along the Aligarh New Bypass road. Several food processing industries, slaughterhouses, and packaging industries have evolved here in the last ten years. These factories have developed by acquiring agricultural lands. Many plots, which were agricultural in the recent past, have been left vacant for sites for industry. These two townships and one industrial zone have developed, acquiring agricultural lands that lie beyond the municipal jurisdiction, where the price of the land is low; taxation of land transfer is also low. These private industrialists and real estate companies got the land outside the city's jurisdiction at a cheaper rate, and farmers got instant compensation, which led to these developments.

5.5 *Issues of Land Acquisition*

The extensive field observation finds three stakeholders in the process of land acquisition in the study area. First is the Aligarh Development Authority. The second stakeholder is the landowners (farmers). The third one is the private property dealers. ADA accomplishes its objective to provide affordable housing to different economic groups through urban housing projects that acquire land from landowners. When the ADA could not meet the housing demand, the private property dealers grabbed the opportunity. They directly buy land from landowners or farmers for different housing projects. To have multi-angle perspectives on the issues of the land acquisition process in the study, these three stakeholders were interviewed, and issues about the land acquisition processes were discussed with them.

5.5.1 Issues Faced by Development Authority

The Aligarh Development Authority is the government authority in the case of land acquisition. ADA prepares the master plan for the city's development and develops affordable housing for different income groups. The development of housing projects requires land to be acquired. To examine the issues and challenges in the land acquisition process in the study area, the land acquisition officer and cartographer in the ADA were interviewed. The outcomes of the interviews are discussed below.

Lengthy Land Acquisition Process

Buying land under LAA is a lengthy and tardy process. Identification of suitable land areas, passing building proposals, notifying landowners, addressing grievances, and transferring land ownership take much time. In many cases, it takes more than a decade. The study finds that it took several years for most of the projects. For example, land acquisition for the Santiniketan colony took twelve years (1985–1997), and in Swarn Jayanti Nagar, the land was acquired in six phases between 1998 and 2007.

Ownership Dispute

In the case of Swarnjayanti Nagar, ADA had to give up around 1.3 ha of acquired land because the *gram samaj* filed a case against the acquisition. The ADA allegedly acquired common land (e.g., graveyards, temples, etc.) owned by *gram samaj,* which is exempt from any acquisition under LAA. The authority believed the farmers did not give the land to ADA, as they were more willing to sell the land to private builders at high selling prices.

Unwillingness of Landowners

In the initial years, land acquisition seemed to be smooth. But, in the late 1990s, when the demand for land for houses increased, private builders started land marketing. The landowners also became aware of the land market and land prices. The reluctance of landowners to sell their land resulted in incomplete land acquisition within the study area. In many instances, the Aligarh Development Authority (ADA) acquired a smaller amount of land than initially proposed. A specific example is the situation in Asadpur Kayam, where villagers were not willing to sell their land to the ADA, despite the authority having already demarcated and fenced the land for the project. But the farmers protested against the acquisition. ADA had to leave the project after acquiring half of the proposed land.

Extra Demands of Owners

The landowners often make extra demands to sell their land to the ADA. In the case of Vikash Nagar Awasiya Yojna, where land acquisition occurred in 1986, a farmer still reportedly visits the ADA asking for a job.

Growth of the Real Estate Market

The land acquisition process is lengthy. When ADA identifies land and publishes the notification in the local newspapers, many private property dealers buy land from farmers to make a profit by selling the land at the time of land acquisition. Besides, when an area is identified for development, by the time the project completes, private buildings, apartments, and housing projects start booming around the area notified for acquisition. This resulted in unplanned sprawl. For example, in the case of Swarn Jayanti Nagar, the completion of the project took more than a decade. Meanwhile, the lands surrounding the project area were developed quickly by private builders. People get houses at a cheaper rate from private builders in a short time. This also led to the rapid growth of apartments in the region.

Landowners Prefer Private Builders

Landowners were more willing to sell the land to private builders. As builders pay a higher rate for land than the authority. Secondly, they offer hand-to-hand cash transactions. As compensation from ADA remains a lengthy process and a lot of paperwork, no owners want to get involved in lengthy paperwork. They prefer hand-to-hand land transfer, as they get instant cash from private builders.

Land Excavation

ADA identifies land excavation by the local mafia as a critical problem in land acquisition and land development. When land is acquired for development, the local land mafia removes the earth to sell illegally and make profits, damaging the land for development. ADA has to infill the land first; then any development can occur. For example, lands were excavated in Swarnjayanti Nagar Extension. In the case of Asadpur Qayam, earth was removed from several plots that ADA had acquired.

Other Issues

There are still cases pending at the court against the land acquisition in Swarn Jayanti Nagar and Vikash Nagar. The authority has to visit the court to hear the cases. In the Asadpur Qayam case, the farmers who had already sold their land got compensation. But as the project was canceled, they did not return the compensation amount yet. When asked about any present or future project, the authority said that they aim to acquire 50 ha of land as directed by the UP Government in 2018 under the ordinance 385/8-3-16-309 for residential and development purposes. But after the incident of Asadpur Qayam, the ADA is not going to get lands following the land acquisition process due to the aforesaid hurdles. The small housing projects were done by taking land from Aligarh Nagar Nigam. ADA is planning to buy land directly from the villagers through direct negotiation instead of following the land acquisition process.

5.5.2 Issues Faced by Land Owners

To get the views of the landowners and their issues and challenges regarding the land acquisition process, villagers in the Kishanpur village were interviewed. The issues discussed with them are mentioned later.

Compensation is Low

Dissatisfaction over the amount of compensation is a common issue stated by the majority of landowners. They blame the government (here, the ADA) for deciding that the compensation amount is not followed by the LAA (as far as they got to know about the mechanism). The ADA decided the amount of compensation based on the circle rate of land value, which often remains outdated. As a result, the compensation amount offered by the authority is low compared to the real market value of the land. One villager said,

> *Muabje ke paise ke liye kayi baar chakkar lagana pada vikash pradhikaran ke office me.*

(To get the compensation, they had to visit the ADA office several times.)

Forced to Give up Land

Most of the land acquisitions occurred before 2011, when the UPA government made some amendments to the old Land Acquisition Act. To date, the LAA 1984 has been in practice, which allowed the government authority to act without farmers' consent (compulsory nature). Therefore, many farmers were forced to give up their land, despite their unwillingness. Also, when a plot of land was sold, the owners of the surrounding land were forced to give up the land.

Lengthy Process and a Lot of Paperwork

Most of the landowners in rural areas where land was acquired are farmers. Many of them were illiterate. They are against the land acquisition process due to the lengthy process and a lot of paperwork.

Marginal Farmers Worst Affected

The city is surrounded by fertile agricultural fields, most of which are smaller in size. Due to the land acquisition process, marginal farmers were the worst affected, as the loss of farms and farming affected them badly. They lost their work, and, on the other hand, they lost their source of food. In a case stated by a farmer, he had to abandon agriculture and find a new job in the city. He works as a laborer on a construction site.

Local People Neglected

When any housing project develops, the local villagers do not get any houses or plots. The villagers remain the villagers. One villager pointed out that during the acquisition process in Swarn Jayanti Nagar, the authority promised to give residential plots to some land givers. However, they did not give any plots to the land-givers after its completion.

Landowners Prefer Selling Land to Private Builders

Private property buyers offered high prices for land near the ADA housing project. The landowners who gave land to ADA found that those who sold the land to the property dealers got much more money than they did. Also, they got hand-to-hand money. So, they were dissatisfied with the ADA land acquisition. Had they known about it earlier, they would have sold the land to private property buyers. When asked about what change occurred in their life after selling land, one said he got his daughter married. One started a grocery shop. Few people said they built their houses, and some bought new land for cultivation, as they were born to farm.

5.5.3 Issues to Private Property Dealer

Private builders and property buyers have been vital stakeholders in the land acquisition process in the study area. To have the perspectives of the private builders on the land acquisition process, the researcher tried to contact some property buyers. Most of them are not aware of the land acquisition processes. It is the demand for residential flats and apartments that they know about. When asked about the ADA projects and their project, one property builder near Swarn Jayanti Nagar said that there is a high demand for houses in these areas. ADA has failed to fulfill the demand, and the ADA projects are empty. The builders and property dealers got into the market and provided affordable houses to the growing population. When asked about the acquisition process of the land they developed, he said that they buy land directly from the landowners. If landowners wish to sell, they sell, and builders buy. No conflict of interest remains in the whole process.

6 Conclusion

Like other Indian cities, the increase in population in Aligarh City led to urban built-up expansion, exceeding the municipal limit in recent times. Urban expansion has brought a change in the socio-economic lives of the people as well as a change in land uses. Urban expansion and land transformation in the fringes are inevitable and need proper control and planning measures. Since its inception, ADA has been engaged in controlling land transformation and planned urban development in the city. The study finds six major and seven minor housing projects where ADA acquired land for urban housing development in the last four decades. The ADA targeted to acquire around 125 ha of land, but could only get 92 ha. Further, 31 ha of land were returned to villagers in Asadpur Qayam village.

The study exposes the poor correlation between the growing demand for urban housing and the government supply mechanisms and finds a huge gap between the two. ADA has failed to provide an adequate number of affordable houses to the increasing city population and housing demands. When the development authority fails, it is the private property dealers and builders who grab the opportunity.

Second, the area identified by the development authority remained undeveloped due to the lengthy and tardy land acquisition process, as observed by Fazal and Ali (2022); meanwhile, the private developers developed the surroundings. It resulted in unplanned urban development in the urban fringe of Aligarh City. This exposes the inefficiency and insincerity of the development authority in dealing with urban housing problems and urban planning.

Third, the competition between development authorities and real estate developers led to an increase in land values. When a development authority starts a project in the fringe, the private property dealers start buying land in the vicinity. Consequently, land prices go up in the fringes. It sidelines the poor migrants, who cannot afford houses at high prices. They start living in slums and unauthorized colonies.

Finally, the case of land acquisition in Aligarh City poses serious questions about the viability and applicability of land acquisition policies in India. The issues discussed above are examples of inadequacies in the urban land acquisition policy, which demands revising the existing policies of urban housing and land acquisition.

While the total area acquired by the ADA for urban housing projects over the past four decades may not be extensive, the case of the land acquisition process holds significance in comprehending the intricacies of land acquisition and disposal policies in smaller cities in India. The study of land acquisition in Aligarh City is crucial for several reasons. First, it unveils the reality of urban housing and development in smaller Indian cities. Second, it prompts a critical examination of the applicability of existing land acquisition policies, indicating a potential need for reform. Lastly, it advocates for the establishment of an effective urban housing policy in India.

References

Ali, M. K. (2021). A GIS based analysis of urban expansion and land use land cover changes in Aligarh city. *Indian Journal of Spatial Science, 12*(2), 1.

Banu, N., & Fazal, S. (2013). Development of infrastructural facilities in public sector on the urban fringe of Aligarh City: A regional perspective from North India. *Journal of Infrastructure Development, 5*(2), 151–168. https://doi.org/10.1177/0974930614527946

Bentinck, J. V. (2000). *Unruly urbanisation on Delhi's fringe*. University of Groningen.

Bhagat, R. B. (2004). Dynamics of urban population growth by size class of towns and cities in India. *Demography India, 33*(1), 47–60. https://doi.org/10.1177/0049085714548537

Bhagat, R. B. (2017). Migration, gender and right to the city: The Indian context. *Economic and Political Weekly, 52*(32), 35–40.

Chadchan, J., & Shankar, R. (2012). An analysis of urban growth trends in the post economic reforms period in India. *International Journal of Sustainable Built Environment, 1*(1), 36–49. https://doi.org/10.1016/j.ijsbe.2012.05.001

Cohen, B. (2006). Urbanization in developing countries: Current trends, future projections, and key challenges for sustainability. *Technology in Society, 28*(1–2), 63–80. https://doi.org/10.1016/j.techsoc.2005.10.005

Datta, P. (2006). Urbanisation in India urbanisation in India. In *Regional and sub-regional population dynamic population process in urban areas* (pp. 1–16). European Population Conference. https://doi.org/10.1121/1.2021445

Diwakar, A. (1991). Urban expansion and rural land acquisition in the fringe villages a case study of union territory of Delhi.

Fazal, S. (2000). Urban expansion and loss of agricultural land—a GIS based study of Saharanpur City, India. *Environment and Urbanization, 12*(2), 133–149. https://doi.org/10.1630/095624700101285343

Fazal, S. (2013). *Land use dynamics in a developing economy regional perspectives from India*. Springer International Publishing. https://doi.org/10.1007/978-94-007-5255-9

Fazal, S., & Ali, K. (2018). Rural–urban transition in India: A perspective on future urbanization. *Annals of the National Association of Geographers, India, 38*(2), 343.

Fazal, S., & Ali, M. K. (2022). The realities of urban land use planning in India: Case study of Aligarh City. *Urban India, 42*(1), 88–112.

Goswami, M., & Khire, M. V. (2016). Land use and land cover change detection for urban sprawl analysis of Ahmedabad City using multitemporal Landsat data. *International Journal of Advanced Remote Sensing and GIS, 5*(1), 1670–1677. https://doi.org/10.23953/cloud.ijarsg.51

He, Q., Liu, Y., Zeng, C., Chaohui, Y., & Tan, R. (2017). Simultaneously simulate vertical and horizontal expansions of a future urban landscape: A case study in Wuhan, Central China. *International Journal of Geographical Information Science, 31*(10), 1907–1928. https://doi.org/10.1080/13658816.2017.1338707

Hui, E. C. M., & Bao, H. (2013). The logic behind conflicts in land acquisitions in contemporary China: A framework based upon game theory. *Land Use Policy, 30*(1), 373–380. https://doi.org/10.1016/j.landusepol.2012.04.001

Johnson, C., & Chakravarty, A. (2013). Re-thinking the role of compensation in urban land acquisition: Empirical evidence from South Asia. *Land, 2*(2), 278–303. https://doi.org/10.3390/land2020278

Kantakumar, L. N., Kumar, S., & Schneider, K. (2016). Spatiotemporal urban expansion in Pune metropolis , India using remote sensing. *Habitat International, 51*, 11–22. https://doi.org/10.1016/j.habitatint.2015.10.007

Kumar, A., Pandey, A. C., Hoda, N., & Jeyaseelan, A. T. (2011). Evaluating the long-term urban expansion of Ranchi urban agglomeration, India using geospatial technology. *Journal of the Indian Society of Remote Sensing, 39*(2), 213–224. https://doi.org/10.1007/s12524-011-0089-z

Kundu, A. (2011). Politics and economics of urban growth. *Economic and Political Weekly, 46*(20), 10–12.

Lata, K. M., Prasad, D. V. K., Badarinath, D. K. V. S., & Raghavaswamy, D. V. (2001). Measuring urban sprawl: A case study of Hyderabad. *GISDevelopment, 5*(12), 26–29. Cited By (since 1996) 10\rExport Date 23 November 2011.

Longyu, S., Guofan, S., Shenghui, C., Xuanqi, L., Tao, L., Kai, Y., & Jingzhu, Z. (2009). Urban three-dimensional expansion and its driving forces—A case study of Shanghai. *China. Chinese Geographical Science, 19*(4), 291–298. https://doi.org/10.1007/s11769-009-0291-x

Mahalingam, A., & Vyas, A. (2011). Comparative evaluation of land acquisition and compensation processes across the world. *Economic and Political Weekly, 46*(32), 94–102.

Mukerji, S. (2017). Land acquisition in contemporary India: The growth agenda, legislation and resistance. *Indian Journal of Public Administration, 63*(1), 85–103. https://doi.org/10.1177/0019556117689851

Narain, V., Banerjee, P., & Anand, P. (2014). *The shadow of urbanization: The Periurban Interface of five Indian cities in transition* (No. 68; environmental change, vulnerability, and governance).

Nielsen, K. B. (2010). Contesting India's development? Industrialisation, land acquisition and protest in West Bengal. *Forum for Development Studies, 37*(2), 145–170. https://doi.org/10.1080/08039411003617252

Population Census. (2023). Aligarh City Population 2023. https://www.census2011.co.in/census/city/112-aligarh.html

Rahman, A., Aggarwal, S. P., Netzband, M., & Fazal, S. (2011). Monitoring urban sprawl using remote sensing and GIS techniques of a fast growing Urban Centre, India. *IEEE Journal of Selected Topics in Applied Earth Observations and Remote Sensing, 4*(1), 56–64. https://doi.org/10.1109/JSTARS.2010.2084072

Ramachandra, T. V., Bharath, H. A., & Sowmyashree, M. V. (2014). Urban footprint of Mumbai—the commercial capital of India. *Journal of Urban and Regional Analysis, 6*(1), 71–94.

Saha, M., Chauhan, D., Patil, S., Kangas, R., Heer, J., & Froehlich, J. E. (2021). Urban accessibility as a socio-political problem. *Proceedings of the ACM on Human-Computer Interaction, 4*(CSCW3), 1–26. https://doi.org/10.1145/3432908

Sarkar, A. (2007). Development and displacement: Land acquisition in West Bengal. *Economic and Political Weekly, 42*, 1435–1442.

Shaw, A. (2005). Peri-urban Interface of Indian cities: Growth, governance and local initiatives. *Economic and Political Weekly, 40*(2), 129–136. https://doi.org/10.2307/4416042

Shaw, R., & Das, A. (2018). Identifying peri-urban growth in small and medium towns using GIS and remote sensing technique: A case study of English bazar urban agglomeration, West Bengal, India. *Egyptian Journal of Remote Sensing and Space Science, 21*(2), 159–172. https://doi.org/10.1016/j.ejrs.2017.01.002

Trivedi, P. K. (2020). The expressway to Agra—Two roads, same destination: Land Acquisition under old and new land acquisition regimes. In *Land and livelihoods in neoliberal India* (pp. 235–248). Palgrave.

Verma, S. (2015). Subverting the land acquisition act, 2013. *Economic and Political Weekly, 50*(37), 18–21.

Verma, K. (2017). Urban sprawl and its changing paradigms a case study of Jaipur City. *Journal for Research, 03*(10), 22–26.

Yun, H., & Kim, S. (2022). Land acquisition methods and housing construction agents for prompt housing supply: A long-term comparative case study of Seoul metropolitan area, South Korea. *Habitat International, 124*, 102577. https://doi.org/10.1016/j.habitatint.2022.102577

Future Urban Slum: A Case of Hyderabad, India

Arijit Das, Sonali Kundu, Tirthankar Basu, Sasanka Ghosh, and Ketan Das

Abstract Slums are considered urban threats due to their overcrowding population, inadequate housing, and services. The social and physical growth of these slum areas is ignored due to the inadequate laws and management policies of urban bodies. From this end, the reason behind the underdevelopment is the absence of sufficient slum maps for the assessment of slums and proper management. In this regard, plenty of methods exist to solve this problem using remote sensing data. In modern days, deep machine learning (DL) approaches in remote sensing are being observed as an important tool for understanding the fundamental structure of physical characteristics situated in the images taken from satellites. This chapter tries to identify the unrecognized slums by using the random forest approach and also to predict the slums by applying cellular automata-Markov chain (CA-MC) techniques. For this, this study has collected training samples using Google Earth, and it is further used to delineate the slum boundaries in Hyderabad. The final output is validated concerning the original slum boundaries of Hyderabad city. The results of the validation process point to a classification process with more than 84.70 percent accuracy. This study has presented a detailed methodology for the delineation of the informal slums in the cities.

Keywords Remote sensing · Image classification · Slum detection · TOC · ROC

A. Das · S. Kundu · T. Basu (✉) · K. Das
Department of Geography, University of Gour Banga, Malda, India

S. Ghosh
Department of Geography, Kazi Najrul University, Asansol, India

P. K. Joshi et al. (eds.), *Geospatial Science for Urban Ecosystems*, The Urban Book Series, https://doi.org/10.1007/978-3-032-07155-2_20

1 Introduction

The literal meaning of slum is "a residential area where a collection of households can be seen living close to one another in a significant number of buildings and have faced one or more types of deprivation, to access proper sanitation facilities, safe drinking water, security of tenure, durability and structural condition of houses, and sufficient space for living" (Vogel et al., 2022). Mapping of the exact slum areas is important to develop proper public policies and development agendas for those who are living in the slums (Kohli et al., 2016; Abascal et al., 2022). Following the Census of India, slum settlements are categorized into three distinct classes: (A) notified slums, (B) recognized slums, and (C) identified slums. Notified slums presently denote areas officially designated under a formal "Slum Act" within a city or urban locality by the appropriate governing body. Known slums, on the other hand, are those acknowledged by housing and slum boards or local authorities but lack formal recognition under any legislative act. Identified slums are characterized by their unhygienic and densely populated nature, typically accommodating a minimum of 300 individuals or comprising 60 to 70 households constructed in extremely congested conditions. These settlements often lack adequate sanitation facilities and suffer from deficient access to clean drinking water. Despite their challenges, identified slums play a significant role in urban dynamics, highlighting the need for targeted interventions intended to enhance living conditions and promote sustainable urban development. The notified slums and the non-notified slums are separated from each other based on their service provisions. For notified slums, the basic services and development needs are provided by the local government (Kundu, 2022), and, for this reason, a huge number of areas remain prohibited from the services and development processes that are provided by the government (Nolan, 2015). Generally, official maps hide the deprived characteristics of the areas (Abascal et al., 2022) and declare them as homogeneous with others (Gupta et al., 2009; Roy et al., 2014). Deprived regions exhibit disparities across various dimensions, including their historical backgrounds, access to services, socioeconomic profiles, physical structures, and living conditions. In the context of India, there has been a notable decline in the urban slum population, plummeting from 26.3% to 17.4% between 2001 and 2011, attributed largely to the success of governmental policies, initiatives, and improved provision of essential services for the impoverished (Patel, 2013). However, it is crucial to recognize that these statistics may not fully capture the true extent of deprivation, as they often exclude marginalized communities residing in areas such as resettlement colonies and those living on footpaths (Office of the Registrar General & Census Commissioner, 2013).

As highlighted by Ooi and Phua (2007), Asia bears the brunt of global slum dwellings, accounting for 60% of such areas, accommodating approximately 554 million individuals in 2001. Moreover, the urban population worldwide continues to surge steadily. Schwab (2021) forecasts a staggering 60% increase in the urban population over the next three decades, with Asia, particularly countries like India, China, the Philippines, Vietnam, Bangladesh, and Pakistan, experiencing significant growth. This expansion predominantly manifests in sprawling informal settlements, with approximately 30% of the urban population residing in slum areas, reflecting prevailing development trends. Despite these challenges, there remains a critical gap in knowledge concerning how to effectively address the diverse array of informal settlements on a global scale. Scholars advocate for the adoption of a "globally integrated method" (Mahabir et al., 2018; Simwanda et al., 2021) capable of grappling with urban expansion while providing actionable insights for planners and policymakers, particularly in developing nations where data availability poses a major hurdle. Closing this gap is essential for promoting sustainable and inclusive urban development that addresses the needs of marginalized communities worldwide.

Numerous studies on slums focus primarily on informal settlements, particularly employing remote sensing techniques to analyze their dynamics (Owusu, 2020; Roy & Lees, 2020). This emphasis arises due to the classification of these settlements as unauthorized, often resulting in government clearance rather than efforts toward improvement or regularization (Badmos et al., 2018). Throughout this paper, the emphasis is on informal settlements and slums. Slum management measures in the past have been ineffectual in preventing new slum growth (Amitava, 2021; Ceppi et al., 2020). Effective management of slums hinges upon prior understanding of their expansion patterns and underlying causes, a task often challenging for many developing nations, particularly in the absence or inaccuracy of census data (Wanie & Akoh, 2020; Surya et al., 2020a, b). Given the unique spatial positioning and characteristics of slums (Takyi et al., 2020), modeling techniques reliant on geographic information systems (GIS) offer a potential solution to bridge this gap. By scrutinizing the temporal and spatial structure of slums, GIS-based approaches facilitate the identification of developmental issues and enable the anticipation of future pathways of slum expansion. Models permit scientists to investigate and evaluate various theories of slums and practices in a controlled environment that is computer-based, gaining a better understanding of the urban-related phenomena (Thomson et al., 2020), which may also aid in the arising of specialized slum management plans. Though these methodologies' usefulness is in its early stages till now, particularly in Asian nations, where the availability of data is a big concern. Hence, the identification and mapping of slums play a crucial role in enhancing living standards and ensuring access to essential amenities for residents of these areas (Kohli et al., 2016; Henninger and Padhi & Mishra, 2021). Throughout the last few years, the government has implemented various policies and provided various facilities for slum areas to ensure proper slum management. But the policies and facilities are only applied to the notified slum areas. The recognized and identified slums, which also have the same characteristics as slum areas (like inadequate access to drinking water, poor sanitation facilities, and nonavailability of sufficient living areas), are not included as slums. Behind these, there are some reasons, which include issues like land tenure and political/legal problems. But the main thing is that the areas suffering from the same problems are not able to get the facilities that are given to the notified slums. So, the accurate mapping of slums is very necessary for the proper management of slums. In this regard, high-resolution satellite images can be useful for the mapping procedure (Kuffer et al., 2016). These images are available at high temporal frequencies that allow researchers to map and also predict the structure and dynamics of urban slums (Kuffer et al., 2016; Roy et al., 2017). Slums can be identified by their distinctive building morphology. In the past, slum research has primarily focused on three areas: socioeconomic and policy (Patel et al., 2014; Omole, 2010; Sola, 2013); material characteristics of slums (Kit et al., 2012; Kohli et al., 2012); and, more recently, modeling of slum areas using sophisticated techniques such as cellular automata in combination with Markov chain (CA-MC) (Augustijn-Beck). There has been very little research in this area. Shekhar (2020) used multi-criteria evaluation (MCE) to estimate the future expansion of areas in Karnataka, India, that were prone to slum construction. The model delves into the identification of potential hotspots for future slum growth within the already urbanized neighborhoods of Karnataka. Roy and Lees (2020) utilized agent-based modeling integrated with geographic information systems (GIS) to delve into the complex dynamics involving diverse stakeholders, including slum residents and governmental bodies, within their spatial context. Their study shed light on the intricate processes contributing to the formation of Indian slums. Similarly, KazeminiaKorrani (2020) developed a vector-based housing model within an agent-based framework, offering insights into the expansion patterns of informal settlements.

In another study, Mutua et al. (2020) employed logistic regression modeling to examine the driving factors behind the emergence of informal settlements and to forecast their potential future locations. Conversely, Wang et al. (2021) introduced a theoretical dynamics model capable of simulating various forms of urban expansion, contributing to a deeper understanding of urban growth dynamics. Furthermore, the realm of remote sensing and image classification methodologies has seen significant advancements. Researchers such as Pati et al. (2020), Li et al. (2021), and Protopapadakis et al. (2021) have extensively explored supervised and unsupervised machine learning techniques tailored for remote sensing applications. These advancements have enriched the array of tools available for analyzing remote sensing data, thereby enhancing our ability to study and manage urban environments effectively. Kohli et al. (2016), for example, developed a technique for detecting slums in Pune, India, depending on satellite image categorization and utilizing specialists' knowledge of the morphology of the surrounding slum region. What is evident, however, is that all of the aforementioned methodologies need complex and sophisticated data, which many developing nations may not have access to. This emphasizes the importance of having a single technique that can be used globally (Thomson et al., 2020).

It is observed that machine learning approaches are very effective in dealing with large datasets without having a detailed perception of the subject (Lary et al., 2016). The potential of machine learning approaches like Random Forests (RF) to classify urban slums is being explored by Duque et al. (2017). Machine learning (ML) techniques leverage an array of features, encompassing textural, spectral, and structural attributes, to analyze and interpret data (Kuffer et al., 2016; Jokar et al., 2013). Notably, the random forest classifier (RF) has emerged as a dominant tool for extracting slum areas from satellite imagery due to its wide-ranging applicability (Wurm et al., 2017; Kuffer et al., 2016). Moreover, Support Vector Machines (SVM) and Artificial Neural Networks (ANN) are also harnessed for slum detection (Kuffer et al., 2016). This study adopts a holistic approach aimed at identifying previously uncharted slum regions and forecasting potential expansion using advanced geospatial techniques and spatial matrices. The resultant predictions undergo rigorous comparison with the original slum boundary map of Hyderabad city to evaluate both accuracy and efficacy, shedding light on the effectiveness of the methodology employed.

2 Study Area

Presently, Hyderabad serves as the capital of Telangana, a recently formed state. Situated on the Deccan Plateau and bordered by the Musi River, it takes up an area of 650 square kilometers (250 square miles) in the northern part of South India (see Fig. 1). As per the 2011 Indian Census, Hyderabad ranks as the nation's fourth-largest city, with a population of 6.9 million residents within city boundaries and 9.7 million within the metropolitan area, positioning it as the sixth-most populous metropolitan region in the country. This city now possesses around 0.5 million slum households where around 2.27 million dwellers live. With a GDP of US $74 billion,

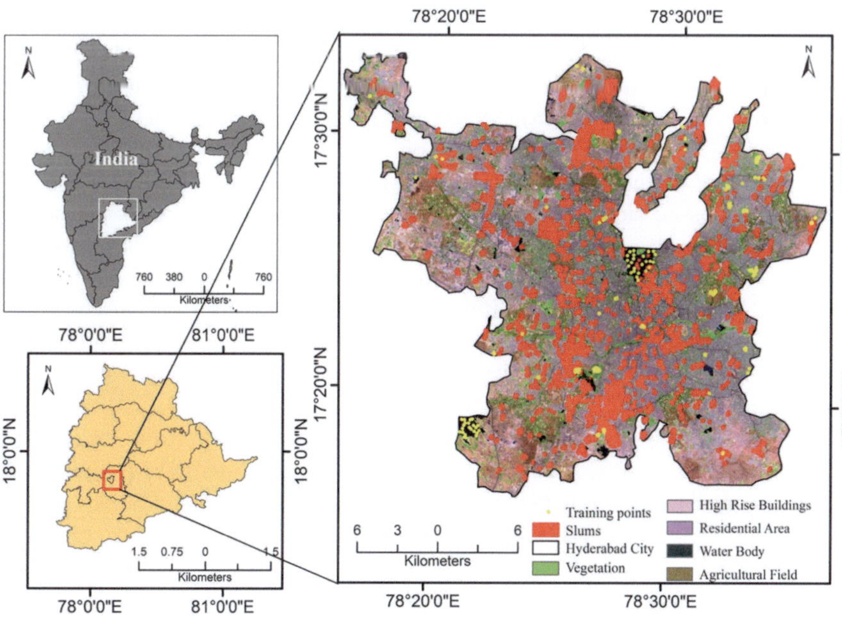

Fig. 1 Location of Hyderabad city

Hyderabad is India's fifth-largest city. Its population increased from over one million in 1951 to 10,269,000 persons in 2021 (https://www.macrotrends.net). It was marked by population growth rates of more than 50% from 1981 to 1991 and 27% from 1991 to 2001 (Kit et al., 2013). Like other growing megacities, Hyderabad also experiences hardship from social inequalities. It is estimated that more than a third of its population lives in slums of varying quality (Uthman et al., 2022). 13% of the population in the broader metropolitan region lives in poverty. Hyderabad has 1476 slums with a population of 1.7 million, with 66 percent of the population living in 985 slums in the "core" of the city and the remaining 34 percent living in 491 suburban slums (Das et al., 2022). Almost 22% of slum-dwelling households relocated from other parts of India in the latter decade of the twentieth century, and 63 percent are said to have lived in the slums for more than 10 years.

3 Materials and Methodology

3.1 Data

The city of Hyderabad has been selected for this study to examine and evaluate the position of slums. Here, the slum-demarked map is collected from the Hyderabad Metropolitan Regional Development Authority (HMRDA). The extension of the slums (spatially) is similar to the area of Hyderabad city taken for this study. LANDSAT-7 and LANDSAT-8 true-color imagery having a spatial resolution of 30 m have been considered for covering the Hyderabad city region dated 07 March 2011 and 29 March 2019. In this study, the pan-sharpened method (band of R-G-B) is taken into consideration. The land-use and land-cover (LULC) classification map is prepared from the satellite imagery to know the different classes in these regions. Classification algorithms obtain information from the pixel level, and then random forest and support vector machine (SVM) are applied to segment the pixels into different classes based on the information. The classes, like built-up areas (resistance or impervious layer), vegetation (trees and bushes), and water bodies (streams, lakes, and wetlands), are simply classified concerning the reflectance values of all the pixels. Following this, the classification algorithm can forecast classes based on the same information as the pixels present in the satellite imagery. Deep learning approaches, like random forest or SVM, estimate hierarchical feature statements, which need the input following the array form (Goodfellow et al., 2016). In this present study, small-sized images have been clipped sequentially from the satellite imagery. It is provided as an input for the random forest and SVM algorithms. Figure 2 represents the entire methodological way of this study.

Fig. 2 Methodological framework of this study

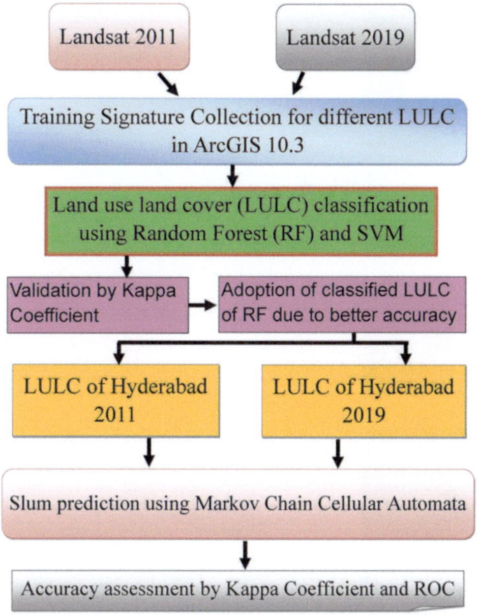

3.2 *Creation of Training Dataset*

In this study, the training dataset consists of four different urban classes: (1) slums, (2) high-rise buildings, (3) vegetation, and (4) water bodies. From the satellite datasets, about 948 samples (points) are obtained randomly, corresponding to the four different urban classes. These optimum numbers of points are obtained following Cochran (1977). 404 samples are collected for the class of slum (204 samples) and built-up (200 samples). Similarly, vegetation (328 samples) and water bodies (216 samples) together include 544 samples, respectively. The training samples are collected from both periods (2011 and 2019) with an equal proportion using the historical Google Earth imagery. Among the different land categories present in Hyderabad, this study has categorized the slums as structures with a low-rise form. In the case of Indian cities, these types of forms generally designate the presence of slums or areas with aged built-up structures (Kotharkar & Bagade, 2018). The rest of the built-up areas are categorized under the "Built-up areas or Non-Slums forms" class. The natural vegetation, forests, agricultural fields, and open field areas are placed under "vegetation." The blue spaces are classified as "Water." The stretching of contrast in the imagery was performed by taking contrast stretched at a further level by employing pixel values at 1–99 percentiles to secure sufficient variation within percentile values. The digitally enhanced satellite imagery is used to obtain the training dataset, testing dataset, and prediction dataset. The dataset corresponding to the four classes is collected randomly in the city (Figs. 3 and 4). The point data is created using Google Earth by employing the "path" tool.

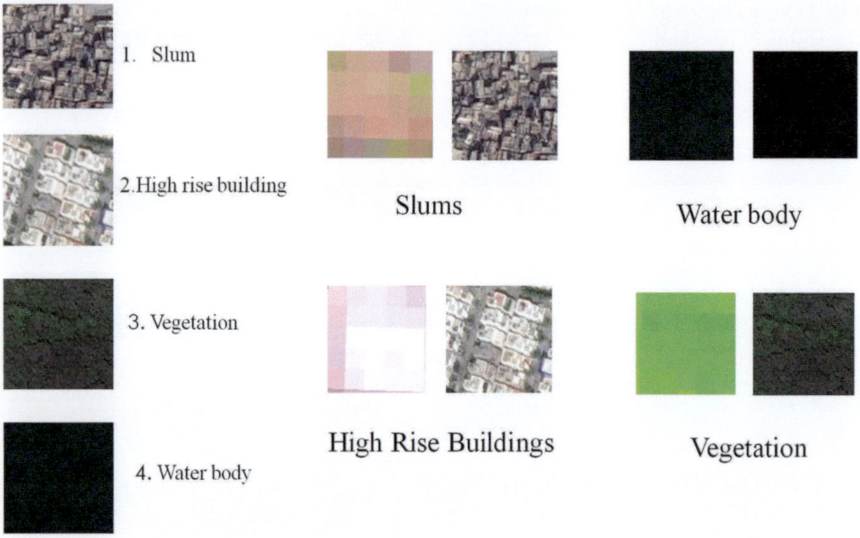

Fig. 3 Land typology in Hyderabad corresponding to the points taken in Google Earth for different classes as the training samples

3.3 Application of Random Forest and Support Vector Machine for LULC Classification

Random forest classifier is a supervised classification algorithm that was suggested by Breiman in the year 2001. It contains several trees and is mainly based on improved bagging and bootstrap techniques. In this technique, every tree is developed from the randomly selected training pixels to train the random forest classification. The two most important parameters of this classification are (1) number of trees (ntree) and (2) the number of variables that need to be divided at every node (mtry) (Nguyen et al., 2018). In the random forest classification, an individual tree is constructed from the original data by applying a different bootstrap sample. It is important to mention that around 1/third of the total cases were skipped out of the bootstrap sample. These are not incorporated into the construction of trees. These left-out cases can be used as a test set for classification. This is known as the out-of-bag (OOB) error estimate (Breiman, 2001a, b). In this regard, after a successful run, suppose "j" is the class that got the majority of votes all the time, while case "n" was OOB. Therefore, the OOB error estimate is the ratio of the number of times that "j" is unequal to the true class of "n," averaged over all cases (Breiman, 2001a, b). This study incorporated all the bands of Landsat images except the thermal band for RF classification. Training areas are obtained throughout the image to ensure sufficient representation and distribution of every class. Finally, the random forest classification is performed in R statistical software using the random forest package. The optimum ntree and mtry are evaluated and obtained following the lowest OOB error.

Support Vector Machine (SVM) is a supervised learning algorithm, and it is generally used for pattern recognition (Wijaya et al., 2008). The SVM algorithm

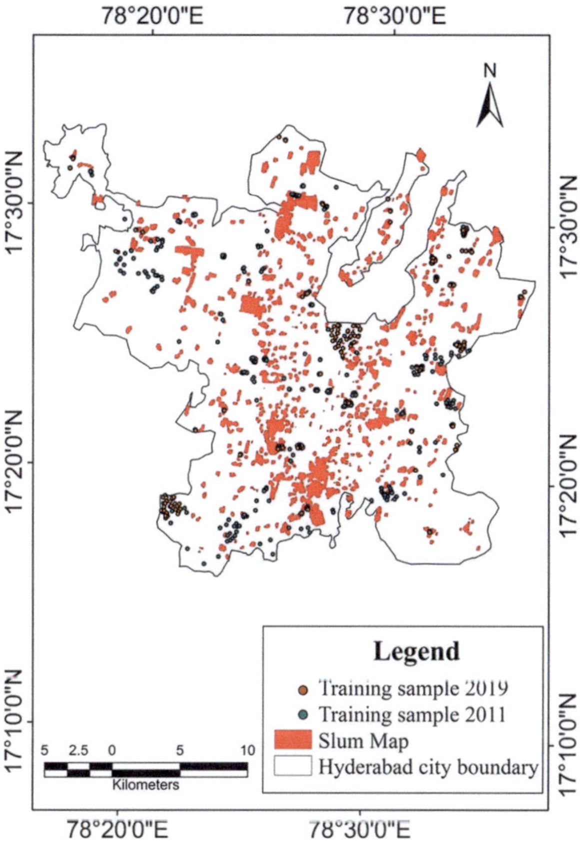

Fig. 4 Map showing the locations of selected point-based samples

separates point data into different classes using hyperplanes. In the original conception stage, this method is described with a labeled data set. Herein, the main objective is to identify a suitable hyperplane that can separate the data set into the classes, which are in accordance with the training samples (Vapnik, 1982). SVM tries to find the optimal separating hyperplane by the application of Lagrange multiplier and quadratic programming methods (Moosaei & Hladík, 2022).In this regard, the proper selection of the Kernel function is essential to project the training data in a larger area to develop a superior separating margin. In this study, the polynomial function kernel is used, which is represented in Eq. 1.

$$K\left(x_{i,}x_{j}\right)=\left(\gamma x_{i}^{\gamma}x_{j}+r\right)^{d},\gamma>0 \tag{1}$$

where x_i, x_j are the feature vectors in the input space, γrepresents the kernel width function, d represents the polynomial degree term, and r represents the bias term in the kernel function.

The signatures of four LULC classes, like slums (Class 1), high-rises (Class 2), water bodies (Class 3), and vegetation (Class 4), are obtained, and then, random forest and SVM are used in Rv4.0.5 software to classify the study areas in different classes based on training signatures. The classified output is validated by the confusion matrix. For this, a separate total set of 601 samples was collected randomly from the study area. The validation output shows an overall accuracy of about 0.91 and 0.88 for the classification by random forest and SVM, respectively (Table 1). Therefore, it can be expressed that the model of random forest performs slightly better for delineating the LULC class of this region. Following this, this study adopted the LULC map by random forest for further analysis.

3.4 *Cellular Automata and Markov Chain for Slum Prediction*

https://www.google.com The Markov chain model plays a pivotal role in assessing the likelihood of transitioning from one land-use type to another. This transformation probability serves as a crucial determinant in forecasting the demand for specific land-use categories, forming an integral component of the cellular automata model. By scrutinizing the Markov chain as a stochastic model, it offers insights into the dynamic evolution of land-use and land-cover (LULC) states. Analyzing the historical and current land-use types of a pixel within a cell, the Markov chain derives a transition likelihood matrix through its internal equation. This matrix provides valuable nonspatial information, elucidating the probability of land-use changes over time.

Table 1 Accuracy assessment for classified LULC

Random forest						
Assigned class	Slum	Built-up	Vegetation	Waterbody	**Total**	**User's accuracy**
Slum	114	9	0	2	125	91.20
Built-up	15	151	0	4	170	88.82
Vegetation	0	0	184	10	194	94.85
Waterbody	0	5	9	98	112	87.50
Total	129	165	193	114	601	
Producer's accuracy	88.37	91.52	95.34	85.96		
Overall accuracy	0.91					
SVM						
Assigned class	Slum	Built-up	Vegetation	Waterbody	**Total**	**User's accuracy**
Slum	108	10	0	7	125	86.40
Built-up	16	149	0	5	170	87.65
Vegetation	0	0	181	10	191	94.76
Waterbody	5	6	12	92	115	80.00
Total	129	165	193	114	601	
Producer's accuracy	83.72	90.30	93.78	80.70		
Overall accuracy	0.88					

Conversely, the cellular automata model computes the evolving land-use patterns within a cell by considering its prior land-use state and the surrounding conditions of neighboring cells. The neighborhood function is determined based on a kernel. In this study, a 5 × 5 cell-based kernel is used, which means the land-use state of a central cell is determined based on the surrounding 24 cells. Although the cellular automata model transforms each cell's land use spatially, it is necessary to integrate both the Markov chain (Eq. 1) and the cellular automata model to get the final results of land-use transformation of a study area based on land-use demand. Development in the cellular automata model makes it more complex by incorporating knowledge-based development suitability area generation for generating user-specific modification of the projected land-use patterns for incorporating the local influence of land-use transformation (Eq. 1).

$$X_{pi,t} = X_{di,t}{}^{*}X_{ni,t}{}^{*}Cnx^{*}Y \quad (2)$$

In this context, $X_{pi,t}$ represents the potential for conversion in cell "*i*," while $X_{di,t}$ denotes the probability of land transformation in cell "*i*," influenced by pertinent variables from the current to the subsequent time step. $X_{Ni,t}$ indicates the probability of land transformation based on neighboring conditions. "*Cnx*" encompasses both global and local constraints, and *Y* represents a stochastic disturbance variable.

3.4.1 Transition Suitability

The cornerstone of predicting future land-use change in an area lies in understanding the transition probability of various land types from time step one to time step two. The proportion of change in different land uses or the demand for land is contingent upon the time interval and the previous state of land use for each land-use type. This transition probability is calculated using Eq. 3.

$$P_{di,t}(a) = \frac{1}{1 + \exp\left(-\left(\mathrm{a}_0 + \sum_{j=1}^{k} a_j \mathrm{x}_j\right)\right)} \quad (3)$$

Here, $a_0$is constant, a_j(j = 1,2….k) is the coefficient that represents the jth factor impact for land-use transformation at "*t*" time.

3.4.2 Neighborhood Condition and Land Conversion Probability

The neighborhood function is a fundamental component of the cellular automata (CA) model, which relies on the land-use states of surrounding cells within a defined kernel. This function assesses the influence of adjacent cells to determine whether a particular cell will change its land-use type or remain in its current state. Essentially, the neighborhood function drives the dynamic behavior of the CA model by simulating how local interactions can lead to broader spatial patterns and transitions in land use. For this experiment, a 5 × 5 neighborhood function is calculated following Eq. 4.

$$P_{ni,t} = \frac{\sum_{j}^{5*5} \left(s_{i,t=urban}\right)\left(j \neq i\right)}{5^{*}5-1} \tag{4}$$

$P_{Ni,t}$ indicates land transformation probability for cell "*i*" between the present and next time step, depending on the neighborhood conditions. For this study, the Moore 5*5 kernel function is used as a neighborhood function.

3.4.3 Stochastic Disturbance Factor

The stochastic disturbance factor is calculated as Eq. 5.

$$R = 1 + \left(\ln r\right)^{a} \tag{5}$$

Here, "*r*" represents a random number between 0 and 1 and represents a real number between 0 and 10, which manages stochastic factor control for urban development.

Following the methods mentioned before, overall land-use transformation prediction is simulated. Land-use prediction is simulated for 2035 based on the 2011 and 2019 land-use patterns of the study area. Based on the subsections of the cellular automata model, the overall prediction is simulated using Eq. 6.

$$P_{gi,t} = \frac{1}{1+\exp\left(-\left(\begin{array}{l}-0.22 - weight_{NH21}\\ +weight_{Main\_rail}\\ -weight_{Minor\_rail}\\ -weight_{CBD\_distance}\end{array}\right)\right)} \times \frac{\sum_{j}^{5\times5}\left(S_{i,t=urban}\right)\left(\mathrm{j} \neq \mathrm{i}\right)}{5\times5-1} \times Bin\left(cell_{i}\left(t\right) \sim \text{available}\right) \times \left(1+\left(-\ln r\right)^{6}\right) \tag{6}$$

The final predicted slum map of Hyderabad city for the year 2035 is represented in Fig. 6.

4 Model Evaluation

The model is validated in two ways. It is done by using the Kappa coefficient and the receiver operating characteristic (ROC) curve. For this, the slum map of Hyderabad is predicted for the year 2021. Later on, a total number of 282 samples for slum areas, mainly for validation purposes were collected separately from the Google Earth imagery of 2021. These samples are collected randomly throughout the study area. Total Operating Characteristics (TOC) and Receiver Operating Characteristic (ROC) curves following Pontius and Si (2014) are used to validate the predicted result of identified slums of Hyderabad city. The ROC curve can be mathematically interpreted as Eq. 7.

$$P(X_1 > X_0) = P(X_1 - X_0 > 0) \tag{7}$$

where X_1 represents a continuous random variable offering "score" output by the binary classifier for the randomly chosen positive sample, and X_0 represents a continuous random variable offering 'score' output by the binary classifier for the randomly chosen negative sample.

The area inside the ROC curve indicates that the model can predict the slum areas with good accuracy. Slum areas of Hyderabad city show an accuracy of 84.7% for the predicted map in 2021 (Fig. 5). The TOC shows that the model has a random diagnostic ability at the origin of the curve. Here, the number of false alarms and

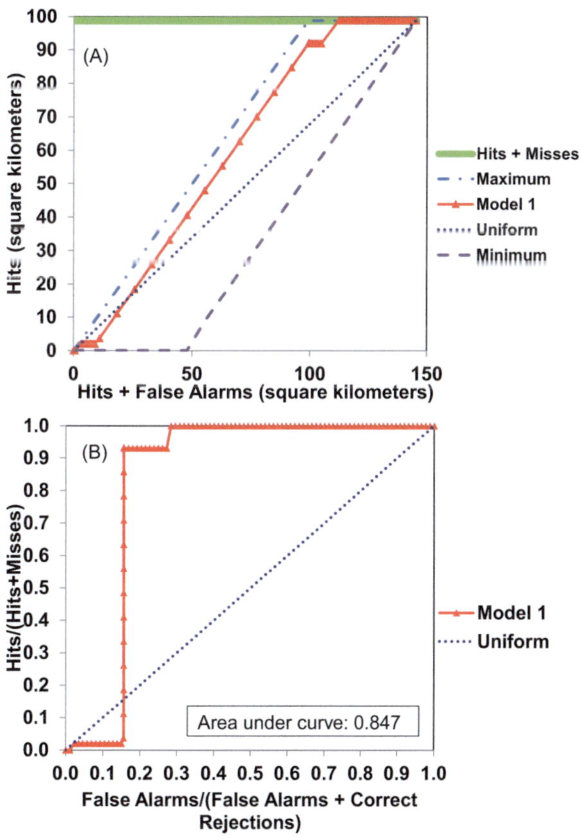

Fig. 5 Validation output for predicted slum (**a**) total operating characteristics curve; (**b**) receiver operating characteristics curve

misses is relatively high. However, the diagnostic ability increases further beyond the threshold value of 15 km^2 with a significant reduction in the number of false alarms and misses. The false alarms and misses become zero toward the upper end of the curve, indicating an accurate diagnostic. Overall, the TOC curve suggests good accuracy in the prediction modeling (Pointus and Si, 2014).

5 Results

Slum identification was conducted using a slum map provided by the local urban development authority, created in 2012. Although this map does not include slums that have emerged recently, it accurately depicts the locations of slums identified by the local authority at that time. In Hyderabad, slums are dispersed throughout the city rather than concentrated in a specific area. These are present in a scattered way throughout the city. The shape and size of the slums differ extensively. The slums are located beside the major roads in linear forms, in small plots, beside the railway lines, and adjacent to the lakes. From the predicted LULC and slum maps by the use of random forest and the cellular automata and Markov chain approach, it is found that the slum areas are mainly along the city boundary, river, and lakes, and they are continuously increasing with time. In Hyderabad, the built-up density of the preexisting slums has increased, and this chapter shows that the new, smaller patches will emerge near the Hussain Sagar areas, the periphery of the city, and in the central part by 2035 (Fig. 6; Table 2). As compared to the existing slums, the newly identified slums are quite smaller. This study has changed the original slum locations (vector form) into the raster form with an equal pixel size as the predicted maps. The variation patterns exhibited in the transition likelihood matrices of 2035 can be used to identify the level of change within the four LULC types (Table 2). The values of the diagonal order in the separate transition matrices show how likely it is for a land-cover type to remain nearly unchanged, but the off-diagonal values show how often it is for a land-cover type to migrate to new classes. Only slums and high-rise structures have high transition likelihood values in the transition matrix, while the other classes have relatively low transition likelihood. This indicates that slums and high-rise structures are likely to stay unaffected, but other classes are likely to alter throughout this period. The lower unaffected transition likelihood score of less than 0.05 demonstrates the development of slums when water bodies and vegetation are converted to slums. During this period, the region around slums and high-rise structures was not transformed into water bodies.

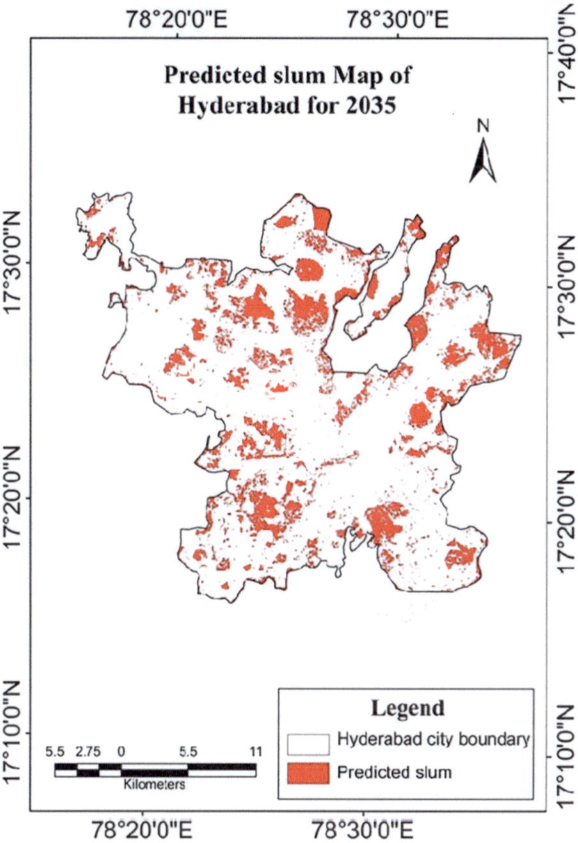

Fig. 6 Predicted slum map of Hyderabad for 2035

Table 2 Transition likelihood of Hyderabad in 2035

	Slums	High-rise buildings	Vegetation	Water bodies
Slums	0.8238	0.0372	0.0019	0.0000
High-rise buildings	0.2184	0.7789	0.0000	0.0000
Vegetation	0.0055	0.0005	0.7768	0.2172
Water bodies	0.0687	0.0006	0.1340	0.7967

6 Discussion

Satellite images capture a moment in time that encompasses the whole city (Hofierka et al., 2020; Nouri et al., 2020). Since it does not rely on previous slum declarations and identification processes, a slum identification approach based on remote sensing data is well-suited to address challenges such as changes in urban morphology resulting from slum upgrading efforts and the rapid emergence of new slums (Owusu, 2020; Surya et al., 2020a, b). Many areas that local governments have designated as slums and reported as such in census data may have lost their slum characteristics over time, even though they retain their slum status (Milbert, 2006). When viewed alongside the city's industrial growth pattern, the higher concentration of slums toward the northwest of Hyderabad becomes more apparent (Kumar et al., 2020). Several industries located on the city's northern outskirts depend on less-skilled, lower-paid labor provided by slum residents. Establishing slums near these employment centers reduces the need for workers to commute, resulting in significant time and cost savings (Kit et al., 2013). Slums possess varied and complex morphological characteristics from place to place, which are generally overlooked in different studies (Kuffer et al., 2016; Taubenböck & Kraff, 2014). This chapter has presented a way to map the slums using an image classification method using machine learning techniques. This includes research on the capability of machine learning techniques to identify slums from satellite imagery (Dahmani et al., 2020; Salem et al., 2020; Ghaffarian & Emtehani, 2021). The model demonstrates optimal performance with an AUC of 0.847 for Hyderabad city. According to Rasyid et al. (2016), an AUC of more than 0.8 is considered good to designate the respective modeling approach as sufficiently accurate. Therefore, the predicted slum map for the study area is found to be reliable. The application of cellular automata for the modeling of slum areas in India is successfully demonstrated in the work of Shekhar (2021). This chapter enhances the knowledge further in this field through the generation of a reliably predicted slum map with the help of the CA-Markov chain. Following the quantity of spatial information in the case of each training sample, the effectiveness of satellite imagery for the detection of slums is quite satisfactory (Dahmani et al., 2020; Owusu, 2020). The classifier model has effectively detected the slums in the city and identified the clusters (Pan et al., 2020; Ghaffarian & Emtehani, 2021). The classification accuracy is improved a lot by employing the magnification of the imagery and combining OOB methods (He et al., 2020; Shackelford, 2020). In deep learning methods, the main concern is the availability of a large training dataset (Li et al., 2020; Rajpurkar et al., 2020), as the model's accuracy heavily relies on the quality of the training data (Kamath et al., 2020; van der Sommen et al., 2020). However, this study circumvents these limitations by utilizing a pretrained CA model. Further, it is also important to mention that the modeling approach also contains some limitations. For instance, Landsat imagery provides a comparatively greater spatial resolution of 30 m. Therefore, it possessed some uncertainties regarding getting the clear-cut boundaries of the slums. The same concern regarding slum modeling is also highlighted in the work of Ibrahim et al. (2019). Therefore, it may lead to some overestimation in the prediction of slums. Besides, this study only tested the cellular automata-Markov chain model for the prediction. Using other models, such as the multilayer perceptron neural network (MLPNN) combined with a Markov chain, may yield slightly different results. Furthermore, the incorporation of data like demographic factors, basic amenities, and infrastructural services into these settlements can provide a detailed overview of the extent, locations, and characteristics of such settlements.

The results of this study indicate that natural areas such as vegetation and water bodies are more likely to be converted into slums in the near future. This finding is consistent with other studies (Das & Basu, 2020; Ghosh & Das, 2019). It was observed that most migrants who come to the city in search of better economic opportunities often find themselves unable to afford land within the city. As a result, they moved to the unfavorable peri-urban region of the city (Chimankar, 2016). This ultimately creates heavy pressure on the natural covers of the peri-urban region (Ali, 2010). The transition probability matrix of this study also highlighted similar facts. More intensive planning is required to restrict the further degradation of the natural covers and the formation of new slums in the region.

7 Conclusion

The requirements of affordable housing space, public transportation facilities, and improved sanitation as part of development can be successfully addressed with rapid and periodic delineation of the slums. This process can highlight changes in the urban extent and slum morphology at regular intervals to provide insight into the facts about the advancement of the slums. With the rapid rate of urbanization, new slums are also forming in medium and small cities. Therefore, extensive spatial-temporal analysis of slum areas is essential for making informed urban policy decisions and assessing city development. The slum prediction modeling approach of this study is found to be sufficiently accurate to demonstrate a general picture regarding the future slum areas in the city. The primary benefit of this modeling approach is its use of a free, publicly available dataset. This makes the approach easy and efficient to implement, especially in developing and underdeveloped countries, where acquiring multiple high-resolution satellite images can be challenging and expensive. Besides, the predicted slums will help to demarcate the areas susceptible to the formation of future slums. Therefore, effective policies can be developed to cope with the situation. It will help to promote sustainable urbanization in the region.

References

Abascal, A., Rothwell, N., Shonowo, A., Thomson, D. R., Elias, P., Elsey, H., & Kuffer, M. (2022). "Domains of deprivation framework" for mapping slums, informal settlements, and other deprived areas in LMICs to improve urban planning and policy: A scoping review. *Computers, Environment and Urban Systems, 93*, 101770.

Ali, S. I. (2010). Alternatives for safe water provision in urban and peri-urban slums. *Journal of Water and Health, 8*(4), 720–734.

Amitava, B. A. S. U. (2021). Smart redevelopment of slums prevent epidemics. *Smart Cities and Regional Development (SCRD) Journal, 5*(2), 11–22.

Badmos, O. S., Rienow, A., Callo-Concha, D., Greve, K., & Jürgens, C. (2018). Urban development in West Africa—Monitoring and intensity analysis of slum growth in Lagos: Linking pattern and process. *Remote Sensing, 10*(7), 1044.

Breiman, L. (2001a). Random forests. *Machine Learning Journal Paper, 45*(1), 5–32.

Breiman, L. (2001b). Random forests. *Machine Learning, 45*, 5–32. https://doi.org/10.1023/A:1010933404324
Ceppi, P., Masera, G., Tadi, M., & Rossi, A. G. (2020). The Rocinha favela as a paradigmatic case study of informal settlements in large cities: Current situation and past slum-upgrading Programmes. In *Environmental performance and social inclusion in informal settlements* (pp. 1–13). Springer.
Chimankar, D. A. (2016). Urbanization and condition of urban slums in India. *The Indonesian Journal of Geography, 48*(1), 28.
Dahmani, R., Belmzoukia, A., Said Tkatek, D., & Fora, A. A. (2020). Automatic slums identification around normal and smart cities: Using machine-learning on VHR satellite imagery. *International Journal, 9*(5), 9071.
Das, A., & Basu, T. (2020). Assessment of peri-urban wetland ecological degradation through importance-performance analysis (IPA): A study on Chatra wetland, India. *Ecological Indicators, 114*, 106274.
Das, B. N., Mohandas, A., & Killi, B. (2022). Prevalence and associated factors of domestic violence among married women in an urban slum in South India. *Journal of Family Medicine and Primary Care, 11*(3), 1006–1011.
Duque, J., Patino, J., & Betancourt, A. (2017). Exploring the potential of machine learning for automatic slum identification from VHR imagery. *Remote Sensing, 9*(9), 895.
Ghaffarian, S., & Emtehani, S. (2021). Monitoring urban deprived areas with remote sensing and machine learning in case of disaster recovery. *Climate, 9*(4), 58.
Ghosh, S., & Das, A. (2019). Urban expansion induced vulnerability assessment of East Kolkata wetland using fuzzy MCDM method. *Remote Sensing Applications: Society and Environment, 13*, 191–203.
Goodfellow, I., Bengio, Y., & Courville, A. (2016). Deep learning. MIT press. Gunter, A. W. (2009). Getting it for free: Using Google earth??? andIlwis to map squatter settlements in Johannesburg. *International Geoscience and Remote Sensing Symposium (IGARSS), 3*, 1–2.
Gupta, K., Arnold, F., & Lhungdim, H. (2009). *Health and living conditions in eight Indian cities. National Family Health Survey (NFHS-3) India 2005–06.*
He, J., Chen, Y., Wu, J., Stow, D. A., & Christakos, G. (2020). Space-time chlorophyll-a retrieval in optically complex waters that accounts for remote sensing and modeling uncertainties and improves remote estimation accuracy. *Water Research, 171*, 115403.
Hofierka, J., Gallay, M., Onačillová, K., & HofierkaJr, J. (2020). Physically-based land surface temperature modeling in urban areas using a 3-D city model and multispectral satellite data. *Urban Climate, 31*, 100566.
Ibrahim, M. R., Titheridge, H., Cheng, T., & Haworth, J. (2019). predictSLUMS: A new model for identifying and predicting informal settlements and slums in cities from street intersections using machine learning. *Computers, Environment and Urban Systems, 76*, 31–56.
Kamath, A., Jia, R., & Liang, P. (2020). Selective question answering under domain shift. *arXiv preprint arXiv:2006.09462*.
KazeminiaKorrani, A. (2020). Introducing a method for the preparation of maps and an automatic property separation tool for urban residential lands. *Town and Country Planning, 12*(2), 547–570.
Kit, O., Lüdeke, M., & Reckien, D. (2012). Texture-based identification of urban slums in Hyderabad, India using remote sensing data. *Applied Geography, 32*(2), 660–667.
Kit, O., Lüdeke, M., & Reckien, D. (2013). Defining the bull's eye: Satellite imagery-assisted slum population assessment in Hyderabad, India. *Urban Geography, 34*(3), 413–424.
Kohli, D., Sliuzas, R., Kerle, N., & Stein, A. (2012). An ontology of slums for image-based classification. *Computers, Environment and Urban Systems, 36*(2), 154–163.
Kohli, D., Sliuzas, R., & Stein, A. (2016). Urban slum detection using texture and spatial metrics derived from satellite imagery. *Journal of Spatial Science, 61*(2), 405–426.
Kotharkar, R., & Bagade, A. (2018). Local climate zone classification for Indian cities: A case study of Nagpur. *Urban Climate, 24*, 369–392.

Kuffer, M., Pfeffer, K., Sliuzas, R., & Baud, I. (2016). Extraction of slum areas from VHR imagery using GLCM variance. *IEEE Journal of Selected Topics in Applied Earth Observations and Remote Sensing, 9*(5), 1830–1840.

Kumar, A., Pandey, A. C., & Khan, M. L. (2020). Urban risk and resilience to climate change and natural hazards: A perspective from million-plus cities on the Indian subcontinent. In *Techniques for disaster risk management and mitigation* (pp. 33–46). Wiley.

Kundu, A. (2022). Housing for all: From rhetoric to a reality check on slum statistics for operationalization of flagship missions in India. In *Accessible housing for South Asia* (pp. 9–27). Springer.

Lary, D. J., Alavi, A. H., Gandomi, A. H., & Walker, A. L. (2016). Machine learning in geosciences and remote sensing. *Geoscience Frontiers, 7*(1), 3–10.

Li, A., Duan, Y., Yang, H., Chen, Y., & Yang, J. (2020, August). TIPRDC: Task-independent privacy-respecting data crowdsourcing framework for deep learning with anonymized intermediate representations. In *Proceedings of the 26th ACM SIGKDD international conference on Knowledge Discovery & Data Mining* (pp. 824–832).

Li, Y., Shi, T., Zhang, Y., Chen, W., Wang, Z., & Li, H. (2021). Learning deep semantic segmentation network under multiple weakly-supervised constraints for cross-domain remote sensing image semantic segmentation. *ISPRS Journal of Photogrammetry and Remote Sensing, 175*, 20–33.

Mahabir, R., Croitoru, A., Crooks, A. T., Agouris, P., & Stefanidis, A. (2018). A critical review of high and very high-resolution remote sensing approaches for detecting and mapping slums: Trends, challenges and emerging opportunities. *Urban Science, 2*(1), 8.

Milbert, I. (2006). Slums, slum dwellers and multilevel governance. *The European Journal of Development Research, 18*(2), 299–318.

Moosaei, H., & Hladík, M. (2022). A lagrangian-based approach for universum twin bounded support vector machine with its applications. *Annals of Mathematics and Artificial Intelligence, 91*, 1–23.

Mutua, M. K., Mohamed, S. F., Kimani-Murage, E. W., Kiarie, L., Janusz, C. B., Aaby, P., & Ravn, H. (2020). Complete and on-time routine childhood immunisation: Determinants and association with severe morbidity in urban informal settlements, Nairobi, Kenya. *Annals of Human Biology, 47*(2), 132–141.

New Delhi, I. (2013). Office of the Registrar General & Census Commissioner (India).

Nguyen, H. T. T., Doan, T. M., & Radeloff, V. (2018). Applying random forest classification to map land use/land cover using Landsat 8 OLI. *The International Archives of the Photogrammetry, Remote Sensing and Spatial Information Sciences, 42*(3), W4.

Nolan, L. B. (2015). Slum definitions in urban India: Implications for the measurement of health inequalities. *Population and Development Review, 41*(1), 59–84.

Nouri, H., Nagler, P., ChavoshiBorujeni, S., BarretoMunez, A., Alaghmand, S., Noori, B., & Didan, K. (2020). Effect of spatial resolution of satellite images on estimating the greenness and evapotranspiration of urban green spaces. *Hydrological Processes, 34*(15), 3183–3199.

Omole, K. F. (2010). An assessment of housing condition and socio-economic life styles of slum dwellers in Akure, Nigeria. *Contemporary Management Research, 6*(4), 1.

Ooi, G. L., & Phua, K. H. (2007). Urbanization and slum formation. *Journal of Urban Health, 84*, 27–34. https://doi.org/10.1007/s11524-007-9167-5

Owusu, M. (2020). *Spatio-temporal slum mapping at citywide scale using high-resolution images* (Master's thesis, University of Twente).

Padhi, B., & Mishra, U. S. (2021). Assessment of living condition of urban slum dwellers in India in the New millennium. *Urban Research & Practice, 15*, 1–23.

Pan, Z., Xu, J., Guo, Y., Hu, Y., & Wang, G. (2020). Deep learning segmentation and classification for urban village using a worldview satellite image based on U-net. *Remote Sensing, 12*(10), 1574.

Patel, S. (2013). Upgrade, rehouse or resettle? An assessment of the Indian government's basic Services for the Urban Poor (BSUP) programme. *Environment and Urbanization, 25*(1), 177–188.

Patel, A., Koizumi, N., & Crooks, A. (2014). Measuring slum severity in Mumbai and Kolkata: A household-based approach. *Habitat International, 41*, 300–306.

Pati, C., Panda, A. K., Tripathy, A. K., Pradhan, S. K., & Patnaik, S. (2020). A novel hybrid machine learning approach for change detection in remote sensing images. *Engineering Science and Technology, an International Journal, 23*(5), 973–981.

Protopapadakis, E., Doulamis, A., Doulamis, N., & Maltezos, E. (2021). Stacked autoencoders driven by semi-supervised learning for building extraction from near infrared remote sensing imagery. *Remote Sensing, 13*(3), 371.

Rajpurkar, P., Park, A., Irvin, J., Chute, C., Bereket, M., Mastrodicasa, D., & Patel, B. N. (2020). AppendiXNet: Deep learning for diagnosis of appendicitis from a small dataset of CT exams using video pretraining. *Scientific Reports, 10*(1), 1–7.

Rasyid, A. R., Bhandary, N. P., & Yatabe, R. (2016). Performance of frequency ration and logistic regression model in creating GIS based landslides susceptibility map at Lompobattang Mountain, Indonesia. *Geoenvironmental Disasters, 3*(19), 1–16.

Roy, D., & Lees, M. (2020). Understanding resilience in slums using an agent-based model. *Computers, Environment and Urban Systems, 80*, 101458.

Roy, D., Lees, M. H., Palavalli, B., Pfeffer, K., & Sloot, M. P. (2014). The emergence of slums: A contemporary view on simulation models. *Environmental Modelling & Software, 59*, 76–90.

Roy, D., Lees, M. H., Pfeffer, K., & Sloot, P. M. (2017). Modelling the impact of household life cycle on slums in Bangalore. *Computers, Environment and Urban Systems, 64*, 275–287.

Salem, M., Tsurusaki, N., Eissa, A., & Osman, T. (2020). Detection of slums from very high-resolution satellite images using machine learning algorithms: A case study of Fustat area in Cairo, Egypt. *Sciences (IEICES), 6*, 219–224.

Schwab, K. (2021). *Stakeholder capitalism: A global economy that works for progress, people and planet*. John Wiley & Sons.

Shackelford, D. (2020). *Comparing Machine Learning Classification Methods for Biological Tracking Data* (Doctoral dissertation).

Shekhar, S. (2020). Effective management of slums-case study of Kalaburagi city, Karnataka, India. *Journal of Urban Management, 9*(1), 35–53.

Shekhar, S. (2021). Slum modeling for growth prediction. In *Slum development in India. The urban book series*. Springer. https://doi.org/10.1007/978-3-030-72292-0_5

Simwanda, M., Murayama, Y., Phiri, D., Nyirenda, V. R., & Ranagalage, M. (2021). Simulating scenarios of future intra-urban land-use expansion based on the neural network–Markov model: A case study of Lusaka, Zambia. *Remote Sensing, 13*(5), 942.

Sola, O. (2013). Pattern of housing expenditure in Ondo state, Nigeria. *OIDA International Journal of Sustainable Development, 6*(2), 25–36.

Surya, B., Saleh, H., Suriani, S., Sakti, H. H., Hadijah, H., & Idris, M. (2020a). Environmental pollution control and sustainability management of slum settlements in Makassar City, South Sulawesi, Indonesia. *Land, 9*(9), 279.

Surya, B., Syafri, S., Hadijah, H., Baharuddin, B., Fitriyah, A. T., & Sakti, H. H. (2020b). Management of Slum-Based Urban Farming and Economic Empowerment of the Community of Makassar City, South Sulawesi, Indonesia. *Sustainability, 12*(18), 7324.

Takyi, S. A., Amponsah, O., Yeboah, A. S., & Mantey, E. (2020). Locational analysis of slums and the effects of slum dweller's activities on the social, economic and ecological facets of the city: Insights from Kumasi in Ghana. *GeoJournal, 86*, 1–15.

Taubenböck, H., & Kraff, N. J. (2014). The physical face of slums: A structural comparison of slums in Mumbai, India, based on remotely sensed data. *Journal of Housing and the Built Environment, 29*(1), 15–38.

Thomson, D. R., Kuffer, M., Boo, G., Hati, B., Grippa, T., Elsey, H., & Kabaria, C. (2020). Need for an integrated deprived area "slum" mapping system (IDEAMAPS) in low-and middle-income countries (LMICs). *Social Sciences, 9*(5), 80.

Uthman, O. A., Ayorinde, A., Oyebode, O., Sartori, J., Gill, P., & Lilford, R. J. (2022). Global prevalence and trends in hypertension and type 2 diabetes mellitus among slum residents: A systematic review and meta-analysis. *BMJ Open, 12*(2), e052393.

van der Sommen, F., de Groof, J., Struyvenberg, M., van der Putten, J., Boers, T., Fockens, K., & Bergman, J. J. (2020). Machine learning in GI endoscopy: Practical guidance in how to interpret a novel field. *Gut, 69*(11), 2035–2045.

Vapnik, V. (1982). *Estimation of dependences based on empirical data*. Springer.

Vogel, W., Hwang, C. D., & Hwang, S. (2022). Gender and sanitation: Women's experiences in rural regions and urban slums in India. *Societies, 12*(1), 18.

Wanie, C. M., & Akoh, N. R. (2020). Strengthening institutional planning framework on urban slum upgrading challenges in Bamenda City, Cameroon. *Journal of Arts and Humanities, 3*(1), 23–40.

Wijaya, A., Marpu, P. R., & Gloaguen, R. (2008). Geostatistics texture classification of tropical rainforest in Indonesia. In S. Stein, W. Shi, & W. Bijker (Eds.), *Quality aspects in spatial data mining* (pp. 199–210). CRC.

Wurm, M., Weigand, M., Schmitt, A., Geiß, C., & Taubenböck, H. (2017, March). Exploitation of textural and morphological image features in sentinel-2A data for slum mapping. In *2017 joint urban remote sensing event (JURSE)* (pp. 1–4). IEEE.

Spatiotemporal Patterns of Land Surface Temperature (LST): A Case of Herat, Afghanistan

Sebghatullah Mahmoodi, Debolina Basu, and Sonam Agrawal

Abstract The Urban Heat Island (UHI) phenomenon is a prevalent environmental issue in urban landscapes, impacting both climatic and ecological processes. This study examines the spatiotemporal patterns of Land Surface Temperature (LST) to analyze the magnitude and characteristics of the UHI effect in Herat, a major metropolitan city in Afghanistan. The seasonal LST and UHI patterns have been correlated to the land use/land cover (LULC) changes that occurred in this city over a decade (winter and summer; 2008–2009 and 2018–2019). The results indicated that the city experienced rapid expansion during the study period. Barren land and vegetation were the main classes that got converted to the built-up class. The study of LULC changes showed that the built-up area has expanded by 61.7% in Herat. The UHI phenomenon was found to be dominant in the summer season. The maximum UHI intensity (UHII) was noted at the city center due to the dense urbanization and high anthropogenic activities. The maximum average seasonal UHI intensity of Herat was observed to be 1.6 °C. The present study is expected to help city planners in framing sustainable land management policies.

Keywords Urban Heat Island (UHI) · Land Surface Temperature (LST) · GIS · Afghanistan

S. Mahmoodi
Greenery Directorate of Kabul Municipality, Kabul, Afghanistan

D. Basu
Department of Civil Engineering, Motilal Nehru National Institute of Technology Allahabad, Prayagraj, Uttar Pradesh, India
e-mail: basud@mnnit.ac.in

S. Agrawal (✉)
Geographic Information System (GIS) Cell, Motilal Nehru National Institute of Technology Allahabad, Prayagraj, Uttar Pradesh, India
e-mail: sonam@mnnit.ac.in

P. K. Joshi et al. (eds.), *Geospatial Science for Urban Ecosystems*, The Urban Book Series, https://doi.org/10.1007/978-3-032-07155-2_21

1 Introduction

Urbanization represents one of the most significant human-induced changes in land use, profoundly transforming natural landscapes into built environments. A well-documented consequence of this transformation is the Urban Heat Island (UHI) effect, a phenomenon where urban areas are significantly warmer than their rural surroundings (Easterling et al., 1997). In most large cities, the temperature at the city center is markedly higher than in the surrounding suburban areas, a disparity referred to as the UHI effect (Akbari et al., 2016; Akbari & Matthews, 2012). This temperature increase occurs because extensive natural land is replaced by artificial surfaces, such as asphalt and concrete, which absorb incoming solar radiation during the day and reradiate it at night. This heat retention by built surfaces leads to the development and intensification of the UHI effect (Dutta et al., 2019). Consequently, urbanization not only alters the physical landscape but also significantly impacts the local climate, contributing to various environmental challenges.

The UHI effect impacts energy consumption, air quality, and urban ecosystems (Grimm et al., 2008). The combined effect of UHI and heat waves could increase mortality risks (Anderson & Bell, 2011). Under synergistic conditions, UHI intensity (UHII) may reach up to 10–15 °C (Oke, 1995). Consequently, the hotter microclimates created by UHI increase the energy demand for cooling buildings (Tursilowati et al., 2012). Furthermore, more power generation is needed to fulfill this demand, resulting in increased greenhouse gas emissions and a decline in climatic conditions. Akbari et al. (2001) found that electricity demand increases by 2–4% for every 1 °C rise in temperature. However, in the winter season, as it is colder in rural or suburban areas, the UHI effect positively impacts city dwellers by providing them with a warmer climate (Voogt & Oke, 2003).

Remote sensing is an efficient technique for estimating the temperature values and UHI intensities over urban environments (Dutta et al., 2021). Initially, the UHI phenomenon was studied using in situ ground-based observations. They were taken from fixed thermometers located at the sampling stations or mobile thermometers mounted on vehicles. These approaches have generated empirical data for UHI studies but for smaller study areas and microenvironments. However, satellite imagery-based remote sensing analysis technique has a very wide areal coverage that would be suitable for measuring the UHI magnitude and its intensity (Dutta et al., 2022). This has provided new avenues for researchers to observe UHI's spatial pattern and its dynamic behavior (Malcoti et al., 2023).

The present study aimed to assess the intensities of the UHI effect in Herat city, Afghanistan, over a decade (2009–2019). The research focused on quantifying the UHI intensity (UHII) at specific urban sites within the city, selected based on dominant land use/land cover (LULC) classifications. By targeting these particular sites, the study sought to establish a direct correlation between UHII and the associated LULC changes. This approach was expected to provide valuable insights into how different land use practices and cover types contribute to the variation in UHI intensities across the urban landscape.

The study aimed to achieve several objectives to substantiate its underlying research hypothesis: First, it aimed to investigate the spatiotemporal variations in daytime Land Surface Temperature (LST) within Herat, Afghanistan, a rapidly evolving metropolitan area, using multisource remotely sensed data. Second, it sought to assess the magnitude and characteristics of the UHI phenomenon in Herat through advanced geospatial techniques. Third, the study aimed to identify patterns of variation in daytime LST and UHII over seasonal and decadal scales within Herat city. Last, it aimed to explore the relationship between UHI and changes in LULC using remote sensing image-based analysis for Herat city.

2 Materials and Methods

2.1 *Description of the Study Area*

Herat is located in western Afghanistan between 34°18′N to 34°24′N latitude and 62°06′E to 62°16′E longitude. Herat is the second most populous city in the country after Kabul. The ancient city of Herat (the capital of Herat Province) lies on the side of the Harirud River at an elevation of 922 meters. Herat is the economic center of western Afghanistan, and it is linked to Mashhad city of Iran through Islam Qala land port and Turkmenistan via Torghundi land port. According to the report of the National Statistics and Information Authority (NSIA) of Afghanistan for the year 2019–2020, the population of Herat city is 632,206. Herat has a hot semiarid climate, with an annual mean temperature of 17 °C. The mean temperature of Herat is changing from a minimum mean of 4.5 °C in January to a maximum mean of 29.2 °C in July. Monsoon winds that flow from June to September are an important climatic feature of Herat (Nasery et al., 2021). The core of the city is surrounded by high residential density. The old city has several heritage and archeological sites in the form of forts, mosques, shrines, etc. (Samimi et al., 2019).

2.2 *Data Use*

To detect changes in urban expansion and evaluate LST variations related to urbanization in the study area, data from Landsat 5 thematic mapper (TM) and Landsat 8 operational land imager (OLI) and thermal infrared sensor (TIRS) were utilized. The analysis considered seasonal variability by obtaining satellite images for both winter and summer months. Specifically, images were collected for November, December, January, and February to represent winter and for May, June, July, and August to represent summer. To ensure accuracy and consistency, only cloud-free images were used, which were downloaded from the United States Geological Survey (USGS) Earth Explorer portal. The specific dates for the imagery

Table 1 Details of the images used

Season	Sensor	Date	Path/row
Winter	Landsat 5 TM	20.11.2008 22.12.2008 23.01.2009 24.02.2009	157/ 36
	Landsat 8 OLI/TIRS	16.11.2018 18.12.2018 19.01.2019 20.02.2019	
Summer	Landsat 5 TM	15.05.2009 16.06.2009 02.07.2009 19.08.2009	
	Landsat 8 OLI/TIRS[a]	27.05.2019 30.07.2019 31.08.2019	

[a] The image for June was unavailable due to high cloud cover

acquisition are detailed in Table 1. This comprehensive dataset allowed for a detailed examination of how urban growth impacts LST over different seasons.

2.3 Methodology

Surface temperature extraction from Landsat thermal band imagery, assessment of the UHI effect, and classification of various LULC types were conducted, following the methodology depicted in Fig. 1. Detailed descriptions of these procedures are provided in the subsequent subsections. This methodological approach ensured a systematic and standardized analysis of thermal data, UHI characteristics, and landscape classifications, facilitating a comprehensive understanding of urban environmental dynamics and their implications.

2.3.1 LST Mapping

For obtaining LST and its mapping, the following steps have been followed in this work (Landsat 8 Handbook, 2019). The digital number (DN) to spectral radiance conversion is carried out with the following formula:

$$L_\lambda = \frac{L_\lambda \max - L_\lambda \min}{Q_{Cal} \max - Q_{Cal} \min} \times \left(Q_{Cal} - Q_{Cal} \min\right) + L_\lambda \min \tag{1}$$

where L_λ = spectral radiance at the sensor's aperture in W/(m^2 sr μm), L_λMax = spectral radiance that is scaled to Q_{Cal}Max in W/(m^2 sr μm),

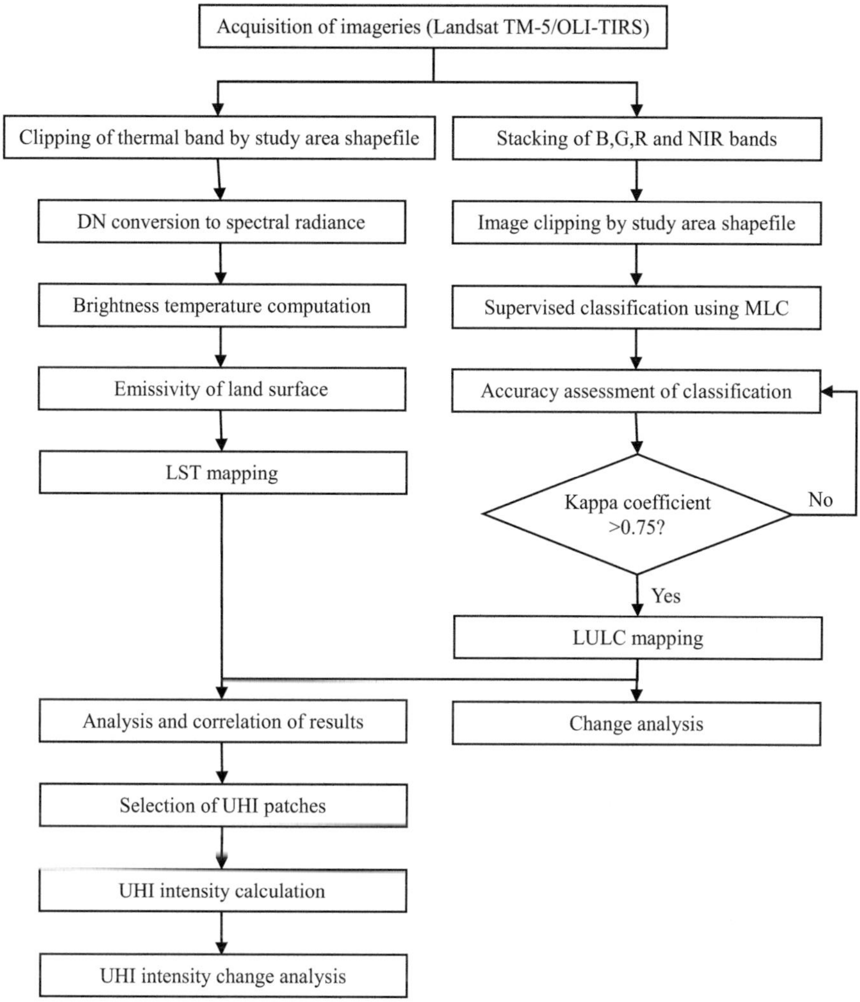

Fig. 1 Flowchart of research methodology

L_λMin = spectral radiance that is scaled to Q_{Cal}Min in W/(m^2 sr μm), Q_{Cal} = quantized calibrated pixel value in DN, Q_{Cal}Max = maximum quantized calibrated pixel value in DN, and Q_{Cal}Min = minimum quantized calibrated pixel value in DN.

L_λ is then used to compute the brightness temperature (i.e., blackbody temperature) as per Eq. (2):

$$\mathrm{BT_K} = \frac{K_2}{\ln \times \left(1 + \frac{K_1}{L_\lambda}\right)} \tag{2}$$

where BT_K = Top of atmosphere brightness temperature in Kelvin, and K_1 and K_2 = prelaunch calibration constants. Their value is given in the metadata file of the image.

BT_K provides the temperature of the blackbody. Therefore, emissivity correction was performed. Land surface emissivity (ε) was estimated considering the Normalized Difference Vegetation Index (NDVI) values. If NDVI <0.2, then ε= 0.98; if NDVI >0.5, then ε= 0.99; otherwise (0.2 ≤ NDVI ≤0.5), the emissivity is computed as per Eq. (3) (Wang et al., 2018).

$$\varepsilon = 0.004 \times \text{PV} + 0.986 \tag{3}$$

where P_v is the vegetation proportion, which is calculated as per Eq. (4).

$$P_v = \left[\frac{\text{NDVI} - \text{NDVI}_{\text{min}}}{\text{NDVI}_{\text{max}} - \text{NDVI}_{\text{min}}} \right]^2 \tag{4}$$

$NDVI_{min}$ = 0.2 and $NDVI_{max}$ = 0.5 were given in the previous studies (Wang et al., 2018). NDVI is computed with Eq. (5):

$$\text{NDVI} = \frac{\text{NIR} - \text{Red}}{\text{NIR} + \text{Red}} \tag{5}$$

Finally, LST was calculated as per Eq. (6).

$$\text{LST} = \frac{\text{BT}_\text{K}}{1 + \lambda \times \text{BT}_\text{K} \times \frac{\ln \varepsilon}{\rho}} \tag{6}$$

where λ = 11.5 µm, the wavelength of emitted radiance; $\rho = h \times c/\sigma = 1.438 \times 10^{-2}$ mK; σ = Stefan Boltzmann's constant = 5.67×10^{-8} $Wm^{-2}K^{-4}$; h = Planck's constant; c = velocity of light; and ε = land surface emissivity.

2.3.2 LULC Mapping

The LULC map illustrates the distribution of different land classes within a specific area. To create the LULC map for Herat, a composite image was generated by stacking the blue, green, red, and near-infrared (NIR) bands. These images were spatially constrained using the administrative boundary shapefile of Herat. Subsequently, supervised image classification was conducted using the Maximum Likelihood Classifier (MLC). The analysis identified three primary LULC classes in Herat: built-up, barren land, and vegetation. Training samples were meticulously chosen from each decade's images to accurately represent these classes. Following classification, the accuracy of the LULC map was assessed, including the computation of the kappa coefficient. A kappa coefficient greater than 0.75 indicates high agreement between classified and reference data, validating the accuracy of the classification results. Change analysis was then performed based on significant kappa coefficients, enabling the study of decadal changes in LULC patterns. This analysis utilized transition probability matrices to understand the dynamics of land cover changes over time, providing valuable insights into urbanization trends and their environmental impacts in Herat.

2.3.3 UHI Mapping

A UHI map was generated from the LST calculated in the previous steps. For understanding the UHI effect in a city, different patches of major densely built-up areas, including commercial areas, industrial areas, and administrative areas were considered for the analysis of UHI. All these patches were compared with a patch of rural area with a natural cover of vegetation and less urbanization, which served as the baseline. The UHI effect is defined as the average temperature difference between an urban area and its adjacent rural surroundings (Chang et al., 2023; Jaiswal et al., 2022). This study adopted a methodological approach similar to that used by Lazzarini et al. (2013) for calculating the UHII in Abu Dhabi, UAE, where specific patches within the urban and rural areas are considered instead of the entire built-up region. This approach allows for a more localized assessment of temperature variations influenced by urbanization. The UHII is computed using the formula developed by Voogt and Oke (2003), which accounts for spatially averaged temperature differences to quantify the intensity of the UHI effect. This methodological framework ensures that variations in LULC types across different patches are taken into account, providing insights into how specific urban features contribute to temperature disparities between urban and rural environments.

$$UHII = LST_U - LST_R \tag{7}$$

where LST_U= LST of the urban patch, and LST_R= LST of the rural patch.

3 Research Findings

This section includes the findings on the satellite-detected UHI effect observed in Herat. Differences in LST were identified in the city area. The growth pattern of heat islands was studied on a spatiotemporal scale. The correlation between the magnitude of UHII and LULC change was also analyzed for Herat for the selected decade.

3.1 Direction of LULC Change

LULC mapping was conducted for the years 2009 and 2019, as illustrated in Figs. 2 and 3. In these maps, red denotes built-up areas, yellow represents vegetation cover, and pink indicates barren land. The built-up areas primarily encompass residential, commercial, and administrative zones. Key landmarks, such as the city center and Taraqi Park (Park-e Taraqi), were prominently marked on the LULC maps to provide spatial context. Confusion matrices were generated to assess the overall accuracy of each LULC classification. For both 2009 and 2019, the accuracy achieved was deemed satisfactory, with kappa coefficients of 0.90 and 0.92, respectively. Over the ten-year period from 2009 to 2019, the built-up area expanded significantly by 2410.5 hectares (Fig. 3), representing a 61.7% increase compared to 2009 (Fig. 2). Concurrently, there was a notable reduction in barren land, characterized by rocky, stony, and sandy terrains, which decreased by 2679.1 hectares. An analysis using transition probability matrices revealed substantial conversions of other land cover classes into built-up areas. Approximately 25% of barren land and 18% of vegetation underwent transformation into built-up areas over the decade. This transformation underscores the rapid urbanization observed in Herat, accompanied by the decline in barren land and vegetative cover. Similar trends of urban growth in Herat have been observed in other studies, highlighting the significant expansion of urban footprints since the early 2000s (Kristy, 2018; Samimi & Maraschin, 2022).

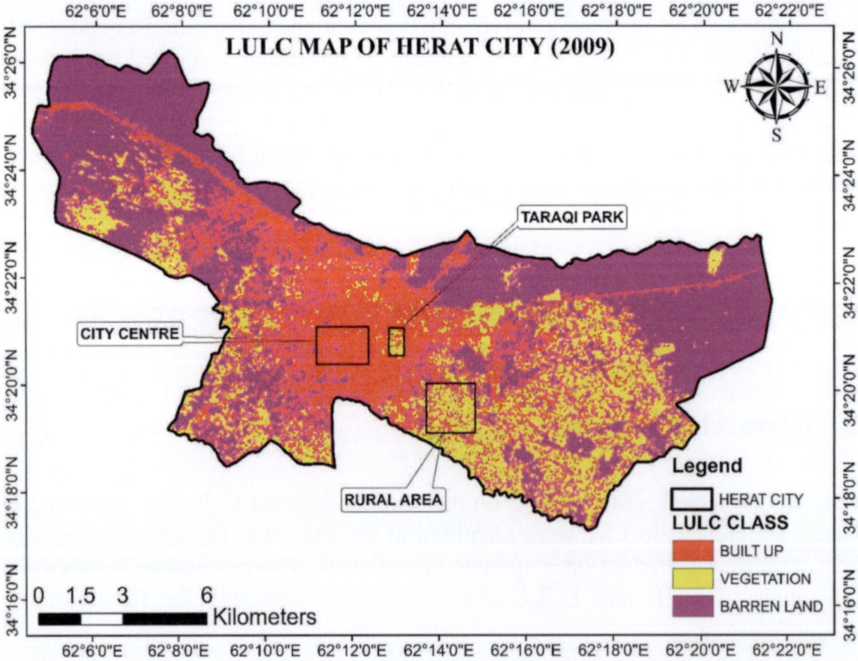

Fig. 2 LULC map of Herat for the year 2009

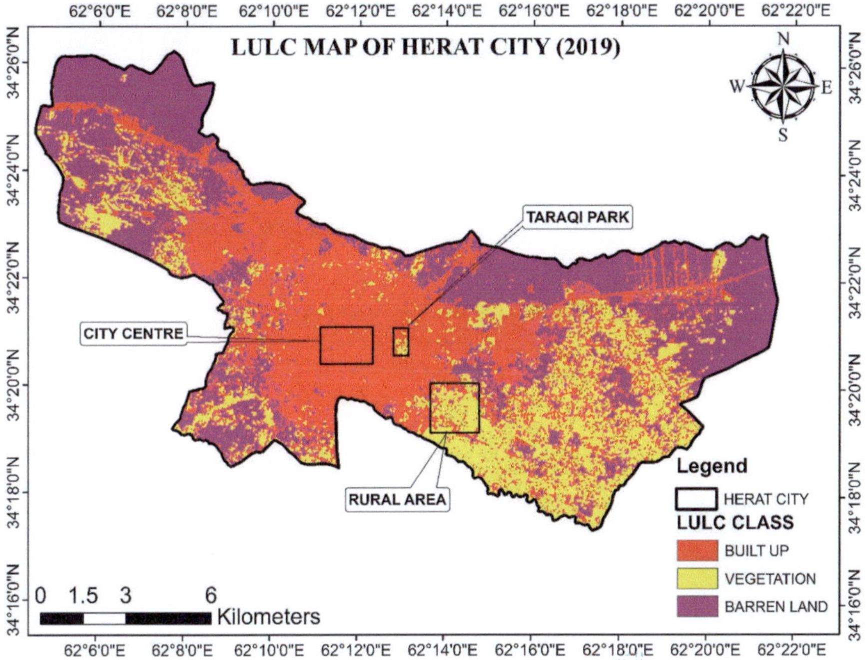

Fig. 3 LULC map of Herat for the year 2019

These findings collectively underscore the dynamic changes in land use and urbanization patterns within Herat, emphasizing the need for sustainable land management practices to mitigate environmental impacts and ensure urban resilience.

3.2 Variation in LST and Its Interrelationship with LULC

LST profiles for Herat were developed to investigate the seasonal and spatiotemporal variability across different LULC classes. Seasonal average LST values were calculated and mapped, as depicted in Fig. 4. During the summer seasons of 2009 and 2019, the average LSTs in Herat were recorded at 36 °C and 38.3 °C, respectively, indicating an increase of 2.3 °C over the decade. An analysis of the summer LST maps revealed a significant rise in minimum temperatures from 26.3 °C in 2009 to 30.5 °C in 2019. This observation underscores the intensification of urban heat within Herat during the summer months. Furthermore, the decadal analysis of summer LST profiles highlighted that the built-up areas experienced an average temperature increase of 2.1 °C, as detailed in Table 2. This trend reflects the thermal impact of urbanization and underscores the need for strategies to mitigate heat stress in rapidly developing urban environments like Herat.

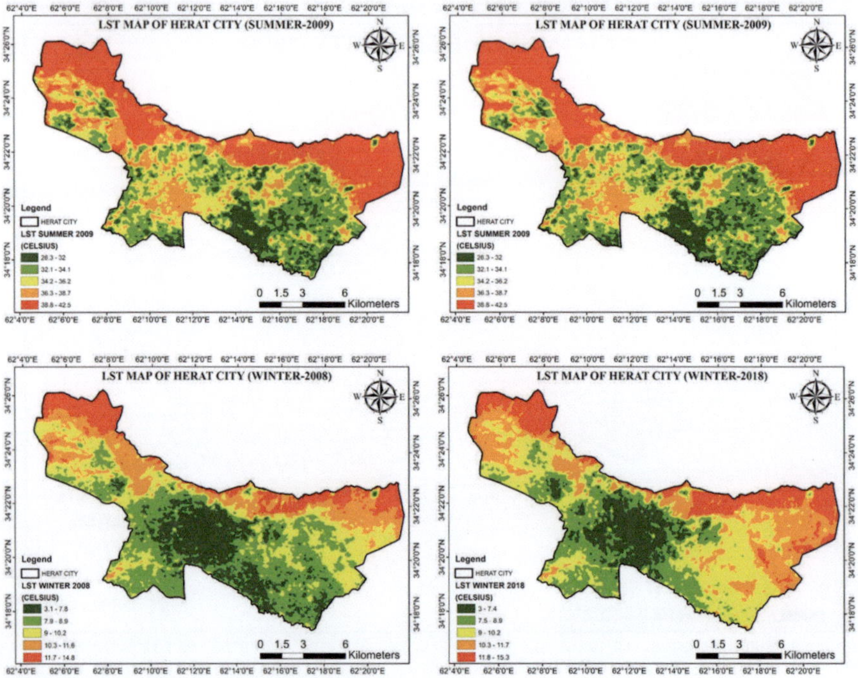

Fig. 4 Seasonal LST profile of Herat at the decadal interval

Table 2 Decadal seasonal changes in LST in different LULC classes (all values are in °C)

	Summer (2009)			Summer (2019)			Winter (2008)			Winter (2018)		
LULC class	Min	Mean	Max	Min	Mean	Max	Min	Mean	Max	Min	Mean	Max
Built-up	27.3	35.5	40.8	30.6	37.6	43.7	3.1	8.3	14.7	3.0	8.3	14.8
Vegetation	26.3	32.6	36.6	30.5	35.3	38.7	3.6	8.5	13.5	5.0	9.6	13.3
Barren land	27.7	37.5	42.5	32.0	40.4	45.8	5.0	10.0	14.8	4.3	10.8	15.3

The impact of LULC changes on the LST of the study area was examined over a decade-long time frame. Table 2 provides a detailed overview of the average LST values associated with each LULC class. Throughout the decadal study, barren lands and built-up areas consistently exhibited the highest average temperatures among all LULC classes. A notable observation was the conversion of 2679 hectares of barren land into built-up areas, indicating significant infrastructural development and increased human activities. This transformation was expected to influence the LST patterns, contributing to localized temperature increases within the urban fabric. Interestingly, areas adjacent to relatively dense vegetation exhibited temperatures lower than the average urban temperature. This suggests that vegetation can mitigate the urban heat effect by providing shade and evaporative cooling, highlighting the importance of green spaces in urban planning and temperature regulation. The findings underscore the complex interplay between land use changes and thermal dynamics in urban environments like the study area. Effective land management strategies and sustainable urban planning are crucial for mitigating the adverse impacts of urbanization on local climate and enhancing urban resilience to heat stress.

The decadal analysis of winter seasons (2008–2018) revealed an overall increase of 0.3 °C in the average LST of the city. Specifically, the average LST during the winter season rose from 9.3 °C in 2008 to 9.6 °C in 2018. Within the built-up class, which includes residential, commercial, and industrial areas and road networks, the LST remained relatively stable at 8.3 °C throughout both 2008 and 2018 (see Table 2). This consistency suggests that thermal properties within these urban infrastructures did not significantly change over the decade. Conversely, the LST of vegetated areas increased from 8.5 °C in 2008 to 9.6 °C in 2018, marking a noticeable rise of 1.1 °C over the decade. This increase highlights potential shifts in vegetation cover or density, which can affect local temperature dynamics during winter months. Distinct regions with elevated temperatures were consistently observed in barren lands during both the 2009 and 2019 seasons. This phenomenon can be attributed to the lower thermal capacity of barren land, which tends to absorb and retain heat more readily compared to vegetated or built-up areas. These findings underscore the complex interplay between land cover types and seasonal temperature variations within urban environments. Understanding these dynamics is crucial for informed urban planning and climate adaptation strategies aimed at mitigating the UHI effect and enhancing urban resilience to temperature extremes.

The LSTs for significant urban areas were analyzed for the summer and winter seasons. In the summer of 2019, the temperatures for all LULC classes were found to increase (by more than +2.1 °C). LSTs of the major urban centers of Herat city were also analyzed, which includes Masjid Jami, commercial area, administrative area, and densely built-up area. The seasonal variation in LST in the major urban centers of Herat has been outlined in Table 3. The LST of the Masjid Jami area had an average increase of 0.9 °C from 36.6 °C to 37.5 °C during the summer of 2009 and 2019, respectively. However, the LST of the commercial areas had shown no growth. The average LST of the administrative area changed from 31.8 °C to 35 °C, with an increase of 3.2 °C. LST of the densely built-up area changed from 34.4 °C to 35.8 °C, with a rise of 1.4 °C from the summer of 2009 to the summer of 2019.

Table 3 Average LST in the major urban centers of Herat (all values are in °C)

Prime points	Winter (2008)	Winter (2018)	Summer (2009)	Summer (2019)
Masjid Jami	7.3	7.0	36.6	37.5
Commercial area	6.9	6.2	35.0	35.0
Administrative area	6.1	7.1	31.8	35.0
Densely built-up area	7.0	6.4	34.4	35.8

3.3 Urban Heat Island Intensity

LULC changes play a significant role in urban climate dynamics, notably contributing to the UHI effect observed in megacity areas such as Herat, Afghanistan. The UHII was assessed by comparing mean and maximum temperatures between urban and rural areas, reflecting both average and peak UHI effects. Spatial analyses were conducted to examine UHI variations within the boundaries of Herat across different seasons. An analysis of LST images revealed distinct spatial patterns of UHI throughout the year. Notably, UHII was found to be the highest in the city's core, where dense urbanization and extensive human activities contribute to increased heat retention. During the summer season, elevated UHII levels were attributed to higher solâr radiation absorption and rapid heating of fallow barren lands and agricultural fields. These areas, lacking vegetation cover, heat up quickly due to their low thermal capacity, exacerbating local temperature disparities. Summer LST maps depicted a notable contrast: Urban areas appeared to be cooler islands amid the surrounding agricultural and barren lands, which functioned as heat islands. This thermal discrepancy is primarily attributed to the lower thermal inertia of open barren lands, composed of sandy soils that absorb and retain heat more readily. To further explore these dynamics, two distinct urban patches were analyzed: one located within the densely built-up area, characterized by intensive urbanization and high human activity, and another patch serving as a control within the rural outskirts of the city. This comparative approach provided insights into how different land use patterns influence local temperature variations and the spatial distribution of UHI effects within Herat. Understanding these complex interactions between land cover, urbanization, and temperature dynamics is essential for developing effective urban planning strategies aimed at mitigating heat stress and enhancing the resilience of cities like Herat to climate change impacts. As seen in the LULC maps of Herat, these patches were located at (1) the city center, consisting of commercial and residential areas; (2) Taraqi Park, comprising a combination of administrative buildings and vegetation; and (3) a rural patch, which was made up of less densely built-up areas and thicker vegetation cover and is farther away from the densely built-up urban area.

The rural area was taken as the baseline for UHII calculation in urban hotspot patches. The LST values were retrieved from Landsat satellite imageries over a decade (2008–2009 to 20,018–2019). Figs. 5, 6, 7 and 8 show the LST maps of two urban patches considered UHI hotspots and one rural patch considered the baseline. It is observed that during the summer of 2009, the temperature ranged between 26.3 °C and 42.5 °C, having a mean LST of 36.0 °C. The overall mean temperature showed an increasing trend during the summer of 2019, with a mean LST of 38.3 °C and the temperature ranging between 30.5 °C and 45.8 °C. Herat is the second fastest developing city in Afghanistan after the capital city of Kabul. The built-up area in the city has increased rapidly due to the construction of roadways, buildings, and other permanent structures. This has contributed to a rise in LST, further leading to the growth of UHII.

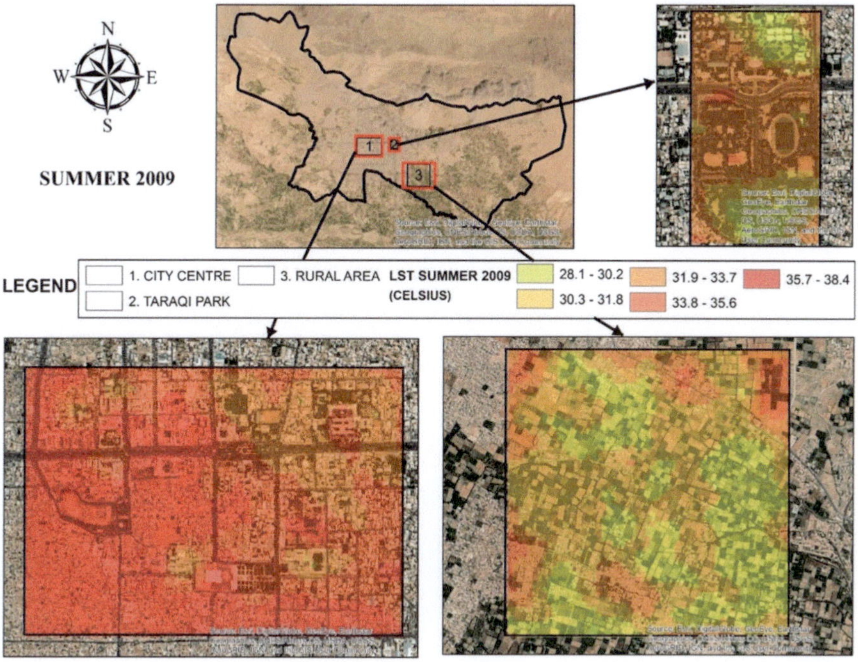

Fig. 5 LST map of urban patches in summer 2009

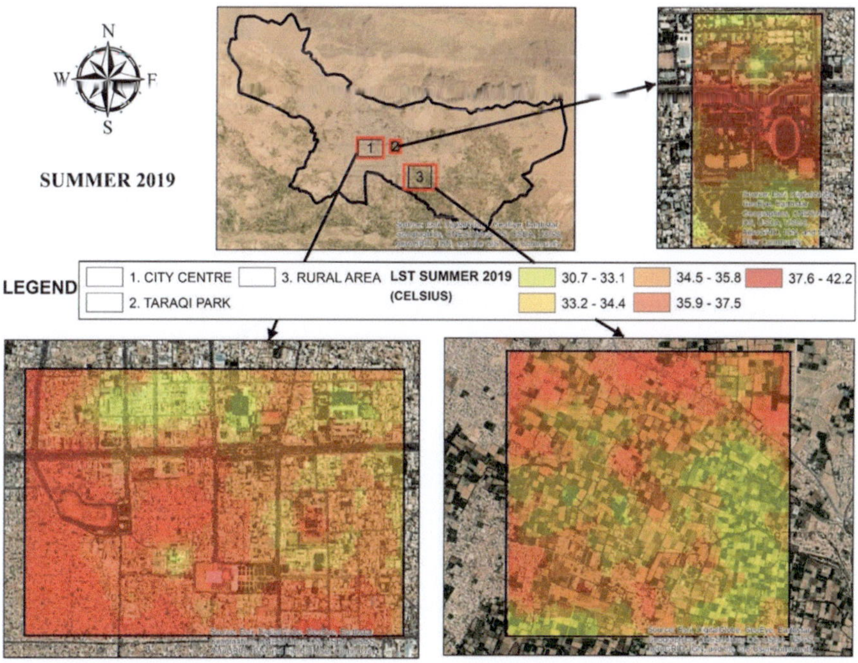

Fig. 6 LST map of urban patches in summer 2019

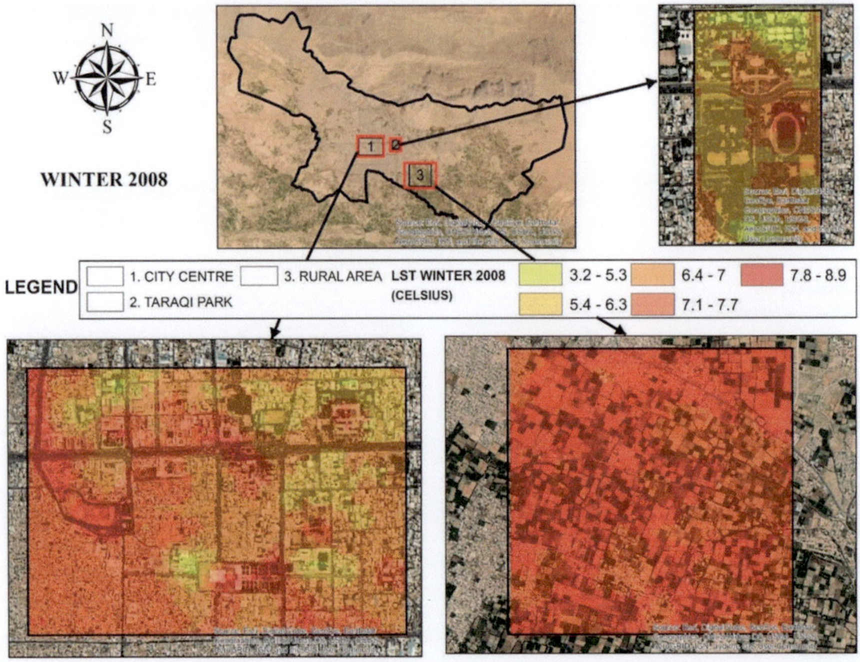

Fig. 7 LST map of urban patches in winter 2008

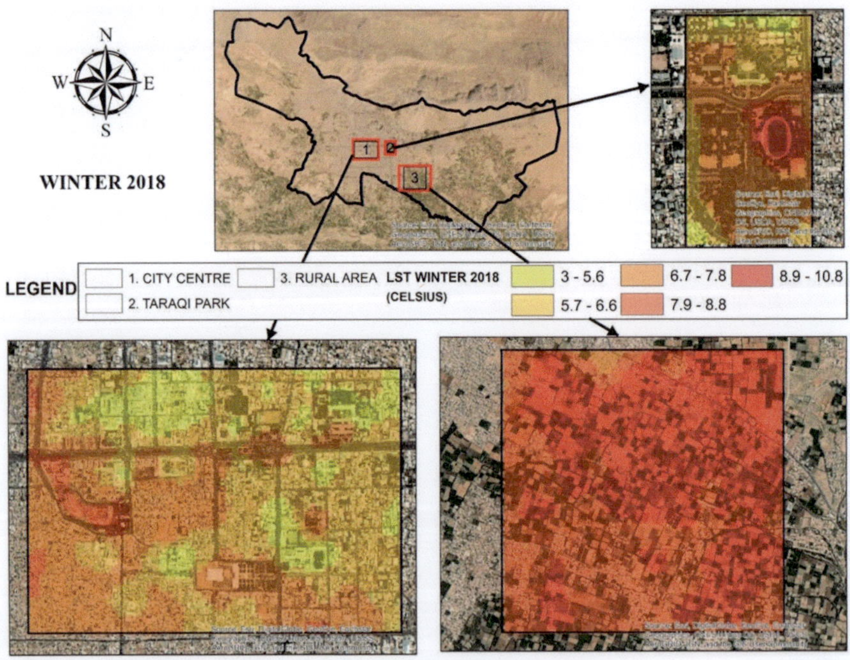

Fig. 8 LST map of urban patches in winter 2018

It was observed that during the winter of 2008, the temperature ranged between 3.1 °C and 14.8 °C, having a mean LST of 9.3 °C. The average LST increased during winter 2018, with a mean value of 9.7 °C. The temperature range in this season was 3.0–15.3 °C. The mean LST of patches showed an average increase of 1.9 °C for Herat from winter 2008 to winter 2018. Table 4 shows the LSTs of all selected urban patches.

The UHII showed seasonal and decadal level variability, as shown in Fig. 9. The UHII in the summer season for the study years (2009 and 2019) has been calculated. In the summer of 2009, the overall maximum UHII was 10.3 °C. The spatial variation in UHII in Herat for all patches was calculated by comparing it with the rural area patch. An extreme average UHII was observed in the city center, with a UHII of 3.6 °C. The average UHII of the administrative area (Taraqi Park) was 1.1 °C. The overall maximum UHII for the summer of 2019 was 11.5 °C. An extreme average UHII was observed in the city center area, with a UHII of 1.6 °C. The mean UHII showed an average increase of 2.3 °C for Herat from the summer of 2009 to the summer of 2019. The rural patch was the baseline for UHII calculation in the temporal study. The average UHII of the city center drastically changed from 4.6 °C to 1.6 °C from the summer of 2009 to the summer of 2019. This decrease in the UHII of the city center is due to the increase in built-up area from 2009 to 2019. The UHII of the administrative and recreational areas of Taraqi Park changed from 1.1 °C to 0.2 °C from the summer of 2009 to the summer of 2019. The decrease in intensity of the UHI of this patch was due to the change of vacant land to built-up areas and vegetation. A report presented by the Ministry of Urban Development Affairs shows that 20.7% of vacant plots are present in the built-up area of Herat (French et al., 2016).

The UHI phenomenon was not prominent in the winter season in Herat. The average LST in winter for each patch has been calculated. The rural patch showed a higher LST in comparison to urban patches. This reversal in temperature difference is known as Urban Cold Island (UCI). It may be formed due to the densely built-up area being colder than sparsely built-up rural areas. As discussed earlier, during winter, there is quicker and higher heating of the fallow agricultural land, compared to the features in the urban section of the city. The presence of vegetative cover within the city could also help normalize the adverse effects of the UHI phenomenon due to the lower albedo value of the vegetation cover. The UCI intensity for the winter season of 2008 was −5.7 °C. The spatial variation of UCI intensity within Herat for selected urban patches was obtained through a comparison with the rural patch. The extreme average UCI intensity was observed in the Taraqi Park area, with a cold island intensity of −1.7 °C. The UCI intensity of the city center was −0.9 °C. The overall maximum UCI intensity of winter 2018 was −7.8 °C. An extreme average UCI intensity was observed in the city center area, with a UCI intensity of −2.2 °C. The rural area was the baseline area for UCI intensity calculation in the temporal study. The average UCI intensity of the city center drastically changed from −0.9 °C to −1.3 °C from winter 2008 to winter 2018, respectively. The UCI intensity of the Taraqi Park area was −1.7 °C during the winter of 2008, which changed drastically to a UHII of 0.8 °C during the winter of 2018.

Table 4 Temporal changes in the summer and winter LSTs for patches of Herat city (all values are in ^{0}C)

Patch	Summer (2009)				Summer (2019)				Winter (2008)				Winter (2018)			
	Min	Mean	Max	Std Dev	Min	Mean	Max	Std Dev	Min	Mean	Max	Std Dev	Min	Mean	Max	Std Dev
City center	30.6	35.3	37.6	1.5	32.6	36.1	39.5	1.8	5.3	6.9	8.5	0.6	4.4	6.6	8.6	0.9
Taraqi Park	28.7	31.8	34.0	1.3	32.2	34.7	42.2	1.9	3.9	6.1	7.4	0.7	5	6.9	10.7	1
Rural area	28.2	30.7	34.8	1.3	31.4	34.5	38.1	1.6	7.1	7.8	8.7	0.3	7.9	8.8	9.9	0.4

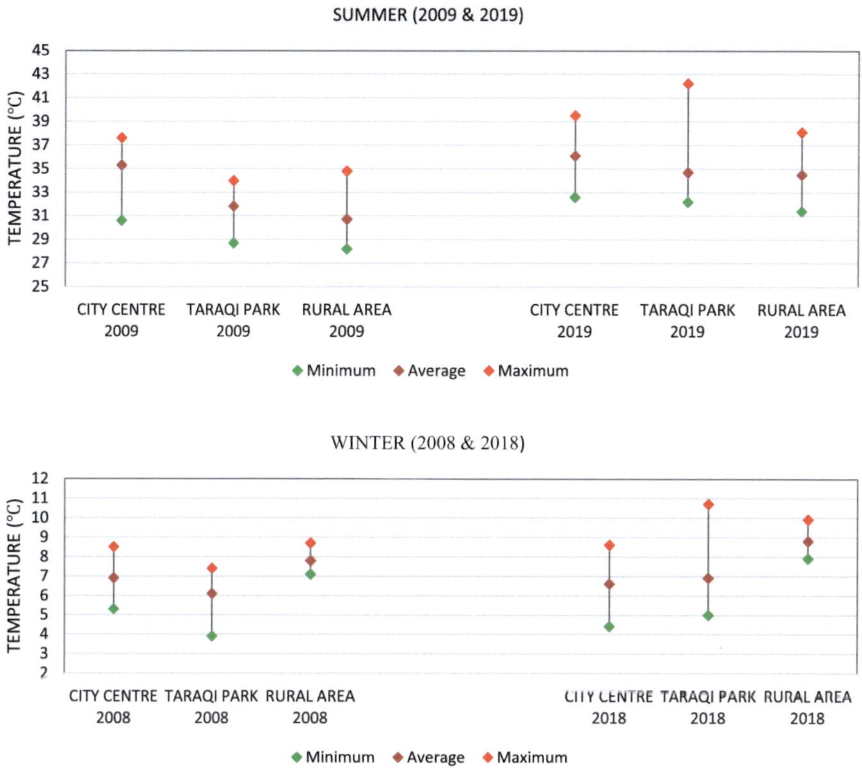

Fig. 9 Stock chart of UHII in UHI patches

Table 5 UHI intensities for different seasons (all values are in °C)

Patch	Winter (2008)	Summer (2009)	Winter (2018)	Summer (2019)
City center	−0.9	4.6	−2.2	1.6
Taraqi Park	−1.7	1.1	−1.9	0.2

To understand the seasonal changes of LST and UHI during the study period, LST (minimum, maximum, and mean) values were calculated for two seasons (Table 4). Table 5 summarizes the UHII for winter and summer. It can be observed from Table 5 that UHII was dominant in the summer of 2009 as compared to the winter of 2008. The UHI intensities changed from −0.9 °C to 4.6 °C for the city center from winter 2008 to summer 2009. A drastic increase of 5.5 °C in the intensity of UHI was observed for the city center. The Taraqi Park area also showed a rise of 2.8 °C in the intensity of UHI from winter 2008 to summer 2009. UHI was dominant in the summer of 2019 compared to the winter of 2018, with the city center showing a maximum intensity of 1.6 °C in the summer of 2019 and − 2.2 °C in the winter of 2018. The Taraqi Park area showed an increase of 2.1 °C from the winter of 2018 to the summer of 2019, with UHII changing from −1.9 °C to 0.2 °C from the winter of 2018 to the summer of 2019, respectively.

3.4 Discussion

The consequences of rapid urbanization and other accompanying anthropogenic activities, such as the increasing use of air conditioning systems, the use of more fossil fuels for energy generation and heating, and the increase in automobiles in the city, have led to the availability of more energy to heat the surface. The mean LST values of the LULC classes in summer 2009 were 32.6 °C in vegetation, 35.5 °C in built-up class, and 37.5 °C in barren land class. The mean LST of all the classes showed an increasing trend. The LST values of the LULC classes in summer 2019 were 35.3 °C in vegetation, 37.6 °C in built-up class, and 40.4 °C in barren land class. The change analysis for LST from summer 2009 to summer 2019 showed that the LST values of built-up class, vegetation, and barren land increased by 2.7 °C, 2.1 °C, and 2.9 °C, respectively. The cause of this increase in LST was the rapid expansion of the city and increased anthropogenic activities.

The effect of anthropogenic activities can be assumed to be an important factor for the overall increase in the surface temperature, as observed through the decadal analysis. A wide cover of land surface has been converted from barren land and vegetation to built-up area, with newly established residential, shopping, and administrative facilities. Therefore, the LST of Herat has changed and gradually increased from 2008 to 2019. UHII shows seasonal cycles and is controlled by weather-related conditions, such as wind, humidity, cloud conditions, and other factors like altitude, topography, and characteristics and concentration of atmospheric aerosols. Overall, the maximum UHII has been recorded as 11.5 °C during the summer season of 2019. The maximum UHII of Herat during different seasons varies from −5.7 °C to 11.5 °C. The maximum UHII varies from 10.3 °C to 11.5 °C during the summer season and from −7.8 °C to −5.7 °C during the winter season.

Tables 4 and 5 show the seasonal and temporal variations of mean UHII between patches of Herat by considering the rural area as the baseline. It can be figured out that the UHI pattern within Herat was not the same in every season considered during the study period. It has been observed that the city center was the hotspot of the city in the built-up area from the summer of 2009 to 2019. The maximum UHII was 4.6 °C during the summer of 2009. The leading causes of high UHII in the city's central area are dense urbanization, overconsumption of energy, and generation of anthropogenic heat.

The UHII of the considered patches of Herat varied seasonally. The variation of UHII was due to vacant plots and barren land inside and outside the city boundary. The UHII of the city center varied from −0.9 °C to −2.2 °C in the winter seasons of 2008 and 2018, respectively. There was an overall decline of −1.3 °C. This decline was due to increased built-up area, replacing barren land and vacant plots. Barren land and vacant plots with low thermal capacity can heat up quickly. So, when replaced by a built-up area, a longer time to heat up is seen due to the high thermal capacity of concrete surfaces. The presence of green cover within planned urban segments also plays an important role in reducing the UHII within the city. A decrease in UHII trend was observed for the city center in the summer season, with a 3 °C change in UHII from 4.6 °C to 1.6 °C in the summer of 2009 and 2019, respectively. The UHII of the Taraqi Park area changed with a change in LULC classes from 2008 to 2019. UHII of Taraqi Park was −1.7 °C in the winter of 2008 and − 1.9 °C in the winter of 2018. A similar decreasing trend was also observed (from 1.1 °C to 0.2 °C) for the summer season of 2009 and 2019. The UHI maps of winter 2008 and summer 2009 show that the locations of UHIs were well-distributed within the patch, but they are relatively concentrated in the UHI maps of winter 2018 and summer 2019. This difference in the pattern of UHI from 2008 to 2019 was due to the change in LULC classes in this particular time period.

4 Conclusion

The study was conducted to assess the spatiotemporal changes in UHII in Herat city of Afghanistan. LST maps were produced for each city using Landsat TM and OLI/TIRS data for 2008–2009 and 2018–2019, respectively. It has been found that UHI was dominant in the summer season throughout the study period. Second, an analysis of the LST pattern was carried out to understand the spatial and temporal changes in UHII. The intensity of UHI varied spatially within different segments of the city. This variation can be attributed to the density of the built-up area and associated anthropogenic activities in such areas. In all the cities, the city centers with densely built-up areas and high commercial activities were seen to have maximum UHI intensities. The temporal variation of UHI could be correlated to the change in LULC due to the depletion of barren/agricultural land or vegetative cover class to the built-up class. In general, the selected urban patches within the dense built-up area with high anthropogenic activities were more than 1.0 °C warmer in 2019 compared to the year 2009.

It was observed that three major classes of land cover were present in Herat, i.e., built-up area, vegetation, and barren land. The built-up area expanded by 61.7% from 2009 to 2019. There was an increase in the built-up area from 3905.8 hectares in 2009 to 6316.3 hectares in 2019. The barren land and vegetation conversion percentages were 25% and 18%, respectively. The UHII was found to be dominant in the summer season in Herat. The UHII in the city center was observed to be 4.6 °C and 1.6 °C in the summer of 2009 and 2019, respectively. The UHII was related to the LULC classes existing in the city and showed temporal changes. The spatial variation of UHI can be related to the density of buildings and the ongoing anthropogenic activities in the study area. The presence of vacant plots in the city also played a role in the formation of UHIs in the city. The temporal variation in UHII could be related to the conversion of vegetation, water, and barren land classes to the built-up class. It has been observed that the UHI changed with the change of barren land and vacant plots in the study period (2009–2019). The increase in anthropogenic activities with time is also expected to play a major role in the temporal changes of UHI.

The work has laid the foundation for quantifying UHII and understanding the pattern of UHI formation within Herat city using geospatial techniques. This would help in the planning and development of sustainable and environmentally friendly zones and the adoption of suitable regulatory measures in order to develop well-planned smart cities in the future. In addition, research may be carried out over a longer period of time to capture the changes in daytime and nighttime UHI. Depending on the existing climatic conditions, the UHI within and between urban micro-locations could be much higher or lower due to multiple intensive land use patterns within a city. If possible, future analysis should also include ground truth data to understand such local changes within selected urban patches during the calculation of UHII.

References

Akbari, H., & Matthews, H. D. (2012). Global cooling updates: Reflective roofs and pavements. *Energy and Buildings, 55*, 2–6. https://doi.org/10.1016/j.enbuild.2012.02.055

Akbari, H., Pomerantz, M., & Taha, H. (2001). Cool surfaces and shade trees to reduce energy use and improve air quality in urban areas. *Solar Energy, 70*, 295–310. https://doi.org/10.1016/S0038-092X(00)00089-X

Akbari, H., Cartalis, C., Kolokotsa, D., Muscio, A., Pisello, A. L., Rossi, F., Santamouris, M., Synnef, A., Wong, N. H., & Zinzi, M. (2016). Local climate change and urban heat Island mitigation techniques—The state of the art. *Journal of Civil Engineering and Management, 22*, 1–16. https://doi.org/10.3846/13923730.2015.1111934

Anderson, G. B., & Bell, M. L. (2011). Heat waves in the United States: Mortality risk during heat waves and effect modification by heat wave characteristics in 43 U.S. Communities. *Environmental Health Perspectives, 119*, 210–218. https://doi.org/10.1289/ehp.1002313

Chang, Y., Xiao, J., Li, X., & Weng, Q. (2023). Monitoring diurnal dynamics of surface urban heat Island for urban agglomerations using ECOSTRESS land surface temperature observations. *Sustainable Cities and Society, 98*, 104833. https://doi.org/10.1016/j.scs.2023.104833

Dutta, K., Basu, D., & Agrawal, S. (2019). Nocturnal and diurnal trends of surface urban heat Island intensity: A seasonal variability analysis for smart urban planning. ISPRS annals of the photogrammetry. *Remote Sensing and Spatial Information Sciences, IV-5/W2*, 25–33. https://doi.org/10.5194/isprs-annals-IV-5-W2-25-2019

Dutta, K., Basu, D., & Agrawal, S. (2021). Synergetic interaction between spatial land cover dynamics and expanding urban heat islands. *Environmental Monitoring and Assessment, 193*, 184. https://doi.org/10.1007/s10661-021-08969-4

Dutta, K., Basu, D., & Agrawal, S. (2022). Evaluation of seasonal variability in magnitude of urban heat islands using local climate zone classification and surface albedo. *International journal of Environmental Science and Technology, 19*, 8677–8698. https://doi.org/10.1007/s13762-021-03602-w

Easterling, D. R., Horton, B., Jones, P. D., Peterson, T. C., Karl, T. R., Parker, D. E., Salinger, M. J., Razuvayev, V., Jamason, P., Folland, C. K., Easterling, D. R., Horton, B., Jones, P. D., Peterson, T. C., Karl, T. R., Parker, D. E., Salinger, M. J., Razuvayev, V., Plummer, N., Jamason, P., & Folland, C. K. (1997). Maximum and minimum temperature trends for the globe. *Science, 277*, 364–367.

French, M., Turkstra, J., & Farid, M. (2016). Vacant land plots in afghan cities: A problem and an opportunity. *Urbanisation, 1*, 79–94. https://doi.org/10.1177/2455747116671825

Grimm, N. B., Faeth, S. H., Golubiewski, N. E., Redman, C. L., Wu, J., Bai, X., & Briggs, J. M. (2008). Global change and the ecology of cities. *Science, 319*, 756–760. https://doi.org/10.1126/science.1150195

Jaiswal, N., Deb, S. K., Panda, S. K., Mandal, A. K., Khan, A. W., & Kishtawal, C. M. (2022). Investigating intra-urban thermal variability of Ahmedabad, India: Heat wave prediction perspective. *Journal of the Indian Society of Remote Sensing, 50*, 1903–1913. https://doi.org/10.1007/s12524-022-01570-3

Kristy, G. (2018). The impact of urban sprawl on cultural heritage in Herat, Afghanistan: A GIS analysis. *Digital Applications in Archaeology and Cultural Heritage, 11*, e00086. https://doi.org/10.1016/j.daach.2018.e00086

Landsat 8 (L8) Data Users Handbook. (2019). 8.

Lazzarini, M., Marpu, P. R., & Ghedira, H. (2013). Temperature-land cover interactions: The inversion of urban heat Island phenomenon in desert city areas. *Remote Sensing of Environment, 130*, 136–152. https://doi.org/10.1016/j.rse.2012.11.007

Malcoti, M. D., Zia, H., Kabre, C., Hang, H. T., Shahfahad, & Rahman, A. (2023). Analysis of urban streets and surface thermal characteristics using thermal imaging camera in residential streets of Gurugram City, India. *Environmental Science and Pollution Research, 30*, 86892–86910. https://doi.org/10.1007/s11356-023-28553-2

Nasery, S., Matci, D. K., & Avdan, U. (2021). GIS-based wind farm suitability assessment using fuzzy AHP multi-criteria approach: The case of Herat, Afghanistan. *Arabian Journal of Geosciences, 14*, 1091. https://doi.org/10.1007/s12517-021-07478-5

Oke, T. R. (1995). The Heat Island of the urban boundary layer: Characteristics, causes and effects. In *Wind climate in cities* (pp. 81–107). Springer. https://doi.org/10.1007/978-94-017-3686-2_5

Samimi, S. A. B., & Maraschin, C. (2022). Built heritage and urban spatial configuration: The case of Herat, Afghanistan. *GeoJournal, 88*, 2101–2120. https://doi.org/10.1007/s10708-022-10738-w

Samimi, S. A. B., Ando, T., & Kawish, K. (2019). Analysis of the transformation of Herat Old City, Afghanistan. *Conservation and Management of Archaeological Sites, 21*, 143–159. https://doi.org/10.1080/13505033.2019.1664626

Tursilowati, L., Sumantyo, J. T. S., Kuze, H., & Adiningsih, E. S. (2012). Relationship between urban Heat Island phenomenon and land use/land cover changes in Jakarta—Indonesia. *Journal of Emerging Trends in Engineering and Applied Sciences, 3*, 645–653. https://doi.org/10.10520/EJC126643

Voogt, J., & Oke, T. (2003). Thermal remote sensing of urban climates. *Remote Sensing of Environment, 86*, 370–384. https://doi.org/10.1016/S0034-4257(03)00079-8

Wang, S., Ma, Q., Ding, H., & Liang, H. (2018). Detection of urban expansion and land surface temperature change using multi-temporal landsat images. *Resources, Conservation and Recycling, 128*, 526–534. https://doi.org/10.1016/j.resconrec.2016.05.011

Environmental Determinants of Land Surface Temperature Variations: A Case of Kolkata Urban Agglomeration

Ratnadeep Dutta, Suvamoy Pramanik, and Milap Punia

Abstract This chapter explores environmental determinants influencing land surface temperature (LST) in the Kolkata Urban Agglomeration (KUA). We categorized the determinants as surface variables (such as fractional vegetation cover, normalized difference vegetation index, built-up height, sky view factors, impervious surfaces, and population density) and atmospheric variables (such as aerosol optical depth (AOD), sulfur dioxide, cardon dioxide, and nitrogen dioxide) for finding the dominant determinants, which are not frequently selected in contemporary LST studies. We unfolded different developmental zones and tried to determine whether there are any differential effects of selected determinants upon LST. Our results show that impervious surfaces, such as roads and buildings, significantly impacted LST, with areas with a higher density of impervious surfaces showing higher LST. Reduced sky view leads to higher net heat storage within buildings and exacerbates the urban heat island (UHI) effect. Air pollutants like NO_2, SO_2, and CO correlate positively with nighttime LST. The findings emphasize the importance of careful urbanization and economic development planning to mitigate the UHI effect and reduce LST.

Keywords Fractional vegetation cover · Air pollutants (NO_2 · SO_2 · CO) · Urban agglomeration · Urban heat island (UHI) · Developmental zones

---

R. Dutta · M. Punia
Centre for the Study of Regional Development, School of Social Sciences, Jawaharlal Nehru University, New Delhi, India

S. Pramanik (✉)
Centre for the Study of Regional Development, School of Social Sciences, Jawaharlal Nehru University, New Delhi, India

Department of Geography, Shaheed Bhagat Singh Evening College, University of Delhi, New Delhi, India
e-mail: suvamo60_ssf@jnu.ac.in; suvamoy.pramanik@sbse.du.ac.in

P. K. Joshi et al. (eds.), *Geospatial Science for Urban Ecosystems*, The Urban Book Series, https://doi.org/10.1007/978-3-032-07155-2_22

1 Introduction

The global trend toward urbanization is accelerating rapidly, projected to grow annually at a rate of 1.85% (United Nations Department of Economic and Social Affairs, 2018). Currently, over half of the world's population resides in urban areas, a statistic poised to persist and expand. According to the United Nations, by 2050, urban dwellers are expected to constitute two-thirds of the global population (United Nations Department of Economic and Social Affairs, 2018). This profound shift underscores the ongoing transformation of human settlement patterns toward increasingly urbanized environments, necessitating careful consideration of urban planning and sustainability practices to meet the evolving needs of urban populations worldwide. Asia and Africa will be the major hotspots. India has recently become the most populous country in the world, surpassing China. India boasts 53 cities, each with populations exceeding one million inhabitants, positioning it among the most urbanized nations globally (Census of India, 2011). As per the Census of India in 2011, Kolkata ranks as the third most populous urban agglomeration in the country. Kolkata Urban Agglomeration (KUA) stands out as one of India's largest, encompassing Kolkata city along with its extensive network of suburbs and surrounding towns. This sprawling metropolitan area not only serves as a major cultural and economic hub but also faces significant urbanization challenges and opportunities, necessitating strategic urban planning and sustainable development initiatives to cater to its burgeoning population and ensure inclusive growth. Although KUA ranks third in the list of Indian megacities after Delhi and Mumbai, it has a larger potential impact on the socio-economic development of eastern India.

Even though rapid urbanization is frequently linked to social and economic advancement, it has also accelerated the exploitation of nature and its resources, leading to environmental degradation (Bai et al., 2017; McKinney, 2002; Rahim et al., 2022). One of the adverse impacts affecting cities is the heightened temperatures of both land and air compared to their surrounding areas, often referred to as the urban heat island (UHI) effect. The UHI effect occurs due to changes made to the natural environment and the use of materials that absorb and store heat (Chakraborty et al., 2021; Deilami et al., 2018; Leal Filho et al., 2021), such as concrete and asphalt, instead of natural vegetation (i.e., high thermal inertia materials). Furthermore, the concentration of human activities and the built environment in urban areas generates heat from various sources, including transportation, residential and business activities, and industries, further exacerbating the UHI effect.

Numerous studies globally have investigated the relationship between the UHI effect, LST, and various spatial and ecological indicators. (Halder et al., 2021; Kayet et al., 2016). One approach has been to examine the influence of vegetation and water bodies on LST using spectral indices such as normalized difference vegetation index (NDVI), normalized difference water index (NDWI), and normalized difference built-up index (NDBI). These studies have found a visible correlation between these indicators and the UHI effect, indicating the cooling effect of vegetation and water bodies (Malik et al., 2019). For instance, high NDVI values, indicating dense green vegetation, have been linked to lower LST values. This suggests that vegetation plays a crucial role in mitigating the UHI effect by offering shade and cooling through transpiration. (Gunawardena et al., 2017; Pramanik & Punia, 2019). Similarly, high NDWI values, which indicate a high density of water bodies, have been associated with lower LST values, suggesting that water bodies can also help mitigate the UHI effect by providing evaporative cooling. Other studies have focused on the landscape metrics as the key component and have tried to understand the LST as an effect of different land use categories (Asadi et al., 2019; Buyantuyev & Wu, 2010; Dai et al., 2018; Shao & Ge, 2020). These studies indicate that urban areas characterized by high densities of impervious surfaces, such as roads and buildings, typically exhibit higher LST values.

In contrast, regions characterized by dense vegetation and water bodies generally exhibit lower LST values. Vegetation provides shade and cooling through transpiration, while water bodies contribute to evaporative cooling, thereby mitigating the UHI effect. Moreover, research indicates that the spatial configuration of land use categories plays a crucial role in influencing LST. Fragmented landscapes, where natural features are interspersed with urban development, often experience higher LST values. This is attributed to the diminished cooling capacity of fragmented green spaces and water bodies, which are essential for regulating local temperature extremes in urban environments (Dahanayake et al., 2022; Zhou et al., 2011). Collectively, these studies underscore the critical role of integrating ecological and spatial indicators into urban planning and design practices. By doing so, cities can effectively address and alleviate the adverse effects of the UHI phenomenon. Strategies that prioritize the preservation and enhancement of green spaces, such as parks, urban forests, and vegetated corridors, contribute significantly to reducing local temperatures by providing shade, enhancing evaporative cooling, and improving air quality. Moreover, incorporating water features like ponds, lakes, and green infrastructure can further enhance cooling effects and resilience to climate impacts. In addition to ecological considerations, optimizing the spatial layout of urban areas is crucial. Compact and mixed-use developments that prioritize pedestrian-friendly streetscapes, reduce the extent of impervious surfaces, and encourage efficient public transportation systems help mitigate heat buildup. Furthermore, promoting green roofs, cool roofs, and reflective surfaces in building designs can significantly reduce surface temperatures and energy consumption for cooling. By adopting these integrated approaches, cities can foster more sustainable and livable urban environments that are resilient to climate change while enhancing the well-being and quality of life for their residents.

The UHI effect is more complex with its determining factors. For example, not only the NDVI and NDWI but also other components like population density, urban morphology, and air pollution are quite responsible for enhancing the UHI effects (Cao et al., 2016; Deilami et al., 2018; Li et al., 2020; Zhou & Chen, 2018). Considering all those cofounding factors and their interactions eventually becomes necessary to understand that complex interrelation. The higher the building height and the lower the sky view factors, the more urban canyon effects are generated, which are found to be most responsible for nocturnal heat islands (Chen et al., 2021;

Loughner et al., 2012; Theeuwes et al., 2014). The air pollution generated from various sources, including vehicles, industries, and agricultural processes, can contribute to the UHI phenomenon by creating greenhouse effects and elevating the temperature of the local environment (Mohajerani et al., 2017; Mohan Kumar et al., 2022; Wu et al., 2017). Urbanization is a dynamic process where the distinct rural-urban boundary is not noticeable; rather, a continuum of gradation or zonation is found from the urban center to its periphery. This continuum represents the human activities (i.e., changes in economic activities, growth in physical infrastructure, and other anthropogenic activities) for different developmental phases. We hypothesize that this developmental zone creates significant differences in influence on the thermal environment or UHI, as different cofounding factors also change with respect to different development zones. With this context, the objectives of the present articles are (1) to demarcate different developmental zones in KUA, (2) to understand the spatial pattern of thermal conditions and various environmental determinants as per developmental zones, and (3) to understand if there is any difference in environmental determinants' association with the thermal environment.

1.1 Database and Methodology

1.1.1 Study Area

Kolkata Urban Agglomeration (KUA) ranks as India's third-largest after Mumbai and Delhi. It encompasses the Kolkata Metropolitan Area (KMA), comprising 4 municipal corporations, 37 municipalities, and 23 Panchayat Samities (PS) across North 24 Parganas, South 24 Parganas, Howrah, and Hooghly districts. The KUA exhibits a diverse urban morphology, with the Kolkata Metropolitan Development Authority (KMDA) playing a pivotal role in its development and management. The region also has several industrial zones, including the Haldia Port and the Kolkata Port. The urban agglomeration has witnessed a moderate growth rate compared to its predecessors, Delhi and Mumbai. According to the 2011 Census, the KUA housed approximately 15.87 million residents, accounting for a significant portion of West Bengal's total urban population of about 29 million. By 2025, the population of KUA is projected to reach 21.1 million. Understanding temperature fluctuations within KUA is crucial for studying the UHI effect, as the region consistently registers higher temperatures compared to its rural surroundings. This temperature disparity is influenced by various factors, including urban morphology and anthropogenic activities.

1.1.2 Data Collection

The LST is the primary parameter under investigation. For our study, the LST for both day and night was derived from the Moderate Resolution Imaging Spectroradiometer (MODIS) daytime and nighttime LST bands. The derivation is given in the methodology section. The data have been acquired for the year 2022, the year of investigation, and the average value of October is used. The environmental determinants explain the spatial variations in LST in the study area. These environmental determinants are mentioned in Table 1.

1.1.3 Methodology

Regression analysis was employed to assess how various surface and atmospheric variables influence LST. A moving window or grid of 1 km × 1 km was taken as the unit of analysis to ensure the coverage of the whole study area. The LST was derived for each unit. The landscape characteristics using the urban morphology were also calculated for each unit, along with different indicators of air pollution. These indicators were taken as independent variables. Multiple linear regression (MLR) was then employed, where each grid (of area 1 km^2) represented the observation points for the dependent and predictor variables. Then, the implications of the regression equation were interpreted and discussed.

Table 1 List of environmental variables used for multiple linear regression (MLR)

Environmental determinants	Source	Spatial resolution (m)
Suface variables		
Fractional vegetation cover (FVC)	Landsat 8 OLI	30
Normalized difference water index (NDWI)	Landsat 8 OLI	30
Building height (BH)	GHSL: Global human settlement layers	38
Sky view factor (SVF)		
Impervous surfaces		
Population density (PD)	WorldPop global project population data	100
Atmospheric variables		
Aerosol optical depth (AOD)	MODIS Terra and aqua combined multi-angle implementation of atmospheric correction (MAIAC)	1000
Carbon monoxide (CO)	Sentinel-5P NRTI CO: Near real-time carbon monoxide	1113.2
Nitrogen dioxide (NO_2)	Sentinel-5P NRTI NO_2: Near real-time nitrogen dioxide	1113.2
Sulfur dioxide (SO_2)	Sentinel-5P NRTI SO_2: Near real-time sulfur dioxide	1113.2

1. Zonation of Economic and Human Activities
   The study area was divided into zones of economic development and human interaction using a unique approach that used satellite-based nighttime light data as a proxy indicator of human activity and development. Specifically, the Visible Infrared Imaging Radiometer Suite (VIIRS) Day/Night Band (DNB) was used to obtain nighttime data, and the K-means clustering technique was employed to statistically divide the study area into three zones. These zones included the urban core area with higher economic development and anthropogenic activities, the peri-urban urban fringe area with less activity, and the peripheral rural areas with the least economic and human activities.
2. Preparation of Environmental Variables of LST
   (a) Fractional Vegetation Cover (FVC)
   FVC represents the percentage of green space within any given grid. To derive the vegetation cover fraction, the NDVI was used. Vegetation cover was extracted from the normalized difference vegetation index (NDVI). The area of the vegetation pixels within each 1 km × 1 km grid was then calculated to determine the FVC using the following equations:

$$NDVI = \frac{NIR - Red}{NIR + Red} \quad (1)$$

where NIR and red refer to the near-infrared (band 5) and red (band 4) bands of Landsat 8 imagery, respectively.

$$\text{FVC} = \frac{\text{Area of vegetation cover}}{\text{Area of the grid}} \times 100\% \quad (2)$$

   (b) Normalized Difference Water Index (NDWI)
   The normalized difference water index (NDWI) was used as an indicator to understand the pattern of LST. NDWI highlights water pixels in an image, with pixel values ranging from −1 to +1, where positive values indicate the presence of water. Average NDWI values were calculated within each grid using the zonal statistics tool. The following formula was used to derive NDWI:

$$\text{NDWI} = \frac{\text{Green} - \text{NIR}}{\text{Green} + \text{NIR}} \quad (3)$$

where green and NIR refer to green (band 3) and near-infrared (band 5) bands of Landsat 8 imagery, respectively.
   (c) Normlized Difference Built-up Index (NDBI)
   The presence of impervious built structures is observed to influence the LST, which is one major cause of UHI. Therefore, the normalized difference built-up index (NDBI) has been used in our study to indicate the amount of built environment within a grid. The average NDBI was aggregated for each grid using the zonal statistics tool, with higher average NDBI values

indicating a higher presence of impervious surfaces. NDBI is derived using the following equation:

$$\mathrm{NDWI} = \frac{\mathrm{SWIR} - \mathrm{NIR}}{\mathrm{SWIR} + \mathrm{NIR}} \tag{4}$$

where SWIR and NIR refer to the shortwave infrared (band 6) and near-infrared (band 5) bands of Landsat 8 imagery, respectively.

(d) Sky View Factor (SVF)

The sky view factor (SVF) is a measure of the proportion of visible sky from a given location, representing the fraction of visible sky. It is crucial for calculating the surface radiation balance, as it affects the amount of solar radiation reaching the ground. Lower SVF values can lead to higher urban heat island (UHI) intensity, as less radiation is reflected and more is absorbed by the built environment. To estimate the SVF from a point in a street canyon, the following formula, developed by De Wolff (Hämmerle et al., 2011), can be used:

$$\mathrm{SVF} = \cos\left(\arctan\left(\left[\frac{H}{0.5W}\right]\right)\right) \tag{5}$$

where H is the height of the obstacle or buildings, and W is the distance between the obstacles. The average building height per grid was derived from the Global Human Settlement Layer (GHSL, 2018).

The remaining variables, namely AOD, NO_2, SO_2, CO, PD, and BH, were readily available in their respective datasets and required no additional processing.

3. Analysis of Environmental Determinants and LST Using Multiple Linear Regression

Multiple linear regression (MLR) was applied to explore the correlation between LST and selected surface and atmospheric variables, including FVC, NDWI, AOD, CO, NO_2, SO_2, PD, BH, and SVF. NDBI was removed from the MLR model due to multicollinearity. Both daytime and nighttime LST values were used as dependent variables to investigate potential differences in the influence of the independent variables on LST between these periods. Mean values were utilized for all variables. The objective was to identify significant relationships between the environmental determinants and LST, evaluating the strength and direction of these relationships. This analysis aims to uncover factors influencing LST, offering insights for developing strategies to mitigate heat stress in urban areas.

2 Results

2.1 Development Zones

The total area of KUA has been divided into three zones using cluster analysis based on satellite-derived nighttime light data (Fig. 1a). These zones give a genuine idea of economic and infrastructural development and human activities.

2.1.1 Zone 1

Zone 1 is the urban core area in the center with high-density built-up. This region mainly comprises the Kolkata Municipal Corporation area and parts of its adjacent urban centers, namely Dumdum and Bidhannagar. Zone 1 has an area of around 217 km^2, the lowest of the three zones. This zone has the highest population within the study area, with a higher density of settlements, a nonagricultural economy, better economic facilities, and enhanced services and living standards.

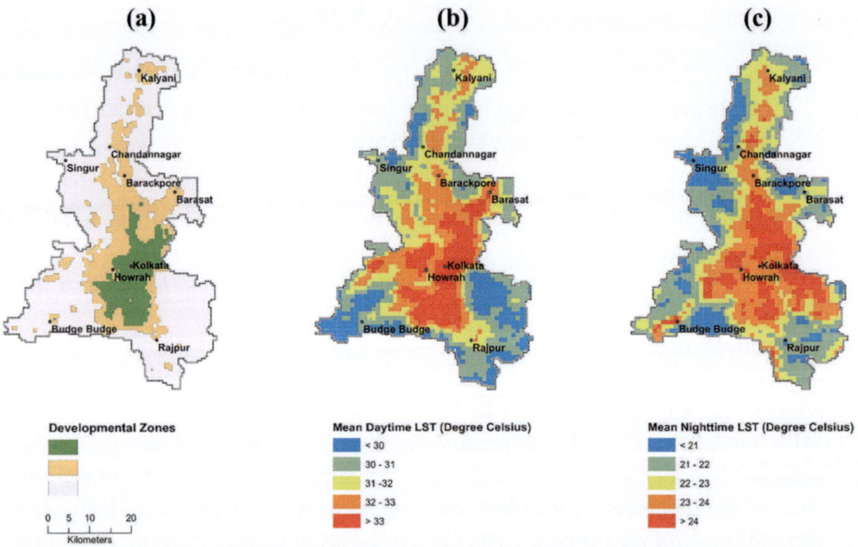

Fig. 1 (**a**) Identified development zones and spatiality of land surface temperature across the zones, (**b**) daytime, and (**c**) nighttime

2.1.2 Zone 2

This zone is characterized by a moderate density of built-up areas comprising the satellite towns of Kolkata. With an area of around 455 km2, this region spreads longitudinally along both banks of the Hooghly River. Here, population density is less than in the core area. Although this region's economic structure comprises small and medium enterprises and administrative units, it largely depends on the state capital, Kolkata.

2.1.3 Zone 3

Beyond Zone 2, there is a visible change in the land use characteristics. This is the boundary region of the KUA with the highest area among the three zones (about 1204 km^2). The region comprises the "less urban" part of KUA with a lesser density of population, sparsely built settlements, agricultural fields, and a less dense transport network. The economic and service sectors in these sections of the study area heavily rely on the smaller towns within the second zone.

2.2 *Characteristics and Spatiality*

A distinctive distribution of land surface temperature (LST) across the development zones (Fig. 1a) is evident across the study area, revealing a visually discernible relationship with development zones and notable diurnal variations. Throughout the study period, average daytime LST was 31.82 degrees Celsius, peaking at 36.42 degrees. Elevated daytime LST values (Fig. 1b) are concentrated in core areas characterized by dense settlement, higher population density, and limited vegetation and water bodies, notably including Kolkata and surrounding cities like Howrah and Barasat. Conversely, maximum nighttime LST (Fig. 1b) was 25.58 degrees Celsius, approximately 10 degrees lower than daytime peaks, with an average nighttime temperature of 22.33 degrees Celsius. During nighttime hours, inner zones adjacent to these core areas experience elevated LST, attributed to residual heat absorption from daytime conditions.

Table 2 summarizes the monitored characteristics of LST and the selected variables. Population density (PD) and building height (BH) show high variations (>100% of the mean) and have relatively few high values (positively skewed). Figure 2 displays the spatial variability of the selected indicators. Variables such as CO concentration, NO_2 concentration, and SO_2 concentration are closely associated with the urban area, depicting the source of these pollutants. The high vehicular and industrial emissions generate these pollutants, mostly concentrated within Kolkata city. Conversely, NDWI, vegetation coverage, and sky view factor exhibit higher values in the surrounding peri-urban and rural areas, where building heights are naturally lower. The distribution of AOD, however, shows a random pattern. The

Table 2 Summary statistics for the dependent and independent variables ($n = 2575$)

Variables	Minimum	Maximum	Median	Mean ± SD
LST_D	27.82684085	36.41625	31.803333	31.8433 ± 1.53
LST_N	19.91584959	25.58765394	22.241623	22.3347 ± 0.93
FVC	0.001752644	0.867306472	0.420895	0.4156 ± 0.19
NDWI	−0.071758519	0.25164978	0.110811	0.10490 ± 0.05
AOD	125.9172176	665.7000471	441.509759	440.5262 ± 60.71
CO	0.034089902	0.039132851	0.036633	0.03671 ± 0.001
NO_2	3.14E-05	8.04E-05	0.000050	0.000051 ± 0.00001
SO_2	0.000139602	0.000391177	0.000255	0.0003 ± 0.00004
PD	4.222898997	1839.644487	25.799016	90.7398 ± 166.71
BH	0	9.249195484	0.739777	1.4912 ± 1.70
SVF	0.975066775	1	0.995786	0.9954 ± 0.004

urban, peri-urban, and rural areas of different compactness are associated with high and low aerosol concentration values. This is because aerosol pollution occurs due to various causes, namely fossil fuel burning, burning of wood, vehicular exhaust, irrigation mist, smoke, dust, etc. The Kolkata region's spatial distribution of environmental variables reflects a complex interplay of urban, peri-urban, and rural dynamics. This spatial heterogeneity underscores the need for comprehensive environmental management strategies tailored to the specific characteristics of each area, taking into account factors such as land use, emission sources, and climate patterns.

2.3 Regression Results

A multiple linear regression was applied where the dependent variable is LST. The independent variables are FVC, NDWI, AOD, CO, NO_2, SO_2, PD, BH, and SVF. The general equation for multiple linear regression used to derive estimation models for LST is presented as:

$$LST(D/N) = \beta_0 + \beta_1.FVC + \beta_2.NDWI + \beta_3.CO \\ +\beta_4.AOD + \beta_5.NO_2 + \beta_6.SO_2 + \beta_7.PD + \beta_8.BH + \beta_9.SVF \tag{6}$$

where β_0 is the intercept, $\beta_{i(1-10)}$ is the slope for the independent variables.

The results of the analysis help to point out the significant factors that have a definite nonzero influence on the LST and link it with the spatially located source and sink points (Pramanik & Punia, 2020).

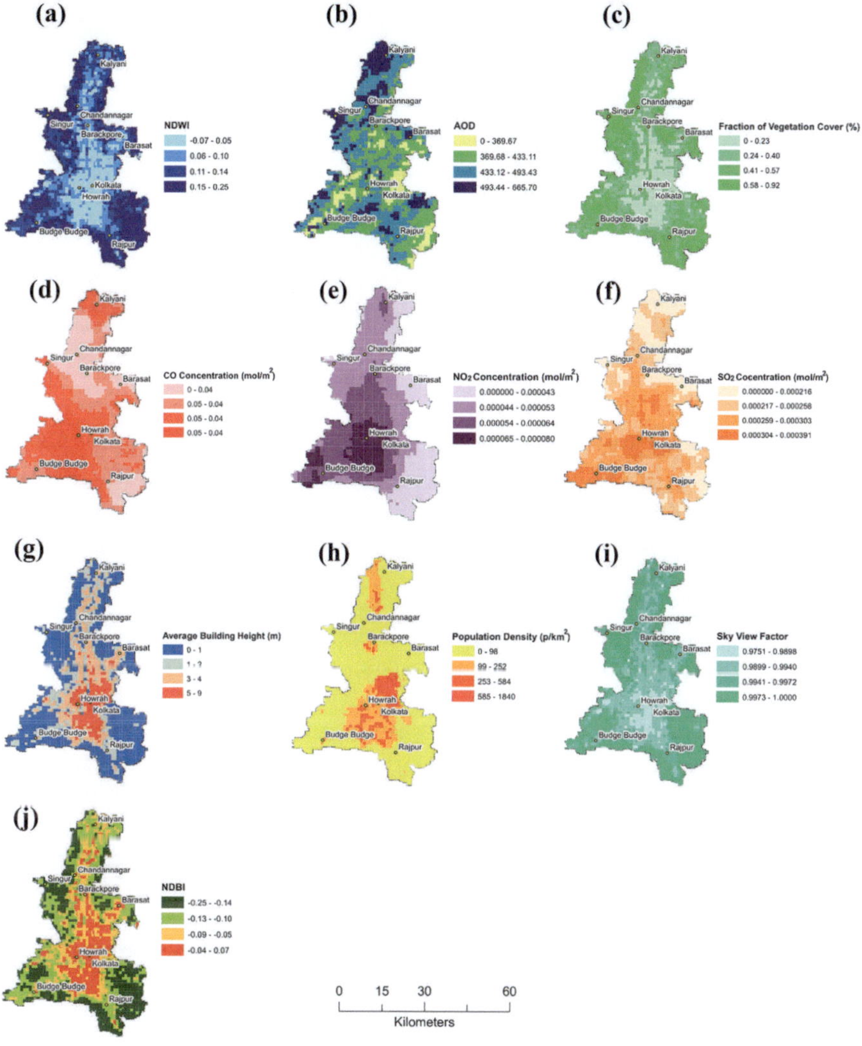

Fig. 2 Spatial distribution of selected variables in KUA: (**a**) NDWI, (**b**) AOD, (**c**) FVC, (**d**) CO concentration, (**e**) NO_2 concentration, (**f**) SO_2 concentration, (**g**) building height (BH), (**h**) population density (PD), (**i**) sky view factor (SVF), and (**j**) NDBI

2.3.1 Zone 1

The results of the regression analysis for Zone 1 (Table 3) are somewhat unexpected, as many of the independent variables turned out to be statistically insignificant. Specifically, NDWI, NO_2, and SO_2 show high significance ($p < 0.001$), while AOD is significant at $p < 0.05$. Variables CO, BH, and SVF did not contribute significantly to explaining daytime LST variability. The overall F-statistic was 21.76

Table 3 Summary of multiple regression analysis for Zone 1

	Zone 1							
	Daytime LST				Nighttime LST			
Variables	Coefficients	Standard error	t-Stat	*P*-value	Coefficients	Standard error	t-Stat	*P*-value
Intercept	0.402	0.1251	3.21	0.00**	0.7055	0.0847	8.321	0.00***
FVC	0.1379	0.1467	0.94	0.34	0.1672	0.0994	1.682	0.09
NDWI	−0.535	0.1428	−3.74	0.00***	−0.3618	0.0967	−3.737	0.00***
AOD	−0.0939	0.0425	−2.2	0.02*	−0.0602	0.0288	−2.09	0.03*
CO	0.0258	0.0718	0.36	0.71	−0.1672	0.0486	−3.437	0.00***
NO_2	0.2521	0.0715	3.52	0.00*	0.0266	0.0484	0.55	0.58
SO_2	−0.2189	0.0529	−4.13	0.00**	0.0339	0.0358	0.945	0.34
PD	0.0381	0.0216	1.76	0.07	0.0323	0.0146	2.21	0.02*
BH	0.0744	0.0744	1.25	0.2	−0.0225	0.04	−0.563	0.57
SVF	−0.0337	0.0464	−0.72	0.46	−0.0997	0.0314	−3.168	0.001**
	Total F (9,280): 21.76 *p*-value: <2.2E-16				Total F (9,280): 7.823 *p*-value: 2.733E-10			

Note: $^{*}p < 0.1$, $^{**}p < 0.05$, $^{***}p < 0.001$

with a p-value less than 0.001, indicating that the overall model does not effectively explain the variability in LST, suggesting that excluding insignificant variables could enhance its predictive power. Nevertheless, the relationship established by the significant variables was robust.

This shows that NDWI, AOD, and SO_2 are negatively correlated with daytime LST, whereas NO_2 and PD aid in increasing the LST and, in turn, the UHI effect. The results of the regression analysis showed that FVC ($\beta = 0.167$, $p = 0.094$), NDWI ($\beta = -0.362$, $p < 0.001$), AOD ($\beta = -0.060$, $p = 0.038$), CO ($\beta = -0.167$, $p < 0.001$), PD ($\beta = 0.032$, $p = 0.028$), and SVF ($\beta = -0.100$, $p = 0.002$) were significantly associated with nighttime LST. These findings indicate that increased levels of air pollution, as measured by NDWI, AOD, CO, and SVF, were associated with decreased LST. Conversely, FVC and PD were positively associated with nighttime LST. However, NO_2 ($\beta = 0.027$, $p = 0.583$), SO_2 ($\beta = 0.034$, $p = 0.345$), and BH ($\beta = -0.023$, $p = 0.574$) were not significantly associated with LST.

2.3.2 Zone 2

The regression results for Zone 2 are presented in Table 4. The adjusted R-squared value for this zone was 0.4016, indicating that the model explained approximately 40% of the variance in lung function. The F-statistic was significant ($F = 56.03$, $p < 0.001$), suggesting that the overall model fit the data well.

The results indicated that FVC ($\beta = 0.28109$, $p < 0.001$), NDWI ($\beta = -0.37133$, $p < 0.001$), AOD ($\beta = -0.06947$, $p < 0.01$), CO ($\beta = 0.14399$, $p < 0.001$), NO_2 ($\beta = 0.12982$, $p < 0.01$), and SO_2 ($\beta = -0.13626$, $p < 0.001$) had significant effects on lung function. In contrast, PD ($\beta = 0.04113$, $p < 0.1$), BH ($\beta = 0.34090$, $p < 0.001$),

Table 4 Summary of multiple linear regression for Zone 2

	Zone 2							
	Daytime LST				Nighttime LST			
Variables	Coefficients	Standard error	t-Stat	*P*-value	Coefficients	Standard error	t-Stat	*P*-value
Intercept	0.279	0.036	7.764	0.000^{***}	0.089	0.043	2.102	0.036^{*}
FVC	0.281	0.083	3.372	0.001^{***}	−0.331	0.099	−3.352	0.001^{***}
NDWI	−0.371	0.083	−4.464	0.000^{***}	0.061	0.098	0.624	0.533
AOD	−0.069	0.024	−2.847	0.005^{***}	−0.122	0.029	−4.237	0.000^{***}
CO	0.144	0.026	5.513	0.000^{***}	−0.123	0.031	−3.976	0.000^{***}
NO_2	0.130	0.041	3.177	0.002^{**}	0.301	0.048	6.219	0.000^{***}
SO_2	−0.136	0.028	−4.885	0.000^{***}	0.042	0.033	1.272	0.204
PD	0.041	0.022	1.912	0.056	0.127	0.025	5.002	0.000^{***}
BH	0.341	0.055	6.172	0.000^{***}	−0.043	0.065	−0.655	0.513
SVF	−0.061	0.039	−1.555	0.120	−0.086	0.047	−1.849	0.065
	Total F (9,729): 56.03 *p*-value: <2.2E-16				Total F (9,729): 27.84 *p*-value: <2.2E-16			

Note: $^{*}p < 0.1$, $^{**}p < 0.05$, $^{***}p < 0.001$
Source: calculated by authors

and SVF (β = −0.06111, not significant) did not significantly influence lung function.

For nighttime LST, the regression analysis revealed significant coefficients for six out of ten variables. FVC, AOD, CO, NO_2, PD, and SVF significantly predicted LST. Higher values of FVC, AOD, CO, and SVF were linked to lower surface temperatures, as indicated by negative coefficients. Conversely, NO_2 and PD were associated with higher LST, reflected in their positive coefficients.

NDWI, SO2, and BH were not significant predictors of LST. The model explained about 26% of the variance in LST, as indicated by the multiple R-squared value of 0.2558. Considering the number of variables in the model, the adjusted R-squared value was 0.2466. The significant F-statistic ($p < 2.2e\text{-}16$) confirmed that the model adequately fit the data.

2.3.3 Zone 3

The regression result for Zone 3 is shown in Table 5. The result shows that FVC has a positive coefficient of 0.09650, indicating that the LST also increases as the fraction of vegetation cover (FVC) increases. NDWI has an obvious negative coefficient of −0.32819, indicating that the LST decreases as the NDWI mean value increases. The coefficient for AOD is 0.02299, indicating a small positive effect that is not statistically significant ($p > 0.05$), implying a weak relationship between this variable and LST. CO has a positive coefficient of 0.15254, indicating that the LST also increases as the carbon monoxide (CO) mean value increases. NO_2 has a negative coefficient of −0.03518, which is not statistically significant at the 0.05 level, suggesting no strong relationship between this variable and the target variable. SO_2 has

Table 5 Summary of multiple linear regression for Zone 3

	Zone 3							
	Daytime LST				Nighttime LST			
Variables	Coefficients	Standard error	t-Stat	*P*-value	Coefficients	Standard error	t-Stat	*P*-value
Intercept	−0.1331	0.0245	−5.43	0.00***	−0.1366	0.0373	−3.661	0.00***
FVC	0.0965	0.0307	3.13	0.00**	−1.1122	0.0468	−23.753	0.00***
NDWI	−0.3281	0.0397	−8.25	0.00***	0.8019	0.0604	13.259	0.00***
AOD	0.0229	0.0128	1.78	0.07	−0.0594	0.0195	−3.034	0.00**
CO	0.1525	0.016	9.53	0.00***	−0.0011	0.0243	−0.048	0.96
NO_2	−0.0351	0.0198	−1.777	0.07	0.0245	0.0301	0.815	0.41
SO_2	−0.1355	0.0157	−8.61	0.00***	−0.0074	0.0239	−0.31	0.75
PD	−0.2164	0.0512	−4.22	0.00***	0.4177	0.0779	5.358	0.00***
BH	0.1142	0.0581	1.96	0.04*	−0.1086	0.0883	−1.229	0.21
SVF	−0.5426	0.0362	−14.96	0.00***	−0.0219	0.0551	−0.398	0.69
	Total F (9,1536): 189 *p*-value: <2.2E-16				Total F (9,1536): 86.54 *p*-value: <2.2E-16			

Note: 1*$p < 0.1$, **$p < 0.05$, ***$p < 0.001$
Source: Calculated by authors

a coefficient of −0.13552, suggesting that higher levels of sulfur dioxide are associated with lower LST. PD has a coefficient of −0.21647, indicating that higher population densities correlate with lower LST. BH has a coefficient of 0.11421, indicating that taller buildings are associated with higher LST. SVF has a coefficient of −0.54261, suggesting that higher SVF values correspond to lower LST.

The model's multiple R-squared value of 0.5255 suggests that approximately 52.55% of the variation in the target variable is explained by the nine predictor variables. The adjusted R-squared value of 0.5228 indicates a good fit of the model to the data, considering the number of predictors used.

The analysis results show that five predictor variables—FVC, NDWI, AOD, PD, and BH—have statistically significant relationships with nighttime LST at a significance level of $p < 0.05$. The intercept term is also significant, indicating a nonzero response variable value when all predictors are zero. However, the coefficients for CO, NO_2, SO_2, and SVF are not statistically significant, suggesting no significant relationship with nighttime LST at this level of significance.

The adjusted R-squared value of 0.3326 indicates that approximately 33.3% of the variation in nighttime LST is explained by the predictor variables in this model. The residual standard error of 0.7479 suggests some unexplained variation in the response variable that the predictors do not account for. Overall, the F-statistic of 86.54 with a p-value of <2.2e-16 indicates that the model is statistically significant and fits the data well.

3 Discussion

This study employed a multidimensional approach to investigate the correlation between the UHI effect, or LST in general, and different natural and man-made conditions. The link between urbanization and LST is quite established through numerous studies done in Indian cities and cities worldwide (Chakraborty et al., 2021; Pramanik & Punia, 2020; Tran et al., 2017; Fu & Weng, 2016). Regarding urban agglomeration within the Kolkata Urban Agglomeration (KUA), urban growth primarily occurred through coalescence, leading to expansion at the cost of the urban fringe and outer zones (Chakraborty et al., 2021). As cities expand and urbanization intensifies, the effect of UHI escalates, resulting in higher surface temperatures in urban areas. Studies have shown that the intensity of UHI is strongly correlated with the number of impervious surfaces in an area, with cities having higher amounts of impervious surfaces experiencing stronger UHI effects (Pramanik et al. 2022; Chakraborty et al., 2021). High LST is observed in and around the dense built-up areas of the urban core, as indicated by our study, with expanding urban areas through coalescence and encroachment on surrounding natural areas potentially reducing vegetation cover and exacerbating UHI effects (Alexander, 2020; Pramanik & Punia, 2019). In addition, the loss of green spaces can also lead to reduced air quality and increased levels of air pollution, which can negatively impact human health. This can be attributed to the fact that there is no incoming solar radiation to heat the surface; the nighttime LST is primarily caused by the heat released by the surface itself. In such cases, the landscape characteristics come into action. Furthermore, as there is little green space within Zone 1, FVC becomes insignificant in explaining the variability of nighttime LST.

3.1 *Urban Morphology*

The relationship between LST and different parts of an urban agglomeration, such as the urban core, urban fringe, and peripheries, can vary depending on various factors such as active land use patterns, dynamics of land cover, and predominant human activities. The results indicate that the urban core tends to experience higher LSTs compared to the urban fringe and peripheries, primarily due to the dense concentration of built structures that have limited heat dissipation capabilities. The urban fringe, on the other hand, may have more green spaces, which can help lower LSTs in those areas. During the day, higher LSTs in urban areas are primarily attributed to the absorption and reemission of solar radiation by impervious surfaces. This is especially true during the summer months when temperatures are already high. In addition, the lack of vegetation cover in urban areas also contributes to higher LSTs during the day. At night, the relationship between LST and urban areas can be reversed, with urban areas having lower LSTs than their surrounding rural areas. This phenomenon can be well understood through our study's spatial distribution of nighttime LST. This occurs because the heat stored in urban surfaces is transmitted into the atmosphere, leading to a cooling effect referred to as the "urban cool island" effect (Pramanik & Punia, 2019; Pramanik & Punia, 2020). This is most pronounced on clear, calm nights when the urban surface can cool rapidly by radiation. However, the extent of the urban cool island effect can fluctuate based on urban area attributes, such as the proportion of impervious surfaces and city size. Larger cities typically exhibit a subdued urban cool island effect due to the dispersion of stored heat across wider areas through wind transport, resulting in diminished cooling within the city center.

3.2 Air Pollution

The correlation between LST and CO levels varies depending on the time of day. During nighttime, LST is negatively correlated with CO levels due to the boundary layer that traps pollutants near the surface, resulting in higher CO concentrations. During the daytime, LST is positively correlated with CO levels due to solar radiation, creating a convective boundary layer that mixes pollutants throughout the lower atmosphere, leading to the dispersion of CO.

NO_2 is primarily released by transportation and industrial activities, contributing to the UHI effect by absorbing solar radiation in the visible and ultraviolet parts of the spectrum, leading to increased absorption of solar radiation and a stronger UHI effect. Additionally, NO_2 can react with other atmospheric pollutants, forming ozone, a potent greenhouse gas that absorbs outgoing thermal radiation, increasing surface temperatures and a stronger UHI effect. SO_2 concentration in the atmosphere is inversely proportional to the UHI effect and LST during the daytime. SO_2 is a type of atmospheric aerosol that can scatter and absorb solar radiation, leading to increased scattering and absorption of solar radiation, resulting in lower daytime LST.

There is a complex relationship between AOD and LST, with studies suggesting a negative correlation between LST and atmospheric aerosol concentration. Atmospheric aerosols can absorb and scatter solar radiation, leading to lower LST. However, they can also contribute to atmospheric warming by absorbing solar energy, resulting in elevated atmospheric temperatures and potentially higher LSTs.

3.3 Economic Development

Urbanization and economic development exert substantial influences on LST and the UHI effect. This study investigates how urbanization and economic activities correlate with LST across various zones within the study area. The study area was segmented into three zones based on economic development levels: the urban core, characterized by high economic activity; the urban fringe, with moderate economic development; and the periphery, displaying the lowest economic development. Using satellite imagery, LST was quantified and analyzed across these zones, revealing that higher economic development zones exhibited elevated LST levels, which gradually decreased toward the periphery. This spatial analysis highlights the complex interplay between urbanization, economic growth, and thermal characteristics, underscoring the need for strategic planning and sustainable development practices to manage urban heat and enhance environmental quality in growing urban areas. These findings suggest that economic development and urbanization should be carefully planned to mitigate the UHI effect and reduce LST. Expanding green spaces and vegetation in urban areas, along with implementing urban design techniques like green roofs and cool pavements, can effectively reduce the UHI effect.

Policy interventions are needed to mitigate the impacts of UHIs, such as implementing green infrastructure and reducing greenhouse gas emissions. Green infrastructure, such as green roofs, green walls, and urban parks, offers cooling benefits, enhances air quality, and contributes to the overall quality of urban living environments (Bartesaghi Koc et al., 2018; Norton et al., 2015). Reduction of greenhouse gas emissions can be achieved through incentivizing renewable energy, promoting public transportation, and encouraging low-carbon industries (Bleviss, 2021; Rissman et al., 2020). Controlling air pollutants through localized emission control measures such as retrofitting diesel vehicles and setting up emissions trading schemes can also reduce the concentration of air pollutants in urban areas (Kosobud et al., 2004; Xie et al., 2021), leading to lower LSTs and improved air quality. The sky view factor (SVF) plays a crucial role in understanding urban climatology and its spatial variations, reflecting street geometry and building density. Reduced SVF contributes to greater heat retention within buildings and exacerbates the UHI effect. Trees can influence SVF by reducing it, thereby affecting the emission of outgoing longwave radiation from urban areas

4 Conclusion

In this chapter, we have analyzed the spatial dynamics of LST about different landscape factors, urban morphology, and air pollution. This article also aimed to explore how economic development and urbanization, including environmental determinants, affect LST and the UHI effect across various zones within the KUA. The study divided KUA into three zones based on economic development: the urban core with high economic development, the urban fringe with moderate economic development, and the periphery with the lowest economic development. Satellite imagery was utilized to measure LST, revealing that LST levels were notably elevated in the high economic development zones and gradually decreased toward the outskirts. The study underscored the substantial influence of impervious surfaces, such as roads and buildings, on LST, with areas featuring denser concentrations of impervious surfaces exhibiting higher LST values. The results also showed that urban areas with high economic development tend to have higher LST than areas with low economic development. This can be attributed to the greater extent of impervious surfaces, which absorb and retain more heat than natural surfaces. The study highlights the need for careful urbanization and economic development planning to mitigate the UHI effect and reduce LST. Expanding green spaces and vegetation within urban areas can effectively lower LST and enhance air quality. Urban planning initiatives, such as integrating green roofs and cool pavements, are also viable strategies to mitigate the UHI effect. Furthermore, adopting green infrastructure and implementing policies aimed at reducing greenhouse gas emissions are essential measures to combat and alleviate the impacts of UHIs. These interventions play a pivotal role in fostering sustainable urban environments that prioritize environmental health and resilience to climate change. Reducing air pollutants such as SO_2, NO_2, CO, and other harmful particles mitigates the UHI effect. Localized emission control measures such as retrofitting diesel vehicles, changing fuels, and setting up emissions trading schemes can reduce the concentration of air pollutants in urban areas, leading to lower LSTs during the night and improved air quality.

The study highlighted the significance of the SVF in characterizing urban climatology and its spatial dynamics. Reduced SVF increases net heat retention within buildings by limiting heat dissipation into the atmosphere, thereby exacerbating the urban heat island (UHI) effect. Trees, while reducing the SVF, play a crucial role in moderating urban temperatures by providing shade and evaporative cooling. However, they also restrict the emission of longwave radiation from the surface, influencing the energy balance within urban areas. Therefore, understanding and managing SVF alongside vegetation cover are essential strategies in mitigating urban heat and enhancing urban climate resilience. In conclusion, this study offers significant insights into how economic development and urbanization influence LST and the UHI effect within the KUA. The results underscore the critical need for thoughtful urban planning and economic development strategies to effectively mitigate the UHI effect and lower LST levels in urban areas. These findings highlight the importance of sustainable development practices aimed at enhancing urban climate resilience and improving overall urban livability. The implementation of green infrastructure, reduction in greenhouse gas emissions, and reduction of air pollutants are crucial policy interventions to mitigate the effects of UHIs.

References

Alexander, C. (2020). Influence of the proportion, height and proximity of vegetation and buildings on urban land surface temperature. *International Journal of Applied Earth Observation and Geoinformation, 95*, 102265. https://doi.org/10.1016/j.jag.2020.102265

Asadi, Y., Ezimand, K., Keshtkar, H., & Alavipanah, S. K. (2019). A survey of landscape metrics and land-use/land-cover structures on urban Heat Islands surface: A case study on Urmia City, Iran. *Desert, 24*(2), 1. https://doi.org/10.22059/jdesert.2019.74952

Bai, X., McPhearson, T., Cleugh, H., Nagendra, H., Tong, X., Zhu, T., & Zhu, Y.-G. (2017). Linking urbanisation and the environment: Conceptual and empirical advances. *Annual*

Review of Environment and Resources, 42(1), 215–240. https://doi.org/10.1146/annurev-environ-102016-061128

Bartesaghi Koc, C., Osmond, P., & Peters, A. (2018). Evaluating the cooling effects of green infrastructure: A systematic review of methods, indicators and data sources. *Solar Energy, 166*, 486–508. https://doi.org/10.1016/j.solener.2018.03.008

Bleviss, D. L. (2021). Transportation is critical to reducing greenhouse gas emissions in the United States. *WIREs Energy and Environment, 10*(2), e390. https://doi.org/10.1002/wene.390

Buyantuyev, A., & Wu, J. (2010). Urban heat islands and landscape heterogeneity: Linking spatiotemporal variations in surface temperatures to land-cover and socio-economic patterns. *Landscape Ecology, 25*(1), 17–33. https://doi.org/10.1007/s10980-009-9402-4

Cao, C., Lee, X., Liu, S., Schultz, N., Xiao, W., Zhang, M., & Zhao, L. (2016). Urban heat islands in China enhanced by haze pollution. *Nature Communications, 7*(1), 12509. https://doi.org/10.1038/ncomms12509

Census of India. (2011). New Delhi: Registrar General & Census Commissioner of India. http://censusindia.gov.in/

Chakraborty, S., Pramanik, S., Follmann, A., Giri, B., Mondal, B., Patel, P. P., Maity, I., Das, J., Punia, M., & Sahana, M. (2021). Dominant urban form and its relation to night-time land surface temperature in the rapidly urbanising National Capital Region of India. *Urban Climate, 40*, 101002. https://doi.org/10.1016/j.uclim.2021.101002

Chen, Q., Cheng, Q., Chen, Y., Li, K., Wang, D., & Cao, S. (2021). The influence of sky view factor on daytime and night-time urban land surface temperature in different spatial-temporal scales: A case study of Beijing. *Remote Sensing, 13*(20), 4117. https://doi.org/10.3390/rs13204117

Dahanayake, H., Wickramasinghe, D., & Dahanayaka, D. (2022). Do natural landscapes contribute to reducing land surface temperature (LST)? A case study from Muthurajawela wetland, Sri Lanka. *Journal of degraded and mining lands management, 9*(2), 3329–3339. https://doi.org/10.15243/jdmlm.2022.092.3329

Dai, Z., Guldmann, J.-M., & Hu, Y. (2018). Spatial regression models of park and land-use impacts on the urban heat Island in Central Beijing. *Science of the Total Environment, 626*, 1136–1147. https://doi.org/10.1016/j.scitotenv.2018.01.165

Deilami, K., Kamruzzaman, M., & Liu, Y. (2018). Urban heat Island effect: A systematic review of spatio-temporal factors, data, methods, and mitigation measures. *International Journal of Applied Earth Observation and Geoinformation, 67*, 30–42. https://doi.org/10.1016/j.jag.2017.12.009

Fu, P., & Weng, Q. (2016). A time series analysis of urbanization induced land use and land cover change and its impact on land surface temperature with Landsat imagery. *Remote Sensing of Environment, 175*, 205–214. https://doi.org/10.1016/j.rse.2015.12.040

Gunawardena, K. R., Wells, M. J., & Kershaw, T. (2017). Utilising green and bluespace to mitigate urban heat Island intensity. *Science of the Total Environment, 584–585*, 1040–1055. https://doi.org/10.1016/j.scitotenv.2017.01.158

Halder, B., Bandyopadhyay, J., & Banik, P. (2021). Evaluation of the climate change impact on urban Heat Island based on land surface temperature and geospatial indicators. *International Journal of Environmental Research, 15*(5), 819–835. https://doi.org/10.1007/s41742-021-00356-8

Hämmerle, M., Gál, T., Unger, J., & Matzarakis, A. (2011). Comparison of models calculating the sky view factor used for urban climate investigations. *Theoretical and Applied Climatology, 105*(3–4), 521–527. https://doi.org/10.1007/s00704-011-0402-3

Kayet, N., Pathak, K., Chakrabarty, A., & Sahoo, S. (2016). Urban heat Island explored by co-relationship between land surface temperature vs multiple vegetation indices. *Spatial Information Research, 24*(5), 515–529. https://doi.org/10.1007/s41324-016-0049-3

Kosobud, R. F., Stokes, H. H., & Tallarico, C. D. (2004). Does emissions trading lead to air pollution hot spots? Evidence from an urban ozone control programme. *International Journal of Environmental Technology and Management, 4*(1/2), 137. https://doi.org/10.1504/IJETM.2004.004626

Leal Filho, W., Wolf, F., Castro-Díaz, R., Li, C., Ojeh, V. N., Gutiérrez, N., Nagy, G. J., Savić, S., Natenzon, C. E., Quasem Al-Amin, A., Maruna, M., & Bönecke, J. (2021). Addressing the urban Heat Islands effect: A cross-country assessment of the role of green infrastructure. *Sustainability, 13*(2), 753. https://doi.org/10.3390/su13020753

Li, Y., Schubert, S., Kropp, J. P., & Rybski, D. (2020). On the influence of density and morphology on the urban Heat Island intensity. *Nature Communications, 11*(1), 2647. https://doi.org/10.1038/s41467-020-16461-9

Loughner, C. P., Allen, D. J., Zhang, D.-L., Pickering, K. E., Dickerson, R. R., & Landry, L. (2012). Roles of urban tree canopy and buildings in urban Heat Island effects: Parameterization and preliminary results. *Journal of Applied Meteorology and Climatology, 51*(10), 1775–1793. https://doi.org/10.1175/JAMC-D-11-0228.1

Malik, M. S., Shukla, J., & Mishra, S. (2019). Relationship of LST, NDBI and NDVI using Landsat-8 data in Kandaihimmat Watershed, Hoshangabad, India. *Indian Journal of Geo-Marine Sciences, 48*, 25–31. https://www.semanticscholar.org/paper/Relationship-of-LST%2C-NDBI-and-NDVI-using-Landsat-8-Malik-Shukla/330257c6a59f76e9c032307e2c0288b294adc346

McKinney, M. L. (2002). Urbanisation, biodiversity, and conservation. *Bioscience, 52*(10), 883. https://doi.org/10.1641/0006-3568(2002)052[0883:UBAC]2.0.CO;2

Mohajerani, A., Bakaric, J., & Jeffrey-Bailey, T. (2017). The urban heat Island effect, its causes, and mitigation, with reference to the thermal properties of asphalt concrete. *Journal of Environmental Management, 197*, 522–538. https://doi.org/10.1016/j.jenvman.2017.03.095

Mohan Kumar, T. M., Suhas, K. B., & Siddegowda. (2022). A case study of urban Heat Island and air pollution mechanism in Bengaluru metropolitan city. *IOP Conference Series: Materials Science and Engineering, 1255*(1), 012014. https://doi.org/10.1088/1757-899X/1255/1/012014

Norton, B. A., Coutts, A. M., Livesley, S. J., Harris, R. J., Hunter, A. M., & Williams, N. S. G. (2015). Planning for cooler cities: A framework to prioritise green infrastructure to mitigate high temperatures in urban landscapes. *Landscape and Urban Planning, 134*, 127–138. https://doi.org/10.1016/j.landurbplan.2014.10.018

Pramanik, S., & Punia, M. (2019). Assessment of green space cooling effects in dense urban landscape: A case study of Delhi, India. *Modeling Earth Systems and Environment, 5*(3), 867–884. https://doi.org/10.1007/s40808-019-00573-3

Pramanik, S., Punia, M., Yu, H., & Chakraborty, S. (2022). Is dense or sprawl growth more prone to heat-related health risks? Spatial regression-based study in Delhi, India. Sustainable Cities and Society, 81, 103808. https://doi.org/10.1016/j.scs.2022.103808

Pramanik, S., & Punia, M. (2020). Land use/land cover change and surface urban heat Island intensity: Source–sink landscape-based study in Delhi, India. *Environment, Development and Sustainability, 22*(8), 7331–7356. https://doi.org/10.1007/s10668-019-00515-0

Rahim, F. U., Niaz, T., Shaheen, S., Asma, M., & Almas, L. K. (2022). A composite impact of urbanisation and liberalisation on environmental degradation in developed and developing economies. *WSEAS Transactions on Environment and Development, 18*, 873–883. https://doi.org/10.37394/232015.2022.18.82

Rissman, J., Bataille, C., Masanet, E., Aden, N., Morrow, W. R., Zhou, N., Elliott, N., Dell, R., Heeren, N., Huckestein, B., Cresko, J., Miller, S. A., Roy, J., Fennell, P., Cremmins, B., Koch Blank, T., Hone, D., Williams, E. D., De La Rue, D., Can, S., & Helseth, J. (2020). Technologies and policies to decarbonise global industry: Review and assessment of mitigation drivers through 2070. *Applied Energy, 266*, 114848. https://doi.org/10.1016/j.apenergy.2020.114848

Shao, J., & Ge, J. (2020). Investigation into relationship between intensive land use and urban Heat Island effect in Shijiazhuang City based on the Tapio decoupling theory. *Journal of Urban Planning and Development, 146*(4), 04020043. https://doi.org/10.1061/(ASCE)UP.1943-5444.0000619

Theeuwes, N. E., Steeneveld, G. J., Ronda, R. J., Heusinkveld, B. G., Van Hove, L. W. A., & Holtslag, A. A. M. (2014). Seasonal dependence of the urban heat Island on the street canyon aspect ratio. *Quarterly Journal of the Royal Meteorological Society, 140*(684), 2197–2210. https://doi.org/10.1002/qj.2289

Tran, D. X., Pla, F., Latorre-Carmona, P., Myint, S. W., Caetano, M., & Kieu, H. V. (2017). Characterizing the relationship between land use land cover change and land surface temperature. *ISPRS Journal of Photogrammetry and Remote Sensing, 124*, 119–132. https://doi.org/10.1016/j.isprsjprs.2017.01.001

United Nations Department of Economic and Social Affairs. (2018). *The sustainable development goals report 2018*. United Nations. 10.18356/7d014b41-en.

Wu, H., Wang, T., Riemer, N., Chen, P., Li, M., & Li, S. (2017). Urban heat Island impacted by fine particles in Nanjing, China. *Scientific Reports, 7*(1), 11422. https://doi.org/10.1038/s41598-017-11705-z

Xie, Y., Wu, D., & Zhu, S. (2021). Can new energy vehicles subsidy curb the urban air pollution? Empirical evidence from pilot cities in China. *Science of the Total Environment, 754*, 142232. https://doi.org/10.1016/j.scitotenv.2020.142232

Zhou, X., & Chen, H. (2018). Impact of urbanisation-related land use land cover changes and urban morphology changes on the urban heat Island phenomenon. *Science of the Total Environment, 635*, 1467–1476. https://doi.org/10.1016/j.scitotenv.2018.04.091

Zhou, W., Huang, G., & Cadenasso, M. L. (2011). Does spatial configuration matter? Understanding the effects of land cover pattern on land surface temperature in urban landscapes. *Landscape and Urban Planning, 102*(1), 54–63. https://doi.org/10.1016/j.landurbplan.2011.03.009

Atmospheric Pollution and Land Surface Temperature Intensity in COVID-19 Pandemic: A Case of Major Indian Cities

Anindita Dasgupta and Uttam Kumar

Abstract The twenty-first century, often referred to as the urban century, witnesses a significant portion of the world's population embracing city life as their preferred residence. With the ongoing global population growth, urbanization is expected to intensify in the coming three decades, shaping cities into even more densely populated and interconnected hubs. Large population size in the cities along with the economic activities, rampant lifestyle, technological interventions, energy use, land use pattern, and climate policies, are some of the important parameters in urban areas that impact sustainability. When natural resources are used to meet both the immediate demands of the current population and those of future generations without significantly compromising the urban ecosystem and environment, sustainable development can be achieved. However, high population density and increased resource utilization have given rise to numerous challenges linked to climate change and environmental degradation, such as increased air pollution, greenhouse gas emissions, waste generation, and ecosystem sustainability. In cities, urban climate is mainly driven by topography, air pollution, urban surface materials, anthropogenic heat, greenhouse gases, and carbon emissions. For example, in India, large cities like New Delhi, Kolkata, Mumbai, Hyderabad, Chennai, and Bangalore exhibit the highest carbon footprints of 69.6, 42.9, 32.1, 26.5, 21.8, and 21.6 Mt. CO_2, respectively. However, in contrary to the general observed trend, these cities exhibited local climate variations during the COVID-19 lockdown. Motivated by this unique observation, we attempt to study the major temperature variations that happened in the above cities between the pandemic years (2019 to 2021). The important factors, such as aspects of pollutant concentration that accounted for such a change in local temperature, were investigated for 4 years, i.e., 2019, 2020, 2021, and 2022. The analysis revealed that all the cities except New Delhi and Kolkata showed lower temperatures than normal during the pandemic in the summer season,

A. Dasgupta · U. Kumar (✉)
Spatial Computing Laboratory, Department of Data Science and Artificial Intelligence (DSAI), International Institute of Information Technology Bangalore (IIIT-B), Bangalore, India
e-mail: anindita.dasgupta@iiitb.ac.in; uttam@iiitb.ac.in

P. K. Joshi et al. (eds.), *Geospatial Science for Urban Ecosystems*, The Urban Book Series, https://doi.org/10.1007/978-3-032-07155-2_23

especially when the urban heat effect was evident. In New Delhi, among all the pollutants, the aerosol index had a positive correlation (0.29) with the land surface temperature, while in Bangalore, the presence of carbon monoxide (CO) and nitrogen dioxide (NO_2) showed the highest correlation of 0.57 and 0.69, respectively, with the land surface temperature.

Keywords Urbanization · Cities · Carbon footprint · Land surface temperature · Urban heat island · Environment

1 Introduction

The majority of people on Earth today live in cities (UN-Habitat, 2008). Cities are the hubs of special economic zones where agglomeration economies and capital investments are the main drivers of growth. The globalization of both national and international economies, facilitated by trade, foreign direct investment, short-term capital flows, the international mobility of employees and individuals, technology transfers, and other factors, has generated fresh opportunities within cities (Najam et al., 2010). Therefore, express urbanization and population growth are the results of the internationalization of national economies, which have led to rapidly growing cities around the world. Increasing urbanization is influencing the way the environment appears and functions. In the year 1900, only 13% of the worldwide population resided in urban areas; however, as of today, more than half of the global population, exceeding 50%, now lives in urban settings (UN-Habitat, 2008). As such, the population of the world will continue to increase over the next few decades, and cities will become even more urbanized. In the process of urbanization, natural resources have been overexploited in order to meet the demands of the expanding population. Due to the depletion of these resources, the environment has deteriorated, which has further led to local and regional climate change issues. Problems related to climate change and environmental degradation, such as increasing air pollution, greenhouse gas (GHG) concentrations, and waste generation, have arisen as a result of the high population density and increased resource use. Pollution, noise, and traffic are a few of the social costs that come with different city sizes. The urban environment is a result of the interaction of numerous natural and man-made elements. Given that air pollution directly affects human health by affecting the population that is exposed, analysis of air pollution has been given high priority for environmental management.

Earth observation through remote sensing (RS) technology has been instrumental in gaining access to the human-environmental interactions necessary for the formulation of environmental decisions and policies. In this regard, satellite RS of the Earth's surface is the most trustworthy and persistent method of monitoring the changes happening in land use and land cover, notwithstanding its inherent susceptibility to atmospheric influence. Thermal scanners in optical sensors in the Earth observation satellites offer a one-of-a-kind opportunity to collect information on the region's land surface temperature (LST) for research on urban climatology. Heat waves that cause increased LST leading to the urban heat island (UHI) effect (Dasgupta & Kumar, 2021; Ramachandra & Kumar, 2009) and contribute to climate change are produced by the sun's incoming radiation that is absorbed and then reradiated off of the dark surfaces (Stone, 2012; Kumar et al., 2021). It has been observed that the central metropolitan areas in cities having large proportions of concrete surface witness the UHI effect with ambient temperatures that are several degrees warmer than neighboring rural areas at a similar altitude (Oke, 1976; Ayanlade et al., 2021). Different radiating surfaces' thermal characteristics and a lower rate of evapotranspiration in cities may be the main causes of UHI (Luo & Wu, 2021). As an illustration, the urban heat island (UHI) phenomenon is initiated by impermeable structures such as buildings, which have expanded at the cost of naturally permeable surfaces and other land use classifications within urban areas (Das et al., 2020; Mwangi et al., 2018). UHI impact is also linked to negative consequences for human health (Heaviside et al., 2017; Pricope et al., 2019). Additional elements that contribute to the UHI include heat generated by human activities like industrial processes, transportation, and the usage of heat-absorbing construction materials like concrete and asphalt, which lead to the emission of GHGs. Apart from GHGs, high pollution levels may also increase urban albedo and reflect long-wave radiation. Additionally, in summer, the UHI effect is most noticeable, leading to higher energy use, health hazards, environmental effects, and poorer air quality (Santamouris et al., 2015).

The use of more energy is often evident in cities, as they are hotter than the surrounding areas because they trap more heat, which necessitates cooling and refrigeration. Actually, energy usage rises as anthropogenic activity in the cities increases (Sovacool & Brown, 2010). In dense and clustered urban areas with multistory buildings, the energy consumption is even higher. For example, for individuals to travel vertically in dense urban settings, more energy is needed to propel them up and down through escalators and lifts. Cities suggest proximity, density, and closeness because there is no near gap between residents and businesses. The more hustle and bustle there is in the cities due to economic activities, the more the formation of carbon footprints. In fact, anthropogenic activities cause carbon emissions, and increased energy usage generates a carbon footprint that has a negative effect on the environment. Moran et al. (2018) studied the carbon footprint of 13,000 cities, and the top 400 cities were analyzed. The study showed that 25 Indian cities were among the top 400 cities in the world with high carbon footprints.

In addition to carbon, air pollutants play a crucial role in shaping the local environment of a city. Any observable or invisible particle or gas present in the air that deviates from its original, natural composition is considered an air pollutant. According to the World Health Organization (WHO), approximately seven million people worldwide succumbed to illnesses linked to air pollution in 2012 alone. Many cities in India grapple with severe pollution, failing to meet the safety thresholds set by the Central Pollution Control Board, Government of India, as well as the WHO guidelines for acceptable levels of air pollution in residential, industrial, and eco-sensitive zones. Gases within the 5–50 μm wavelength range, capable of heat absorption, often fall into the category of greenhouse gases (GHGs). These gases, frequently present in concentrations significant enough to absorb a notable portion of Earth's emitted radiation, contribute to the broader environmental picture. Among the GHGs, sulfur oxides (SOx) such as SO_2 are commonly used as indicators of primary concern. SO_3 is another gas that is present in the atmosphere, but its concentration is generally much lower compared to SO_2. The largest source of SO_2 is the burning of fossil fuels by the power sector and other industries (Ramachandra et al., 2014). Similarly, nitrogen oxides (NOx) are highly reactive gases, for example, nitrogen dioxide (NO_2), that are generally formed by the combustion of fossil fuels (Pramanik, 2017). Carbon monoxide (CO) is another colorless, odorless, chemically stable gas that is produced when there is incomplete combustion of fossil fuel (https://www.epa.gov/environmental-topics/air-topics). In the urban centers, it is mainly released by vehicular emissions, and in the rural areas, the reason behind its high concentration is primarily biomass burning. Exposure to high levels of CO can also lead to various health hazards.

Aerosols present in the atmosphere are solid/liquid particles suspended in the atmosphere, whose size ranges from 0.1 to 100 μm. They are small solid particles or liquid droplets (excluding clouds and rain) and can be either man-made (anthropogenic) or naturally occurring (like ocean salt, dust, and emissions from industrial plants). Particles suspended in the atmosphere play a crucial role in the formation of clouds, with water droplets nucleating around them. Certain aerosols are not only beneficial but essential for precipitation to take place. Aerosols contribute significantly to the Earth-atmosphere system's radiation budget and exert an influence on the climate.

During late 2019, the COVID-19 outbreak prompted governments to enact stringent regulations in their states and nations in an effort to curtail the virus's spread. To maintain social distancing, a nationwide lockdown was implemented in India in March 2019. Educational institutions, schools, and shopping malls were shut down; a majority of industrial activities came to a standstill; nonessential commercial and industrial operations were put on hold, and venturing outside one's residence was strictly forbidden. Meanwhile, essential information technology (IT) services largely shifted to online platforms. This lessened people's outdoor activities, decreased vehicular movements, and lowered energy consumption in offices, commercial, and public buildings. As a result, a change in the overall effect on local climate within large cities was noticeable. Researchers have also attempted to analyze changes in air quality under unusual circumstances due to the nationwide shutdowns (Nakada & Urban, 2020).

Inspired by the noteworthy insights mentioned earlier, this study seeks to examine variations in land surface temperature (LST) during the COVID-19 lockdown in major Indian cities, namely New Delhi, Kolkata, Mumbai, Hyderabad, Chennai, and Bangalore. The study aims to identify potential factors, such as pollution levels (specifically sulfur dioxide (SO_2), nitrogen dioxide (NO_2), carbon monoxide (CO), and aerosols), that may contribute to the observed temperature differences before and during the pandemic, as explored by Kaneko et al. (2008). The objectives of this study include analysis of LST variations during the pandemic (2019 and 2021) in major Indian cities such as New Delhi, Kolkata, Mumbai, Hyderabad, Chennai, and Bangalore, and identification and study of major pollutants that correlate with LST in the New Delhi and Bangalore regions during 2019, 2020, 2021, and 2022.

The analysis revealed that all the cities except New Delhi and Kolkata showed lower temperatures than normal during the pandemic in the summer season, especially when the UHI effect was apparent. Further, the presence of aerosol had a positive correlation (0.29) with the LST in New Delhi, whereas other pollutants were negatively correlated. While in Bangalore, the presence of CO and NO_2 density in the atmosphere had the highest correlation of 0.57 and 0.69, respectively, with the LST. This suggests that different pollutants were responsible for increased LST in major cities of India during the COVID-19 pandemic.

2 Data, Methodology and Study Area

2.1 Data

In this section, the dataset and computing platform used in the study are mentioned.

1. The carbon footprint data for six major cities in India were retrieved from the Eora Global Supply Database at https://worldmrio.com/.
2. Sector-specific carbon emissions data were sourced from the World Bank database available at https://databank.worldbank.org/.
3. MODIS LST cloud-free data (MOD11A2 Version 6) products were utilized, offering an average 8-day per-pixel land surface temperature (LST) and Emissivity (LST&E) with a 1 km spatial resolution (1200 × 1200 grid). The data can be accessed at https://lpdaac.usgs.gov/products/mod11a2v061/.
4. The Sentinel 5P dataset with a spatial resolution of 1000 m was employed, and the data can be found at https://developers.google.com/earth-engine/datasets/catalog/sentinel-5p.
5. Geospatial datasets were processed and analyzed using GEE (Google Earth Engine), a cloud-based geospatial processing platform that integrates a multi-petabyte catalog of satellite imagery and geospatial datasets. Additionally, the GRASS (Geographic Resources Analysis Support System) GIS software, available at https://grass.osgeo.org/, was employed for the analysis.

2.2 Methodology

In this section, the overall methodology is briefed, as also depicted in Fig. 1.

Initially, MODIS LST cloud-free data (MOD11A2 Version 6) were used for analysis in GEE. The administrative city boundaries of the six Indian metropolitan regions, along with a buffer area to accommodate the new urban growth and urban

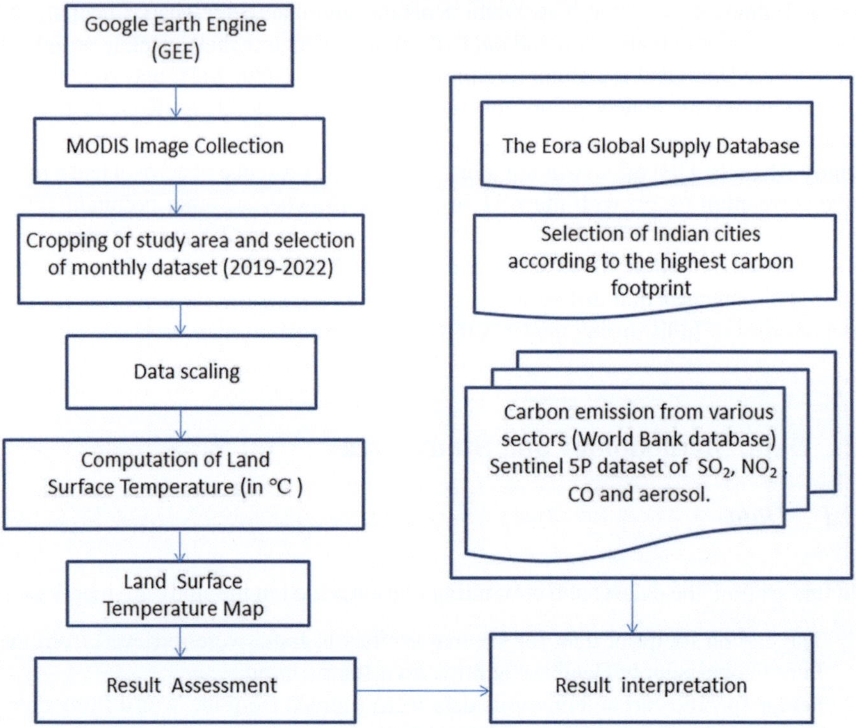

Fig. 1 Pictorial representation of the overall methodology

sprawl that happened outside the existing administrative boundaries, were cropped and filtered in the time domain to obtain the monthly datasets. A scaling factor (0.02) was applied to the dataset, and LST was derived in °C. The carbon footprint of each city was obtained from the Eora Global Supply Database, and the World Bank dataset was used to plot emissions from each sector. The pollution datasets for yearly assessment were obtained from Sentinel 5P. Yearly correlation and aggregate correlation analysis were performed between pollutants and LST.

2.3 Study Area

India is one of the fastest-growing economies in the world, and most Indian cities exhibit high energy consumption rates, leading to greater natural resource reserve depletion with consequent detrimental environmental impact. Major Indian metropolises are characterized by their high population densities and the proximity of people and businesses to one another (UN-Habitat, 2008). Typically, in developing countries, including India, cities have shown high emissions due to changes in lifestyle, such as vehicle ownership, dietary plans, the transformation of the joint family structure to nuclear family, more purchasing power leading to more sprawl outgrowth in the cities, etc. As such, Indian metropolitan areas today are becoming increasingly crowded as more people flock to education, employment opportunities, tourism, etc. In this research, six metropolitan regions of India, namely, New Delhi, Kolkata, Mumbai, Hyderabad, Chennai, and Bangalore, have been considered for analysis, as shown in Fig. 2.

New Delhi: New Delhi, the capital of the Republic of India, is geospatially located between the latitudes 28°40′ and 28°67′ N and the longitude 77°14′ and 77°22′ E at 216 meters above mean sea level in the northern part of the country. Its latitudinal extent from east to west is around 51 km long, while its longitudinal stretch from north to south is approximately 54 km, covering an area of 1483 sq. km. Due to its large distance from the sea, the region experiences a tropical climate. The seasonal temperature fluctuations between summer and winter are extremely wide. Rainfall occurs regularly in the monsoon season, with an average annual measurement ranging from 750 millimeters to 1500 millimeters (Punia et al., 2011). The city boasts a rich and extensive history, having played a pivotal role in India's struggle for independence. New Delhi, as part of the extended urban agglomeration

Fig. 2 Location of cities

within the National Capital Region (NCR), encompasses Delhi, Faridabad, Ghaziabad, Gurugram, and Noida. The NCR hosts key service industries such as information technology, telecommunication, hotels, banking, media, and tourism. Despite accounting for only 2% of the country's population, New Delhi is a significant contributor to India's overall gross domestic product (GDP) and emissions. Buildings, automobiles, and waste generated from the city are the main sources of emissions, with household and industrial activities also contributing significantly.

Kolkata: Formerly known as Calcutta, Kolkata is the capital city of the state of West Bengal in India. Positioned on the eastern bank of the Hooghly River within the Ganges Delta, Kolkata is situated between 22°55′ and 23°03′ N latitude and 88°39′ and 88°42′ E longitude, with an elevation of approximately 9 meters above mean sea level, located 154 km (96 miles) upstream from the Bay of Bengal, as documented by Bhatta (2009). The Kolkata Metropolitan Area, the city's metropolitan agglomeration, has a substantially wider area, totaling 1750 sq. km. Geographically, Kolkata's layout stretches along the Hooghly River's bank in a north-to-south direction. The Hooghly River's western shoreline extends all the way to the Eastern Metropolitan Bypass, which is where the city's eastern boundary begins. Playing a vital role in economic development, this city serves as the primary hub for trade, commerce, and finance in eastern India and the northeastern states. With more than 300 years of history, Kolkata once served as India's capital through the British colonial era. Ranked among the largest metropolitan regions globally, it stands as an exemplar showcasing both the potential and challenges encountered by megacities in developing nations (Banerjee, 2005).

Mumbai: Mumbai, previously known as Bombay, is a significant megacity, the economic capital of India, and the capital of the western state of Maharashtra. Geographically, Mumbai covers an area of 4355 sq. km., extending from 18°31′ to 19°9.6′ N latitude and 72 to 72°35′ E longitude. The area has a population of more than 20 million with a population density of 4445 people per sq. km. (Kaneko et al., 2008). The Mumbai metropolitan area contains a complex medley of natural elements. This consists of rivers, water bodies, and an important terrestrial harbor, coupled with a network of canals that connect creeks, coastal lowlands, linear ridges, and peninsulas—a blending of geographical components as illustrated in Anbazhagan and Ghosh's work (2002). This city plays an essential role as the country's commercial and financial hub, often referred to as the "Gateway to India," and the engine of the nation's activities. Mumbai's significant contribution to a number of economic metrics, including 40% of international trade, 60% of customs duty revenue, and 33% of the country's income tax revenue (Vishwas & Joshi, 2013), is an excellent measure of the city's economic importance.

Hyderabad: Hyderabad is the capital of Telangana state in southern India, located between 17°22′ to 17°26′ N latitude and 78°22′ to 78°33′ E longitude. Historically, Hyderabad has played a significant role as a political and economic hub, and the city reached its one millionth resident in the 1940s. By 2025, the population is predicted to increase to 11.64 million (UN, 2011). Presently, Hyderabad stands as one of India's major hubs for information technology, biotechnology, pharmaceuticals, manufacturing, and various service sectors, a transformation that gained momentum in the aftermath of the economic reforms of the 1990s (Greater Hyderabad Municipal Corporation, 2011; Wakode et al., 2014; Krank, 2007).

Chennai: Chennai (Madras) is a coastal city on the Bay of Bengal, located between latitude 12°50′ to 13°17′ N and longitude 79°59′ to 80°20′ E. The city has a long shoreline that stretches 45 km from north to south. The land is mostly flat, with scattered hillocks that slant slightly eastward toward the Bay of Bengal. The highest point in Chennai is 60 m above mean sea level, with an average height of roughly 6.7 m. According to the 2011 census, this metropolis'metropolitan agglomeration contained close to 8.7 million people with a population density of 26,553 people per sq. km. Chennai's tremendous urban growth over the previous two decades has resulted in an area of 1189 sq. km, or a nearly 17-fold rise in spatial growth, which is the result of significant industrialization. Three significant rivers, namely, Coovam, Adayar, and Kosathalayar, cross the city and join the Bay of Bengal. The climate is generally hot and humid with a tropical wet and dry climate (Jeganathan et al., 2021).

Bangalore: Bangalore (officially known as Bengaluru) is the political, economic, knowledge, and industrial capital of the state of Karnataka in southern India, located spatially between latitude 12°50′ and 12°59′ N and longitude 77°27′ and 77°47′ E. Bangalore has experienced rapid growth and urbanization in recent decades. Renowned for its information technology, biotechnology, manufacturing, aerospace, and aviation sectors, the city is commonly referred to as the Silicon Valley of India. In the twenty-first century, Bangalore has emerged as a pivotal hub for innovative ideas, akin to Mumbai's role in the nineteenth century. The growth and success of newer cities like Bangalore can be attributed in part to technological innovations that have increased returns to knowledge production and proximity to others. Bangalore, the third-largest Indian city, is the greatest success story of the subcontinent. Bangalore's strength comes as the city of ideas rather than industrial might, even though it has plenty of textile industries (Ramachandra et al., 2014). Bangalore is no different from many other growing cities facing challenges related to sustainability, increasing vehicular traffic, high air pollution, and waste management, including increased carbon emissions and per capita resource consumption.

3 Results and Discussion

Figures 3, 4 and 5 show the LST maps of New Delhi and Kolkata, Mumbai and Hyderabad, and Chennai and Bangalore for selected months (January, February, April, May, November, and December) for the years 2019 and 2021. Other months' LST maps have not been shown because either they had cloud coverage due to the monsoon season or they showed normal trends of LST variation, which were obvious.

Figure 3a shows that certain areas of NCR, such as the southwest comprising Gurugram, the southeast, which is Faridabad, and the northwestern parts adjacent to Haryana, are urban hotspots having higher LST in the summer months. Pusa Hill

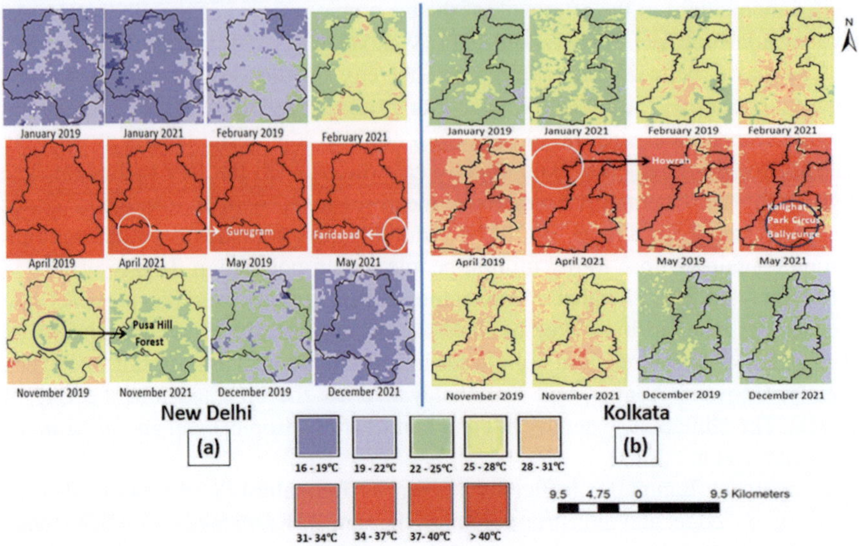

Fig. 3 Land surface temperature of (**a**) New Delhi and (**b**) Kolkata

Forest showed a relatively cooler temperature. In the year 2021, January, November, and December showed mean temperatures of 17.80 °C, 25.46 °C, and 18.68 °C, which were lower than the year 2019 (see Fig. 6). However, data analysis showed that there were not many variations in LST in and around New Delhi before and during COVID as a result of the pandemic lockdown.

Northwest of Kolkata, which is popularly known as Howrah, was found to be hotter, especially in summer. The area encircled in the May 2021 image in Fig. 3b, consisting of Kalighat, Park Circus, and Ballygunge in Kolkata, is an example of UHI in the midst of the city. Overall, the average temperature in summer (April and May) ranged between 31 °C and 34 °C, while in winter (December and January), the mean temperature ranged between 22 °C and 25 °C in 2019 and 2021, respectively. Overall, the LST in 2021 did not show much variation from the year 2019, as also observed in NCR.

Figure 4a shows the LST map consisting of the Mumbai region, including Navi Mumbai, Thane, Kalyan, Bhwandi, Tungareshwar Wildlife Sanctuary, etc. In Mumbai (highlighted in the January 2021 image in Fig. 4a), the southern parts, such as Dadar, Parel, Bandra, Chembur, etc., showed a relatively higher temperature throughout all the seasons. Sanjay Gandhi National Park, which is located in the northern part of the city (see April 2021 image in Fig. 4a), showed a relatively lower temperature than the rest of the region because of the presence of dense vegetation. Adjacent areas of Thane (see May 2019 image in Fig. 4a), Kalyan, and Bhwandi, etc., were also warm regions near Mumbai. Navi Mumbai, which is adjacent to Mumbai and located in the Konkan division of Maharashtra (see May 2021 image in Fig. 4a), was relatively cooler in May 2021, as this region is also characterized by undulating terrain surrounded by forest and waterfalls. May and December in

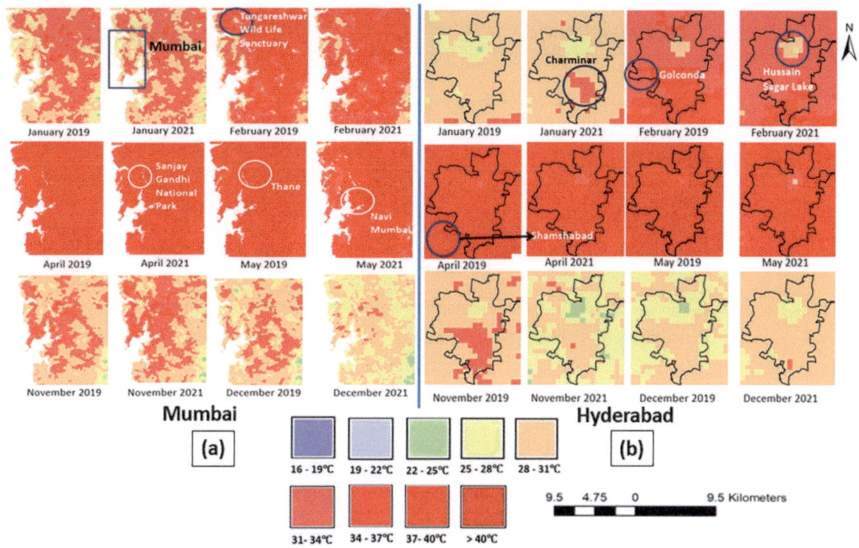

Fig. 4 Land surface temperature of (**a**) Mumbai and (**b**) Hyderabad

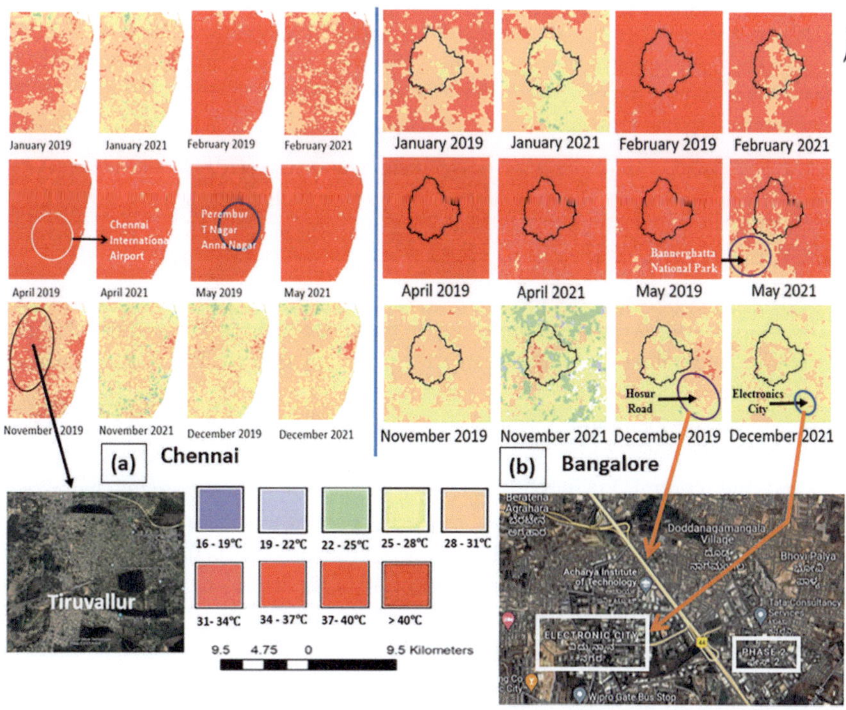

Fig. 5 Land surface temperature of (**a**) Chennai and (**b**) Bangalore

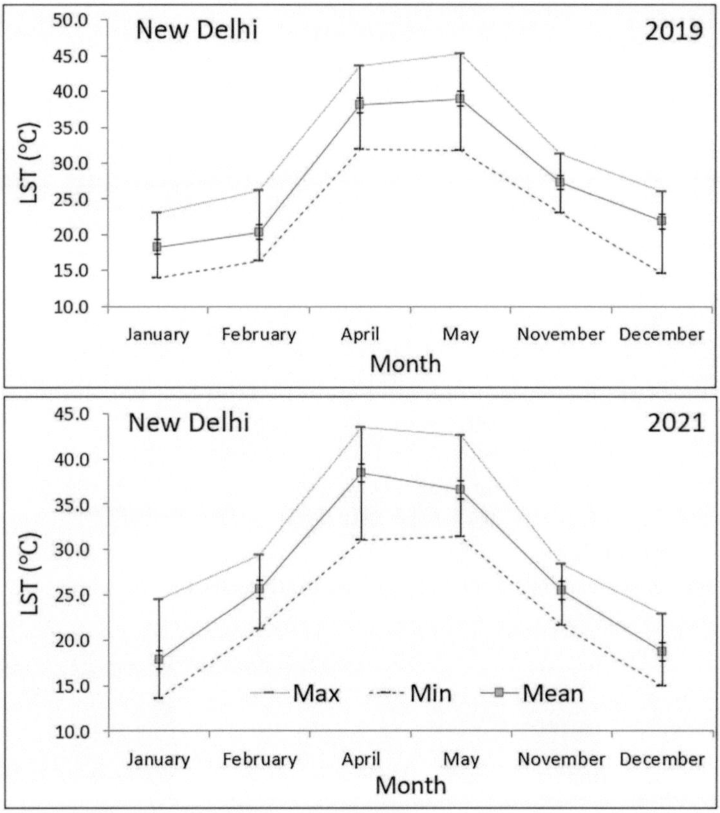

Fig. 6 Minimum, maximum, and mean LST (°C) of the New Delhi region in 2019 and 2021

particular showed overall lower temperatures in 2021 when compared to pre-pandemic observations.

The old city of Hyderabad consists of Charminar and Golkonda Fort, and the adjacent Secunderabad was initially an army cantonment area. The northern part of the city, comprising Hussain Sagar Lake (see February 2021 image in Fig. 4b), showed lower temperatures throughout the year. Southern parts of Hyderabad, including Shamshabad, Rajiv Gandhi International Airport, Gandiguda, Badangpet, etc., and the central business district encompassing Mehdipatnam, Shaikpet, Pujjagatta, and various commercial areas, illustrate the UHI phenomenon. The mean temperature of summer months was approximately 37 °C in 2021, which was found to be lower than the summer average of 40 °C in 2019. To conclude, the overall mean LST in 2021 was lower by ~1 °C compared to 2019.

Central parts of Chennai (highlighted in the May 2019 LST image by a circle in Fig. 5a), such as Perambur, T. Nagar, Anna Nagar, Koyambedu, etc., showed higher temperatures in all the seasons. Areas of dense built-up consisting of commercial and residential layouts, and Chennai International Airport and its surroundings, also had high LST. In the summer season (April and May), the mean temperature in 2021 was lower (35 °C) than in 2019 (37 °C). Similar observations were also made in regions such as Tiruvallur (highlighted in the November 2019 image in Fig. 5a for illustration) and adjoining open vacant areas that were found to have significantly lower temperatures in November 2021 as compared to November 2019 (see LST maps of November 2019 and November 2021 in Fig. 5a).

It is interesting to see that in Bangalore, the minimum and maximum temperatures were lower by 2–3 °C and 1 °C in 2021 compared to the year 2019. On average, the city experienced lower temperature of 2.5 °C in 2021. Overall, LST was observed to be lower in south-eastern Bangalore, including Bannerghatta National Park and along Hosur Road leading to Electronics City in south-western Bangalore in 2021, compared to 2019 (highlighted by circles in Fig. 5b in May 2021 and December 2019, 2021 images). In January 2019, northern Bangalore was found to be hotter than southern Bangalore, which is due to large-scale building construction happening near the Kempegowda International Airport and its surroundings in the northern part of the City near Devanahalli, thereby radiating more latent heat and increasing land surface temperature. November and December experienced the lowest temperature in both years (2019 and 2021), while February to May were the hottest months (summer season) with temperatures ranging from 32 °C to 36 °C (see Fig. 7). November 2021 was the coolest month among all the observations. It is to be noted that all the months' LST maps have not been shown due to either cloud coverage or similar results.

One of the potential limitations of this study was the use of the MODIS 1 Km LST product, which is a low/coarse spatial resolution to capture high LST details and variations in urban areas (compared to the Landsat thermal band at 30 m spatial resolution). However, MODIS has a very high temporal resolution of 1 day, and due to this advantage, it is possible to get frequent observations useful for many applications, which is a pressing limitation of the Landsat series of satellites. The MODIS11A2 v061 product furnishes daytime surface temperature bands, inclusive of associated quality control assessments, observation times, view zenith angles, and clear-sky coverages with emissivities for bands 31 and 32 from different land cover types. Calibration of MODIS LST data involved comparison with observations from sources such as the World Weather API and Weather Forecast (https://www.worldweatheronline.com/search-weather.aspx?q=India), Open City Urban Data Portal (https://data.opencity.in/dataset/daily-temperature-70-years-data-for-major-indian-cities), Climate Data Service Portal - Indian Meteorological Department (IMD), Pune, India (https://cdsp.imdpune.gov.in/), and Open Government Data (OGD) Platform India (https://data.gov.in/). The final LST maps were derived from this calibration process. Version 6.1 Level-1B products underwent enhancements through various calibration stages, including modifications to the response-versus-scan angle (RVS) approach affecting reflectance bands for Aqua and Terra MODIS, corrections for optical crosstalk in Terra MODIS infrared (IR) bands, and updates to the Terra MODIS forward look-up table (LUT). Additionally, a polarization correction is applied to the L1B Reflective Solar Bands (RSB) (https://lpdaac.usgs.gov/products/mod11a2v061/).

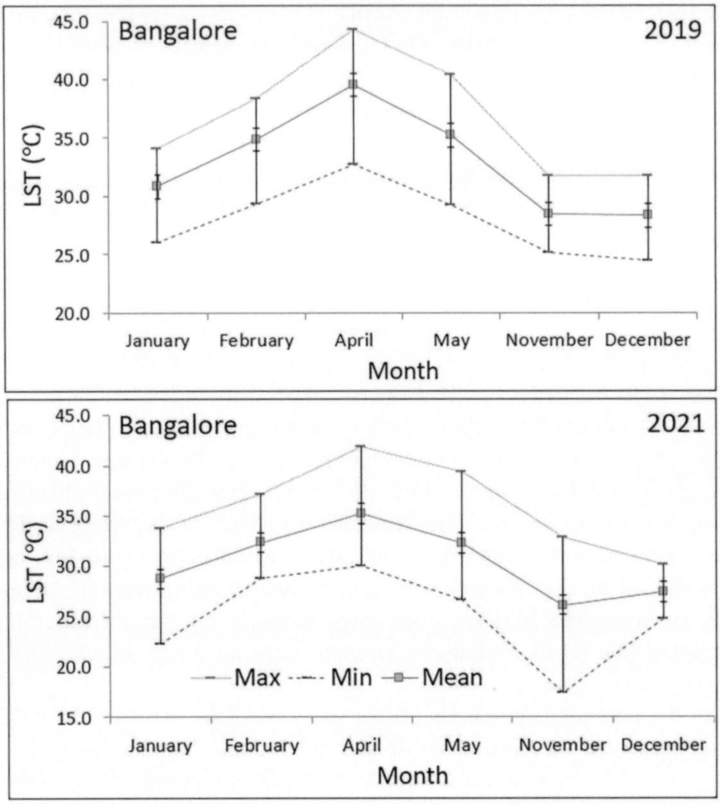

Fig. 7 Minimum, maximum, and mean LST (°C) of the Bangalore City region in 2019 and 2021

Figure 8a, b shows the CO_2 emission from various sectors such as transport, electricity, residential, manufacturing, and "others" (indicating remaining sources of CO_2 emissions) for the entire of India. It is evident that the emissions from electricity production are highest, followed by the manufacturing sector. Emissions from the transport sector attribute to the third largest in terms of carbon emissions.

Figure 9 shows the total highest carbon footprint (Mt CO_2) of the 25 major Indian cities based on the Eora Global Supply Database [https://worldmrio.com/]. It is evident that New Delhi has the highest carbon footprint (approximately 70 Mt. CO_2), and the curve almost flattens after Bangalore (having the lowest carbon footprint of 21.6 Mt. CO_2 among the top six Indian cities accounting for the total carbon footprint).

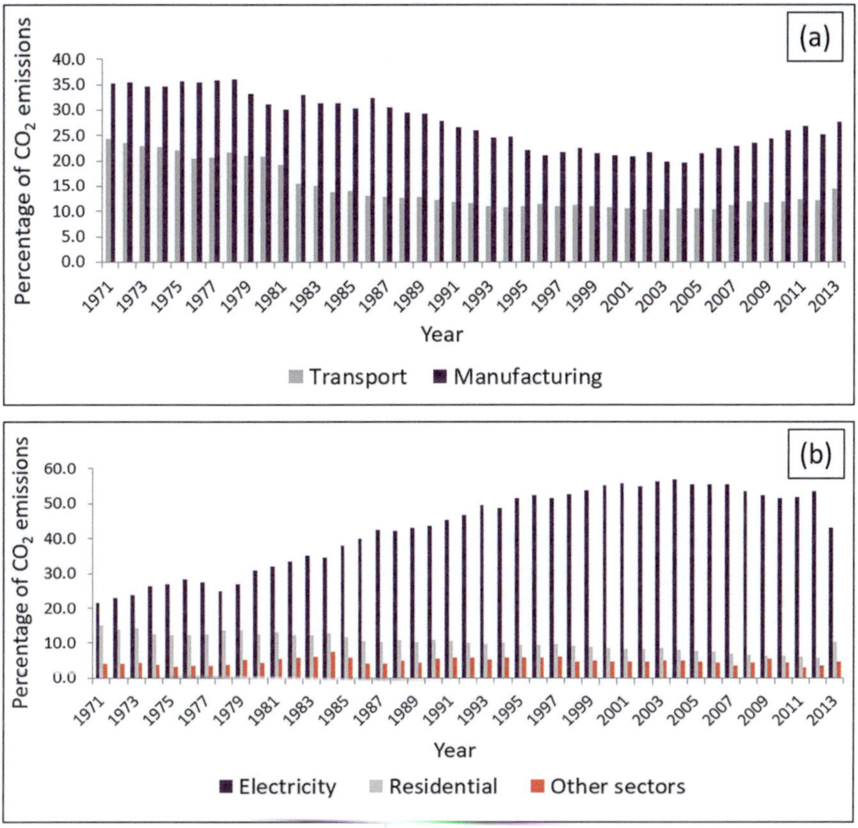

Fig. 8 CO_2 emission from different sectors: (**a**) transport and manufacturing and (**b**) electricity, residential, and other sectors in India

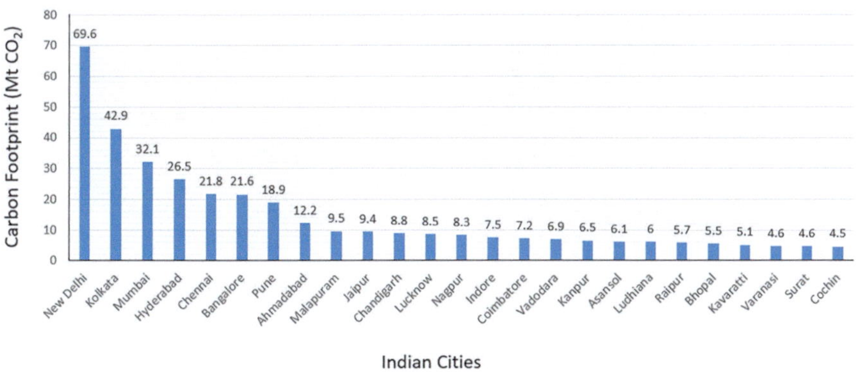

Fig. 9 25 Indian cities having the highest total carbon footprint (Mt CO_2)

3.1 Correlation Analysis of Pollution Datasets

In this work, 4 years of pollutant datasets (2019, 2020, 2021, and 2022) were used for computing the relationship between LST and pollutants for the two major cities, that is, New Delhi, having the highest carbon footprint, and Bangalore, having the lowest carbon footprint among all six Indian cities studied earlier. In New Delhi, aerosols had a higher correlation with LST than any other pollutant. Beginning from the pre-pandemic year 2019 to 2020 and 2021, the correlation between LST and aerosol index was found to be 0.71, 0.56, and – 0.07, respectively. Interestingly, as shown in Fig. 10, the correlation was high again at 0.54 in the year 2022. Figure 11 highlights the yearly aggregated correlation.

In Bangalore, during the pre-pandemic year of 2019, the yearly correlation between LST and CO density was 0.56, LST and NO_2 was 0.87, and the least correlation of 0.26 was observed between LST and SO_2. However, in the years 2020 and 2021, the correlation between LST and CO decreased to 0.37 and eventually increased to 0.55, while the correlation between LST and NO_2 was 0.54 and 0.79, respectively (see Fig. 12). To summarize, the correlation between both LST and CO, and LST and NO_2 showed a decreasing pattern in 2020 and an increasing trend in 2021 to similar values as observed in 2019.

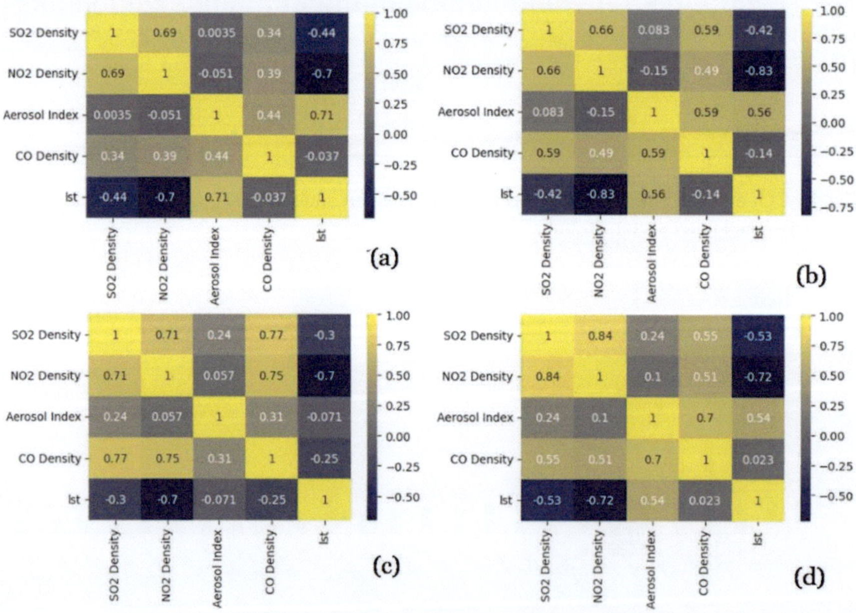

Fig. 10 Yearly correlation between LST and pollutants during (**a**) 2019, (**b**) 2020, (**c**) 2021, and (**d**) 2022 in New Delhi

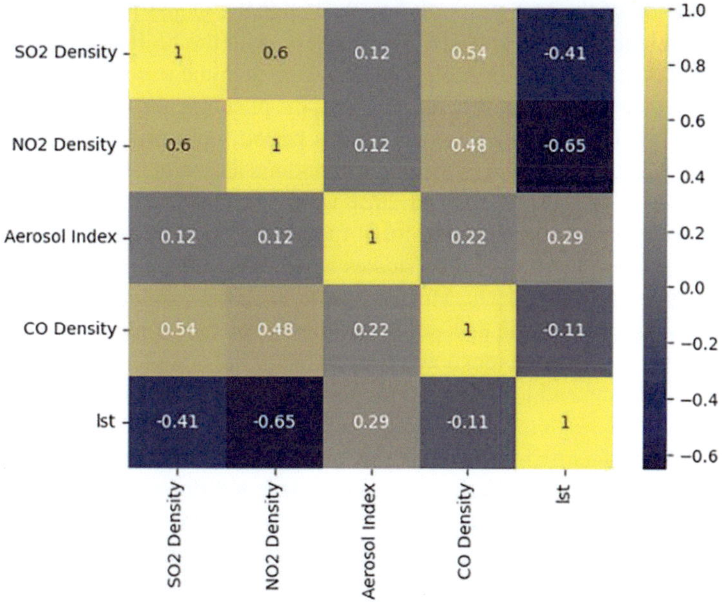

Fig. 11 Aggregated correlation between LST and pollutants in New Delhi

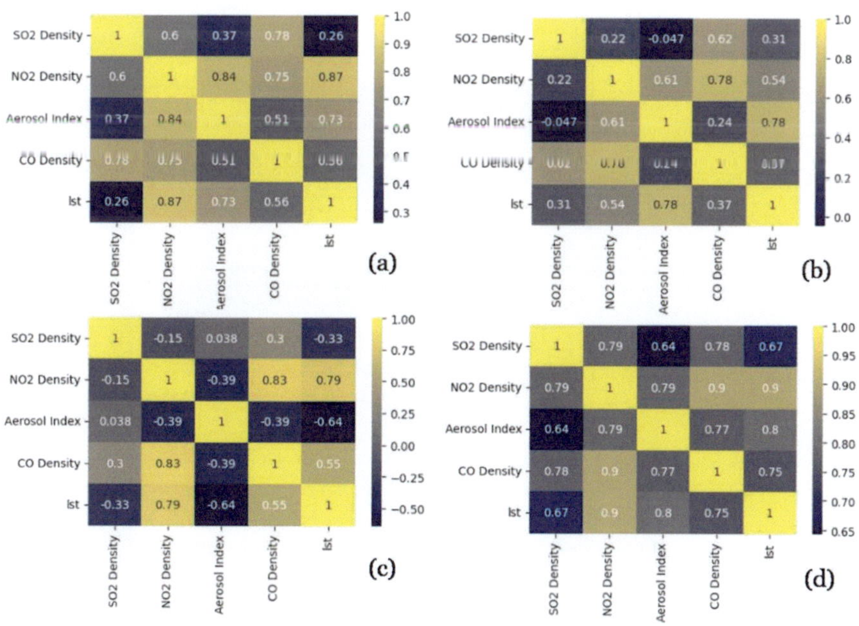

Fig. 12 Yearly correlation between LST and pollutants during (**a**) 2019, (**b**) 2020, (**c**) 2021, and (**d**) 2022 in Bangalore

Aggregation of the four years of data showed that CO and NO_2 density had the highest correlation of 0.57 and 0.69, respectively, with the LST (Fig. 13).

The temperature of the Earth is gradually rising with the rise in the concentration of carbon. However, as seen from this analysis, the phenomenon underwent a change during the COVID-19 pandemic, making this period more important to study and analyze. During the shutdown of the major economic activities, the seasonal changes in LST between the years 2019 and 2021 were measured for assessment. Human activities, such as the combustion of fossil fuels and the modification of natural surface cover, produce aerosols. While studies have proved that GHG warming occurs globally (Lashof & Ahuja, 1990; Meinshausen et al., 2009), pollution aerosol cooling is regionally dependent, with uncertain outcomes in some regions.

Since there were several lockdown phases (including complete lockdown, partial lockdown, and weekend lockdown) during the pandemic (2019–2022) with restrictions on people's movement and economic activities, low pollution levels were prevalent during this period. The analysis results indicated that in New Delhi, the presence of aerosols has effects on the environment at different levels, from local to regional. At the local level, there is an increasing recognition of aerosols and NO_2 as significant health concerns, especially leading to breathing problems. Aerosols also have a noticeable effect on visibility, causing haze and reducing clarity in the atmosphere. Regionally, aerosols play a role in the development of smog and air pollution, particularly in urban areas characterized by elevated industrial and vehicular emissions. The combination of aerosols with other pollutants can give rise to

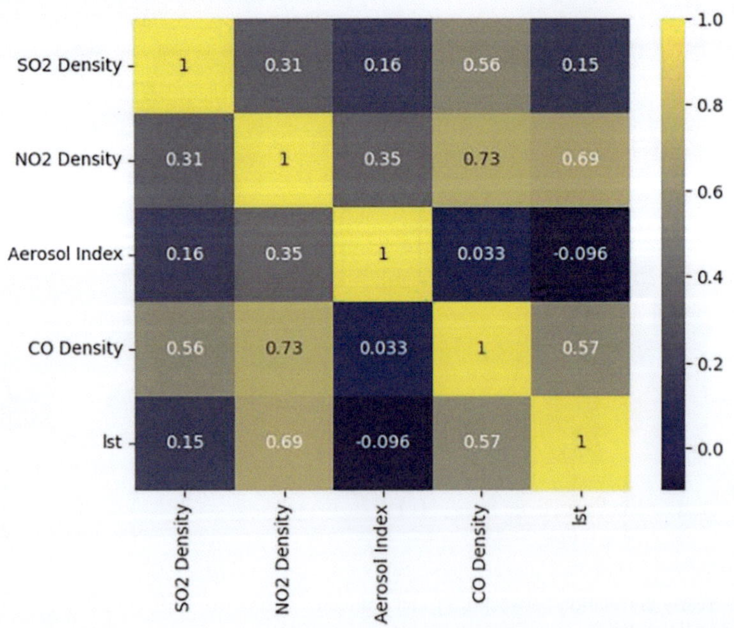

Fig. 13 Aggregated correlation between LST and pollutants in Bangalore

secondary pollutants, such as ozone, with detrimental impacts on both human health and ecosystems (Clemente et al., 2022). Depending on the composition of the aerosols and their sizes, they can either cool or warm the Earth's surface. Most aerosols cool the surface; for example, sulfate particles from industrial sources have a cooling effect by reflecting sunlight back into space, reducing GHG-induced warming (Kumar et al., 2008). Conversely, an aerosol such as black carbon, which is produced from fossil fuel and biomass burning, absorbs sunlight, heats the atmosphere, and contributes to warming. Aerosols also influence cloud formation and precipitation patterns, impacting regional and global climate dynamics.

Many Tier-1 Indian cities, including Bangalore, the third-largest city in terms of population in India, have undergone substantial alterations in their land use and land cover due to urban expansion. In Bangalore City, the presence of high levels of CO and NO_2 in the atmosphere, which have a positive correlation with LST, is a matter of concern, as they are harmful for the human respiratory system when they aggravate respiratory diseases. The city has tremendous pressure on the road infrastructure because of the ever-increasing number of motorized vehicles running on fossil fuels. The existing public transportation system is unable to cater to the needs of more than 12 million people. Vehicular transportation in the urban centers of the city contributes to harmful gas emissions, and in the surrounding areas (the outskirts of the city), high concentrations of these gases are attributed to biomass burning. Exposure to high levels of CO in humans reduces the oxygen that is transported to the bloodstream to vital organs such as the heart and brain, thus posing a great threat to the life and health of people.

The COVID-19 pandemic and the following lockdown measures implemented by governments worldwide provided a unique opportunity for researchers to study the impact of reduced human activity. There was a significant decrease in air pollution levels in many areas with fewer vehicles on the roads, reduced industrial activity, and limited air travel. In India, the strict nationwide lockdown implemented in March 2020 resulted in noticeably improved air quality. With nonessential, commercial, and industrial activities put to a halt, there was a substantial reduction in pollutants during the pandemic. The restrictions on mobility also led to a significant decrease in vehicular emissions, particularly in urban areas. This study underscores the favorable impact of the lockdown on air quality in Bangalore. Additionally, the land surface temperature (LST) maps of cities, with a few exceptions, exhibited comparatively lower temperatures during the pandemic lockdown year.

The four key factors leading to higher emissions in major Indian cities are transportation, buildings, industries, and waste generated from the region. Transport is a substantial source of carbon emissions due to the high rate of car ownership and traffic congestion in cities. The majority of cities have public transportation such as buses and metros, but these have had difficulty keeping up with the city's increasing population and demand. It is necessary to adopt strategies for encouraging environmentally friendly transportation, such as creating designated bike lanes, boosting electric automobiles, and taking proper actions to create pedestrian lanes. Secondly, buildings are another significant source of carbon emissions due to the high need for energy-intensive air cooling and heating in the summer. Additionally, many structures are inadequately insulated, which increases energy usage. Growing cities must put policies into place to encourage building energy efficiency, such as energy audits, for instance. Thirdly, although the city's extensive industrial sectors produce items for both domestic and foreign markets, industries are key contributors to carbon emissions. Strict emission requirements and promoting the use of renewable energy are actions that should be taken to encourage cleaner production processes and lower industrial emissions. Finally, last but not least, waste generation due to the growing population and increased consumption in the cities has given rise to emissions. As a result, landfills are being used to dispose of a lot of garbage, which increases GHG emissions. The majority of the food consumed in the big cities is imported from other areas and nations. As such, emissions from agriculture, food processing, and distribution are influenced by the production and transportation of food materials. To cut emissions, the cities have to take actions to support regional food production and minimize food waste. Actions are required to promote trash reduction, recycling, composting, and a ban on single-use plastics.

"Think globally and act locally" is apt for good environmentalism, as a worldwide perspective and local actions are required. Technology based on nonrenewable resources leads to more utilization of natural resources like fossil fuels, which have indirectly led to higher concentrations of GHGs like carbon dioxide, nitrous oxide, methane, etc. Reducing emissions involves implementing measures such as promoting the increased use of renewable energy sources (wind, solar, biomass, etc.) through combined heat and power installations, enhancing energy efficiency in buildings, industry, and household appliances, and minimizing CO_2 emissions from new passenger cars in numerous cities and urban areas. To sustain low levels of pollutants and minimize land surface temperature (LST) in cities, various initiatives can be undertaken. For instance, creating inclusive spaces for recreational activities, such as public green spaces and easily accessible sports areas, promotes low-carbon living where there is minimal pressure to consume. In most populous cities, the number of vehicles operating on the roads is ever-increasing, resulting in traffic congestion, environmental pollution, and increasing incidents of traffic accidents. Energy-efficient buildings, urban green spaces, improved transportation systems, etc. are important aspects of lowering carbon emissions (Pey et al., 2008). The implementation of a low-emission zone (LEZ) serves as a significant component of emission control strategies, delineating specific areas where the entry of polluting vehicles is either restricted or penalized based on emission standards (Holman et al., 2015). New technology of electric vehicles in place of existing automobiles that cause more pollution by the emission of GHGs is currently being promoted. However, their wider usage and long-term benefits are yet to be realized. As such, electric cars are propelled by electric motors that require rechargeable batteries using energy stored in them. Solar power and wind energy are the most sustainable green technologies that can be used for power generation, and they are competitive with fossil-based energy options. Indeed, economic activity and energy consumption exhibit a positive correlation with increased emissions stemming from the utilization of fossil fuels like coal, etc. (Najam et al., 2010). However, technological advancements and increased productivity can help reduce pollution levels by making energy production and consumption more efficient. Managing environmental impact in cities requires a comprehensive approach that considers the interdependent relationships between consumption, production, and the environment. This could entail the adoption of policies aimed at fostering sustainable consumption and production, improving energy efficiency, and expanding the utilization of renewable energy sources.

Green building technology, including geothermal heating to supply energy to appliances, has already proved to be very efficient. Buildings with zero energy consumption generate their own power from renewable energy sources, resulting in minimal carbon emissions. For example, Heating Ventilation and Air Conditioning (HVAC) contains high-performing, eco-friendly refrigerant, which reduces the electricity consumption in energy-efficient buildings (Gupta, 2018). Utilizing cool roofs made of special tiles and reflective paints, which absorb less heat and reflect solar radiation, can contribute to a decreased reliance on air conditioners. The United Nations Framework Convention on Climate Change (UNFCCC) asserts that achieving energy efficiency can lead to cost-effective emission reduction. While modern electrical appliances such as fans, air conditioners, washing machines, ovens, etc., consume significantly less energy than their older counterparts, ongoing efforts are essential to further diminish carbon emissions in urban areas and advocate for sustainable urban development practices.

4 Conclusions

A rapidly globalizing world will bring about incoming city growth both in terms of size and population. Therefore, to accommodate the incoming changes, proper planning of the city is required. As a result of the increasing concrete and impervious surfaces in the city, local LST has increased. Minimum winter temperatures and maximum summer temperatures have witnessed to increase, thereby creating extreme temperature events. In cities, building density also hinders the air flow and reduces the cooling effect of wind, particularly when there is more anthropogenic heat generation in the urban ecosystem. The pollutant concentration also accounts for changes in local temperature. This study revealed that six major Indian cities, namely, New Delhi, Kolkata, Mumbai, Hyderabad, Chennai, and Bangalore, exhibited the highest carbon footprint chronologically. Mumbai, Hyderabad, Chennai, and Bangalore showed lower temperatures than normal during the pandemic in the summer season, especially with the urban heat island effect. It was also observed that there were noticeable differences in the local temperature across all the pre- and during-pandemic seasons. Disagreeing with the general observed pattern before the pandemic year, all six cities showed local climatic deviations during the COVID-19 lockdown. In New Delhi, aerosol had a positive correlation with the land surface temperature, while in Bangalore, the presence of carbon monoxide (CO) and nitrogen dioxide (NO_2) showed the highest correlation of 0.57 and 0.69, respectively, with the surface temperature. The existence of aerosols, carbon monoxide (CO), and nitrogen dioxide (NO_2) has detrimental impacts on both human health and urban ecosystems.

In order to circumvent the adverse impacts of increasing temperature and pollutants, zero/low-emission zones can be created in individual cities to reduce the carbon footprint. Compact urban development consists of accessibility, affordability, and safety of transport. A mass transport system helps in the optimization of fuel, as there is less fuel usage per passenger. Appropriate urban planning and further green building design offer ways to lessen heat exhaustion for city dwellers and can lower the urban heat island effect. From the design of the individual buildings to regional planning, climatic factors should be taken into account in order to maximize thermal comfort in urban settings. We belong to a system where any change in the system affects us too. Therefore, when alternative technology is made available to the people then the objectives of a cleaner environment and sustainability are also accomplished. Sustainable cities provide a high quality of living to people without much use of resources.

References

Anbazhagan, S., & Ghosh, A. (2002). Coastal geomorphology and land use/land cover mapping in and around Manori creek, Mumbai. *IAFRS & SIS, 34*(7), 420.

Ayanlade, A., Aigbiremolen, M. I., & Oladosu, O. R. (2021). Variations in urban land surface temperature intensity over four cities in different ecological zones. *Scientific Reports, 11*(1), 20537.

Banerjee, A. (2005, October). Population growth, environment and development: Some issues in sustainability of the mega city of Kolkata (Calcutta), West Bengal. In *Proceedings of the National Seminar on population environment and nexus, population environment Centre, IIPS, Deonar, Mumbai* (Vol. 21).

Bhatta, B. (2009). Analysis of urban growth pattern using remote sensing and GIS: A case study of Kolkata, India. *International Journal of Remote Sensing, 30*(18), 4733–4746.

Clemente, Á., Yubero, E., Nicolás, J. F., Caballero, S., Crespo, J., & Galindo, N. (2022). Changes in the concentration and composition of urban aerosols during the COVID-19 lockdown. *Environmental Research, 203*, 111788.

Das, D. N., Chakraborti, S., Saha, G., Banerjee, A., & Singh, D. (2020). Analysing the dynamic relationship of land surface temperature and landuse pattern: A city level analysis of two climatic regions in India. *City and Environment Interactions, 8*, 100046.

Dasgupta, A., & Kumar, U. (2021). Urban heat island and its impact on impervious surfaces during two seasons: A case study of Bangalore. In *Proceedings of the IEEE international India geoscience and remote sensing symposium (InGARSS 2021), Ahmedabad, India, 2021* (pp. 250–253). https://doi.org/10.1109/InGARSS51564.2021.9792024

Gupta, M. (2018, 9 24). *Green building technology-concept, need, and types*. https://gosmartbricks.com/green-building-technology-concept-need-and-types/

Heaviside, C., Macintyre, H., & Vardoulakis, S. (2017). The Urban Heat Island: Implications 615 for health in a changing environment. *Current Environmental Health Reports, 4*(3), 296–305. 616. https://doi.org/10.1007/s40572-017-0150-3

Holman, C. H., Harrison, R., & Querol, X. (2015). Review of the efficacy of low emission zones to improve urban air quality in European cities. *Atmospheric Environment, 111*, 161–169.

Jeganathan, A., Ramachandran, A., & Palanivelu, K. (2021). Challenges in Chennai City to cope with changing climate. *European Journal of Climate Change, 3*(1), 33–43.

Kaneko, Shinji, Dhakal, & Shobhakar. (2008). Comparison of urban energy use and carbon emission in Tokyo, Beijing, Seoul and Shanghai. In *Presentation to the International Workshop on Urban Energy and Carbon Modeling, February 5–6, 2008, AIT Centre, Asian Institute of Technology, Pathumthani, Thailand.*

Krank, S. (2007). *Cultural, spatial, and socio-economic fragmentation in the Indian megacity Hyderabad.*

Kumar, G. M., Sampath, S., Jeena, V. S., & Anjali, R. (2008). Carbon monoxide pollution levels at environmentally different sites. *Journal of Indian Geophysical Union, 12*(1), 31–40.

Kumar, A., Agarwal, V., Pal, L., Chandniha, S. K., & Mishra, V. (2021). Effect of land surface temperature on Urban Heat Island in Varanasi City, India. *J, 4*(3), 420–429.

Lashof, D. A., & Ahuja, D. R. (1990). Relative contributions of greenhouse gas emissions to global warming. *Nature, 344*(6266), 529–531.

Luo, H., & Wu, J. (2021). Effects of urban growth on the land surface temperature: A case study in Taiyuan, China. *Environment, Development and Sustainability, 23*(7), 10787–10813.

Meinshausen, M., Meinshausen, N., Hare, W., Raper, S. C., Frieler, K., Knutti, R., Frame, D. J., & Allen, M. R. (2009). Greenhouse-gas emission targets for limiting global warming to 2 C. *Nature, 458*(7242), 1158–1162.

Moran, D., Kanemoto, K., Jiborn, M., Wood, R., Többen, J., & Seto, K. C. (2018). Carbon footprints of 13 000 cities. *Environmental Research Letters, 13*(6), 064041.

Mwangi, P. W., Karanja, F. N., & Kamau, P. K. (2018). Analysis of the relationship between land surface temperature and vegetation and built-up indices in upper-hill, Nairobi. *Journal of Geoscience and Environment Protection, 6*(1), 1–16.

Najam, A., Runnalls, D., & Halle, M. (2010). Environment and globalization: Five propositions (2010). In *The globalization and environment reader* (Vol. 94). Wiley.

Nakada, L. Y. K., & Urban, R. C. (2020). COVID-19 pandemic: Impacts on the air quality during the partial lockdown in São Paulo state, Brazil. *Science of the Total Environment, 730*, 139087.

Oke, T. R. (1976). The distinction between canopy and boundary-layer urban heat islands. *Atmosphere, 14*(4), 268–277.

Pey, J., Rodríguez, S., Querol, X., Alastuey, A., Moreno, T., Putaud, J. P., & Van Dingenen, R. (2008). Variations of urban aerosols in the western Mediterranean. *Atmospheric Environment, 42*(40), 9052–9062.

Pramanik, M. K. (2017). Impacts of predicted sea level rise on land use/land cover categories of the adjacent coastal areas of Mumbai megacity, India. *Environment, Development and Sustainability, 19*, 1343–1366.

Pricope, G. N., Mapes, L. K., & Woodward, D. K. (2019). Remote sensing of human–environment interactions in global change research: A review of advances, challenges and future directions. *Remote Sensing, 11*(23), 2783.

Punia, M., Joshi, P. K., & Porwal, M. C. (2011). Decision tree classification of land use land cover for Delhi, India using IRS-P6 AWiFS data. *Expert Systems with Applications, 38*(5), 5577–5583.

Ramachandra, T. V., & Kumar, U. (2009). Land surface temperature with land cover dynamics: Multi-resolution, Spatio-temporal data analysis of greater Bangalore, India. *International Journal of Geoinformatics, 5*(3), 1.

Ramachandra, T. V., Sreejith, K., & Bharath, H. A. (2014). Sector-wise assessment of carbon footprint across major cities in India. *Assessment of Carbon Footprint in Different Industrial Sectors, 2*, 207–267.

Santamouris, M., Cartalis, C., Synnefa, A., & Kolokotsa, D. (2015). On the impact of urban 684 heat Island and global warming on the power demand and electricity consumption of 685 buildings—A review. *Energy and Buildings, 98*(119–124), 686. https://doi.org/10.1016/j.enbuild.2014.09.052

Sovacool, B. K., & Brown, M. A. (2010). Twelve metropolitan carbon footprints: A preliminary comparative global assessment. *Energy Policy, 38*(9), 4856–4869.

Stone, B. (2012). *The City and the coming climate*. Cambridge University Press.

UN-Habitat. (2008). *State of the world's cities 2008/2009: Harmonious cities*. Earthscan. http://www.iclei.org/

Vishwas, S., & Joshi, V. U. (2013). Coastal environmental resources and users in Mumbai and Chennai metropolitan regions, India. In *Sustainability in the coastal Urban environment: Thematic profiles of resources and their users* (Vol. 3, p. 249). Sapienza Università Editrice.

Wakode, H. B., Baier, K., Jha, R., & Azzam, R. (2014). Analysis of urban growth using Landsat TM/ETM data and GIS—A case study of Hyderabad, India. *Arabian Journal of Geosciences, 7*, 109–121.

Index

P. K. Joshi et al. (eds.), *Geospatial Science for Urban Ecosystems*, The Urban Book Series, https://doi.org/10.1007/978-3-032-07155-2

MIX
Papier aus verantwortungsvollen Quellen
Paper from responsible sources
FSC® C105338

If you have any concerns about our products,
you can contact us on
ProductSafety@springernature.com

In case Publisher is established outside the EU,
the EU authorized representative is:
Springer Nature Customer Service Center GmbH
Europaplatz 3, 69115 Heidelberg, Germany

Printed by Libri Plureos GmbH
in Hamburg, Germany